Control Engineering

Stability and Control of Dynamical Systems with Applications

A Tribute to Anthony N. Michel

Derong Liu
Panos J. Antsaklis
Editors

Springer Science+Business Media, LLC

Derong Liu
University of Illinois at Chicago
Department of Electrical
and Computer Engineering
Chicago, Illinois 60607
USA

Panos J. Antsaklis
University of Notre Dame
Department of Electrical Engineering
Notre Dame, IN 46556
USA

Library of Congress Cataloging-in-Publication Data

Stability and control of dynamical systems with applications : a tribute to Anthony N. Michel / Derong Liu, Panos J. Antsaklis, editors.
p. cm.
Selected papers presented at the workshop at the University of Notre Dame on Apr. 5, 2003.
Includes bibliographical references and index.
ISBN 978-1-4612-6583-2 ISBN 978-1-4612-0037-6 (eBook)
DOI 10.1007/978-1-4612-0037-6
1. Stability–Congresses. 2. Differentiable dynamical systems–Congresses. 3. Control theory—Congresses. I. Michel, Anthony N. II. Liu, Derong, 1963- III. Antsaklis, Panos J.

QA871.S7634 2004
003'.5–dc21

2003045390
CIP

Printed on acid-free paper.

ISBN 978-1-4612-6583-2 SPIN 10952331

Typeset by the editors.

9 8 7 6 5 4 3 2 1

To Anthony N. Michel

Anthony N. Michel

Contents

Foreword

It is with great pleasure that I offer my reflections on Professor Anthony N. Michel's retirement from the University of Notre Dame. I have known Tony since 1984 when he joined the University of Notre Dame's faculty as Chair of the Department of Electrical Engineering.

Tony has had a long and outstanding career. As a researcher, he has made important contributions in several areas of systems theory and control theory, especially stability analysis of large-scale dynamical systems. The numerous awards he received from the professional societies, particularly the Institute of Electrical and Electronics Engineers (IEEE), are a testament to his accomplishments in research. He received the IEEE Control Systems Society's Best Transactions Paper Award (1978), and the IEEE Circuits and Systems Society's Guillemin-Cauer Prize Paper Award (1984) and Myril B. Reed Outstanding Paper Award (1993), among others. In addition, he was a Fulbright Scholar (1992) and received the Alexander von Humboldt Forschungspreis (Alexander von Humboldt Research Award for Senior U.S. Scientists) from the German government (1997). To date, he has written eight books and published over 150 archival journal papers.

Tony is also an effective administrator who inspires high academic standards. He has the vision and determination to implement his plans. During his tenure at the University of Notre Dame, first as Chair of the Department of Electrical Engineering (1984-1988) and then as Dean of the College of Engineering (1988-1998), he has made positive impacts on both the Department of Electrical Engineering and the College of Engineering at Notre Dame.

It is humbling for me to step into the Department Chair position that he once held. I can see clearly how he contributed to the success that our department now enjoys. Specifically, our success in the nano-electronics area is attributable to Tony's vision and determination back in the 1980s. He hired key faculty members who are now leaders in nano-electronics and who accomplished the pioneering work in quantum-dot cellular automata, a novel paradigm for quantum computing. I believe Tony's impact will be long-lasting.

I consider it a privilege to have been a colleague of Anthony N. Michel and I wish him the best.

Yih-Fang Huang
Department of Electrical Engineering
University of Notre Dame

Preface

On April 5, 2003, the University of Notre Dame hosted a one-day workshop in honor of Anthony N. Michel on the occasion of his retirement. Tony has been on the faculty at Notre Dame since 1984. He served as the Chair of the Department of Electrical Engineering from 1984 to 1988 and as the Matthew H. McCloskey Dean of Engineering from 1988 to 1998. Prior to that, he was on the faculty at Iowa State University for sixteen years and he was with General Motors' AC Electronics Division for six years.

Tony has been a pioneer in many research fields including large-scale dynamical systems, nonlinear systems stability analysis, recurrent neural networks, and recently, qualitative analysis of hybrid and discontinuous dynamical systems. He has also rendered extraordinary service to our community as General Chair of the IEEE Conference on Decision and Control, General Chair of the IEEE International Symposium on Circuits and Systems, Distinguished Lecturer of the IEEE Circuits and Systems Society, Editor of the *IEEE Transactions on Circuits and Systems*, President of the IEEE Circuits and Systems Society, and Vice-President of the IEEE Control Systems Society, just to name a few.

Tony has sustained a high level of significant research accomplishments, mostly in systems and controls, throughout his career. His work is characterized by great depth, as exemplified by his contributions to stability theory of dynamical systems, and by great breadth, as demonstrated by the wide range of problems that he addresses.

Tony has proved to be an excellent teacher and mentor, he has demonstrated to be an effective administrator, and he has rendered more than his share of service to his profession. He has received several lifetime career awards, including the 1995 Technical Achievement Award of the IEEE Circuits and Systems Society, the Distinguished Member Award of the IEEE Control Systems Society in 1998, the IEEE Centennial Medal in 1984, the 1999 Golden Jubilee Medal of the IEEE Circuits and Systems Society, and the IEEE Third Millennium Medal in 2000.

The workshop provided a venue for researchers, colleagues, friends, and students to honor Tony's contributions while discussing contemporary issues in systems stability and control. Speakers at the workshop discussed topics ranging from stability analysis of dynamical systems to many application areas.

The workshop featured two keynote addresses and 16 technical presentations. The two keynote addresses were delivered by Irwin Sandberg and Alfred Fettweis. The 16 technical presentations were given by nine of Tony's friends and colleagues and by seven of Tony's former Ph.D. students. Due to the limited number of time slots available for technical presentations during the day of the workshop, not all papers accepted by the workshop were presented at the workshop. In particular, three of the papers included in this book were not presented at the workshop.

This book is divided into three parts. Part I contains seven chapters that deal with issues in stability analysis of dynamical systems. Part II contains six chapters about artificial neural networks as well as signal processing. Part III contains eight chapters that deal with power systems and control systems.

Chapter 1 introduces the wave-digital concepts and shows some interesting relationships to the relativity theory through some modifications to Newton's second law. Chapter 2 investigates the notion of time and establishes the so-called consistent Lyapunov methodology. The Physical Continuity and Uniqueness Principle and the Time Continuity and Uniqueness Principle are introduced as crucial for adequate modeling of physical systems. Chapter 3 develops a mathematical model for a multibody attitude system and provides asymptotic stability analysis results for such a system. The results of this chapter are motivated by a 1980 publication of R. K. Miller and A. N. Michel. Chapter 4 studies robust regulation of uncertain hybrid systems, provides a method for checking attainability, and applies the results to networked control systems. Chapter 5 uses Lyapunov stability theory to characterize and analyze swarm dynamics. In particular, it is shown that the swarm can achieve cohesion even under very noisy measurements. Chapter 6 establishes a necessary and sufficient stability condition for discrete time-varying uncertain delay systems and applies the result to communication network control problems. Chapter 7 conducts stability analysis and $\mathcal{L}_2$ gain analysis for switched symmetric systems. It is established that when all the subsystems are stable, the switched systems are exponentially stable under arbitrary switching.

Chapter 8 shows that some interesting families of shift-varying input-output maps can be uniformly approximated using Gaussian radial basis functions over an infinite time or space domain. Chapter 9 presents a generalized state-space formulation for blind source recovery where natural gradient neural learning is used. Various state-space demixing network structures are exploited to develop learning rules. Chapter 10 discusses the relationships, results, and challenges of various approaches under the theme of approximate dynamic programming. It also presents direct neural dynamic programming and its application to a challenging continuous state control problem of helicopter command tracking. Chapter 11 derives algorithms for estimating the aircraft state vector and for approximating the nonlinear forces and moments acting on the aircraft. Convergence of the present algorithms is shown using a Lyapunov-like function in this chapter. Chapter 12 proposes a novel evolutionary approach to multiobjective optimization problems called the dynamic multiobjective evolutionary algorithm. Qualitative analysis of the proposed algorithm is also performed. Chapter 13 introduces set-membership adaptive filtering which offers a viable alternative to traditional filtering techniques. Novel features of the set-membership adaptive filtering such as data-dependent selective update are highlighted.

Chapter 14 introduces trajectory sensitivity analysis techniques. Theoretical and practical applications of trajectory sensitivity analysis to power system dynamic security analysis are discussed in this chapter. Chapter 15 addresses the design of a corrective control strategy after large disturbances in large electric power systems where a system is separated into smaller islands at a slightly reduced capacity. Chapter 16 reviews control methods developed in the past for the purpose of maintaining the stability of the electric power generation-transmission-distribution grid. It also presents some new stability control possibilities. Chapter 17 uses data fusion modeling for groundwater systems identification. Kalman filtering methods are generalized using information filtering methods coupled with a Markov random field representation for spatial variations. Chapter 18 develops parameterizations of nominal (control, output) responses and provides results for feedback synthesis in an algebraic framework. Chapter 19 presents the adaptive dynamic programming algorithm and provides a complete proof for the adaptive dynamic programming theorem. Chapter 20 estimates the reliability of the supervisory control and data acquisition system used in offshore oil and gas platforms. Probabilistic risk assessment and fault tree analysis are used in obtaining these results. Chapter 21 develops call admission control algorithms for signal-to-interference ratio (SIR)-based power-controlled CDMA cellular networks based on calculated power control setpoints for all users in the network.

We would like to express our sincere thanks to the Department of Electrical Engineering, the Graduate School, and the Center for Applied Mathematics of the University of Notre Dame for providing the financial support that made this workshop possible. We are grateful to all contributing authors for their timely and professional response to our numerous requests. Without their hard work, the workshop and this book would not have been possible.

Finally, we would like to thank Anthony N. Michel for his friendship, guidance, and support throughout the past many years. Please join us in wishing Tony the best of retirements!

Derong Liu
Chicago, Illinois

Panos J. Antsaklis
Notre Dame, Indiana

Anthony N. Michel: A Friend, a Scholar, and a Mentor

Derong Liu and Panos J. Antsaklis

The son of a grade school principal and a homemaker, Anthony N. Michel was born in Rekasch, Romania, on November 17, 1935. He and his mother left Communist Romania (at great personal risk) in 1947 to join his father in Linz, Austria. While in Austria, he and his parents lived for five years in a refugee camp. In 1952, he emigrated with his mother to the United States, settling in Milwaukee, Wisconsin. While in Austria, even though times were very difficult, he was fortunate to be able to attend the Gymnasium in Linz, a superb secondary school. After finishing high school in Milwaukee, one year after arriving there, he attended Marquette University to study electrical engineering under a five year co-op program. While at Marquette, he met his future wife, Leone Flasch, who was attending Alverno College, also in Milwaukee. They were married in 1957, the year Leone graduated, and one year before Tony received his B.S. degree in Electrical Engineering. Subsequently, they had five children: Mary, Katherine, John, Anthony, and Patrick. They now have five grandchildren.

After graduation, Tony spent one year with the U.S. Army, Corps of Engineers, and six years in the aerospace industry at AC Electronics, a Division of General Motors, in Milwaukee. While at AC Electronics, he also worked at Marquette toward an M.S. degree in Mathematics, which he received in 1964. In 1965 he was awarded an NSF fellowship to study for a Ph.D. degree in Electrical Engineering at Marquette. After receiving this degree in 1968, he joined the faculty of Electrical Engineering at Iowa State University as an Assistant Professor. He was promoted one year later to Associate Professor with tenure and in 1974, he was promoted to Full Professor. In 1972–1973, while on a sabbatical leave, he worked under the direction of Professor Wolfgang Hahn at the Technical University in Graz, Austria, where he received the D.Sc. degree in Applied Mathematics in 1973.

In 1984 he joined the faculty of Electrical Engineering at the University of Notre Dame as Department Chair and Professor. He served as Chair until 1988. In 1987 he was named Frank M. Freimann Professor of Engineering, and in 1988, he was appointed Matthew H. McCloskey Dean of Engineering. He served two terms as the Dean of the College of Engineering, from 1988 to 1998. From 1998 to December 31, 2002, he was Frank M. Freimann Professor in the Department of Electrical Engineering. Since January 1, 2003, he is Frank M. Freimann Professor of Engineering Emeritus and Matthew H. McCloskey Dean of Engineering Emeritus.

Summary of the Research Accomplishments of Anthony N. Michel

In a distinguished career spanning over 40 years, Anthony N. Michel has made seminal contributions in the qualitative studies of dynamical systems, with an emphasis on stability theory. Specific areas in which he has contributed include finite-time and practical stability, Lyapunov stability of interconnected (resp., large-scale) dynamical systems, input-output properties of interconnected (resp., large-scale) systems, artificial neural networks with applications to associative memories, robust stability analysis, stability preserving mapping theory, and stability theory for hybrid and discontinuous dynamical systems. Throughout, he has demonstrated the significance of his work with specific applications to signal processing, power systems, artificial neural networks, digital control systems, systems with state saturation constraints, and other areas.

On the topic of finite-time and practical stability, in contrast to other workers, Michel uses prespecified time-varying sets in formulating a notion of set stability. His Lyapunov-like results for set stability yield estimates for system trajectory behavior, obtained from the boundaries of the prespecified sets [9]. A radical departure from existing practices, this approach was subsequently adopted and extended by others.

To circumvent difficulties encountered in the analysis of large-scale systems with complex structure, Michel and his colleagues view such systems as interconnections of several simpler subsystems. The analysis is then accomplished in terms of the qualitative properties of the subsystems and the interconnecting structure. Michel advocates the use of scalar Lyapunov functions consisting of weighted sums of Lyapunov functions for the free subsystems. This approach has resulted in significantly less conservative results than the weak-coupling M-matrix results obtained by others who employ vector Lyapunov functions [1, 10, 12]. These results in turn are used by Michel in the analysis and synthesis of artificial neural networks [8, 14], and he and his colleagues also use them as the basis of further results involving computer generated norm-Lyapunov functions which then are applied with good success in the analysis of interconnected power systems and digital filters [13].

Using the same philosophy as in the preceding paragraph, Michel and his coworkers discovered the first results for the input-output stability of interconnected systems [1, 11]. These results make possible the systematic analysis of multi-loop nonlinear feedback systems (consisting of interconnections of subsystems that satisfy, e.g., the small gain theorem, the circle criterion, the passivity theorem, or Popov-like conditions). In the same vein, Michel also established results for the response (due to periodic inputs) of nonlinear single-loop and multi-loop feedback systems, and for the existence, nonexistence, and stability of limit cycles for such systems [1].

For his work on qualitative analysis of interconnected systems, Michel has received substantial recognition. In response to an invitation by Professor Richard Bellman, Michel co-authored with R. K. Miller the first book on the qualitative analysis of large-scale dynamical systems [1], which appeared in the Bellman Series in Mathematics in Science and Engineering (Academic Press). This book is widely cited and has had an impact on other areas of large-scale systems (e.g., power systems). Also, he received with R. D. Rasmussen the 1978 Best Transactions Paper Award of the IEEE Control Systems Society (currently called the Axelby Award) [12], and with R. K. Miller and B. H. Nam the 1984 Guillemin-Cauer Prize Paper Award of the IEEE Circuits and Systems Society [13]. Moreover, he was elected Fellow of the IEEE for "contributions in the qualitative analysis of large-scale systems."

Michel and his co-workers have also conducted extensive research in artificial neural networks with applications to associative memories [8, 14, 15]. This work, which addresses network architectures, qualitative analysis, synthesis procedures, and implementation issues for several classes of continuous and discrete recurrent neural networks, is widely cited and one of their paradigms [15], "LSSM-linear systems in a saturated mode," has been used in the software tool MATLAB's Neural Network Toolbox.

Michel and his co-workers have contributed significantly to robust stability analysis, most notably, for perturbed systems with perturbed equilibria [16] and for systems with interval matrices [17]. Michel and his co-workers have further extended the results in [16] to the robust stability analysis of recurrent neural networks. The work in [17] was possibly the first to provide necessary and sufficient conditions for the Hurwitz and Schur stability of interval matrices with a practical computer algorithm.

Michel has conducted fundamental research in qualitative analysis involving stability preserving mappings of dynamical systems. In doing so, he has developed an equation-free comparison theory that is applicable to virtually all types of deterministic dynamical systems. Some of this work has been published in Russian (in *Avtomatika i Telemekhanika*) and in a highly original book [5] (co-authored with K. Wang). For his contributions in this area, Michel was elected Foreign Member of the Russian Academy of Engineering.

Michel's more recent research addresses stability analysis of hybrid and discontinuous dynamical systems. For such systems, he and his co-workers formulate a general model, suitable for stability analysis (involving a notion of generalized time), which contains most of the hybrid and discontinuous systems considered in the literature as special cases. For this model, he established the principal Lyapunov and Lagrange stability results, including Converse Theorems [7, 18–20] and he applied these results in the analysis of several special classes of systems, including switched systems, digital control systems, impulsive systems, pulse-width-modulated feedback control systems, systems with saturation constraints, and others.

Michel's scholarly work has been made public in eight books, 30 chapters in books, 165 journal papers, and 250 conference papers. A measure of the high quality of his work is indicated by approximately 1400 citations of his publications (since 1976) in the *Science Citation Index*.

Michel has played a significant role as an educator. His eight books ([1–8]) which have been well received in the systems and controls community around the world and in many instances have blazed new trails when first introduced, demonstrate his contributions as a teacher. Furthermore, his record of maintaining a highly productive research program while simultaneously serving as an effective administrator at Notre Dame, first as Department Chair (1984–1988) and then as Dean (1988–1998), puts him in rare company. Michel has served as mentor to many outstanding graduate students. Equal numbers of these are in academe and in industry, attesting to the fine balance Michel maintains in his research program between theory and practice. These former students have all outstanding careers. (For example, one of them is the current Dean of Engineering and Architecture at Washington State University.)

Theses Supervised by Anthony N. Michel

Anthony N. Michel graduated 13 Ph.D. students from Iowa State University and 11 Ph.D. students from the University of Notre Dame before 2001. He also graduated a total of 10 Master students.

List of Doctoral Dissertations Supervised by Anthony N. Michel

(1) D. E. Cornick, "Numerical Optimization of Distributed Parameter Systems by Gradient Methods," Ph.D. Dissertation, Iowa State University, 1970.

(2) D. W. Porter, "Stability of Multiple-Loop Nonlinear Time-Varying Systems," Ph.D. Dissertation, Iowa State University, 1972.

(3) A. B. Bose, "Stability and Compensation of Systems with Multiple Nonlinearities," Ph.D. Dissertation, Iowa State University, 1974.

(4) E. P. Oppenheimer, "Application of Interval Analysis to Problems of Linear Control Systems," Ph.D. Dissertation, Iowa State University, 1974.

(5) E. L. Lasley, "The Qualitative Analysis of Composite Systems," Ph.D. Dissertation, Iowa State University, 1975.

(6) R. D. Rasmussen, "Lyapunov Stability of Large-Scale Dynamical Systems," Ph.D. Dissertation, Iowa State University, 1976.

(7) W. R. Vitacco, "Qualitative Analysis of Interconnected Dynamical Systems Containing Algebraic Loops," Ph.D. Dissertation, Iowa State University, 1976.

(8) R. L. Gutmann, "Input-Output Stability of Interconnected Stochastic Systems," Ph.D. Dissertation, Iowa State University, 1976.

(9) W. Tang, "Structure and Stability Analysis of Large Scale Systems using a New Graph-Theoretic Approach," Ph.D. Dissertation, Iowa State University, 1978.

(10) J. N. Peterson, "Wind Generator Network Methodology and Analysis," Ph.D. Dissertation, Iowa State University, 1980.

(11) N. R. Sarabudla, "Stability Analysis of Complex Dynamical Systems: Some Computational Methods," Ph.D. Dissertation, Iowa State University, 1981.

(12) B. H. Nam, "Asymptotic Stability of Large-Scale Dynamical Systems using Computer Generated Lyapunov Functions," Ph.D. Dissertation, Iowa State University, 1983.

(13) K. T. Erickson, "Stability Analysis of Fixed-Point Digital Filters using a Constructive Algorithm," Ph.D. Dissertation, Iowa State University, 1983.

(14) J.-H. Li, "Qualitative Analysis and Synthesis of a Class of Neural Networks," Ph.D. Dissertation, University of Notre Dame, 1988.

(15) J. A. Farrell, "Analysis and Synthesis Techniques for Two Classes of Nonlinear Dynamical Systems: Digital Controllers and Neural Networks," Ph.D. Dissertation, University of Notre Dame, 1989.

(16) H.-F. Sun, "Two Problems in Finite Dimensional Dynamical Systems: Qualitative Analysis and Synthesis of a Class of Neural Networks and Linear Systems Subject to Parameter Variations," Ph.D. Dissertation, University of Notre Dame, 1990.

(17) D. L. Gray, "New Paradigms for Feedforward and Feedback Artificial Neural Networks," Ph.D. Dissertation, University of Notre Dame, 1990.

(18) J. Si, "Analysis and Synthesis of Discrete-Time Recurrent Neural Networks with High Order Nonlinearities," Ph.D. Dissertation, University of Notre Dame, 1991.

(19) G. Yen, "Learning, Forgetting, and Unlearning in Associative Memories: The Eigenstructure Method and the Pseudo Inverse Method with Stability Constraints," Ph.D. Dissertation, University of Notre Dame, 1991.

(20) C.-H. Kuo, "Robust Control Strategies for a Class of Large Scale Dynamical Systems: Contaminated Groundwater Remediation," Ph.D. Dissertation, University of Notre Dame, 1993.

(21) D. Liu, "Qualitative Theory of Dynamical Systems with Saturation Nonlinearities," Ph.D. Dissertation, University of Notre Dame, 1993.

(22) H. Ye, "Stability Analysis of Two Classes of Dynamical Systems: General Hybrid Systems and Neural Networks with Delays," Ph.D. Dissertation, University of Notre Dame, 1996.

(23) B. Hu, "Qualitative Analysis of Hybrid Dynamical Systems," Ph.D. Dissertation, University of Notre Dame, 1999.

(24) L. Hou, "Qualitative Analysis of Discontinuous Deterministic and Stochastic Dynamical Systems," Ph.D. Dissertation, University of Notre Dame, 2000.

List of Master Theses Supervised by Anthony N. Michel

(1) J. D. Arend, "Finite Time Stability of Systems," M.S. Thesis, Iowa State University, 1969.

(2) E. P. Oppenheimer, "Numerical Solution of the Transient Currents of the Townsend Discharge," M.S. Thesis, Iowa State University, 1969.

(3) H. T. Zyduck, "Numerical Solution of Linear Systems Containing Widely Separated Eigenvalues," M.S. Thesis, Iowa State University, 1970.

(4) D. W. Porter, "Qualitative and Quantitative Absolute Stability of Systems," M.S. Thesis, Iowa State University, 1970.

(5) D. D. Havenhill, "Limit Cycle Oscillations in Digital Filters," M.S. Thesis, Iowa State University, 1979.

(6) C. P. Lee, "Absolute Stability Analysis by a Constructive Algorithm," M.S. Thesis, Iowa State University, 1984.

(7) C. H. Hsu, "Application of Intel 2920 to a Digital Controller: A Study of Quantization Effects," M.S. Thesis, Iowa State University, 1984.

(8) N. P. Sarang, "Designing a Controller for Contaminated Groundwater Remediation using Computer Simulation Methods," M.S. Thesis, University of Notre Dame, 1987.

(9) D. L. Gray, "Synthesis Procedures for Neural Networks," M.S. Thesis, University of Notre Dame, 1988.

(10) D. T. Flaherty, "Feedback and Redundancy in the Back-Propagation Algorithm," M.S. Thesis, University of Notre Dame, 1991.

Bibliography

[1] A. N. Michel and R. K. Miller, *Qualitative Analysis of Large Scale Dynamical Systems*, Academic Press, New York, 1977. (Written at the invitation of Professor Richard Bellman; vol. 134 in Bellman Series on Mathematics in Science and Engineering. First book on this subject. This book has been translated into Chinese.)

[2] A. N. Michel and C. J. Herget, *Mathematical Foundations in Engineering and Science: Algebra and Analysis*, Prentice-Hall, Englewood Cliffs, NJ, 1981. (Republished after going out of print as *Applied Algebra and Functional Analysis*, Dover Publications, New York, 1993.)

[3] R. K. Miller and A. N. Michel, *Ordinary Differential Equations*, Academic Press, New York, 1982. (Graduate math text, widely used as a text and reference book by students in control and applied math.)

[4] D. Liu and A. N. Michel, *Dynamical Systems with Saturation Nonlinearities: Analysis and Design*, Lecture Notes in Control and Information Sciences, vol. 195, Springer-Verlag, Berlin, 1994. (Treatise on systems with state saturation constraints. This book has inspired many works by other researchers.)

[5] A. N. Michel and K. Wang, *Qualitative Theory of Dynamical Systems: The Role of Stability Preserving Mappings*, Marcel Dekker, New York, 1995. (Develops entire Lyapunov stability theory by use of stability preserving mappings. Highly original. Michel was elected Foreign Member of the Russian Academy of Engineering for his contributions in this area.)

[6] P. J. Antsaklis and A. N. Michel, *Linear Systems*, McGraw-Hill, New York, 1997. (Adopted as text by several good academic departments. Reviews of this book have appeared in *IEEE Trans. Automatic Control*, vol. 44, pp. 1320–1321, June 1999 by T. E. Djaferis, and in *Automatica*, vol. 36, pp. 783–787, May 2000 by S. P. Bhattacharyya.)

[7] A. N. Michel, K. Wang, and B. Hu, *Qualitative Theory of Dynamical Systems: The Role of Stability Preserving Mappings*, 2nd ed., Marcel Dekker, New York, 2001. (Revised and expanded. Includes latest work on stability theory of hybrid and discontinuous dynamical systems.)

[8] A. N. Michel and D. Liu, *Qualitative Analysis and Synthesis of Recurrent Neural Networks*, Marcel Dekker, New York, 2002. (A great part of the book is based on research conducted by Michel and his co-workers.)

[9] A. N. Michel, "Quantitative analysis of systems: Stability, boundedness and trajectory behavior," *Archive for Rational Mechanics and Analysis*, vol. 38, no. 2, pp. 107–122, 1970. (Paper was based on Michel's Ph.D. dissertation. Radical departure from practices used at the time. Approach in the paper was subsequently adopted and expanded by others.)

[10] A. N. Michel, "Stability analysis of interconnected systems," *SIAM J. Control*, vol. 12, pp. 554–579, Aug. 1974. (Paper was based on Michel's D.Sc. dissertation.)

[11] D. W. Porter and A. N. Michel, "Input-output stability of time-varying nonlinear multiloop feedback systems," *IEEE Trans. Automatic Control,* vol. 19, pp. 422–427, Aug. 1974. (The first results on input-output stability of interconnected systems were presented in this paper.)

[12] R. D. Rasmussen and A. N. Michel, "Stability of interconnected dynamical systems described on Banach spaces," *IEEE Trans. Automatic Control*, vol. 21, pp. 464–471, Aug. 1976. (For this paper, the authors received the 1978 Best Transactions Paper Award of the IEEE Control Systems Society, currently called the Axelby Award.)

[13] A. N. Michel, R. K. Miller, and B. H. Nam, "Stability analysis of interconnected systems using computer generated Lyapunov functions," *IEEE Trans. Circuits and Systems*, vol. 29, pp. 431–440, July 1982. (For this paper, the authors received the 1984 Guillemin-Cauer Prize Paper Award of the IEEE Circuits and Systems Society.)

[14] A. N. Michel, J. A. Farrell, and W. Porod, "Qualitative analysis of neural networks," *IEEE Trans. Circuits and Systems*, vol. 36, pp. 229–243, Feb. 1989. (The theory in [1] is applied. This paper received more than 80 citations in the *Science Citation Index*.)

[15] J. H. Li, A. N. Michel, and W. Porod, "Analysis and synthesis of a class of neural networks: Linear systems operating on a closed hypercube," *IEEE Trans. Circuits and Systems*, vol. 36, pp. 1405–1422, Nov. 1989. (The synthesis procedure in this paper has been incorporated as a tool into MATLAB.)

[16] A. N. Michel and K. Wang, "Robust stability: Perturbed systems with perturbed equilibria," *Systems and Control Letters*, vol. 21, pp. 155–162, 1993. (This paper has led to several important robust stability results for recurrent neural networks.)

[17] K. Wang, A. N. Michel, and D. Liu, "Necessary and sufficient conditions for the Hurwitz and Schur stability of interval matrices," *IEEE Trans. Automatic Control*, vol. 39, pp. 1251–1255, June 1994. (See also *IEEE Trans. Automatic Control*, vol. 41, p. 311, Feb. 1996.) (This paper was possibly the first to give necessary and sufficient conditions for the stability of interval matrices with a practical computer algorithm.)

[18] H. Ye, A. N. Michel, and L. Hou, "Stability theory for hybrid dynamical systems," *IEEE Trans. Automatic Control*, vol. 43, pp. 461–474, Apr. 1998. (Invited paper for Special Issue on Hybrid Dynamical Systems.)

[19] A. N. Michel and B. Hu, "Towards a stability theory of general hybrid dynamical systems," *Automatica,* vol. 35, pp. 371–384, Apr. 1999. (Invited paper for Special Issue on Hybrid Dynamical Systems.)

[20] A. N. Michel, "Recent trends in the stability analysis of hybrid dynamical systems," *IEEE Trans. Circuits and Systems–Part I: Fundamental Theory and Applications*, vol. 46, pp. 120–134, Jan. 1999. (Invited paper in Special Issue honoring Sid Darlington.)

List of Contributors

Panos J. Antsaklis
H. C. and E. A. Brosey Professor
Department of Electrical Engineering
University of Notre Dame
Notre Dame, IN 46556
Phone: (574) 631-5792
Fax: (574) 631-4393
antsaklis.1@nd.edu

Peter H. Bauer
Professor
Department of Electrical Engineering
University of Notre Dame
Notre Dame, IN 46556
Phone: (574) 631-8015
Fax: (574) 631-4393
pbauer@mars.ee.nd.edu

Anjan Bose
Professor and Dean
College of Engineering & Architecture
Washington State University
Pullman, WA 99164-2714
Phone: (509) 335-5593
Fax: (509) 335-9608
bose@wsu.edu

Egemen Cetinkaya
NTAC Engineer
Sprint PCS
15500 West 113th Street
Lenexa, KS 66219
Phone: (913) 227-1887
EgemenCetinkaya@
NMCC.SprintSpectrum.com

Chadwick J. Cox
Member of the Technical Staff
Accurate Automation Corporation
7001 Shallowford Rd.
Chattanooga, TN 37421
Phone: (423) 894-4646
Fax: (423) 894-4645
ccox@accurate-automation.com

Shari Dunn-Norman
Associate Professor
Dept. Geological & Petroleum Engn
University of Missouri-Rolla
Rolla, MO 65409-0040
Phone: (573) 341-4840
Fax: (573) 341-6935
caolila@umr.edu

Russell Enns
The Boeing Company
Helicopter Systems
5000 East McDowell Road
Mesa, AZ 85215
russell.enns@boeing.com

Kelvin T. Erickson
Professor and Chair
Dept. Electrical & Computer Engn
University of Missouri-Rolla
Rolla, MO 65409-0040
Phone: (573) 341-6304
Fax: (573) 341-4532
kte@umr.edu

Jay Farrell
Professor
Department of Electrical Engineering
University of California, Riverside
Riverside, CA 92521
Phone: (909) 787-2159
Fax: (909) 787-2425
farrell@ee.ucr.edu

Alfred Fettweis
Professor
Lehrstuhl fuer Nachrichtentechnik
Ruhr-Universitaet Bochum
D-44780 Bochum, Germany
Phone: +49 234 322 2497 or 3063
Fax: +49 234 321 4100
fettweis@nt.ruhr-uni-bochum.de

Lyubomir T. Gruyitch
University Professor
University of Technology Belfort
- Montbeliard
90010 Belfort Cedex, France
Phone/Fax: +33 384 583 348
lyubomir.gruyitch@utbm.fr

Sanqing Hu
Dept. Electrical & Computer Engn
University of Illinois at Chicago
Chicago, IL 60607-7053
Phone: (312) 413-2407
Fax: (312) 996-6465
shu@cil.ece.uic.edu

Yih-Fang Huang
Professor and Chair
Department of Electrical Engineering
University of Notre Dame
Notre Dame, IN 46556
Phone: (574) 631-5350
Fax: (574) 631-4393
Yih-Fang.Huang.2@nd.edu

Hai Lin
Department of Electrical Engineering
University of Notre Dame
Notre Dame, IN 46556
Phone: (574) 631-6435
Fax: (574) 631-4393
hlin1@nd.edu

Derong Liu
Associate Professor
Dept. Electrical & Computer Engn
University of Illinois at Chicago
Chicago, IL 60607-7053
Phone: (312) 355-4475
Fax: (312) 996-6465
dliu@ece.uic.edu

N. Harris McClamroch
Professor
Department of Aerospace Engineering
The University of Michigan
Ann Arbor, MI 48109-2140
Phone: (734) 763-2355
Fax: (734) 763-0578
nhm@engin.umich.edu

Ann Miller
C. Tang Missouri Dist. Professor
Dept. Electrical & Computer Engn
University of Missouri-Rolla
Rolla, MO 65409-0040
Phone: (573) 341-6339
Fax: (573) 341-4532
millera@umr.edu

John J. Murray
Associate Professor
Dept. Electrical & Computer Engn
SUNY at Stony Brook
Stony Brook, NY 11794-2350
Phone: (631) 632-8413
Fax: (631) 632-8494
John.Murray@sunysb.edu

Trong B. Nguyen
Energy Science and
Technology Directorate
Pacific Northwest National Lab
P. O. Box 999 / MS K5-20
Richland, WA 99352
Phone: (509) 375-816
tony.nguyen@pnl.gov

M. Anantha Pai
Professor
Dept. Electrical & Computer Engn
Univ. Illinois at Urbana-Champaign
Urbana, IL 61801
Phone: (217) 333-6790
Fax: (217) 244-7075
pai@ece.uiuc.edu

Kevin M. Passino
Professor
Department of Electrical Engineering
The Ohio State University
Columbus, OH 43210-1272
Phone: (614) 292-5716
Fax: (614) 292-7596
passino@ee.eng.ohio-state.edu

Marios Polycarpou
Associate Professor
Department of ECE and CS
University of Cincinnati
Cincinnati, OH 45219-0030
Phone: (513) 556-4763
Fax: (513) 556-7326
polycarpou@uc.edu

David W. Porter
Senior Professional Staff
Applied Physics Laboratory
Johns Hopkins University
Laurel, MD 20723-6099
Phone: (443) 778-7622
Fax: (443) 778-1313
david.w.porter@jhuapl.edu

Richard E. Saeks
Chief Technical Officer and
Vice President Engineering
Accurate Automation Corporation
Chattanooga, TN 37421
Phone: (423) 894-4646
Fax: (423) 894-4645
richard@saeks.org

Michael K. Sain
Frank M. Freimann Professor
Department of Electrical Engineering
University of Notre Dame
Notre Dame, IN 46556
Phon:e (574) 631-6538
Fax: (574) 631-4393
sain.1@nd.edu

Fathi M. Salem
Professor
Department of Electrical Engineering
Michigan State University
East Lansing, MI 48824-1226
Phone: (517) 355-7695
Fax: (517) 353-1980
salem@egr.msu.edu

Irwin W. Sandberg
Cockrell Family Regents Professor
Dept. Electrical & Computer Engn
The University of Texas at Austin
Austin, TX 78712-0240
Phone: (512) 471-6899
Fax: (512) 471-5532
sandberg@uts.cc.utexas.edu

Amit K. Sanyal
Department of Aerospace Engineering
The University of Michigan
Ann Arbor, MI 48109-2140
Phone: (734) 763-1305
Fax: (734) 763-0578
asanyal@engin.umich.edu

Manu Sharma
Research Scientist
Barron Associates, Inc.
1160 Pepsi Place, Suite 300
Charlottesville, VA 22902-0807
Phone: (434) 973-1215
Fax: (434) 973-4686
sharma@barron-associates.com

Jennie Si
Professor
Department of Electrical Engineering
Arizona State University
Tempe, AZ 85287
Phone: (480) 965-6133
Fax: (480) 965-0461
si@asu.edu

E. Keith Stanek
Finley Distinguished Professor
Dept. Electrical & Computer Engn
University of Missouri-Rolla
Rolla, MO 65409-0040
Phone: (573) 341-4545
Fax: (573) 341-4532
stanek@umr.edu

Khurram Waheed
Department of Electrical Engineering
Michigan State University
East Lansing, MI 48824-1226
Phone: (517) 432-1650
Fax: (517) 353-1980
waheedkh@msu.edu

Bostwick F. Wyman
Professor
Department of Mathematics
Ohio State University
Columbus, OH 43210
Phone: (614) 292-2560
Fax: (614) 292-1479
wyman@math.ohio-state.edu

Jinglai Shen
Post Doctoral Fellow
Department of Aerospace Engineering
The University of Michigan
Ann Arbor, MI 48109-2140
Phone: (734) 763-1305
Fax: (734) 763-0578
jinglais@engin.umich.edu

Mihail L. Sichitiu
Assistant Professor
Dept. Electrical & Computer Engn
NC State University
Raleigh, NC 27695-7911
Phone: (919) 515-7348
Fax: (919) 515-2285
mlsichit@eos.ncsu.edu

Vijay Vittal
Harpole Professor
Dept. Electrical & Computer Engn
Iowa State University
Ames, IA 50011
Phone: (515) 294-8963
Fax: (515) 294-4263
vittal@ee.iastate.edu

Yu-Tsung Wang
Senior Engineer
Scientific Monitoring, Inc.
4801 S. Lakeshore Drive, Suite 103
Tempe, AZ 85282
jim@scientificmonitoring.com

Lei Yang
Department of Electrical Engineering
Arizona State University
Tempe, AZ 85287
Phone: (480) 965-3257
Fax: (480) 965-0461
leiyang@asu.edu

Gary G. Yen
Associate Professor
School of Electrical & Computer Engn
Oklahoma State University
Stillwater, OK 74078-5032
Phone: (405) 744-7743
Fax: (405) 744-9198
gyen@ceat.okstate.edu

Guisheng Zhai
Assistant Professor
Faculty of Systems Engineering
Wakayama University
Wakayama 640-8510, Japan
Phone: +81 73 457 8187
Fax: +81 73 457 8201
zhai@sys.wakayama-u.ac.jp

Yi Zhang
Dept. Electrical & Computer Engn
University of Illinois at Chicago
Chicago, IL 60607-7053
Phone: (312) 413-2407
Fax: (312) 996-6465
yzhang@cil.ece.uic.edu

Stability and Control of Dynamical Systems with Applications

PART I

ISSUES IN STABILITY ANALYSIS

Chapter 1

Wave-Digital Concepts and Relativity Theory

Alfred Fettweis

Abstract: The wave-digital approach to numerical integration owes its advantageous behavior partly to the use of wave concepts, but mostly to the use of passivity and losslessness properties that occur naturally in physical systems. For handling such nonlinear systems, one is naturally led to certain formulations that turn out to be of fundamental physical significance, yet are violated by some basic relations in special relativity theory. By starting from classical relativistic kinematics and making some assumptions that are at least not a priori physically unreasonable, however, one is led to a modified version of relativistic dynamics that (1) is in complete accord with the formulations just mentioned, (2) yields expressions of appealing elegance (including a four-vector, thus a Lorentz-invariant, quadruplet that is of immediate physical significance and coincides with a four-vector already considered by Minkowski), and (3) is, at least at first sight, in good agreement with some reasonable analytic expectations. In this alternative approach, Newton's second law is altered in a slightly different way than in classical relativity, and, as a consequence, Newton's third law, which is taken over untouched in classical theory, must also be subjected to some modification. For problems concerning collisions of particles or action of fields (electromagnetic, gravitational) upon particles, the alternative approach yields exactly the same dynamic behavior as the classical theory. Corresponding experiments are thus unable to differentiate, and the same holds for some other available experimental results. This chapter builds on the same basic concepts as those that have previously been published [1] and, in some respect, expands them. On the other hand, an unnecessary additional earlier requirement that had led to an unavoidable factor 1/2 in the expression for the equivalence between mass and energy is abandoned. This way, for example, a remarkable agreement with certain results in electromagnetics is obtained. To further test the validity, the crucial issue to be considered now appears to be the kinetic energy of fast particles. A classical experiment by Bertozzi addresses this issue, but it is not yet sufficiently clear how the results obtained there should properly be interpreted in the present context. It is hoped that the present chapter can contribute to clarifying some of the issues involved, even if the conventional theory should in the end be confirmed, by accurate and unequivocal measurements, to be the one with the closet connection to reality.

1.1 Introduction

In a previous paper [1] it was shown that certain ideas originally developed for the wave-digital methods of filtering and numerical integration can also throw some interesting light on certain aspects of the theory of special relativity [2–10]. Surprisingly, the analogy, which involves in particular the expression for nonlinear lossless inductances, suggests an expression for the force in terms of mass and velocity that is somewhat different than the classical one. This alternative expression leads to noticeably elegant but puzzling conclusions. What is even more puzzling is that, ignoring all relationships to wave-digital principles, the same force expression had been shown to be obtainable from the Lorentz transformation if at the outstart one places some assumptions that are at least not a priori unreasonable.

These ideas are further developed later. Some of these, e.g., the extension to three-dimensional movements and added mathematical rigor in deriving the alternative force expression, have also been planned, although in condensed form, for a companion paper [11] of [1]. On the other hand, a general assumption about proportionality between energy and mass increase, which implicitly had fixed the value of an integration constant in the energy expression, is abandoned. That assumption had been retained formerly in order to preserve as much as possible the features of the classical approach, where it is indeed rather natural, but it is nowhere necessary in the present context, while the added freedom offers challenging prospects. One of these is the possibility of allowing for elementary cells (particles) whose dynamic behavior is characterized by a mass that is not simply a scalar, as usual, but a matrix, in analogy to the way of characterizing coupled inductances in generalization of individual inductances. Another prospect is that the previous claim for the equivalence between mass and energy to be $E = mc^2/2$ does not have to be upheld.

Of the three fundamental Newton laws, the first one is of a purely qualitative nature. The second one is modified in classical relativity theory and also in the alternative approach [1, 11], albeit in a somewhat different way. The combination of the alternative force and the resulting power expressions yields in fact a four-vector identical to what is sometimes called the Minkowski force [2] and has thus Lorentz invariant form. Newton's third law is taken over untouched in the classical theory but has to be modified in the alternative approach. One of the consequences is that in the case of a collision between particles, conservation of momentum holds in the classical sense. Another consequence is that in the case of the action of a field (electromagnetic, gravitational, etc.) upon a moving particle, the classical expression of the force has also to be corrected, which altogether results in exactly the same dynamic behavior of the particle. Thus, in many situations the combined effect of the changes in Newton's second and third laws leads in the alternative approach to results that are the same as in the classical one.

Also of immediate interest is the expression of the energy density of a field that is electrostatic in one inertial frame but is observed in another one moving with constant velocity with respect to the first. Using the well-known formulas for transforming the electromagnetic field, the resulting energy density is not consistent with what we would expect it to be according to the classical approach. However, the result is in full agreement with the alternative approach, provided one characterizes the dynamic behavior of a cell by a matrix rather than a scalar mass, as is possible according to that alternative approach. This is altogether astonishing since Maxwell's equations have a Lorentz invariant form, i.e., they should implicitly satisfy the requirements of relativity theory.

As this theoretical result confirms, both approaches should definitely lead to a different behavior concerning the kinetic energy of fast particles. Unfortunately, the published results of a classical experiment by Bertozzi [12] do not allow us to draw a definite conclusion.

All these results, obviously, are quite puzzling. It is not claimed in the least that the alternative approach has any actual validity, but it is sufficiently interesting to deserve further attention from theoreticians as well as experimentalists.

1.2 Nonlinear Passive Kirchhoff Circuits and the Wave-Digital Method

We first consider a nonlinear inductance, thus a lossless (and therefore a fortiori passive) device, and we assume that its voltage, u, can be appropriately expressed in terms of its current, i. It can be described by either one of the equations

$$u = \mathrm{D}\left(L_g i\right) \qquad u = L_l \mathrm{D} i \tag{1.2.1}$$

where

$$\mathrm{D} = \mathrm{d}/\mathrm{d}t, \tag{1.2.2}$$

$L_g = L_g(i)$ being the global inductance and $L_l = L_l(i)$ the local inductance. $L_g(i) \geq 0 \ \forall i$ is necessary for guaranteeing passivity but not sufficient while $L_l(i) \geq 0 \ \forall i$ is sufficient but not necessary.

As recalled from [1] an alternative way of writing (1.2.1) is

$$u = \sqrt{L}\mathrm{D}\left(\sqrt{L}i\right) = \frac{1}{2}\left(\mathrm{D}\left(Li\right) + L\mathrm{D}i\right), \tag{1.2.3}$$

and for the power absorbed and the stored energy W_L one finds

$$ui = \mathrm{D}W_L, \qquad W_L = \frac{1}{2}Li^2. \tag{1.2.4}$$

$L(i) \geq 0\ \forall i$ is now necessary and sufficient for passivity. Writing (1.2.3) in the form

$$x = \mathrm{D}y, \qquad x = u\big/\sqrt{L}, \qquad y = i\sqrt{L},$$

approximating D by means of the trapezoidal rule with step size T, and designating the discretized time variable by t_n, the result can be written in the form

$$b\,(t_n) = -a\,(t_n - T) \tag{1.2.5}$$

where the so-called waves (wave quantities) a and b are given by

$$a = \frac{u + iR}{2\sqrt{R}}, \quad b = \frac{u - iR}{2\sqrt{R}}, \quad R = \frac{2L}{T}. \tag{1.2.6}$$

We obviously also have $ui = a^2 - b^2$. The simplicity of expressions such as (1.2.5) and (1.2.6) and their combination with passivity and losslessness aspects are the essential reasons for the advantageous properties obtained by making use of the wave-digital principles for filtering [13] and numerical integration of ordinary and partial differential equations [14–17]. The wide importance that wave and scattering concepts have in physics suggests that relations such as (1.2.3) and (1.2.4) could also have some fundamental physical importance. This, however, is not the case in the classical theory of special relativity.

1.3 Nonlinear Relativistic Mass

1.3.1 Scalar and matrix case

Consider a particle of rest mass m_0. Let

$$\mathbf{f} = (f_x,\ f_y,\ f_z)^T, \quad \mathbf{p} = (p_x,\ p_y,\ p_z)^T, \quad \mathbf{v} = (v_x,\ v_y,\ v_z)^T \tag{1.3.1}$$

be the force acting upon it, its momentum, and its velocity, respectively, the superscript T designating transposition. Define v, β, and α by

$$\mathbf{v}^T\mathbf{v} = v^2, \quad \beta = v/c, \quad \alpha = \sqrt{1-\beta^2} \tag{1.3.2}$$

where c is the speed of light. According to the basic principles adopted in classical relativity theory, the following holds (cf. (1.2.2)),

$$\mathbf{f} = \mathrm{D}\mathbf{p}, \quad \mathbf{p} = m_g\mathbf{v}, \quad m_g = m_0/\alpha. \tag{1.3.3}$$

Since the force, as used here, multiplied by the displacement yields the work done, one finds for the power delivered to the particle $\mathbf{v}^T\mathbf{f} = \mathrm{D}\left(m_g c^2\right)$. The quantity $m_g c^2$ is then interpreted as total energy while $(m_g - m_0)\,c^2$ is the kinetic energy.

Clearly, m_g as defined by (1.3.3) corresponds to L_g in (1.2.1), thus not to L in (1.2.3). This is somewhat surprising and, as explained in [1], the reason for examining whether an expression such as

$$\mathbf{f} = \sqrt{m}\mathrm{D}\left(\sqrt{m}\mathbf{v}\right) = \frac{1}{2}\left(\mathrm{D}\,(m\mathbf{v}) + m\mathrm{D}\mathbf{v}\right) \tag{1.3.4}$$

where m is some function of v, say $m = m(\beta)$, could be of interest. In the affirmative, we could write

$$\mathbf{v}^T\mathbf{f} = \mathrm{D}E_k = \mathrm{D}E \tag{1.3.5}$$

where

$$E = E_i + E_k, \quad E_k = mv^2/2, \quad E_i = \text{const.} \tag{1.3.6}$$

Since $E_k = 0$ for $v = 0$, E_k would be the kinetic energy, thus E_i the rest energy (internal energy) and E the total energy. One could then let m assume a role somewhat similar to the former one of m_g. In particular, defining a rest mass

$$m_0 = m(0) \tag{1.3.7}$$

one would want to preserve the essential former properties and thus require that E_k is equal to $m - m_0$ times a constant and that $m = \infty$ for $v = c$. It then follows from the above expression for E_k (c.f. (1.3.6) and (1.3.7)) that

$$m = \frac{m_0}{1-\beta^2} = \frac{m_0}{\alpha^2}, \tag{1.3.8}$$

and thus that

$$E_k = \frac{mc^2}{2} - \frac{m_0 c^2}{2}. \tag{1.3.9}$$

In view of $E_k = E - E_i$, the validity of (1.3.9) had suggested in the earlier paper [1] the identification of E with $mc^2/2$ and thus E_i with $m_0c^2/2$. This, however, leads to theoretical difficulties (e.g., when applying it to the inelastic collision of bodies) and also to certain experimental incompatibilities (e.g., for nuclear reactions). Nothing contradicts the above equations, however, if one does not impose a requirement on E_i, i.e., on the integration constant implied by (1.3.5), in which case E and E_k are still given by (1.3.6), thus E_k by

$$E_k = \frac{1}{2}mv^2 = E_0\frac{\beta^2}{1-\beta^2}, \quad E_0 = \frac{m_0c^2}{2}, \tag{1.3.10}$$

with m as defined in (1.3.8). Note that E_k can be decomposed into the sum of kinetic energy components according to

$$E_x = \frac{1}{2}mv_x^2, \quad E_y = \frac{1}{2}mv_y^2, \quad E_z = \frac{1}{2}mv_z^2,$$

$$v_x f_x = \mathrm{D}E_x, \quad v_y f_y = \mathrm{D}E_y, \quad v_z f_z = \mathrm{D}E_z.$$

The case discussed in [1] amounted to choosing $E_i = E_0$. Adopting the present point of view, however, amounts to saying that a particle is, in general, characterized by two constants, E_i and E_0. For $E_i = 2E_0$, the rest energy is $E_i = m_0c^2$, as in the classical case. In general, however, no simple relation has to exist between E_i and E_0. As an example, for a particle traveling at the speed of light (photon, neutrino), one could have $E_0 = 0$ and $E_i > 0$. On the other hand, one can apply the general case to the inelastic collision of two equal bodies, i.e., both characterized by the same E_i and E_0. It can be shown that if the resulting fused body is characterized by E_i' and E_0' and if Q is the annihilated kinetic energy, one has

$$E_i' = 2E_i + Q, \quad E_0' = 2E_0 + 2Q,$$

provided the actual kinetic nature of the heat Q is ignored.

In fact, one can go even a step further and consider a particle characterized, in addition to E_i, by a constant positive-definite symmetric 3×3 matrix $\mathbf{m}_0$ instead of a scalar m_0. The scalar m has then also to be replaced by the matrix $\mathbf{m}$ given by (cf. (1.3.8)),

$$\mathbf{m} = \frac{1}{1-\beta^2}\mathbf{m}_0 = \frac{1}{\alpha^2}\mathbf{m}_0. \tag{1.3.11}$$

In (1.3.4), for example, the second expression for $\mathbf{f}$ becomes

$$\mathbf{f} = \frac{1}{2}\left(\mathrm{D}(\mathbf{mv}) + \mathbf{m}\mathrm{D}\mathbf{v}\right).$$

For the power delivered, (1.3.5) remains unchanged, but in (1.3.6) and (1.3.10) the expression for E_k has to be replaced by

$$E_k = \frac{1}{2}\mathbf{v}^T\mathbf{mv} = \frac{1}{2\alpha^2}\mathbf{v}^T\mathbf{m}_0\mathbf{v}. \tag{1.3.12}$$

The matrix case reduces to the scalar one if $\mathbf{m}_0 = \mathbf{1}m_0$ where $\mathbf{1}$ is the unit matrix and m_0 a scalar. All relevant new equations then indeed become as discussed previously. An irreducible example for the matrix case will be discussed in Section 1.5.1.

1.3.2 Four-vector

By means of $\mathbf{f}$ and E_k one can form the quadruple

$$\begin{pmatrix} \mathbf{f} \\ \frac{1}{c}\mathbf{v}^T\mathbf{f} \end{pmatrix} = \begin{pmatrix} \mathbf{f} \\ \frac{1}{c}\mathrm{D}E_k \end{pmatrix}.$$

In the case of a scalar mass one can compare this quadruple to well-known results in relativity theory [2–10]. It can then easily be verified that this quadruple is identical to the four-vector (world vector) originally introduced by Minkowski [2]. It has a Lorentz invariant form. One immediate conclusion is that a Lorentz transformation changes only that component of $\mathbf{f}$ that is oriented in the direction of the relative movement of the two reference frames with respect to each other, i.e., the components of $\mathbf{f}$ perpendicular to that direction remain unchanged, contrary to what holds in classical relativity theory.

It is likely that similar properties also hold in the matrix case, provided $\mathbf{m}_0$ satisfies some restrictions that have still to be examined.

1.4 Direct Derivation of the Alternative Results

1.4.1 Some useful general relations

In view of the puzzling results obtained in Section 1.3, it is of interest to examine whether these same results and related ones could not be obtained by some direct approach. For this we consider two reference frames, S and S', with S' moving with constant velocity v_0, say, in the x-direction of S. For S we use unprimed notation as in Section 1.3 (coordinates, velocities, forces, differential operators, etc.) and for S' the corresponding primed notation. Between S and S' we thus have the Lorentz transformation

$$x' = \frac{1}{\alpha_0}(x - v_0 t), \quad y' = y, \quad z' = z, \quad t' = \frac{1}{\alpha_0}\left(t - \beta_0 \frac{x}{c}\right) \tag{1.4.1}$$

where

$$\beta_0 = v_0/c, \qquad \alpha_0 = \sqrt{1 - \beta_0^2}. \tag{1.4.2}$$

We consider a particle, P, moving as discussed in Section 1.3. For P it is appropriate, however, to introduce vectors of position coordinates $\mathbf{r} = \mathbf{r}(t) = (x, y, z)^T$ and $\mathbf{r}' = \mathbf{r}'(t') = (x', y', z')^T$, while t' can be uniquely expressed in terms of t, say $t' = t'(t)$ and vice versa. While (1.3.1) and (1.3.2) refer to S, we have, with respect to S'

$$\mathbf{f}' = \left(f'_x, f'_y, f'_z\right)^T, \qquad \mathbf{v}' = \left(v'_x, v'_y, v'_z\right)^T, \tag{1.4.3}$$

$$\mathbf{v}'^T\mathbf{v}' = v'^2, \quad \beta' = v'/c, \quad \alpha' = \sqrt{1 - \beta'^2}, \tag{1.4.4}$$

where (cf. (1.2.2)) $\mathbf{v}' = \mathrm{D}'\mathbf{r}'$ and $\mathrm{D}' = \mathrm{d}/\mathrm{d}t'$. As is known and as can be verified by means of (1.4.1),

$$\frac{\mathrm{d}t}{\mathrm{d}t'} = \frac{\alpha'}{\alpha} = \frac{\alpha'_x}{\alpha_x}, \qquad \alpha'\mathrm{D} = \alpha\mathrm{D}' \tag{1.4.5}$$

where

$$\alpha_x = \sqrt{1 - \beta_x^2}, \quad \alpha'_x = \sqrt{1 - \beta'^2_x}, \quad \beta_x = \frac{v_x}{c}, \quad \beta'_x = \frac{v'_x}{c}, \tag{1.4.6}$$

and furthermore,

$$v'_x = \frac{v_x - v_0}{1 - \beta_x\beta_0}, \quad v'_y = \frac{\alpha'}{\alpha}v_y, \quad v'_z = \frac{\alpha'}{\alpha}v_z. \tag{1.4.7}$$

The following expressions can, in turn, be derived from (1.4.2) to (1.4.7),

$$\frac{1}{\alpha^3}\mathrm{D}v_x = \frac{1}{\alpha'^3}\mathrm{D}'v'_x \tag{1.4.8}$$

$$\frac{1}{\alpha_x^3}\mathrm{D}v_x = \frac{1}{{\alpha'_x}^3}\mathrm{D}'v'_x \tag{1.4.9}$$

$$\alpha'^2\mathrm{D}v_y - \alpha^2\mathrm{D}'v'_y = \alpha' v'_y \mathrm{D}\alpha - \alpha v_y \mathrm{D}'\alpha' \tag{1.4.10}$$

$$\alpha'^2\mathrm{D}v_z - \alpha^2\mathrm{D}'v'_z = \alpha' v'_z \mathrm{D}\alpha - \alpha v_z \mathrm{D}'\alpha' \tag{1.4.11}$$

$$\frac{1}{\alpha'^3}\mathrm{D}'v'_x = \mathrm{D}\frac{v_x}{\alpha} + \frac{1-\alpha^2}{\alpha^3}\mathrm{D}v_x + \frac{v_x}{2\alpha^3}\mathrm{D}\alpha^2$$

$$= \mathrm{D}\frac{v_x}{\alpha} + \frac{\beta_y^2+\beta_z^2}{\alpha^3}\mathrm{D}v_x - \frac{v_x}{\alpha^3}\left(\beta_y\mathrm{D}\beta_y + \beta_z\mathrm{D}\beta_z\right) \tag{1.4.12}$$

$$\mathrm{D}'v'_y = \frac{\alpha'^2}{\alpha}\mathrm{D}\left(\frac{v_y}{\alpha}\right) + v'_y\frac{1}{\alpha}\mathrm{D}\alpha' \tag{1.4.13}$$

$$\mathrm{D}'v'_z = \frac{\alpha'^2}{\alpha}\mathrm{D}\left(\frac{v_z}{\alpha}\right) + v'_z\frac{1}{\alpha}\mathrm{D}\alpha' \tag{1.4.14}$$

and furthermore, from (1.4.9),

$$\mathrm{D}\frac{v_x}{\alpha_x} = \mathrm{D}'\frac{v'_x}{\alpha'_x}. \tag{1.4.15}$$

1.4.2 Modification of Newton's second law

Newton's first law is purely qualitative and thus irrelevant for our purpose, but his second law concerns the expression of the force. For deriving the corresponding alternative expression, we consider two reference frames S and S' as discussed in Section 1.4.1. In order to obtain a mathematically accurate proof we consider two time instants, t_1 and t_2, that are arbitrary close, yet distinct. Quantities referring to t_1 and t_2 will be given subscripts 1 and 2, respectively, thus, e.g.,

$$t'_1 = t'(t_1), \qquad t'_2 = t'(t_2),$$

$$\mathbf{r}_1 = \mathbf{r}(t_1), \qquad \mathbf{r}_2 = \mathbf{r}(t_2), \qquad x_1 = x(t_1), \qquad x_2 = x(t_2),$$

$$\mathbf{v}_1 = \mathbf{v}(t_1), \qquad \mathbf{v}_2 = \mathbf{v}(t_2), \qquad v_{x1} = v_x(t_1), \qquad v_{x2} = v_x(t_2).$$

We also define $\Delta t = t_2 - t_1$, $\Delta t' = t'_2 - t'_1$, $\Delta\mathbf{r} = \mathbf{r}_2 - \mathbf{r}_1$, $\Delta\mathbf{r}' = \mathbf{r}'_2 - \mathbf{r}'_1$, etc. and use a simplified notation defined by

$$\mathbf{O}^n = \mathbf{O}\left((\Delta t)^n\right) = \text{order of } (\Delta t)^n \text{ for } n \geq 0.$$

The proof does indeed involve small quantities of different degrees of smallness, and we have to clearly distinguish between those. Due to $\alpha dt = \alpha' dt'$ (cf. (1.4.5)) we have

$$\Delta t' = t'(t_1 + \Delta t) - t_1' = \Delta t\,(\mathrm{d}t'/\mathrm{d}t)_1 + \mathbf{O}^2 = \frac{\alpha_1}{\alpha_1'}\Delta t + \mathbf{O}^2$$

and therefore

$$\alpha_1' \Delta t' = \alpha_1 \Delta t + \mathbf{O}^2, \qquad \alpha_1'^2 (\Delta t')^2 = \alpha_1^2 (\Delta t)^2 + \mathbf{O}^3 \tag{1.4.16}$$

and thus also

$$\mathbf{O}\left((\Delta t')^n\right) = \mathbf{O}\left((\Delta t)^n\right) = \mathbf{O}^n.$$

By Taylor series expansion we obtain

$$\Delta \mathbf{r} = \mathbf{v}_1 \Delta t + \frac{1}{2}(\Delta t)^2 (\mathrm{D}\mathbf{v})_1 + \mathbf{O}^3 \tag{1.4.17}$$

$$\Delta \mathbf{r}' = \mathbf{v}_1' \Delta t' + \frac{1}{2}(\Delta t')^2 \left(\mathrm{D}'\mathbf{v}'\right)_1 + \mathbf{O}^3 \tag{1.4.18}$$

For further analyzing these expressions we have to distinguish between the individual components of $\mathbf{r}$ and $\mathbf{r}'$. Taking into account (1.4.8) and (1.4.16), we derive from (1.4.17) and (1.4.18) for the x and x' coordinates,

$$\alpha_1 \Delta x_r' = \alpha_1' \Delta x_r + \mathbf{O}^3 \tag{1.4.19}$$

where we have used the definitions

$$\Delta x_r = \Delta x - v_{x1}\Delta t, \qquad \Delta x_r' = \Delta x' - v_{x1}'\Delta t'. \tag{1.4.20}$$

Furthermore, taking into account (1.4.16) we derive from (1.4.17) and (1.4.18) and then (1.4.10) and (1.4.11),

$$\Delta' y_r - \Delta y_r = \frac{1}{2}\left(\left(\mathrm{D}'v_y'\right)_1 - \left(\frac{\alpha_1'}{\alpha_1}\right)^2 (\mathrm{D}v_y)_1\right)(\Delta t')^2 + \mathbf{O}^3$$

$$= \frac{1}{2}\left(\alpha_1 v_{y1}\left(\mathrm{D}'\alpha'\right)_1 - \alpha_1' v_{y1}'(\mathrm{D}\alpha)_1\right)\left(\frac{\Delta t'}{\alpha_1}\right)^2 + \mathbf{O}^3 \tag{1.4.21}$$

$$\Delta' z_r - \Delta z_r = \frac{1}{2}\left(\alpha_1 v_{z1}\left(\mathrm{D}'\alpha'\right)_1 - \alpha_1' v_{z1}'(\mathrm{D}\alpha)_1\right)\left(\frac{\Delta t'}{\alpha_1}\right)^2 + \mathbf{O}^3 \tag{1.4.22}$$

where, correspondingly to (1.4.20),

$$\Delta y_r = \Delta y - v_{y1}\Delta t, \qquad \Delta y_r' = \Delta y' - v_{y1}'\Delta t', \tag{1.4.23}$$

$$\Delta z_r = \Delta z - v_{z1}\Delta t, \qquad \Delta z'_r = \Delta z' - v'_{z1}\Delta t'. \tag{1.4.24}$$

For ease of reasoning we temporarily consider a third reference frame, S'', which travels with velocity $\mathbf{v}_1$ with respect to S, thus with velocity $\mathbf{v}'_1$ with respect to S'. During Δt, S'' moves by $\mathbf{v}_1\Delta t$ with respect to S, and, equivalently, by $\mathbf{v}'_1\Delta t'$ with respect to S', and the particle P will be observed in S as moving by, say, $\Delta\mathbf{r}$ and, equivalently, in S' by $\Delta\mathbf{r}'$. If no force is acting on P we simply have $\Delta\mathbf{r} = \mathbf{v}_1\Delta t$ and, equivalently, $\Delta\mathbf{r}' = \mathbf{v}'_1\,\Delta t'$, i.e., P simply shares the movement of S''. If, however, a force is acting, this simple situation no longer holds, but the movement of P can then be decomposed into a movement identical to the one just explained and a residual movement that is exclusively due to the movement of P with respect to S''. The residual movement results in a displacement that is of size $\Delta\mathbf{r}_r$ if observed in S and, equivalently, $\Delta\mathbf{r}'_r$ if observed in S', where

$$\Delta\mathbf{r}_r = \Delta\mathbf{r} - \mathbf{v}_1\Delta t, \;\; \Delta\mathbf{r}'_r = \Delta\mathbf{r}' - \mathbf{v}'_1\Delta t' \tag{1.4.25}$$

and where in view of (1.4.20), (1.4.23), and (1.4.24),

$$\Delta\mathbf{r}_r = (\Delta x_r, \Delta y_r, \Delta z_r)^T, \qquad \Delta\mathbf{r}'_r = (\Delta x'_r, \Delta y'_r, \Delta z'_r)^T.$$

As follows from (1.4.17) and (1.4.18), both $\Delta\mathbf{r}_r$ and $\Delta\mathbf{r}'_r$ are $\mathbf{O}^2$ and they are related by (1.4.19), (1.4.21), and (1.4.22).

Associated with the residual movement just explained is a work done by the force acting on P. Let $\mathbf{f}$ be the value of that force as observed in S, and, equivalently, $\mathbf{f}'$ its value as observed in S'. Within the interval of interest $t_1 \le t \le t_2$, we may write

$$\mathbf{f} = \mathbf{f}_1 + \mathbf{O}^1, \qquad \mathbf{f}' = \mathbf{f}'_1 + \mathbf{O}^1$$

where $\mathbf{f}_1$ and $\mathbf{f}'_1$ refer to t_1 and t'_1, respectively. For the work done as just explained we can then write, if it is observed in S,

$$\int_{\mathbf{0}}^{\Delta\mathbf{r}_r} \mathbf{f}^T \mathrm{d}\,(\Delta\mathbf{r}_r) = \left(\mathbf{f}_1^T + \mathbf{O}^1\right)\Delta\mathbf{r}_r = \mathbf{f}_1^T\Delta\mathbf{r}_r + \mathbf{O}^3, \tag{1.4.26}$$

and if it is observed in S',

$$\int_{\mathbf{0}}^{\Delta\mathbf{r}'_r} \mathbf{f}'^T \mathrm{d}\,(\Delta\mathbf{r}'_r) = \left(\mathbf{f}_1'^T + \mathbf{O}^1\right)\Delta\mathbf{r}'_r = \mathbf{f}_1'^T\Delta\mathbf{r}'_r + \mathbf{O}^3, \tag{1.4.27}$$

where we have made use of the observation subsequent to (1.4.25) that $\Delta\mathbf{r}_r$ and $\Delta\mathbf{r}'_r$ are $\mathbf{O}^2$. In classical mechanics, the left-hand sides of (1.4.26) and (1.4.27) are equal.

We now concentrate on S and S'. We may choose v_0 and the orientation of S and S' such that

$$\mathbf{v}'_1 = \mathbf{0}, \;\; \text{thus } \alpha'_1 = 1. \tag{1.4.28}$$

We then say that the displacement in S' is *basal*, and we have (cf. (1.4.25))

$$\Delta \mathbf{r}'_r = \Delta \mathbf{r}'. \tag{1.4.29}$$

On the other hand, it is not unreasonable to associate something like an invariance of work done with conservation of energy. In order to explain that notion we consider the evaluation of the work done in S' (cf. left-hand side of (1.4.27)) under the assumption of a basal displacement in that frame. If that work done is evaluated in S, thus by means of the left-hand side of (1.4.26), we require that expression to become at least asymptotically correct if $\Delta t' \to 0$, thus if $\Delta t \to 0$. This asymptotic coincidence of the left-hand sides of (1.4.26) and (1.4.27) implies for the right-hand sides that the value of

$$\mathbf{f}_1^T \Delta \mathbf{r}_r - \mathbf{f}_1'^T \Delta \mathbf{r}'_r + \mathbf{O}^3$$

goes to zero faster than $\Delta \mathbf{r}_r$ and $\Delta \mathbf{r}'_r$, thus, in view of (1.4.17), (1.4.18), (1.4.25), and (1.4.29) that

$$\mathbf{f}_1^T \Delta \mathbf{r}_r - \mathbf{f}_1'^T \Delta \mathbf{r}' = \mathbf{O}^n, \quad n > 2. \tag{1.4.30}$$

Since the components of $\Delta \mathbf{r}_r$ are independent of each other, (1.4.30) must hold for each component separately, i.e., with the other two components set equal to zero.

Taking into account (1.4.19), (1.4.28), and (1.4.29), we thus deduce from (1.4.30),

$$(f'_{x1} - \alpha_1 f_{x1})\,\Delta x' = \mathbf{O}^n, \quad n > 2. \tag{1.4.31}$$

But as follows from (1.4.18) and (1.4.28), $\Delta x'$ not only is $\mathbf{O}^2$ but in fact decreases precisely with $(\Delta t)^2$ if $\left(\mathrm{D}' v'_x\right)_1 \neq 0$, as will in general be the case if a force is acting. Hence (1.4.31) assures us that $f'_{x1} - \alpha_1 f_{x1}$ is of order > 0 in Δt, and since in fact it is independent of Δt, we conclude that

$$f'_{x1} - \alpha_1 f_{x1} = 0. \tag{1.4.32}$$

For the other force components we observe that in view of (1.4.7) and (1.4.28), (1.4.21) and (1.4.22) yield

$$\Delta y'_r = \Delta y_r + \mathbf{O}^3, \quad \Delta z'_r = \Delta z_r + \mathbf{O}^3.$$

Proceeding as before, we then find

$$f'_{y1} = f_{y1}, \quad f'_{z1} = f_{z1}. \tag{1.4.33}$$

Due to (1.4.28) the situation in S' becomes Newtonian for $t' = t'_1$, and we can therefore write

$$\mathbf{f}'_1 = m_0 \left(\mathrm{D}'\mathbf{v}'\right)_1 \tag{1.4.34}$$

where m_0 is some positive constant. Using (1.4.12) to (1.4.14), and (1.4.32) to (1.4.34) we obtain

$$\mathbf{f}_1 = \frac{m_0}{\alpha_1} \left(\mathrm{D}\frac{\mathbf{v}}{\alpha}\right)_1. \tag{1.4.35}$$

Let us next assume that the reference frame S is in fact a rotated version of the one designated so far by S. All properties that have so far been considered as holding for $\mathbf{v}_1$ and $\mathbf{f}_1$ then in fact hold for $\mathbf{U}\mathbf{v}_1$ and $\mathbf{U}\mathbf{f}_1$ where $\mathbf{U}$ is some constant 3×3 orthogonal matrix. In view of (1.4.35) we may thus write

$$\mathbf{U}\mathbf{f}_1 = \frac{m_0}{\alpha_1}\left(\mathrm{D}\left(\frac{1}{\alpha}\mathbf{U}\mathbf{v}\right)\right)_1 = \mathbf{U}\frac{m_0}{\alpha_1}\left(\mathrm{D}\left(\frac{1}{\alpha}\mathbf{v}\right)\right)_1$$

whence (1.4.35) follows again, but this time without having to require that the x-axis of S is aligned with $\mathbf{v}_1$. But t_1 is arbitrary, so that we can finally replace (1.4.35) by

$$\mathbf{f} = \frac{m_0}{\alpha}\mathrm{D}\left(\frac{1}{\alpha}\mathbf{v}\right) = \sqrt{m}\mathrm{D}\left(\sqrt{m}\mathbf{v}\right) \tag{1.4.36}$$

where, confirming (1.3.4) and (1.3.8),

$$m = m_0/\alpha^2, \quad \alpha^2 = 1 - \beta^2, \quad \beta = v/c, \quad v = |\mathbf{v}|,$$

and where the orientation of S is not subject to any restriction. Correspondingly we have in a frame S' that, although differing from the old S', is related to the new frame S by (1.4.1),

$$\mathbf{f}' = \frac{m_0}{\alpha'}\mathrm{D}'\left(\frac{1}{\alpha'}\mathbf{v}'\right), \tag{1.4.37}$$

in accordance with the Lorentz invariance mentioned in Section 1.3.2.

If $v_y \equiv v_z \equiv 0$, i.e., if all phenomena take place exclusively along the x- and x'-axes, we have $\alpha_x = \alpha$, $\alpha'_x = \alpha'$, and we then derive from (1.4.15), (1.4.36), and (1.4.37)

$$\alpha\mathbf{f} = \alpha'\mathbf{f}'. \tag{1.4.38}$$

This corresponds to the result obtained in [1].

1.4.3 Newton's third law and the conservation of momentum

While in classical relativity Newton's third law remains untouched, it has to be modified if one adopts the alternative viewpoint. In order to show this, we consider action and reaction between two particles P_1 and P_2, the subscripts 1 and 2 being used hereafter consistently to distinguish between quantities referring either to P_1 or P_2, respectively. We assume P_1 and P_2 to be traveling with velocities $\mathbf{v}_1$ and $\mathbf{v}_2$. Let there be an interaction between P_1 and P_2 and let $\mathbf{f}_1$ and $\mathbf{f}_2$ be the resulting forces acting upon P_1 and P_2, respectively. We first consider the case that $\mathbf{f}_1$, $\mathbf{f}_2$, $\mathbf{v}_1$, and $\mathbf{v}_2$, are all located on the same straight line, which we may assume to be the x-axis of S, thus the x'-axis of S' (defined as before); for S' we thus have to consider $\mathbf{f}'_1$, $\mathbf{f}'_2$, $\mathbf{v}'_1$ and $\mathbf{v}'_2$ (and similarly primed quantities used hereafter). Defining

$$\beta_1 = v_{x1}/c, \quad \beta'_1 = v'_{x1}/c, \quad \beta_2 = v_{x2}/c, \quad \beta'_2 = v'_{x2}/c,$$

we can write (cf. (1.4.7)),

$$\beta_1' = \frac{\beta_1 - \beta_0}{1 - \beta_1\beta_0}, \quad \beta_2' = \frac{\beta_2 - \beta_0}{1 - \beta_2\beta_0}, \quad \beta_0 = v_0/c. \tag{1.4.39}$$

Assume P_1 and P_2 to touch each other at $t = t_0$, which implies $x_1(t_0) = x_2(t_0)$ and thus (cf. (1.4.1))

$$x_1'(t_0') = x_2'(t_0'), \quad t_0' = t_1'(t_0) = t_2'(t_0).$$

To simplify the writing we add a subscript zero to specify evaluations at $t = t_0$, thus at $t' = t_0'$. As it can be shown to be always feasible (cf. (1.4.39)), we choose v_0 such that we obtain the symmetric situation

$$\beta_{10}' = -\beta_{20}', \quad \text{thus} \quad \alpha_{10}' = \alpha_{20}'. \tag{1.4.40}$$

Under the assumption concerning the orientation of the velocities and forces and the reference frames, we may make use of (1.4.38) and thus write in particular,

$$\alpha_{10}\mathbf{f}_{10} = \alpha_{10}'\mathbf{f}_{10}', \qquad \alpha_{20}\mathbf{f}_{20} = \alpha_{20}'\mathbf{f}_{20}'. \tag{1.4.41}$$

But due to the symmetry of the velocities in S' we must have $\mathbf{f}'_{10} = -\mathbf{f}_{20}'$. Hence, we obtain from (1.4.40) and (1.4.41),

$$\alpha_{10}\mathbf{f}_{10} = -\alpha_{20}\mathbf{f}_{20}. \tag{1.4.42}$$

Furthermore if, while maintaining the assumption about the alignment of $\mathbf{v}_1$, $\mathbf{v}_2$, $\mathbf{f}_1$, $\mathbf{f}_2$, we abandon the assumption about the coincidence of that common direction with the one of the x- and x'-axes, a rotation argument as used in Section 1.4.2 shows that (1.4.42) remains valid even then.

Consider now the general case of the relationship between action and reaction. In view of (1.4.42) it is clear that it could not be $\mathbf{f}_1 = -\mathbf{f}_2$ as in the classical theory. For modifying Newton's third law, however, the simplest way compatible with (1.4.42) is

$$\alpha_1\mathbf{f}_1 = -\alpha_2\mathbf{f}_2. \tag{1.4.43}$$

An important consequence can immediately be drawn from (1.4.43). Indeed, if the relevant masses of P_1 and P_2 are m_{10} and m_{20}, it follows from (1.4.36) that

$$\mathrm{D}(\mathbf{p}_1 + \mathbf{p}_2) = \mathbf{0}$$

where

$$\mathbf{p}_1 = m_{10}\mathbf{v}_1/\alpha_1, \quad \mathbf{p}_2 = m_{20}\mathbf{v}_2/\alpha_2$$

are the momenta mentioned in Section 1.3.1. More generally, if there are n particles P_1 to P_n and if forces $\mathbf{f}_{\nu 1}$ to $\mathbf{f}_{\nu k}$ are acting upon P_ν, which is moving with velocity $\mathbf{v}_\nu$, we have

$$\mathrm{D}\mathbf{p}_\nu = \alpha_\nu(\mathbf{f}_{\nu 1} + \cdots + \mathbf{f}_{\nu k}) \tag{1.4.44}$$

where

$$\mathbf{p}_\nu = m_{\nu 0}\mathbf{v}_\nu/\alpha_\nu, \quad \alpha_\nu = \sqrt{1-\beta_\nu^2}, \quad \beta_\nu = v_\nu/c, \quad v_\nu^2 = \mathbf{v}_\nu^T\mathbf{v}_\nu.$$

Summing (1.4.44) over all particles, there will be pairwise cancellations in the right-hand side due to (1.4.43). Hence,

$$\mathrm{D}\,(\mathbf{p}_1 + \cdots + \mathbf{p}_n) = \mathbf{0}. \tag{1.4.45}$$

Conservation of momentum thus holds exactly as in classical relativity. Vice versa, if in the alternative theory we require (1.4.45) to apply in all cases, one is forced to modify Newton's third law as stated in (1.4.43).

1.4.4 Forces and energy in fields

A further consequence can be drawn from the result in Section 1.4.3. Assume that a field (electromagnetic, gravitational, etc.) is exerting a force $\mathbf{f}$ upon a particle P traveling with velocity $\mathbf{v}$, that the equipment, Eq, producing the field is at rest ($\mathbf{v}_1 = \mathbf{0}$), and that the reaction upon Eq is $\mathbf{f}_1$. Assume furthermore that all relevant distances are sufficiently small so that we can state (cf. (1.4.43)),

$$\alpha\mathbf{f} = -\alpha_1\mathbf{f}_1 = -\mathbf{f}_1 \tag{1.4.46}$$

with α and α_1 as in (1.3.2) and as used in Section 1.4.3 (here thus $\alpha_1 = 1$). But $\mathbf{f}_1$ is independent of the velocity of P and must thus be equal to the force existing for $\mathbf{v} = \mathbf{0}$. Hence, we conclude from (1.4.46) that

$$\mathbf{f} = \frac{1}{\alpha}\mathbf{f}_0 \tag{1.4.47}$$

where $\mathbf{f}_0$ is that force that the field exerts upon P if $\mathbf{v} = \mathbf{0}$ and thus is equal to $-\mathbf{f}_1$, i.e., $\mathbf{f}_0$ is the force as used in all classical expressions (e.g., $\mathbf{f}_0 = e\mathbf{E}$ if e is a charge and $\mathbf{E}$ the electric field). Furthermore, if in (1.4.47) we replace $\mathbf{f}$ by (1.4.36) we can write

$$m_0\mathrm{D}\left(\frac{1}{\alpha}\mathbf{v}\right) = \mathbf{f}_0. \tag{1.4.48}$$

Hence, the alternative theory leads to exactly the same dynamic behavior of a particle in a field as the classical relativity theory.

This, however, does not include energy. Consider indeed a relation such as

$$-\rho\mathbf{v}^T\mathbf{E} = \frac{1}{2}\mathrm{D}\left(\varepsilon\mathbf{E}^T\mathbf{E} + \mu\mathbf{H}^T\mathbf{H}\right) + \mathrm{div}\,(\mathbf{E}\times\mathbf{H})$$

from classical electrodynamics, where $\rho = \varepsilon\cdot\mathrm{div}\mathbf{E}$ is the charge density and where ε and μ are assumed constant (e.g., $\varepsilon = \varepsilon_0$, $\mu = \mu_0$). The alternative theory obviously requires us to multiply the left-hand side by $1/\alpha$ to obtain a proper expression for the energy supplied, and the right-hand side thus has to be multiplied in the same way.

We consider here only the case of a constant α (see also Section 1.5.1). Clearly, the field energy density then turns out to be

$$e = \frac{1}{2\alpha}\left(\varepsilon \mathbf{E}^T\mathbf{E} + \mu \mathbf{H}^T\mathbf{H}\right) \tag{1.4.49}$$

and the Poynting vector has to be replaced similarly by $(\mathbf{E} \times \mathbf{H})/\alpha$. The classical expression

$$\frac{\varepsilon \mathbf{E}^T\mathbf{E} + \mu \mathbf{H}^T\mathbf{H}}{2}$$

may then be considered to be something like an apparent energy density.

1.5 Some Applications

1.5.1 Moving electrostatic field

We define S and S' as done so far and consider a field that in S' is electrostatic, thus at rest. In S' we designate its strength by $\mathbf{E}'$, which is thus independent of t', while in S it comprises $\mathbf{E}$ and $\mathbf{H}$. Since we assume ε and μ constant we can ignore $\mathbf{D}$ and $\mathbf{B}$. In S, the field is moving with constant velocity $\mathbf{v} = (v_0, 0, 0)^T$, i.e., using the notation as before, we have $v = v_0$. Making use of standard results for transforming an electromagnetic field from S' to S as well as of (1.3.2) we can write

$$E_x = E'_x, \quad E_y = \frac{1}{\alpha}E'_y, \quad E_z = -\frac{1}{\alpha}E'_z,$$

$$H_x = 0, \quad H_y = -\varepsilon\frac{v}{\alpha}E'_z, \quad H_z = \varepsilon\frac{v}{\alpha}E'_y.$$

For the field energy density in S at any fixed time t we thus obtain according to (1.4.49),

$$e = \frac{1}{\alpha}e' + \frac{\beta^2}{\alpha^3}e'_0 \tag{1.5.1}$$

where

$$e' = \frac{1}{2}\varepsilon\left(E'^2_x + E'^2_y + E'^2_z\right), \quad e'_0 = \varepsilon\left(E'^2_y + E'^2_z\right) \tag{1.5.2}$$

e' being the field energy density in S'. Due to the electrostatic hypothesis, the right-hand side of (1.5.1) is independent of t'. Let

$$\mathrm{d}V = \mathrm{d}x \cdot \mathrm{d}y \cdot \mathrm{d}z, \quad \mathrm{d}V' = \mathrm{d}x' \cdot \mathrm{d}y' \cdot \mathrm{d}z'$$

be the elementary volumes in S and S', respectively. As is known, it follows from (1.4.1) that

$$\mathrm{d}x = \alpha\mathrm{d}x', \quad \mathrm{d}y = \mathrm{d}y', \quad \mathrm{d}z = \mathrm{d}z',$$

so that $\mathrm{d}V = \alpha \mathrm{d}V'$, thus from (1.5.1),

$$e\mathrm{d}V = e'\mathrm{d}V' + \frac{\beta^2}{\alpha^2} e'_0 \mathrm{d}V'. \tag{1.5.3}$$

This expression is precisely of the form determined by (1.3.6) and (1.3.8), with E, E_i, and $m_0 c^2/2$ replaced by $e\mathrm{d}V$, $e'\mathrm{d}V'$, and $e'_0\mathrm{d}V'$, respectively, and no strict relationship existing between e' and e'_0. We may also integrate (1.5.3) over the entire volume, which, using W instead of E for designating energy to avoid confusion, leads to

$$W = W' + W_k,$$

$$W = \int_V e\mathrm{d}V, \quad W' = \int_{V'} e'\mathrm{d}V', \quad W_k = \frac{\beta^2}{\alpha^2} \int_{V'} e'_0 \mathrm{d}V'.$$

Since e' and e'_0 are independent of t', W' and W_k may be considered to be evaluated at a constant t'. Furthermore, V and V' may be the entire space. If then the total contribution by each one of the components E_x, E_y, and E_z is the same, we obtain from (1.5.2)

$$\int_{V'} e'_0 dV' = \frac{4}{3} W'.$$

In all cases we thus obviously obtain a perfect correspondence with the results developed in the preceding sections. The same can be shown to hold if a static field (electron ?) is combined electrostatic and magnetostatic and has a total field momentum equal to zero. Similarly elegant interpretations would not be possible if we use the decomposition into rest energy and kinetic energy as occurs in classical relativity theory. This is remarkable since Maxwell's equations are inherently compatible with the Lorentz transformation.

The second expression (1.5.2) can also be written as

$$e'_0 = \varepsilon \mathbf{E}'^T \mathbf{E}' - \varepsilon E'^2_x.$$

But E'_x is actually the component of $\mathbf{E}'$ in the direction of $\mathbf{v}'$, $\mathbf{v}' = -\mathbf{v}$ being the velocity of S with respect to S', and we have $\mathbf{v}'^T \mathbf{E}' = -\mathbf{v}^T \mathbf{E}' = -vE'_x$. Hence, (1.5.3) can also be written in the more general form

$$e\mathrm{d}V = \left(e' + \frac{1}{2\alpha^2} \mathbf{v}^T \mathbf{m}_0 \mathbf{v} \right) \mathrm{d}V', \tag{1.5.4}$$

$$\mathbf{m}_0 = \frac{2\varepsilon}{c^2} \left(\left(\mathbf{E}'^T \mathbf{E}' \right) \mathbf{1} - \mathbf{E}' \mathbf{E}'^T \right), \tag{1.5.5}$$

$\mathbf{1}$ being the unit matrix of order 3. These expressions hold for any orientation of $\mathbf{v}$ and have the form of the matrix case discussed in relation with (1.3.12).

1.5.2 Kinetic energy

As we have seen, tests based on conservation of momentum or movement of particles in fields cannot differentiate between the classical and the alternative approach. Tests based on the rest energy are also inconclusive because the latter occurs in the alternative approach as an arbitrary integration constant. For fast moving particles, however, the kinetic energy is distinct. In 1964, Bertozzi [12] published a few results of an experiment in which he had determined β^2 in terms of the kinetic energy of fast electrons. He had found that for appreciably large velocities the data he had measured were located far away from the Newton straight line (as had obviously been expected) but were reasonably close to the curve

$$\beta^2 = 1 - \frac{1}{(1+\gamma)^2}; \quad \gamma = E_k/m_0c^2 \tag{1.5.6}$$

that follows from the classical Einstein formula. If one plots into the same diagram the curve

$$\beta^2 = 2\gamma/(1+2\gamma) \tag{1.5.7}$$

which follows from (1.3.10) and which has the same general tendency as (1.5.6), one notices (Figure 1.5.1) that the measured data are located roughly evenly between (1.5.6) and (1.5.7).

1.6 Conclusions

Stimulated by results in wave-digital theory, it has been shown that some reasonable assumptions about work done led (for the force exerted on a particle) to an expression that differs from the one in classical relativity theory. This alternative way of modifying Newton's second law implies that Newton's third law has to be modified correspondingly. The combination of these two changes leads in some situations to the same result as the classical theory (conservation of momentum, movement of particles in fields). The rest energy occurs as an unspecified integration constant, and there is thus no incompatibility with the classical expression either. A definite difference exists for the kinetic-energy expressions. For a uniformly moving electrostatic field, the kinetic term of the field energy is in agreement with the alternative approach, not with the classical one. The few measured results published by Bertozzi and aimed at differentiating between the Newton and the Einstein predictions are in fact inconclusive because they are essentially located between the latter curve and the one for the alternative approach.

Although this has not been addressed more specifically, the alternative theory is compatible with some further-reaching ideas that say that elementary particles are in fact compressed (thus high-density) fields. A complete field composed of individual fields such as electromagnetic and gravitational fields is indeed highly nonlinear (electromagnetic forces proportional to products of charges, which are nonvanishing divergences, masses proportional to the square of charges, gravitational forces

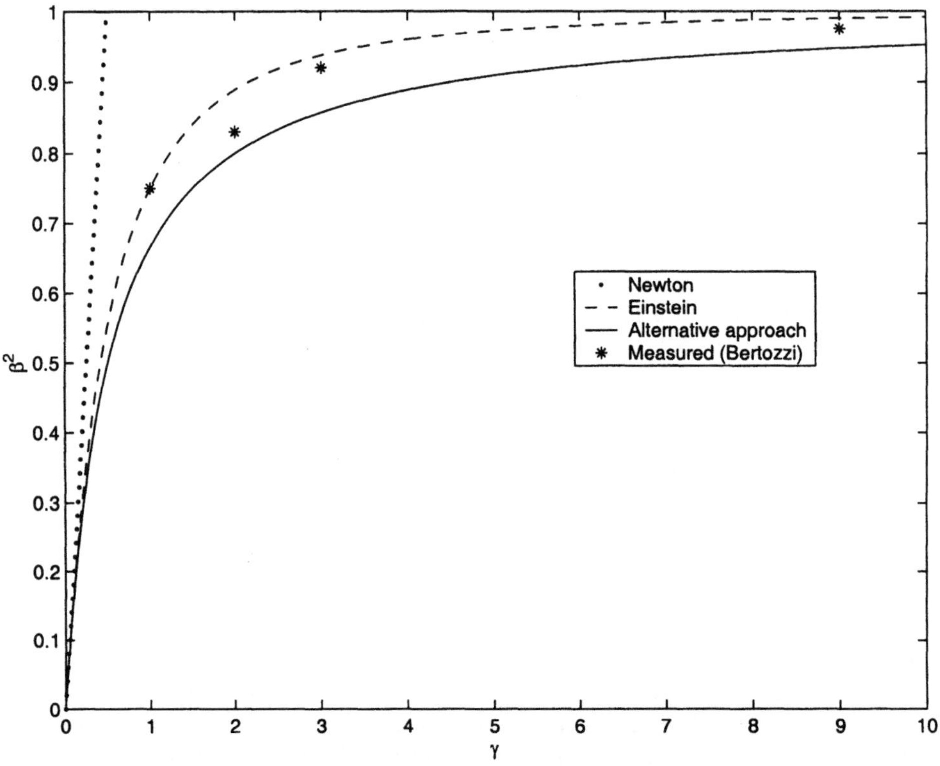

Figure 1.5.1: Results of the Bertozzi experiment in comparison with the curves resulting from the classical and the alternative approach

to the product of masses, etc.) and therefore can exist in a stable fashion only for a discrete set of configurations, reminiscent of quantum states. Due to the varying influence of outside forces acting upon the particle, an equilibrium configuration could not be rigidly maintained but there would be continuous fluctuations around it, and this might more or less frequently result in jumps to other stable configurations, thus giving the appearance of a purely probabilistic behavior. As a further example, under the influence of strong surrounding fields, elementary particles could split (also, e.g., for passing through holes located closely to each other) and then recombine. Other paradoxes of quantum physics could also be explained, at least qualitatively. There would thus be a complicated dynamic behavior behind quantum phenomena. Quantum jumps would be like electronic flip-flops switching from one stable state to another. Such switching is indeed carried out by complicated, nonlinear electric phenomena, but all one is interested in is the final outcome and the settling time. All this could explain why, for example, the use of the formalism of delta functions and the concept of group delay are of similar importance in communication systems and electronics as in physics. In addition to particles there might also exist uncompressed, say vagabonding fields (dark matter ?), which would be undetectable by conventional instruments based on interactions of particles. It could in fact be that all basic laws are quite simple, but that the observed complexity comes into play because of the gigantic variability that is involved.

Despite these positive comments, it is very possible that the ideas developed in this chapter do not correspond to physical reality. Even then, however, it is hoped they could contribute to clarifying some of the issues addressed.

Acknowledgments

The author is greatly indebted to M. Bouten, M. Depenbrock, P. Fettweis, A. Hahn, K.-A. Owenier, H. Rollnick, and R. Staufenbiel for many discussions on the subject of this chapter and to F.-J. Reich for substantial help in preparing the manuscript.

Bibliography

[1] A. Fettweis, "The wave-digital method and some of its relativistic implications," *IEEE Trans. Circuits and Systems I*, vol. 49, no. 6, pp. 862–868, June 2002, and no. 10, p. 1521, Oct. 2002.

[2] H. Minkowski, "Raum und Zeit," published in *Das Relativitätsprinzip*, Teubner, Leipzig, 1913; English version "Space and Time," in *The Principle of Relativity*, Methuen, 1923, and Dover, New York, 1952.

[3] P. G. Bergmann, *Introduction to the Theory of Relativity*, Prentice-Hall, Englewood Cliffs, NJ, 1942, and Dover, New York, 1976.

[4] M. Born, *Die Relativitätstheorie Einsteins*, 5th ed., Springer-Verlag, Berlin, 1969, American translation of an earlier German edition: *Einstein's Theory of Relativity*, Dover, New York, 1965.

[5] S. Weinberg, *Gravitation and Cosmology*, Wiley, New York, 1972.

[6] R. Becker and F. Sauter, *Theorie der Elektrizität*, vol. 1, Teubner, Stuttgart, 1973.

[7] J. D. Jackson, *Classical Electrodynamics*, 2nd ed., Wiley, New York, 1975.

[8] W. Panofsky and M. Phillips, *Classical Electricity and Magnetism*, Addison-Wesley, Reading, MA, 1977.

[9] C. Kacsor, "Relativity, special theory," in *Encyclopedia of Physics*, 2nd ed., R. G. Lerner and G. L. Triggs, Eds., pp. 1052–1058, Wiley-VCH, New York, 1991.

[10] H. and M. Ruder, *Die spezielle Relativitätstheorie*, Vieweg, Braunschweig, Germany, 1993.

[11] A. Fettweis, "A mathematical and physical complement to the paper 'The wave-digital method and some of its relativistic implications'," submitted for publication.

[12] W. Bertozzi, "Speed and energy of relativistic electrons," *American J. Physics*, vol. 32, pp. 551–555, 1964.

[13] A. Fettweis, "Wave digital filters: Theory and practice," *Proc. IEEE*, vol. 74, pp. 270–327, Feb. 1986, and vol. 75, p. 729, May 1987.

[14] H. D. Fischer, "Wave digital filters for numerical integration," *ntz-Archiv*, vol. 6, pp. 37–40, 1984.

[15] A. Fettweis and G. Nitsche, "Transformation approach to numerically integrating PDEs by means of WDF principles," *Multidimensional Systems and Signal Processing*, vol. 2, pp. 127–159, 1991.

[16] A. Fettweis, "Discrete passive modelling of physical systems described by partial differential equations," in *Multivariate Analysis: Future Directions*, C. R. Rao, Ed., pp. 115–130, Elsevier, Amsterdam, 1993.

[17] A. Fettweis, "Improved wave-digital approach to numerically integrating the PDEs of fluid dynamics," *Proc. IEEE International Symposium on Circuits and Systems*, vol. III, pp. 361–364, Scottsdale, AZ, May 2002.

Chapter 2

Time, Systems, and Control: Qualitative Properties and Methods

Lyubomir T. Gruyitch

Abstract: This chapter presents some new results in the areas in which Professor Anthony N. Michel has been a world renowned researcher. Properties of time are summarized. These properties are linked to the features of physical variables expressed by the Physical Continuity and Uniqueness Principle and are important for modeling physical systems and for studies of their qualitative properties. The complete transfer function matrix is defined for multi-input multi-output time-invariant linear systems. This matrix is crucial for zero-pole cancellation, system minimal realization, and synthesis of stabilizing, tracking, and/or optimal control for the systems. A new Lyapunov methodology for nonlinear systems, called the "consistent Lyapunov methodology," enables us to establish the necessary and sufficient conditions for (1) asymptotic stability, (2) direct construction of a Lyapunov function for a given nonlinear dynamical system, and (3) a set to be the exact domain of asymptotic stability. They are not expressed in terms of the existence of a Lyapunov function. The extended concepts of definite vector functions and of vector Lyapunov functions open new directions for studies of complex nonlinear dynamical systems and for their control synthesis.

2.1 Introduction

Anthony N. Michel and co-workers contributed remarkably to the qualitative theory of general and large-scale dynamical systems (see [8, 19, 20, 36–41, 46–53, 55–59, 63–67, 69–72], or papers collected and/or summarized in the books [1, 54, 60, 61]). Scientific results by Michel and colleagues had a significant impact on the research of the author of this chapter.

Michel et al. [52, 72] defined time as an independent mathematical variable, which is in complete disagreement with Einstein's relativity theory. By referring to the new results in relativity theory, we can see why Michel's explanation of time, in the Newtonian sense, is correct not only in the mathematical sense but also in the physical sense. Other properties of time are also explained. These properties, combined with the crucial properties of physical variables that led to the Physical Continuity and Uniqueness Principle (PCUP), result in the Time Continuity and Uniqueness Principle (TCUP). These principles are crucial for an adequate modeling of physical systems. Their application is shown in the analysis of asymptotic stability of the zero equilibrium state of time-invariant nonlinear dynamical systems. The complete transfer function matrix is defined and determined for continuous-time MIMO time-invariant linear systems. It enables us to obtain new results on zero-pole cancellation, system minimal realization, stability analysis, and stabilizing, tracking, and/or optimal control synthesis.

The chapter shows how the new Lyapunov methodology for nonlinear systems, called the "consistent Lyapunov methodology," permits us to resolve all three unsolved fundamental problems of the Lyapunov stability theory: the necessary and sufficient conditions, which are not expressed in terms of the existence of a Lyapunov function, for (1) asymptotic stability, (2) direct construction of a Lyapunov function for a given nonlinear dynamical system, and (3) a set to be the exact domain of asymptotic stability.

2.2 Physical Variables, Time, and Continuity

2.2.1 Physical variables

"Physical variable" is defined as follows.

Definition 2.2.1.

(a) Physical variable. A variable is a physical variable if and only if its value uniquely expresses and characterizes some physical situation, hence some physical phenomenon, of matter (including any material object and any being) and/or of energy.

(b) The existence of a physical variable. Every physical variable has existed as long as the corresponding physical situation, physical phenomenon, of matter (including any material object and any being) and/or of energy has existed. It cannot be created from nothing or destroyed into nothing.

2.2.2 Time

Michel et al. ([52, 72]) defined time as an independent mathematical variable (Definition 3.1 in [52]), which agrees with the characterization of time in mathematics, classical mechanics, systems theory, control science, and control engineering. After Einstein [4–7], time has been considered dependent of space. Recent results in relativity theory [23, 27–29] verify the essential sense of Newton's characterization of time [62]. These results agree with our experience and with the physical facts, which permits us to summarize as follows:

Axiom 2.2.1.

(a) *Time. Time denoted by t (or by τ) is an independent and unique physical variable, the value of which is strictly monotonously continuously increasing, equally in all directions and independently of beings, objects, matter, energy, space, and of all other variables, processes, and events.*

(b) *Time value. Time value is called "moment" (or "instant"). A total zero moment has not existed. A moment (or instant) expresses the age of a being, an object, a particular matter, or a particular energy, and it is used to determine uniquely the order of events happening. A time value difference called "time interval" is used to measure the duration of the existence, i.e., age of a process, a movement, a rest, or the existence of somebody or something.*

A fixed moment (instant) is usually denoted by t (or by τ) and a subscript, e.g., t_2 (or τ_b). An arbitrary moment (instant) will be denoted as time itself by t (or τ). Any moment can be conventionally accepted for the zero moment $t = 0$.

Axiom 2.2.1 explains time in the Newtonian sense as an independent variable [62, p. 8, Scholium I]. It is independent of spatial coordinates. Einstein considered time relativity in the same sense [4–7] as Newton [62, p 8, Scholium I]. Axiom 2.2.1 is the basis for the new general relationships in the relativity theory, which show that the Lorentz–Einstein formulae hold only for a singular case [23, 27–29].

2.2.3 Physical principles

The PCUP expresses some of intrinsic features of physical variables [24].

Claim 2.2.1. *PCUP. A physical variable (a matrix/vector of physical variables) can change its (matrix/vector) value from one (matrix/vector) value to another one only by passing (elementwise) through all intermediate (matrix/vector) values. It possesses a unique local instantaneous real (matrix/vector) value at any place (in any being or in any object) at every moment.*

Claim 2.2.2. *System form of PCUP. Variables of a physical system can change their values from one value to another one only by passing through all intermediate values. They possess unique local instantaneous real values at any place at every moment.*

The term "physical vector (or matrix) variable" means in the sequel "a vector (or matrix) of which all entries are physical variables."

Corollary 2.2.1. *Mathematical model and PCUP. For a mathematical model of a physical system to be an adequate description of a physical system, it is necessary that its system variables obey the PCUP, that is, the mathematical model obeys the PCUP.*

Claim 2.2.3. *TCUP. Any physical variable (any vector/matrix of physical variables) can change its (vector/matrix) value from one (vector/matrix) value to another one only continuously in time by passing (elementwise) through all intermediate (vector/matrix) values. It possesses a unique local instantaneous real (vector/matrix) value at any place (in any being or in any object) at every moment.*

2.3 Complete Transfer Function Matrix

Antsaklis and Michel [1] noted that zero-state system equivalence and zero-input system equivalence do not necessarily imply system equivalence (p. 171 in [1]). Different state-space realizations of the system transfer function matrix yield the same zero-state system response, but the corresponding zero-input (hence, complete) system responses can be quite different (p. 387 in [1]).

Problem 2.3.1. *What is a mathematical description of the system transfer of a simultaneous influence of all inputs and arbitrary initial conditions on its motion $x(\cdot)$ and on its output response $y(\cdot)$?*

2.3.1 Input-output (IO) system description

Problem 2.3.1 was solved by introducing the complete (or full) transfer function matrix of a system described by a vector IO differential equation (2.3.1) [13, 21, 25],

$$\sum_{k=0}^{k=\nu} A_k y^{(k)}(t) = \sum_{k=0}^{k=\nu} B_k i^{(k)}(t), A_\nu \neq O_N \in R^{N\times N}, (\cdot)^{(k)}(t) = \frac{d^k(\cdot)(t)}{dt^k}, \quad (2.3.1)$$

where the input vector $i = (i_1\ i_2 \cdots\ i_M)^T \in R^M$, the output vector $y = (y_1\ y_2\ \cdots\ y_N)^T \in R^N$, $A_k \in R^{N\times N}$ and $B_k \in R^{N\times M}$, $k = 0, 1, \cdots, \nu$, where R^k is the k-dimensional real vector space, $k \in \{M, N\}$. C^k denotes the k-dimensional complex vector space. Let

$$\begin{aligned} A^{(\nu)} &= (A_0\ A_1 \cdots A_\nu) \in R^{N\times(\nu+1)N}, \\ B^{(\nu)} &= (B_0\ B_1 \cdots B_\nu) \in R^{N\times(\nu+1)M}, \end{aligned} \quad (2.3.2)$$

$$i_{0(\mp)}^{\nu-1} = i^{\nu-1}(0^{(\mp)}) = \left(i_{0(\mp)}^{T}\ i_{0(\mp)}^{(1)T} \cdots i_{0(\mp)}^{(\nu-1)T}\right)^T \in R^{\nu M}, \quad (2.3.3)$$

$$y_{0(\mp)}^{\nu-1} = y^{\nu-1}(0^{(\mp)}) = \left(y_{0(\mp)}^{T}\ y_{0(\mp)}^{(1)T} \cdots y_{0(\mp)}^{(\nu-1)T}\right)^T \in R^{\nu N}. \quad (2.3.4)$$

Definition 2.3.1.

(a) The complete transfer function matrix (or the complete matrix transfer function) of system (2.3.1), denoted by $T_{IO}(s)$,

$$T_{IO}(\cdot): C \to C^{N\times[M+\nu(M+N)]},$$

is a matrix function of complex variable s, $s \in C$, such that it determines uniquely the (left, right) Laplace transform $Y^{(\mp)}(s)$ of $y(t)$ as a homogeneous linear function of the (left, right) Laplace transform $I^{(\mp)}(s)$ of $i(t)$, for arbitrary variation of $i(t)$, of arbitrary initial vector values $i^{\nu-1}_{0(\mp)}$ and $y^{\nu-1}_{0(\mp)}$ of the extended input vector $i^{\nu-1}(t)$ and of the extended output vector $y^{\nu-1}(t)$ at $t = 0^{(\mp)}$, respectively,

$$Y^{(\mp)}(s) = T_{IO}(s)\begin{pmatrix} (I^{(\mp)}(s))^T & (i^{\nu-1}_{0\mp})^T & (y^{\nu-1}_{0\mp})^T \end{pmatrix}^T. \tag{2.3.5}$$

(b) The transfer function matrix relative to $y^{\nu-1}_{0(\mp)}$ (or the matrix transfer function relative to $y^{\nu-1}_{0(\mp)}$) of system (2.3.1), denoted by $T_{IOY}(s)$, $T_{IOY}(\cdot): C \to C^{N\times\nu N}$, is a matrix function of complex variable s such that it determines uniquely the (left, right) Laplace transform $Y^{(\mp)}(s)$ of $y(t)$ as a homogeneous linear function of arbitrary initial vector $y^{\nu-1}_{0(\mp)}$ of the extended output vector $y^{\nu-1}(t)$ at $t = 0^{(\mp)}$ for the system in a free regime (i.e., for $i(t) \equiv 0_M$):

$$Y^{(\mp)}(s) = T_{IOY}(s)y^{\nu-1}_{0(\mp)},\quad i(t) \equiv 0_M. \tag{2.3.6}$$

The transfer function matrix $W(s)$ can be set in the following compact form:

$$W(s) = \left(A^{(\nu)}S_N^{(\nu)}(s)\right)^{-1}\left(B^{(\nu)}S_M^{(\nu)}(s)\right), \tag{2.3.7}$$

$$S_k^{(\nu)}(s) = \begin{pmatrix} s^0I_k & s^1I_k & s^2I_k & \cdots & s^\nu I_k \end{pmatrix} \in C^{k\times k(\nu+1)}, k\in\{M,N\}, \tag{2.3.8}$$

in which I_k is the kth-order identity matrix. O_k is the kth-order zero matrix. Let $Z_k^{(\nu-1)}(\cdot): C \to C^{k(\nu+1)xk\nu}$:

$$Z_k^{(\nu-1)}(s) = \begin{pmatrix} O_k & O_k & O_k & \cdots & O_k \\ s^0I_k & O_k & O_k & \cdots & O_k \\ \cdots & \cdots & \cdots & \cdots & \cdots \\ s^{\nu-1}I_k & s^{\nu-2}I_k & s^{\nu-3}I_k & \cdots & s^0I_k \end{pmatrix}, k\in\{M,N\}. \tag{2.3.9}$$

Theorem 2.3.1.

(a) *The complete transfer function matrix $T_{IO}(s)$ of system (2.3.1) has the following form:*

$$T_{IO}(s) = \left(A^{(\nu)}S_N^{(\nu)}(s)\right)^{-1}\begin{pmatrix} B^{(\nu)}S_M^{(\nu)}(s) & -B^{(\nu)}Z_M^{(\nu-1)}(s) & A^{(\nu)}Z_N^{(\nu-1)}(s) \end{pmatrix}.$$

(b) *The transfer function matrix $T_{IOY}(s)$ of system (2.3.1) relative to $y^{\nu-1}_{0\mp}$ has the following form:*

$$T_{IOY}(s) = \left(A^{(\nu)}S_N^{(\nu)}(s)\right)^{-1}\left(A^{(\nu)}Z_N^{(\nu-1)}(s)\right).$$

Proof. The (left, right) Laplace transforms of (2.3.1)–(2.3.4), (2.3.8), and (2.3.9) yield

$$Y^{(\mp)}(s) = \underbrace{\left(A^{(\nu)}S_N^{(\nu)}(s)\right)^{-1}\left(B^{(\nu)}S_M^{(\nu)}(s)\ \left(-B^{(\nu)}Z_M^{(\nu-1)}(s)\right)\ A^{(\nu)}Z_N^{(\nu-1)}(s)\right)}_{T_{IO}(s)}$$
$$\cdot\left(\ \left(I^{(\mp)}(s)\right)^T\ \ \left(i_{0\mp}^{\nu-1}\right)^T\ \ \left(y_{0\mp}^{\nu-1}\right)^T\ \right)^T.$$

The definition of $T_{IO}(s)$ and this equation prove statement (a) of the theorem. Statement (b) results directly from (a) and the definition of $T_{IOY}(s)$. □

Claim 2.3.1. *In order for a triple* $\left(\nu, A^{(\nu)}, B^{(\nu)}\right)$ *to determine the minimal IO system realization (2.3.1), it is necessary and sufficient that*

$$\left(A^{(\nu)}S_N^{(\nu)}(s)\right)^{-1}\left(\ B^{(\nu)}S_M^{(\nu)}(s)\quad -B^{(\nu)}Z_M^{(\nu-1)}(s)\quad A^{(\nu)}Z_N^{(\nu-1)}(s)\ \right)$$

equals the irreducible form of the system complete transfer function matrix $T_{IO}(s)$.

2.3.2 Input-state-output (ISO) system description

Problem 2.3.1 was solved by introducing the complete (or full) transfer function matrix of a system described by an ISO mathematical model (2.3.10), (2.3.11) [13, 21, 25],

$$\begin{aligned}\frac{dx(t)}{dt} &= Ax(t) + Bi(t), A \in R^{n\times n}, B \in R^{n\times M}, x \in R^n, &(2.3.10)\\ y(t) &= Cx(t) + Di(t), C \in R^{N\times n}, D \in R^{N\times M}. &(2.3.11)\end{aligned}$$

The system transfer function matrix $W(s)$ does not depend on the form of the system mathematical model. Initial conditions, although arbitrary, have in general different meanings for the IO model (2.3.1) and for the ISO model (2.3.10), (2.3.11) of the same physical system.

Definition 2.3.2.

(a) The complete transfer function matrix of the system (2.3.10), (2.3.11), denoted by $T_{ISO}(s)$,

$$T_{ISO}(\cdot)\colon C \to C^{N\times(M+n)},$$

is a matrix function of the complex variable s such that it determines uniquely the (left, right) Laplace transform $Y^{(\mp)}(s)$ of $y(t)$ as a homogeneous linear function of the (left, right) Laplace transform $I^{(\mp)}(s)$ of $i(t)$, for arbitrary variation of $i(t)$, and of arbitrary initial vector values $x_{0(\mp)}$ of the state vector $x(t)$ at $t = 0^{(\mp)}$, respectively:

$$Y^{(\mp)}(s) = T_{ISO}(s)\left[\ (I(s))^T\ \ \left(x_{0(\mp)}\right)^T\ \right]^T. \tag{2.3.12}$$

(b) The transfer function matrix relative to $x_{0(\mp)}$ of system (2.3.10), (2.3.11), denoted by $T_{ISOX}(s)$, $T_{ISOX}(\cdot)\colon C \to C^{N\times n}$, is a matrix function of the complex variable s such that it determines uniquely the (left, right) Laplace transform $Y^{(\mp)}(s)$ of $y(t)$ as a homogeneous linear function of arbitrary initial vector value $x_{0(\mp)}$ of the state vector $x(t)$ at $t = 0^{(\mp)}$ for the system in a free regime (i.e., for $i(t) \equiv 0_M$):

$$Y^{(\mp)}(s) = T_{ISOX}(s)x_{0(\mp)}, i(t) \equiv 0_M. \tag{2.3.13}$$

Theorem 2.3.2.

(a) *The complete transfer function matrix $T_{ISO}(s)$ of the system (2.3.10), (2.3.11) has the following form:*

$$T_{ISO}(s) = \begin{pmatrix} C(sI_n - A)^{-1}B + D & C(sI_n - A)^{-1} \end{pmatrix}. \tag{2.3.14}$$

(b) *The transfer function matrix $T_{ISOX}(s)$ relative to x_0 of system (2.3.10), (2.3.11) has the following form:*

$$T_{ISOX}(s) = C(sI_n - A)^{-1}. \tag{2.3.15}$$

Proof. The (left, right) Laplace transforms of (2.3.10), (2.3.11) yields

$$Y^{(\mp)}(s) = \underbrace{\begin{pmatrix} C(sI_n - A)^{-1}B + D & C(sI_n - A)^{-1} \end{pmatrix}}_{T_{ISO}(s)} \begin{pmatrix} I^{(\mp)}(s) \\ x_{0(\mp)} \end{pmatrix}.$$

This equation and (2.3.12) prove statement (a) of the theorem. Statement (b) results directly from (a) and the definition of $T_{ISOX}(s)$. □

Claim 2.3.2. *In order for a quintuple (n, A, B, C, D) to determine the minimal ISO system realization (2.3.10), (2.3.11), it is necessary and sufficient that*

$$\begin{pmatrix} C(sI_n - A)^{-1}B + D & C(sI_n - A)^{-1} \end{pmatrix}$$

equals the irreducible form of the system complete transfer function matrix $T_{ISO}(s)$.

Claim 2.3.3. *In order for the cancellation of zeros and poles with nonnegative real parts in the transfer function matrix W(s) of system (2.3.1) (respectively, system (2.3.10), (2.3.11)) not to influence results*

(a) *on stability or attraction of the equilibrium point (in the Lyapunov sense), it is necessary and sufficient that the same cancellation is possible in the transfer function matrix $T_{IOY}(s)$ (respectively, $T_{ISOX}(s)$) of the system.*

(b) *on system IO stability if the system was not at rest at the initial moment, on the system complete response, on system output tracking, and on system optimization under non-zero initial conditions, it is necessary and sufficient that the same cancellation is possible in the complete transfer function matrix $T_{IOY}(s)$ (respectively, $T_{ISO}(s)$) of the system.*

2.4 Consistent Lyapunov Methodology for Nonlinear Systems

2.4.1 Notation

A will denote a compact connected invariant (relative to motions of system (2.4.1)) subset of R^n, which can be a singleton: $A = \{\overset{\circ}{x}\} \subset R^n$. If $\overset{\circ}{x} = \mathbf{0} \in R^n$, then $A = O, O = \{\mathbf{0}\} \subset R^n$. $N(A)$ and $S(A)$ are open connected neighborhoods of the set A. An α-neighborhood of A is $N_\alpha(A)$, $N_\alpha(A) = \{x \colon x \in R^n, d(x, A) < \alpha\}$, where $\alpha > 0$, $d(x, A) = \inf\{\|x - x^*\| : x^* \in A\}$ and $\|\cdot\| : R^n \to R_+$ is any norm on R^n, $R_+ = [0, \infty[$ and $R^+ =]0, \infty[$. The boundary, closure, and interior of A are denoted by ∂A, ClA, and InA, respectively. $D_a(A)$, $D_s(A)$, and $D(A)$ will mean, respectively, the attraction domain of A, the stability domain of A, and the asymptotic stability domain of A. For their definitions, see [3, 10, 11, 18, 33, 34, 60, 61, 74].

Let $\zeta \in R^+$, and $p(\cdot) \colon R^n \to R$. Then $P_\zeta(A)$ is the largest open connected neighborhood of A such that the function $p(\cdot)$ obeys the following conditions: (1) $p(x) = \zeta$ if and only if $x \in \partial P_\zeta(A)$, (2) $p(x) < \zeta$ if and only if $x \in InP_\zeta(A)$, and (3) $p(x) > \zeta$ if and only if $x \in (R^n - ClP_\zeta(A))$. The function $p(\cdot)$ is the generating function of the set $P_\zeta(A)$ if and only if it obeys these conditions. Additionally if $p(\cdot)$ is globally positive-definite with respect to A and radially unbounded, then its value $p(x)$ may be used as a measure of the distance of x from A, where these notions are used in the well-known sense [3, 34, 35, 42, 73, 74].

A motion of system (2.4.1), which passes through an initial state x_0 at an initial moment t_0, is denoted by $x(\cdot; t_0, x_0)$. Its vector value at a moment t is $x(t; t_0, x_0)$, $x(t; t_0, x_0) \equiv x(t)$. We accept $t_0 = 0$ so that $x(t; t_0, x_0) \equiv x(t; x_0)$. If $v(\cdot) \colon R^n \to R^k$ is continuous then its total upper right-hand derivative along a system motion $x(\cdot; t_0, \mathbf{x})$ at $\mathbf{x} \in R^n$ is its Dini derivative

$$D^+v(\mathbf{x}) = \limsup \left\{ \left(v\left[x(\theta; \mathbf{x})\right] - v(\mathbf{x}) \right) \theta^{-1} : \theta \to 0^+ \right\}.$$

Other notation is explained at its first usage.

2.4.2 Lyapunov methods and methodologies

A. Lyapunov methods

Lyapunov (1892) developed two methods: the first and the second method [42]. The first Lyapunov method comprises all approaches enabling us to solve differential equations for given initial conditions. Hence, the first method has not been appropriate for analysis of qualitative dynamical properties. The second Lyapunov method (called also the "direct Lyapunov method"), for short, the Lyapunov method, comprises all the approaches that enable us to deduce a stability (or another qualitative dynamical) property of a given system directly from its mathematical model without determining its solutions (or motions). The basis for the method is the use of a sign definite function and of its total time derivative along system motions without a determination of the motions themselves. The function is called a "Lyapunov function of the system" if its derivative is semidefinite with the sign either opposite to the sign of the function itself or identically equal to zero. The method opens the following problems in order to solve the considered stability problem.

Problem 2.4.1. *What is a (an effective) direct procedure for constructing a Lyapunov function for a given nonlinear system?*

Problem 2.4.2. *What are necessary and sufficient conditions for Lyapunov stability properties, which are not expressed in terms of the existence of a system Lyapunov function?*

Problem 2.4.3. *What are (effective) necessary and sufficient conditions for a set to be the (exact) domain of a Lyapunov stability property, which are not expressed in terms of the existence of a system Lyapunov function?*

B. Lyapunov's methodology for time-invariant linear systems

In the framework of time-invariant continuous-time linear systems (2.3.10) and (2.3.11), Lyapunov introduced a methodology that enables the determination of a system Lyapunov function $v(\cdot)\colon R^n \to R$ in one step. The main characteristics of Lyapunov's methodology for time-invariant linear systems are

- The stability conditions are not expressed in terms of the existence of a function $v(\cdot)$ with the prespecified properties.
- The procedure starts with an arbitrary choice of a globally positive-definite function $h(\cdot)$ from a functional family.
- The conditions determine the functional family H (of all mth-order homogeneous forms) from which a (global) positive-definite function $h(\cdot)$ may be chosen arbitrarily.
- The procedure determines completely how to find a system Lyapunov function $v(\cdot)$.
- The next step is the test of (global) positive-definiteness of the solution function $v(\cdot)$. We usually use $h(\cdot)$ as a quadratic form ($m = 2$) and then obtain also $v(\cdot)$ as a quadratic form.
- The test result provides a final reply to the question whether the system equilibrium state $x = 0$ is (globally) asymptotically (exponentially) stable or not.
- The positive test result determines exactly the domain of asymptotic (exponential) stability of the equilibrium, which is the whole state space due to the system linearity.

C. Lyapunov's methodology for nonlinear systems

Lyapunov's methodology for nonlinear systems [42] will be essentially summarized in the next theorem for systems described by (2.4.1)

$$\frac{dx(t)}{dt} = f[x(t)],$$

$$f(\cdot)\colon R^n \to R^n, x = (\xi_1\ \xi_2 \cdots \xi_n)^T, n \in \{1, 2, 3, \cdots\}. \tag{2.4.1}$$

Condition 2.4.1. *System (2.4.1) satisfies the PCUP.*

Under this condition, system motions and their first derivative obey the principle on R_+ for every initial state $x_0 \in R^n$.

Condition 2.4.2. *System (2.4.1) has the unique equilibrium state at* $x = 0$.

Theorem 2.4.1. *Let Conditions 2.4.1 and 2.4.2 hold. In order for the equilibrium state* $x = 0$ *of system (2.4.1) to be asymptotically stable, it is necessary and sufficient that there exists a positive-definite function* $v(\cdot)$ *with negative-definite total time derivative along system motions.*

The classical Lyapunov methodology for nonlinear systems left the problems 2.4.1 through 2.4.3 unsolved. It is inverse to, hence inconsistent with Lyapunov's methodology for time-invariant linear systems.

2.4.3 Consistent Lyapunov methodology for nonlinear systems

A. Introduction

The consistent Lyapunov methodology was first developed for asymptotic stability of the equilibrium state $x = 0$ of time-invariant nonlinear systems. The corresponding results and references published until 1996 are presented in [33]. The methodology was further broadened to asymptotic stability and exponential stability of sets of different classes of time-invariant and time-varying nonlinear dynamical systems, to energetic stability, to absolute stability of Lurie–Posntnykov systems, and to Hopfield neural networks [9, 14–17, 26, 30, 32]. A functional family P_i, $i \in \{1, 2\}$ replaces and generalizes the Lyapunov functional family H in the framework of nonlinear systems.

Definition 2.4.1. Definition of functional families P_i.

(a) $P_1 = P_1(f; A, S)$ is a family of all functions $p(\cdot)\colon R^n \to R$ obeying

(i) $p(\cdot)$ is continuous on $S(A)\colon p(x) \in C\left[S(A)\right]$,

(ii) there is an arbitrarily small $\mu \in R^+$, $\mu = \mu(f; p; A)$ such that there exists a unique solution function $v(\cdot)$ to (2.4.2),

$$D^+v(x) = -p(x), v(x) = 0 \; \forall x \in \partial A, \tag{2.4.2}$$

which is continuous in $x \in ClN_\mu(A)\colon v(x) \in C\left[ClN_\mu(A)\right]$.

(b) $P_2 = P_2(f; A, S)$ is a family of all the functions $p(\cdot) \in P_1$ such that for every $\alpha > 0$ obeying $ClN_\alpha(A) \subset S(A)$ there exists $\beta > 0$, $\beta = \beta(\alpha; p; A, S)$, satisfying (2.4.3)

$$\inf\left\{p(x)\colon x \in \left[S(A) - N_\alpha(A)\right]\right\} = \beta. \tag{2.4.3}$$

B. Asymptotic stability of a set

Theorem 2.4.2. *Let Condition 2.4.1 hold. In order for a compact connected invariant set A of system (2.4.1) to be asymptotically stable, it is necessary and sufficient* *that $P_1(f; A, S)$ is nonempty and contains at least one positive-definite function $p(\cdot)$ relative to A, and for any $p(\cdot) \in P_1(f; A, S)$ that is positive-definite relative to A, the equations (2.4.2) have a unique solution function $v(\cdot)$ that is also positive-definite function relative to A.*

Proof. Let Condition 2.4.1 hold. All the conditions of Theorem 1 (a) in [22] hold, which proves the above statement.

Remark 2.4.1. If we can determine both the system power $P(\cdot)$ and the system energy $E(\cdot)$ from physical properties of the physical system represented by (2.4.1), then the function $v(\cdot)$ represents the system energy $E(\cdot)$ and it is not necessary to solve (2.4.2) [32].

C. Asymptotic stability domain of a set

The attraction domain has been widely used notion ([3, 34, 60, 61, 74]). The notions of the stability domain and the asymptotic stability domain were introduced in [10] and further developed and used in [11, 18, 33].

Theorem 2.4.3. *Let Condition 2.4.1 hold. In order for a compact connected invariant set A of system (2.4.1) to have the domain of asymptotic stability and for a subset N of $S(A)$, $N \subseteq S(A)$ to be its domain of asymptotic stability, it is necessary and sufficient that both (1) and (2) hold:*

(1) *the set N is an open connected neighborhood of the set A,*

(2) *for any global positive-definite function $p(\cdot)$ with respect to A obeying $p(\cdot) \in P_1(f; A, R^n)$, (2.4.2) has a unique solution function $v(\cdot)$ with the following properties:*

(i) *$v(\cdot)$ is a positive-definite function relative to A on N,*

(ii) *if the boundary ∂N of N is nonempty then $x \to \partial N$, $x \in N$, implies $v(x) \to \infty$.*

Proof. Let Condition 2.4.1 hold. Hence, all the conditions of Theorem 9 in [22] are satisfied, which proves the above statement.

2.5 Vector Lyapunov Function Concept and Complex Systems

2.5.1 Introduction

The concept of vector Lyapunov functions was coincidentally introduced by Bellman [2] and Matrosov [43]. It became the basic mathematical tool for studying stability properties of complex dynamical systems [18, 44, 54, 68]. We will extend the concept of vector Lyapunov functions.

2.5.2 Definitions

Let i, j, k, l, m, n, and $p \in \{1, 2, 3, \cdots\}$, $\min\{i_l, j_l\} \geq 1$, $\max\{i_l, j_l\} \leq n$, $m_{ll} = 1, \forall l \in \{1, 2, \cdots, k\}$, $i_1 = j_1$, $i_k = j_k = n$, and $\sum_{l=1}^{l=k} m_{i_l j_l} = n$ if $k \leq n$.

Otherwise, the sum ends with $l = n$. All vector and matrix equalities, inequalities, and powers hold elementwise. Let for $n = 2p$, $x_p = (\xi_1\ \xi_2 \cdots \xi_p)^T \in R^p$, $x_p^1 = (x_p^T\ x_p^{(1)^T})^T = (x_1^T x_2^T)^T = x$, $x_i \in R^p$, $i = 1, 2$, $\mathbf{x}_{ij} = (\xi_1 \cdots \xi_{i-1}\ \widetilde{\xi}_i\ \widetilde{\xi}_{i+1} \cdots \widetilde{\xi}_j\ \xi_{j+1} \cdots \xi_p)^T \in R^p$, $\mathbf{x}_{ij}^1 = (\mathbf{x}_{ij}^T\ \ \mathbf{x}_{ij}^{(1)^T})^T$, and let

$$S_{ij} = \left\{\widetilde{\mathbf{x}}_{ij} : \widetilde{\mathbf{x}}_{ij} = (\widetilde{\xi}_i\widetilde{\xi}_{i+1} \cdots \widetilde{\xi}_j)^T \in R^{p_{ij}} \Longrightarrow \exists x = \left(\mathbf{x}_{ij}^T\, x_p^{(1)^T}\right)^T \in S\right\},$$

$$S_{ij}^1 = \left\{\widetilde{\mathbf{x}}_{ij}^1 : \widetilde{\mathbf{x}}_{ij}^1 = \left(\widetilde{\mathbf{x}}_{ij}^T\ \widetilde{\mathbf{x}}_{ij}^{(1)^T}\right)^T \in R^{2p_{ij}} \Longrightarrow \exists x = \mathbf{x}_{ij}^1 = \left(\mathbf{x}_{ij}^T\, \mathbf{x}_{ij}^{(1)^T}\right)^T \in S\right\}.$$

Let $S_{(\cdot)} \in \{S, S_b\}$, where the set S_b is defined as follows.

Definition 2.5.1. A vector function $v(\cdot)\colon R^n \to R^k$, $k \in \{1, 2, \cdots, n\}$, $v(x) = [v_1(x)\ \ v_2(x) \cdots\ v_k(x)]^T$, is

(a) positive (negative)-definite relative to a set S, $S \subset R^n$, if and only if (i) through (iv) hold respectively over some neighborhood $N(S)$ of S:

(i) $v(\cdot)$ is defined and continuous on $N(S)$: $v(x) \in C[N(S)]$,

(ii) $v(x) \geq 0$, $(v(x) \leq 0)$, $\forall x \in [N(S) - InS]$,

(iii) $v(x) = 0$ for $x \in [N(S) - InS]$ if and only if $x \in \partial S$,

(iv) $v(x) \leq 0$, $(v(x) \geq 0)$, $\forall x \in ClS$.

(b) pairwise positive (negative)-definite relative to a set S_b, $S_b = S_{11}^1 \times S_{22}^1 \times \cdots \times S_{pp}^1 \subset R^n$, if and only if (i), (ii), and (iv) of (a) hold for $k = p$, $n = 2p$, $\xi_{p+l} = \xi_l^{(1)}$, $\forall l = 1, 2, \cdots, p$, $S = S_b$,

(i) $v_l(x) = 0$ for $x = \mathbf{x}_{ll}^1 = (\mathbf{x}_{ll}^T\ \ \mathbf{x}_{ll}^{(1)^T})^T \in [N(S_b) - InS_b]$ if and only if $\widetilde{\mathbf{x}}_{ll}^1 \in \partial S_{ll}^1$, $\forall l = 1, 2, \cdots, p$,

(ii) $v_l(x) \equiv v_l(\xi_l^1)$, $\xi_l^1 = (\xi_l\ \ \xi_l^{(1)})^T$, $\forall l = 1, 2, \cdots, p$.

The preceding properties are

(c) global if and only if they hold for $N(S_{(\cdot)}) = R^n$, $S_{(\cdot)} \in \{S, S_b\}$,

(d) on a set A, $A \subset R^n$, if and only if the corresponding conditions hold for $N(S_{(\cdot)}) = A$.

The expression "relative to a set $S_{(\cdot)}$, $S_{(\cdot)} \subset R^n$" should be omitted if and only if the set $S_{(\cdot)}$ is singleton O:

$$S_{(\cdot)} = O = \{x\colon x = 0\},\ S_{(\cdot)} \in \{S, S_b\}.$$

This definition is compatible with Lyapunov's original definition of scalar definite functions [42], as well as with the concept of matrix definite functions introduced in [12].

The vector $e^1 = (x_d - x) \in R^n$ is the error vector of a system real motion $x(\cdot)\colon R \to R^n$ relative to a system desired motion $x_d(\cdot)\colon R \to R^n$ that is well defined, twice differentiable, $x_d(t) \in C^{(2)}(R)$, and fixed. The output error vector $\varepsilon = (y_d - y) \in R^N$ is the error vector of a system real output response $y(\cdot)\colon R \to R^N$ relative to a system desired output response $y_d(\cdot)\colon R \to R^N$ that is fixed, twice differentiable, and well defined, $y_d(t) \in C^{(2)}(R)$.

Remark 2.5.1. Notice that $\varepsilon(t) = y_d(t) - y(t; q_0^1) = \varepsilon(t; q_0^1)$, implying $\varepsilon_0 = \varepsilon(0; q_0^1) = \varepsilon_0(q_0^1)$. Analogously, $e_0 = e(0; q_0^1) = e_0(q_0^1)$.

Definition 2.5.2. A vector function $v(\cdot)\colon R^n \to R^k$, $k \in \{1, 2, \cdots, n\}$, is

(a) a vector Lyapunov function of system (2.4.1) (on A) with respect to a set S, $S \subset R^n$, if and only if both (i) and (ii) hold respectively:

(i) $v(\cdot)$ is positive-definite with respect to the set S (on A),

(ii) there is a neighborhood $N_D(S)$, $N_D(S) \subseteq N(S)$, Definition 2.5.1, $(N_D(S) = A)$, of the set S such that (2.5.1) is valid,

$$D^+v(e^1) \leq 0, \forall e^1 \in N_D(S). \tag{2.5.1}$$

If and only if additionally there is a positive-definite vector function $\Psi(\cdot)\colon R^n \to R^k$ with respect to the set S (on A) such that

$$D^+v(e^1) \leq -\Psi(e^1), \forall e^1 \in N_D(S), \tag{2.5.2}$$

then the function $v(\cdot)$ is a strict vector Lyapunov function of the system (on A) with respect to the set S.

(b) a pairwise vector Lyapunov function of system (2.4.1), (2.5.3) (on A) with respect to a set S_b, $S_b \subset R^n$, if and only if $n = 2p$, $\xi_{p+l} = \xi_l^{(1)}$, $\forall l = 1, 2, \cdots, p$, and both (i) and (ii) hold respectively:

(i) $v(\cdot)$ is pairwise positive-definite with respect to the set S_b (on A),

(ii) there is a neighborhood $N_D(S_b)$, $N_D(S_b) \subseteq N(S_b)$, Definition 2.5.1, $(N_D(S_b) = A)$, of the set S_b such that (2.5.1) is valid for $S = S_b$.

If and only if additionally there is a pairwise positive-definite vector function

$$\Psi(\cdot)\colon R^n \to R^p$$

with respect to the set S_b (on A) such that (2.5.2) holds for $S = S_b$, then the function $v(\cdot)$ is a strict pairwise vector Lyapunov function of the system (on A) with respect to the set S.

This definition is compatible with the concept of vector Lyapunov functions by Bellman [2] and Matrosov [43], as well as with the concept of matrix Lyapunov functions introduced in [12]. Let ϕ be the empty set.

A large class of mechanical systems with neglected dynamics of electrical actuators and sensors can be described by (2.5.3):

$$A[q(t)]q^{(2)}(t) + n[q^1(t)] = B[q^1(t)]u(t), \tag{2.5.3}$$

$$y(t) = g[q(t)], \tag{2.5.4}$$

where $q = x_1 \in R^p$, $q^1 = (q^T \;\; q^{(1)^T})^T = (x_1 \;\; x_2)^T = x \in R^n$ $(n = 2p)$ is the internal dynamics vector, the inertia matrix function $A(\cdot)\colon R^p \to R^{p\times p}$ is known and nonsingular, $\det A(q) \neq 0 \; \forall q \in R^p$, the matrix function $B(\cdot)\colon R^{2p} \to R^{p\times r}$ is known and has full (row) rank, $\text{rank}B(q^1) = p \leq r, \forall q^1 \in R^{2p}$, the nonlinear vector function $n(\cdot)\colon R^p \to R^p$ is known and contains all other terms of the system internal dynamics which are not in $A(q)q^{(2)}$, and the control vector function $u(\cdot)\colon R^{\cdots} \to R^r$ should be synthesized. The output vector dimension N obeys $N+1 \leq p$, the output vector nonlinear function $g(\cdot) = (g_1(\cdot)\ g_2(\cdot) \cdots g_N(\cdot))^T\colon R^p \to R^N$ is known and twice differentiable with Jacobian $J(q) = [\partial g_i(q)/\partial q_j]$. In this setting, $x_{1d}(t) \equiv q_d(t)$, $e = (x_{1d} - x_1) = (q_d - q)$ and $g[q_d(t)] \equiv y_d(t)$.

2.5.3 Vector generalizations of the classical stability theorems

Let $c \in R^{+^k}$, $k \in \{1, 2, \cdots, n\}$. The set $V_c(S)$, $V_c(S) \subseteq R^n$, is the largest open connected neighborhood of the set S such that a vector function $v(\cdot)$ and the set $V_c(S)$ obey (2.5.5):

$$v(e^1) < c, \forall e^1 \in V_c(S). \tag{2.5.5}$$

Condition 2.5.1. *The sets $V_{c_i}(S)$, $c_i \in R^{+^k}$, $i = 1, 2$, satisfy both (a) and (b):*

(a) $ClV_{c_1}(S) \subset ClV_{c_2}(S)$, $\partial V_{c_1}(S) \cap \partial V_{c_2}(S) = \phi$, $\forall c_i \in R^{+^k}$, $i = 1, 2$, $0 < c_1 < c_2$,

(b) $c_i \to \infty \mathbf{1}, \mathbf{1} = (1\ 1\ \cdots\ 1)^T \in R^{+^k} \Longrightarrow V_{c_i}(S) \to R^n, i = 1, 2.$

The function $v(\cdot)$ is the generating function of the sets $V_{c_i}(S)$.

Theorem 2.5.1. *Let Conditions 2.4.1 and 2.5.1 hold. In order for a set S, $S \subseteq R^n$, $[S = S_b]$, to be, respectively, [pairwise] asymptotically stable set of system (2.4.1) [(2.4.1), (2.5.3), (2.5.4)], it is sufficient that there is a strict [pairwise] vector Lyapunov function $v(\cdot)$ of the system with respect to the set S.*

Conclusion 2.5.1. *The vector definite functions and the vector Lyapunov functions introduced in the preceding definitions enable us to solve various stability problems of complex (interconnected and/or large-scale) systems without using a scalar Lyapunov function of the overall system. Moreover, the introduced vector Lyapunov functions enable us to ensure a high quality of stability properties such as pairwise asymptotic stability of a desired motion together with (output) tracking, both with a finite reachability time.*

2.5.4 Control synthesis

Let

$$W = W^1 \times W^2,\ W^1 = \left\{\varepsilon^1 \colon \lambda_\varepsilon \varepsilon^{(1)} + \alpha_\varepsilon w_\varepsilon(\varepsilon) = \mathbf{0}\right\},$$

$$W^2 = \left\{e^1 \colon \lambda_e e^{(1)} + \alpha_e w_e(e) = \mathbf{0}\right\}, \tag{2.5.6}$$

where $\alpha_\varepsilon \in R^+$, $\alpha_e \in R^+$, $\lambda_\varepsilon \in [1, \infty[$, and $\lambda_e \in [1, \infty[$ are design parameters. The vector functions $w_\varepsilon(\cdot)$ and $w_e(\cdot)$ are defined by (2.7.1) and (2.7.5) (see Appendix).

Notice that the application of the vector Lyapunov function $v(\cdot)$ determined by (2.7.1) through (2.7.6), (2.7.13) (Appendix) permits us to use the product set $W =$

$W^1 \times W^2 \subset R^{2N} \times R^{2p}$. The vector Lyapunov function $v(\cdot)$ and the product set W enable us to synthesize control vector function $u(\cdot)$ so that it guarantees simultaneously (output) tracking and asymptotic stability of the desired motion $q_d^1(t)$, both with the same finite reachability time $\tau_{RO}(q_0^1)$:

$$\tau_{RO}(q_0^1) = \tau_{RO}(\varepsilon_0^1(q_0^1), \tau_{RO}(q_0^1)) = \tau_{RW} + \tau_{RWO}(\varepsilon_0^1(q_0^1), e_0^1(q_0^1)), \quad (2.5.7)$$

$$\tau_{RWO}(\varepsilon_0^1, e_0^1) = \max\left\{ \frac{\lambda_\varepsilon}{\alpha_\varepsilon} \frac{\delta_\varepsilon}{\delta_\varepsilon - \gamma_\varepsilon} |\varepsilon_0|^{\frac{\delta_\varepsilon - \gamma_\varepsilon}{\delta_\varepsilon}}, \frac{\lambda_e}{\alpha_e} \frac{\delta_e}{\delta_e - \gamma_e} |e_0|^{\frac{\delta_e - \gamma_e}{\delta_e}} \right\}, \quad (2.5.8)$$

$$\gamma_{(\cdot)} \text{ and } \delta_{(\cdot)} \in R^+, 2\gamma_{(\cdot)} \geq \delta_{(\cdot)} \geq \gamma_{(\cdot)}, (\cdot) = \varepsilon, e. \quad (2.5.9)$$

Theorem 2.5.2. *Let (2.5.6) through (2.7.17) hold (see Appendix). If system (2.4.1), (2.5.3), is controlled by the control $u(\cdot)$ determined by (2.5.6) through (2.7.17) then*

(a) *the set W, (2.5.6), is globally asymptotically stable with the prespecified finite reachability time $\tau_{RW} \in R^+$ for every mixed initial error vector $(\varepsilon_0^{1^T} \;\; e_0^{1^T})^T$:*

$$\begin{bmatrix} \varepsilon^{1^T}(t;\varepsilon_0^1) & e^{1^T}(t;e_0^1) \end{bmatrix}^T \in W, \forall [t, (\varepsilon_0^{1^T} \; e_0^{1^T})^T] \in [\tau_{RW}, \infty[\times R^{2N+n}.$$

(b) *the desired system motion $q_d^1(\cdot)$ is globally asymptotically stable with the finite reachability time $\tau_{RO} = \tau_{RO}(\varepsilon_0^1, e_0^1) = \tau_{RO}(q_0^1)$ determined by (2.5.7), (2.5.8) (see Remark 2.5.1), for every initial state vector $x_0 = q_0^1$,*

$$q^1(t;q_0^1) = q_d^1(t), \forall (t, q_0^1) \in [\tau_{RO}(q_0^1), \infty[\times R^n.$$

(c) *the system exhibits global (output) tracking with a finite reachability time $\tau_{RO}(q_0^1)$, that is, that the real output response $y^1(t;q_0^1)$ tracks and reaches pairwise its desired output response $y_d^1(t)$ at the latest at the moment $\tau_{RO}(q_0^1)$ for every initial state vector $q_0^1 \in R^n$, and rests equal to it after $\tau_{RO}(q_0^1)$ has elapsed:*

$$y^1(t;q_0^1) = y_d^1(t), \forall (t, q_0^1) \in [\tau_{RO}(q_0^1), \infty[\times R^n.$$

The proof of this theorem is given in the Appendix.

2.6 Conclusions

The notion of the system transfer function matrix is broadened to the system complete transfer function matrix that is valid under arbitrary input vector variations and any initial conditions. Both are determined from IO and ISO models of time-invariant continuous-time linear systems. They enable unification of the linear dynamical system theory and various extensions including discrete-time systems [31].

Time is the basic independent variable of dynamical (control) systems. Its properties linked with those of physical variables, summarized in the PCUP, show that every physical variable is a continuous function of time. This is expressed by the

TCUP. The validity of the PCUP, hence of the TCUP, is a necessary condition for an adequate mathematical modeling of physical systems.

The consistent Lyapunov methodology for nonlinear dynamical systems enables us to use the necessary and sufficient conditions for asymptotic stability of $x = 0$, for a set to be its domain and for a direct construction of a Lyapunov function of a given nonlinear dynamical system. The conditions are not expressed in terms of the existence of a system Lyapunov function.

The extended concept of vector Lyapunov functions introduced in this chapter opens new possibilities for studies of qualitative dynamical properties of complex (interconnected and/or large-scale systems) and for synthesis of their control.

2.7 Appendix

2.7.1 Control characterization

The real valued parameters $\gamma_{(\cdot)}$ and $\delta_{(\cdot)}$ (2.5.9) determine $\chi(\cdot)\colon R \to R$ defined by Matyukhin [45]:

$$\chi(\varepsilon_i^1) = |\varepsilon_i|^{-\gamma_\varepsilon/\delta_\varepsilon}\, \varepsilon_i^{(1)} \text{ if } \varepsilon_i \neq 0 \text{ for every } \varepsilon_i^{(1)} \in R,$$

$$\chi(\varepsilon_i^1) = \infty \text{ if } \varepsilon_i = 0 \text{ for every } \varepsilon_i^{(1)} \in R,$$

$$\chi\left(\varepsilon^1\right) = \left(\chi(\varepsilon_1^1)\chi(\varepsilon_2^1)\cdots\chi(\varepsilon_N^1)\right)^T \Longrightarrow \chi\left(e\right) = \left(\chi(e_1^1)\chi(e_2^1)\cdots\chi(e_p^1)\right)^T. \quad (2.7.1)$$

Another subsidiary function also by Matyukhin [45] is defined by

$$\psi_{(\cdot)}(\chi) = (\alpha_{(\cdot)}/2)[1 + \cos(\omega_{(\cdot)}\,|\chi| + \varphi_{(\cdot)})], \alpha_{(\cdot)} \in R^+$$

$$\Psi_{(\cdot)}(\chi) = \text{diag}\{\psi_{(\cdot)}(\chi)\ \psi_{(\cdot)}(\chi)\ \cdots\ \psi_{(\cdot)}(\chi)\} \in R^{k\times k},$$

$$\psi'_{(\cdot)}(\chi) = \frac{d\psi_{(\cdot)}(\chi)}{d\chi} = -(\pi\alpha_{(\cdot)}/2\eta_{(\cdot)})\text{sign}(\chi)\sin[\omega_{(\cdot)}(|\chi| + 2(\eta_{(\cdot)} - \alpha_{(\cdot)}))],$$

$$\text{if } |\chi| \in [2\alpha_{(\cdot)}, 2\alpha_{(\cdot)} + \eta_{(\cdot)}], (\cdot) = \varepsilon, e, \quad (2.7.2)$$

where $k = p$ if $(\cdot) = e$ and $k = N$ if $(\cdot) = \varepsilon$. Further, $\eta_{(\cdot)} \in]\pi\alpha_{(\cdot)}/2, \infty[$, $\omega_{(\cdot)} = \pi/\eta_{(\cdot)}$ and $\varphi_{(\cdot)} = 2(\eta_{(\cdot)} - \alpha_{(\cdot)})/\eta_{(\cdot)}$. The preceding parameters and functions well define the Matyukhin function $\mu(\cdot)\colon R \to R$ [45]:

$$\mu_{(\cdot)}(\chi) = \alpha_{(\cdot)} \text{ if } \chi \in [-2\alpha_{(\cdot)}, 2\alpha_{(\cdot)}],$$

$$\mu_{(\cdot)}(\chi) = \psi_{(\cdot)}(\chi) \text{ if } |\chi| \in [2\alpha_{(\cdot)}, 2\alpha_{(\cdot)} + \eta_{(\cdot)}],$$

$$\mu_{(\cdot)}(\chi) = 0 \text{ if } |\chi| \in [2\alpha_{(\cdot)} + \eta_{(\cdot)}, \infty[,$$

$$M_{(\cdot)}(\chi) = \text{diag}\{\mu_{(\cdot)}(\chi)\mu_{(\cdot)}(\chi)\cdots\mu_{(\cdot)}(\chi)\} \in R^{k\times k},$$

$$k = p \text{ if } (\cdot) = e \text{ and } k = N \text{ if } (\cdot) = \varepsilon, \tag{2.7.3}$$

which is globally differentiable and leads to the following partition of the set R of real numbers:

$$X_{(\cdot)1} = \{\chi\colon\ |\chi| \leq \alpha_{(\cdot)}\}, X_{(\cdot)2} = \{\chi\colon\ |\chi| \in [\alpha_{(\cdot)}, 2\alpha_{(\cdot)} + \eta_{(\cdot)}]\},$$

$$X_{(\cdot)3} = \{\chi\colon\ |\chi| \geq 2\alpha_{(\cdot)} + \eta_{(\cdot)}\}, X_{(\cdot)1} \cup X_{(\cdot)2} \cup X_{(\cdot)3} = R, (\cdot) \in \{e, \varepsilon\}. \tag{2.7.4}$$

Let

$$w(\varepsilon_i) = \text{sign}(\varepsilon_i)\, |\varepsilon_i|^{\gamma_e \delta_e^{-1}} \text{ and } w(e_i) = \text{sign}(e_i)\, |e_i|^{\gamma_e \delta_e^{-1}},$$
$$w_\varepsilon(\varepsilon) = (w(\varepsilon_1)\, w(\varepsilon_2) \cdots w(\varepsilon_N))^T,$$

$$w_e(e) = (w(e_1)\, w(e_2) \cdots w(e_p))^T, \tag{2.7.5}$$

$$z_\varepsilon(\varepsilon^1) = [z_{\varepsilon 1}(\varepsilon_1^1)\ z_{\varepsilon 2}(\varepsilon_2^1)\ \cdots\ z_{\varepsilon N}(\varepsilon_N^1)]^T = [\lambda_\varepsilon \varepsilon^{(1)} + M_\varepsilon(\chi(\varepsilon^1)) w_\varepsilon(\varepsilon)],$$
$$Z_\varepsilon(\varepsilon^1) = \text{diag}\{z_{\varepsilon 1}(\varepsilon_1^1)\ z_{\varepsilon 2}(\varepsilon_2^1)\ \cdots\ z_{\varepsilon N}(\varepsilon_N^1)\},\ \lambda_\varepsilon \in [1, \infty[,$$
$$z_e(e^1) = [z_{e1}(e_1^1)\ z_{e2}(e_2^1)\ \cdots\ z_{ep}(e_p^1)]^T = [\lambda_e e^{(1)} + M_e(\chi(e^1)) w_e(e)],$$

$$Z_e(e^1) = \text{diag}\{z_{e1}(e_1^1)\ z_{e2}(e_2^1)\ \cdots\ z_{ep}(e_p^1)\},\ \lambda_e \in [1, \infty[. \tag{2.7.6}$$

Hence,

$$z_\varepsilon^{(1)}(\varepsilon^1) = [\lambda_\varepsilon I_N + R_\varepsilon(\varepsilon, \chi)]\varepsilon^{(2)} + r(\varepsilon, \chi), \tag{2.7.7}$$

$$\varepsilon^{(2)} = y_d^{(2)} - J(x_p) x_p^{(2)} - J^{(1)}(x_p) x_p^{(1)},$$

$$z_e^{(1)}(e^1) = [\lambda_e I_p + R_e(e, \chi)]e^{(2)} + r(e, \chi), e^{(2)} = x_{pd}^{(2)} - x_p^{(2)}, \tag{2.7.8}$$

$$R_\varepsilon(\varepsilon, \chi) = \text{diag}\{R_{\varepsilon 1}(\varepsilon_1, \chi)\ R_{\varepsilon 2}(\varepsilon_2, \chi)\ \cdots R_{\varepsilon N}(\varepsilon_N, \chi)\},$$
$$R_{\varepsilon i}(\varepsilon_i, \chi) = 0 \text{ if } (\varepsilon_i, \chi) \in R \times (X_{\varepsilon i1} \cup X_{\varepsilon i3}),$$

$$R_{\varepsilon i}(\varepsilon_i, \chi) = \psi_i'(\chi)\text{sign}(\varepsilon_i) \text{ if } (\varepsilon_i, \chi) \in R \times X_{\varepsilon i2}, \tag{2.7.9}$$

$$R_e(e, \chi) = \text{diag}\{R_{e1}(e_1, \chi)\ R_{e2}(e_2, \chi)\ \cdots R_{ep}(e_p, \chi)\},$$
$$R_{ei}(e_i, \chi) = 0 \text{ if } (e_i, \chi) \in R \times (X_{ei1} \cup X_{ei3}),$$

$$R_{ei}(e_i, \chi) = \psi_i'(\chi)\text{sign}(e_i) \text{ if } (e_i, \chi) \in R \times X_{ei2}, \tag{2.7.10}$$

$$r(\varepsilon, \chi) = [r_1(\varepsilon_1, \chi)\ r_2(\varepsilon_2, \chi)\ \cdots\ r_N(\varepsilon_N, \chi)]^T,$$
$$r_i(\varepsilon_i, \chi) = \alpha_{i\varepsilon} \gamma_{i\varepsilon} \delta_{i\varepsilon}^{-1}\, |\varepsilon_i|^{(2\gamma_{i\varepsilon} - \delta_{i\varepsilon})} \chi \text{ if } (\varepsilon_i, \chi) \in R \times X_{\varepsilon i1},$$

$$r_i(\varepsilon_i,\chi)=\gamma_{i\varepsilon}\delta_{i\varepsilon}^{-1}[\psi_{i\varepsilon}(\chi)-\psi'_{i\varepsilon}(\chi)\chi]\,|\varepsilon_i|^{(2\gamma_{i\varepsilon}-\delta_{i\varepsilon})}\,\chi$$
$$\text{if } (\varepsilon_i\neq 0,\chi)\in R\times X_{\varepsilon i2},$$

$$r_i(\varepsilon_i,\chi)=0 \text{ if } (\varepsilon_i,\chi)\in R\times X_{\varepsilon i3}. \tag{2.7.11}$$

$$r(e,\chi)=[r_1(e_1,\chi)\ r_2(e_2,\chi)\ \cdots\ r_p(e_p,\chi)]^T,$$
$$r_i(e_i,\chi)=\alpha_{ie}\gamma_{ie}\delta_{ie}^{-1}\,|e_i|^{(2\gamma_{ie}-\delta_{ie})}\,\chi \text{ if } (e_i,\chi)\in R\times X_{ei1},$$
$$r_i(e_i,\chi)=\gamma_{ie}\delta_{ie}^{-1}[\psi_{ie}(\chi)-\psi'_{ie}(\chi)\chi]\,|e_i|^{(2\gamma_{ie}-\delta_{ie})}\,\chi$$
$$\text{if } (e_i\neq 0,\chi)\in R\times X_{ei2},$$

$$r_i(e_i,\chi)=0 \text{ if } (e_i,\chi)\in R\times X_{ei3}. \tag{2.7.12}$$

The control is now determined by the next equations in which $F_2\equiv F_2(\varepsilon^1,e^1,x)$:

$$v(\varepsilon^1,e^1)=\begin{bmatrix} v^1(\varepsilon^1)\\ v^2(e^1)\end{bmatrix}=\begin{bmatrix}(1/2)Z_\varepsilon(\varepsilon^1)z_\varepsilon(\varepsilon^1)\\ (1/2)z_e^T(e^1)z_e(e^1)\end{bmatrix},$$

$$v(\varepsilon^1,e^1)=\ [v_1(\varepsilon_1^1)\ v_2(\varepsilon_2^1)\ \cdots\ v_N(\varepsilon_N^1)\ v^2(e^1)],\ v_0=v(\varepsilon_0^1,e_0^1),$$

$$V_0^{1/2}=\operatorname{diag}\left\{v_1^{1^{1/2}}(\varepsilon_{10}^1)\ v_2^{1^{1/2}}(\varepsilon_{20}^1)\ \cdots v_N^{1^{1/2}}(\varepsilon_{N0}^1)\ v^{2^{1/2}}(e_0^1)\right\}, \tag{2.7.13}$$

$$F_1(\varepsilon^1,e^1,x)=\begin{bmatrix} Z_\varepsilon(\varepsilon^1)\left\{\begin{matrix}[\lambda_\varepsilon I_N+R_\varepsilon(\varepsilon,\chi(\varepsilon^1))]\cdot\\ \cdot[y_d^{(2)}-J^{(1)}(x_p)x_p^{(1)}]+r(\varepsilon,\chi(\varepsilon^1))\end{matrix}\right\}\\ z_e^T(e^1)\left\{[\lambda_e I_p+R_e(e,\chi(e^1))]x_{pd}^{(2)}+r(e,\chi(e^1))\right\}\end{bmatrix}, \tag{2.7.14}$$

$$F_2(\varepsilon^1,e^1,x)=\begin{bmatrix} Z_\varepsilon(\varepsilon^1)[\lambda_\varepsilon I_N+R_\varepsilon(\varepsilon,\chi(\varepsilon^1))]J(x_p)\\ z_e^T(e^1)[\lambda_e I_p+R_e(e,\chi(e^1))]\end{bmatrix},$$

$$\operatorname{rank}F_2(\varepsilon^1,e^1,x)=N+1,\forall(\varepsilon^1,e^1,x)\in R^{2N}\times R^n\times R^n, \tag{2.7.15}$$

$\beta\in\{0,1\}$,

$$u^*(\varepsilon^1,e^1,x)=A(x_p)F_2^T(F_2F_2^T)^{-1}$$

$$\cdot\left[F_1(\varepsilon^1,e^1)+\tau_{RW}^{-1}\left\langle 2\beta V_0^{1/2}v^{1/2}(\varepsilon^1,e^1)+(1-\beta)v_0\right\rangle\right], \tag{2.7.16}$$

$$u(\varepsilon^1,e^1,x)=B^T(x)[B(x)B^T(x)]^{-1}[u^*(\varepsilon^1,e^1)+n(x)]. \tag{2.7.17}$$

Remark 2.7.1.

$$
\begin{aligned}
z_\varepsilon(\varepsilon^1) &= [\lambda_\varepsilon \varepsilon^{(1)} + \alpha_\varepsilon w_\varepsilon(\varepsilon)] = 0, \forall \varepsilon^1 \in W^1, \\
z_e(e^1) &= [\lambda_e e^{(1)} + \alpha_e w_e(e)] = 0, \forall e^1 \in W^2.
\end{aligned}
$$

2.7.2 Proof of theorem on control synthesis (Theorem 2.5.2)

Proof. The vector function $v^1(\cdot)$, $v^1(z_\varepsilon) = (1/2)Z_\varepsilon z_\varepsilon$, $z_\varepsilon \in R^N$, is globally positive-definite in $z_\varepsilon \in R^N$. The scalar function $v^2(\cdot)$, $v^2(z_e) = (1/2)\|z_e\|^2$, $z_e \in R^p$, is globally positive-definite in $z_e \in R^p$. Both are radially unbounded. Hence, the vector function $v(\cdot)$, (2.7.13), is globally positive-definite radially unbounded relative to the set W, (2.5.6), in which $v^1(z_\varepsilon(\cdot))$ is global pairwise positive-definite radially unbounded vector function relative to the set W^1, and $v^2(z_e(\cdot))$ is global positive-definite radially unbounded scalar function relative to the set W^2, (2.5.6). The Matyukhin function, (2.7.1)–(2.7.3), and (2.7.5) guarantee global differentiability of the vector function $v(\cdot)$ on R^{2N+n} [45]. Its derivative along motions of system (2.4.1), (2.5.3) controlled by control determined by (2.5.6) through (2.7.17) has the next form (Remark 2.7.1):

$$
v^{(1)}(\varepsilon^1, e^1) = -\tau_{RW}^{-1}[\beta V_0^{1/2} v^{1/2}(\varepsilon^1, e^1) + (1-\beta)v_0].
$$

Hence,

$$
\left[\varepsilon^{1^T}(t;\varepsilon_0^1)\; e^{1^T}(t;e_0^1)\right]^T \in W, \forall t \in [\tau_{RW}, \infty[, \forall(\varepsilon_0^1, e_0^1) \in R^{2N+n}.
$$

This proves global asymptotic stability with finite reachability time τ_{RW} of the set W and its positive invariance. The definition of the function $w(\cdot)$, (2.7.5), guarantees

$$
\forall t \in [\tau_{RWO}, \infty[\Longrightarrow \left[\varepsilon^{1^T}(t;\varepsilon_0^1)\; e^{1^T}(t;e_0^1)\right]^T = 0, \forall(\varepsilon_0^1, e_0^1) \in W. \tag{2.7.18}
$$

$$
\forall t \in [\tau_{RO}, \infty[\Longrightarrow \left[\varepsilon^{1^T}(t;\varepsilon_0^1)\; e^{1^T}(t;e_0^1)\right]^T = 0, \forall(\varepsilon_0^1, e_0^1) \in R^{2N+n}.
$$

This proves the theorem.

Bibliography

[1] P. J. Antsaklis and A. N. Michel, *Linear Systems*, McGraw-Hill, New York, 1997.

[2] R. Bellman, "Vector Lyapunov functions," *SIAM J. Control*, Ser. A, vol. 1, no. 1, pp. 32–34, 1962.

[3] N. P. Bhatia and G. P. Szegö, *Dynamical Systems Stability Theory and Applications*, Springer-Verlag, Berlin, 1967.

[4] A. Einstein, *La théorie de la relativité*, Gauthier-Villars et Cie, Paris, 1921.

[5] A. Einstein, *The Meaning of Relativity*, Methuen, London, 1950.

[6] A. Einstein, *La théorie de la relativité restreinte et générale*, Gauthier-Villars et Cie, Paris, 1954.

[7] A. Einstein, *Relativity*, Methuen, London, 1960.

[8] D. L. Gray and A. N. Michel, "A training algorithm for binary feedforward neural networks," *IEEE Trans. Neural Networks*, vol. 3, no. 2, pp. 176–194, 1992.

[9] Ly. T. Grouyitch and G. Dauphin-Tanguy, "Stabilité énergétique exponentielle," *Actes de la Conférence Internationale Francophone d'Automatique*, pp. 75–80, Lille, France, July 2000.

[10] Lj. T. Grujić, "Novel development of Lyapunov stability of motion," *Int. J. Control*, vol. 22, no. 4, pp. 525–549, 1975.

[11] Lj. T. Grujić, "Stability domains of general and large-scale stationary systems," in *Applied Modelling and Simulation of Technological Systems*, P. Borne and S. G. Tzafestas, Eds., pp. 317–327, Elsevier (North Holland), Amsterdam, 1987.

[12] Lj. T. Grujić, "On large-scale systems stability," in *Computing and Computers for Control Systems*, P. Borne et al., Eds., pp. 201–206, Baltzer Scientific Publishing, Bussum, the Netherlands, 1989.

[13] Lj. T. Grujić, *Automatique–Dynamique Linéaire*, Lecture Notes, Ecole Nationale d'Ingénieurs de Belfort, Belfort, France, 1994–1998.

[14] Lj. T. Grujić, "New Lyapunov methodology: Absolute stability of Lurie-Postnykov systems," *Preprints of the IFAC-IFIP-IMACS Conference on Control of Industrial Systems*, vol. 1, pp. 533–538, Belfort, France, May 1997.

[15] Lj. T. Grujić, "Novel Lyapunov stability methodology for nonlinear systems: Complete solutions," *Nonlinear Analysis, Theory, Methods & Applications*, vol. 30, no. 8, pp. 5315–5325, 1997.

[16] Lj. T. Grujić, "Exponential quality of time-varying dynamical systems: Stability and tracking," in *Advances in Nonlinear Dynamics, Stability and Control: Theory, Methods and Applications*, S. Sivasundaram and A. A. Martynyuk, Eds., vol. 5, pp. 51–61, Gordon and Breach, Amsterdam, 1997.

[17] Lj. T. Grujić, "New approach to asymptotic stability: Time-varying nonlinear systems," *Int. J. Math. and Math. Sci.*, vol. 20, no. 2, pp. 347–366, 1997.

[18] Lj. T. Grujić, A. A. Martynyuk, and M. Ribbens-Pavella, *Large Scale Systems Stability under Structural and Singular Perturbations*, Springer-Verlag, Berlin, 1987.

[19] Lj. T. Grujić and A. N. Michel, "Exponential stability and trajectory bounds of neural networks under structural variations," *IEEE Trans. Circuits and Systems*, vol. 38, no. 10, pp. 1182–1192, Oct. 1991.

[20] Lj. T. Grujić and A. N. Michel, "Modelling and qualitative analysis of discrete-time neural networks under pure structural variations," in *Mathematics of the Analysis and Design of Process Control*, P. Borne, S. G. Tzafestas, and N. E. Radhy, Eds., pp. 559–566, Elsevier (North-Holland), Amsterdam, 1992.

[21] Ly. T. Gruyitch, *Continuous Time Control Systems*, Lecture notes for the course DNEL4CN2: Control Systems, Department of Electrical Engineering, University of Natal, Durban, South Africa, 1993.

[22] Ly. T. Gruyitch, "Consistent Lyapunov methodology for time-invariant nonlinear systems," *Avtomatika i Telemehanika*, no. 12, pp. 35–73, Dec. 1997 (in Russian).

[23] Ly. T. Gruyitch, "Time, relativity and physical principle: Generalizations and applications," *Proc. V International Conference: Physical Interpretations of Relativity Theory*, pp. 134–170, London, Sept. 1998.

[24] Ly. T. Gruyitch, "Physical continuity and uniqueness principle. Exponential natural tracking control," *Neural, Parallel & Sci. Comput.*, vol. 6, pp. 143–170, 1998.

[25] Ly. T. Gruyitch, *Conduite des Systèmes*, Lecture Notes, University of Technology Belfort-Montbeliard, Belfort, 2000, 2001.

[26] Ly. T. Gruyitch, "Consistent Lyapunov methodology, time-varying nonlinear systems and sets," *Nonlinear Analysis*, vol. 39, pp. 413–446, 2000.

[27] Ly. T. Gruyitch, "Gaussian generalisations of the relativity theory fundaments with applications," *Preprints of the VI International Conference: Physical Interpretations of Relativity Theory*, pp. 125–136, London, Sept. 2000.

[28] Ly. T. Gruyitch, "Systems approach to the relativity theory fundaments," *Nonlinear Analysis*, vol. 47, pp. 37–48, 2001.

[29] Ly. T. Gruyitch, "Time and uniform relativity theory fundaments," *Problems of Nonlinear Analysis in Engineering Systems*, vol. 7, no. 2, pp. 1–29, 2001.

[30] Ly. T. Gruyitch, "Consistent Lyapunov methodology: Non-differentiable nonlinear systems," *Nonlinear Dynamics and Systems Theory*, vol. 1, no. 1, pp. 1–22, 2001.

[31] Ly. T. Gruyitch, *Contrôle Commande des Processus Industriels*, Lecture Notes, University of Technology Belfort-Montbeliard, Belfort, 2002.

[32] Ly. T. Gruyitch and J. Dauphin-Tanguy, "Strict monotonous asymptotic energetic stability. Part I: Theory," *Pro. 14th Triennial World Congress*, vol. D, pp. 437–442, Beijing, P. R. China, 1999.

[33] Ly. T. Gruyitch, J.-P. Richard, P. Borne, and J.-C. Gentina, *Stability Domains*, Taylor and Francis, London, in press.

[34] W. Hahn, *Stability of Motion*, Springer-Verlag, Berlin, 1967.

[35] N. N. Krasovskii, *Stability of Motion*, Stanford University Press, Stanford, CA, 1963.

[36] J. H. Li, A. N. Michel, and W. Porod, "Qualitative analysis and synthesis of a class of neural networks," *IEEE Trans. Circuits and Systems*, vol. 35, no. 8, pp. 976–986, Aug. 1988.

[37] J. H. Li, A. N. Michel, and W. Porod, "Analysis and synthesis of a class of neural networks: Variable structure systems with infinite gain," *IEEE Trans. Circuits Syst.*, vol. 36, no. 5, pp. 713–731, May 1989.

[38] J. H. Li, A. N. Michel, and W. Porod, "Analysis and synthesis of a class of neural networks: Linear systems operating on a closed hypercube," *IEEE Trans. Circuits Syst.*, vol. 36, no. 11, pp. 1405–1422, Nov. 1989.

[39] D. Liu and A. N. Michel, "Asymptotic stability of discrete-time systems with saturation nonlinearities with applications to digital filters," *IEEE Trans. Circuits and Systems–I: Fundamental Theory and Applications*, vol. 39, no. 10, pp. 798–807, Oct. 1992.

[40] D. Liu and A. N. Michel, "Null controllability of systems with control constraints and state saturation," *Systems and Control Letters*, vol. 20, pp. 131–139, 1993.

[41] D. Liu and A. N. Michel, "Stability analysis of state-space realizations for two-dimensional filters with overflow nonlinearities," *IEEE Trans. Circuits and Systems–I: Fundamental Theory and Applications*, vol. 41, no. 2, pp. 127–137, Feb. 1994.

[42] A. M. Lyapunov, *The General Problem of the Stability of Motion*, Kharkov Mathematical Society, Kharkov, 1892, published in *Academician A. M. Lyapunov. Collected Works. vol. II*, (in Russian), Academy of Sciences of USSR, Moscow, pp. 5–263, 1956. English translation: *Int. J. Control*, vol. 55, pp. 531–773, 1992.

[43] V. M. Matrosov, "To the theory of stability of motion," *Prikl. Math. Mekh.*, vol. 25, no. 5, pp. 885–895, 1962 (in Russian).

[44] V. M. Matrosov, *Vector Lyapunov function method: Analysis of dynamical properties of nonlinear systems*, FIZMATLIT, Moscow, 2001 (in Russian).

[45] V. I. Matyukhin, "Strong stability of motions of mechanical systems," *Automation and Remote Control*, vol. 57, no. 1, pp. 28–44, 1996.

[46] A. N. Michel, "On the bounds of the trajectories of differential systems," *Int. J. Control*, vol. 10, no. 5, pp. 593–600, 1969.

[47] A. N. Michel, "Stability, transient behavior and the trajectories bounds of interconnected systems," *Int. J. Control*, vol. 11, no. 4, pp. 703–715, 1970.

[48] A. N. Michel, "Qualitative analysis of simple and interconnected systems: Stability, boundedness, and trajectory behavior," *IEEE Trans. Circuit Theory*, vol. CT-17, no. 3, pp. 292–301, 1970.

[49] A. N. Michel and J. A. Farrell, "Associative memories via artificial neural networks," *IEEE Control Systems Magazine*, vol. 10, no. 3, pp. 6–17, Apr. 1990.

[50] A. N. Michel, J. A. Farrell, and W. Porod, "Qualitative analysis of neural networks," *IEEE Trans. Circuits and Systems*, vol. 36, no. 2, pp. 229–243, 1989.

[51] A. N. Michel and Lj. T. Grujić, "Modelling and qualitative analysis of continuous-time neural networks under pure structural variations," in *Mathematics of the Analysis and Design of Process Control*, P. Borne, S. G. Tzafestas and N. E. Radhy, Eds., pp. 549–558, Elsevier (North-Holland), Amsterdam, 1992.

[52] A. N. Michel and L. Hou, "Modeling and qualitative theory for general hybrid dynamical and control systems," *Preprints of the IFAC-IFIP-IMACS Conference on Control of Industrial Systems*, vol. 1, pp. 173–183, Belfort, France, May 1997.

[53] A. N. Michel, D. Liu, and K. Wang, "Stability analysis of a class of systems with parameter uncertainties and with state saturation nonlinearities," *Int. J. Robust and Nonlinear Control*, vol. 5, pp. 505–519, 1995.

[54] A. N. Michel and R. K. Miller, *Qualitative Analysis of Large-Scale Dynamical Systems*, Academic Press, New York, 1977.

[55] A. N. Michel, R. K. Miller, and B. H. Nam, "Stability analysis of interconnected systems using computer generated Lyapunov functions," *IEEE Trans. Circuits and Systems*, vol. CAS-29, no. 7, pp. 431–440, 1982.

[56] A. N. Michel, B. H. Nam, and V. Vittal, "Computer generated Lyapunov functions for interconnected systems: Improved results with application to power systems," *IEEE Trans. Circuits Syst.*, vol. CAS-31, no. 2, pp. 189–198, 1984.

[57] A. N. Michel and D. W. Porter, "Practical stability and finite-time stability of discontinuous systems," *IEEE Trans. Circuit Theory*, vol. CT-19, no. 2, pp. 123–129, 1972.

[58] A. N. Michel, N. R. Sarabudla, and R. K. Miller, "Stability analysis of complex dynamical systems," *Circuits, Systems, Signal Processing*, vol. 1, no. 2, pp. 171–202, 1982.

[59] A. N. Michel and K. Wang, "Robust stability: Perturbed systems with perturbed equilibria," *Systems and Control Letters*, vol. 21, pp. 155–162, 1993.

[60] A. N. Michel and K. Wang, *Qualitative Theory of Dynamical Systems: The Role of Stability Preserving Mappings*, Marcel Dekker, New York, 1994.

[61] R. K. Miller and A. N. Michel, *Ordinary Differential Equations*, Academic Press, New York, 1982.

[62] I. Newton, *Mathematical Principles of Natural Philosophy—Book I. The Motion of Bodies*, William Benton, Publisher, Encyclopedia Britannica, Chicago, 1952.

[63] K. M. Passino, A. N. Michel, and P. Antsaklis, "Lyapunov stability of a class of discrete-event systems," *IEEE Trans. Automatic Control*, vol. 39, no. 2, pp. 269–279, Feb. 1994.

[64] M. S. Radenkovic and A. N. Michel, "Robust adaptive systems and self stabilization," *IEEE Trans. Automatic Control*, vol. 37, no. 9, pp. 1355–1369, Sept. 1992.

[65] M. Radenkovic and A. N. Michel, "Verification of the self-stabilization mechanism in robust stochastic adaptive control using Lyapunov function arguments," *SIAM J. Control Optim.*, vol. 30, no. 6, pp. 1270–1294, Nov. 1992.

[66] M. Radenkovic and A. N. Michel, "Stochastic adaptive control of nonminimum phase systems in the presence of unmodelled dynamics," *SIAM J. Control Optim.*, vol. 14, no. 3, pp. 317–349, 1995.

[67] J. Si and A. N. Michel, "Analysis and synthesis of a class of discrete-time neural networks with multilevel threshold neurons," *IEEE Trans. Neural Networks*, vol. 6, no. 1, pp. 105–116, 1995.

[68] D. D. Šiljak, *Large-Scale Dynamic Systems: Stability and Structure*, North-Holland, New York, 1978.

[69] K. Wang and A. N. Michel, "Qualitative analysis of dynamical systems determined by differential inequalities with applications to robust stability," *IEEE Trans. Circuits Syst.–I: Fundamental Theory and Applications*, vol. 41, no. 2, pp. 377–386, May 1994.

[70] K. Wang and A. N. Michel, "On the stability of nonlinear time-varying systems," *IEEE Trans. Circuits Syst.–I: Fundamental Theory and Applications*, vol. 43, no. 7, pp. 517–531, July 1996.

[71] K. Wang and A. N. Michel, "Stability analysis of differential inclusions in Banach space with applications to nonlinear systems with time delays," *IEEE Trans. Circuits Syst.–I: Fundamental Theory and Applications*, vol. 43, no. 8, pp. 617–626, Aug. 1996.

[72] H. Ye, A. N. Michel, and L. Hou, "Stability theory for hybrid dynamical systems," *Proc. 34th IEEE Conf. Decision and Control*, pp. 2670–2684, New Orleans, LA, Dec. 1995.

[73] T. Yoshizawa, *Stability Theory by Liapunov's Second Method*, The Mathematical Society of Japan, Tokyo, 1966.

[74] V. I. Zubov, *Methods of A. M. Liapunov and Their Applications*, Noordhoff, Groningen, 1964.

Chapter 3

Asymptotic Stability of Multibody Attitude Systems*

Jinglai Shen, Amit K. Sanyal, and N. Harris McClamroch

Abstract: A rigid base body, supported by a fixed pivot point, is free to rotate in three dimensions. Multiple elastic subsystems are rigidly mounted on the rigid body; the elastic degrees of freedom are constrained relative to the rigid base body. A mathematical model is developed for this multibody attitude system that exposes the dynamic coupling between the rotational degrees of freedom of the base body and the deformation or shape degrees of freedom of the elastic subsystems. The models are used to assess passive dissipation assumptions that guarantee asymptotic stability of an equilibrium solution. These results are motivated and inspired by a 1980 publication of R. K. Miller and A. N. Michel [6].

3.1 Introduction

A photograph of the triaxial attitude control testbed (TACT) in the Attitude Dynamics and Control Laboratory at the University of Michigan is shown in Figure 3.1.1. Its physical properties are described in detail in [1], and a detailed derivation of mathematical models for the TACT is given in [2, 3]. The TACT is based on a spherical air bearing that provides a near-frictionless pivot for the base body. The stability problem treated in this chapter is motivated by possible set-ups of the TACT.

This chapter presents results of a study of asymptotic stability properties of an abstraction of the TACT. This abstraction consists of a rigid base body that is free to

*This research has been supported in part by the National Science Foundation under Grant ECS-0140053.

Figure 3.1.1: Triaxial attitude control testbed

rotate in three dimensions. Multiple elastic subsystems are rigidly mounted on the base body. This chapter discusses the uncontrolled motion of this multibody attitude system. Results are obtained that guarantee asymptotic stability of an equilibrium.

This chapter is motivated by a 1980 publication of R. K. Miller and A. N. Michel [6]. This Miller and Michel paper studied an elastic multibody system consisting of an interconnection of ideal mass elements and elastic springs. Lyapunov function arguments, based on the system Hamiltonian, were used to develop sufficient damping assumptions that guaranteed asymptotic stability of the equilibrium. A key insight was the use of observability properties to guarantee asymptotic stability. The paper by Miller and Michel provided a clear and direct exposition of these issues. It was one of the earliest papers to make clear connections between properties of Hamiltonian systems and their control theoretical properties. During the last 23 years, these issues have been extensively studied. However, the Miller and Michel paper remains an important resource for researchers on dynamics and control of mechanical systems.

3.2 Equations of Motion

Consider the following class of multibody attitude systems: the base body rotates about a fixed pivot point (see Figure 3.2.1). A base body fixed coordinate frame is chosen with its origin located at the pivot point. We assume that the center of mass of the system is always at the pivot point, and thus does not depend on the shape. This is a restrictive assumption, but we demonstrate that it represents an interesting class of multibody attitude systems. Thus gravity does not affect the dynamics and its effects are irrelevant in the subsequent analysis.

The configuration manifold is given by $Q = SO(3) \times Q_s, \widehat{\omega} \in \mathfrak{so}(3)$ with $\omega \in \mathbb{R}^3$ representing the base body angular velocity expressed in the base body frame. We use $r \in Q_s$ to denote n-dimensional generalized shape coordinates or deformation

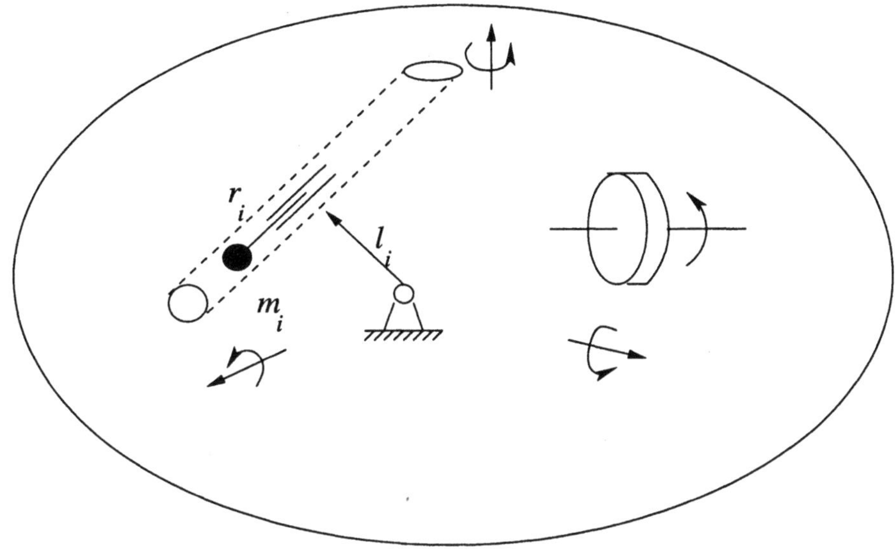

Figure 3.2.1: Schematic configuration of a multibody attitude system with a translational elastic degree of freedom and a rotational elastic degree of freedom

of the elastic systems. Assuming that the kinetic and potential energy are invariant to $SO(3)$ action, the dynamics are only dependent on $(\omega, r, \dot{r})$. This leads to the reduced Lagrangian and the reduced equations of motion.

The reduced kinetic energy is given by

$$T(\omega, r, \dot{r}) = \frac{1}{2} \begin{pmatrix} \omega \\ \dot{r} \end{pmatrix}^T \underbrace{\begin{bmatrix} M_{11}(r) & M_{12}(r) \\ M_{21}(r) & M_{22}(r) \end{bmatrix}}_{M(r)} \begin{pmatrix} \omega \\ \dot{r} \end{pmatrix},$$

where $M(r)$ is symmetric and positive-definite for all $r \in Q_s$. Let $V_s(r)$ denote the potential energy of the elastic subsystems that depends only on the shape coordinates. Throughout the subsequent analysis, we assume that $V_s(r)$ has a local minimum at the shape r_e, i.e.,

$$\frac{\partial V_s(r_e)}{\partial r} = 0 \text{ and } \frac{\partial^2 V_s(r_e)}{\partial r^2} > 0.$$

We obtain the reduced Lagrangian on $\mathfrak{so}(3) \times TQ_s$ as

$$L(\omega, r, \dot{r}) = \frac{1}{2} \begin{pmatrix} \omega \\ \dot{r} \end{pmatrix}^T \begin{bmatrix} M_{11}(r) & M_{12}(r) \\ M_{21}(r) & M_{22}(r) \end{bmatrix} \begin{pmatrix} \omega \\ \dot{r} \end{pmatrix} - V_s(r).$$

Therefore, the equations of motion for the multibody attitude system are in the form of Euler-Poincare equations

$$M(r) \begin{bmatrix} \dot{\omega} \\ \ddot{r} \end{bmatrix} = -\frac{dM(r)}{dt} \begin{bmatrix} \omega \\ \dot{r} \end{bmatrix} + \begin{bmatrix} \Pi \times \omega \\ \frac{\partial [T(\omega, r, \dot{r}) - V_s(r)]}{\partial r} \end{bmatrix}, \qquad (3.2.1)$$

where

$$\Pi = \frac{\partial L}{\partial \omega} = M_{11}(r)\omega + M_{12}(r)\dot{r}$$

is the conjugate angular momentum. It is easy to verify that the spatial angular momentum is conserved and $||\Pi||_2$ is a conserved quantity (see [8]). In addition, $(\omega, r, \dot{r}) = (0, r_e, 0)$ is an isolated equilibrium of (3.2.1).

We now add dissipation to (3.2.1). We consider linear passive damping at the pivot point of the base body attitude dynamics and in the shape dynamics so that (3.2.1) becomes

$$M(r)\begin{bmatrix} \dot{\omega} \\ \ddot{r} \end{bmatrix} = -\frac{dM(r)}{dt}\begin{bmatrix} \omega \\ \dot{r} \end{bmatrix} + \begin{bmatrix} \Pi \times \omega \\ \frac{\partial [T(\omega,r,\dot{r}) - V_s(r)]}{\partial r} \end{bmatrix} - \begin{bmatrix} C_a\omega \\ C_r\dot{r} \end{bmatrix}, \tag{3.2.2}$$

where $C_a \in \mathbb{R}^{3\times3}$, $C_r \in \mathbb{R}^{n\times n}$ denote constant damping matrices that satisfy $C_a = C_a^T \geq 0$, $C_r = C_r^T \geq 0$, i.e., both are symmetric and positive semidefinite. As before, $(\omega, r, \dot{r}) = (0, r_e, 0)$ is an isolated equilibrium of (3.2.2).

3.3 Conservation of Energy and Lyapunov Stability

Define a noncanonical Hamiltonian based on the reduced Lagrangian $H(\omega, r, \dot{r}) = T(\omega, r, \dot{r}) + V_s(r) = L(\omega, r, \dot{r}) + 2V_s(r)$. This Hamiltonian can be viewed as the total energy of the system. We now verify that $\dot{H} \equiv 0$ along the solutions of the undamped dynamics in (3.2.1). In fact,

$$\frac{dH}{dt} = \frac{\partial H}{\partial \omega}\dot{\omega} + \frac{\partial H}{\partial \dot{r}}\ddot{r} + \frac{\partial H}{\partial r}\dot{r} = \frac{\partial L}{\partial \omega}\dot{\omega} + \frac{\partial L}{\partial \dot{r}}\ddot{r} + \left(\frac{\partial T}{\partial r} + \frac{\partial V_s}{\partial r}\right)\dot{r}.$$

Note that

$$\begin{aligned} \frac{\partial L}{\partial \omega}\dot{\omega} + \frac{\partial L}{\partial \dot{r}}\ddot{r} &= (\omega \;\; \dot{r})^T M(r)\begin{pmatrix} \dot{\omega} \\ \ddot{r} \end{pmatrix} = (\omega \;\; \dot{r})^T \left\{-\dot{M}(r)\begin{pmatrix} \omega \\ \dot{r} \end{pmatrix} + \begin{bmatrix} \Pi \times \omega \\ \frac{\partial T}{\partial r} - \frac{\partial V_s}{\partial r} \end{bmatrix}\right\} \\ &= -\left(\frac{\partial T}{\partial r} + \frac{\partial V_s}{\partial r}\right)\dot{r}, \end{aligned}$$

where we use (3.2.1) and $(\omega \;\; \dot{r})^T \dot{M}(r)\begin{pmatrix} \omega \\ \dot{r} \end{pmatrix} = 2\frac{\partial T}{\partial r}\dot{r}$. Consequently, $\frac{dH}{dt} \equiv 0$. This result agrees with the fact that the total energy is conserved if there is no dissipation.

In the case of linear damping, $\dot{H}$ along the solutions of (3.2.2) is given by

$$\dot{H} = -\omega^T C_a \omega - \dot{r}^T C_r \dot{r} \leq 0.$$

Therefore, $\dot{H}$ is negative semidefinite. Moreover, since the Hamiltonian is a positive-definite function of $(\omega, r, \dot{r})$ in any small neighborhood of the equilibrium satisfying $H(0, r_e, 0) = 0$, it is a Lyapunov function. Thus by the property that $\dot{H} \leq 0$ along the solutions of (3.2.2), we conclude that the equilibrium $(\omega, r, \dot{r}) = (0, r_e, 0)$ is stable in the sense of Lyapunov.

Since both C_a and C_r are symmetric and positive semidefinite, it is easy to show that the set where $\dot{H} = 0$ is the set $S = \{(\omega, \dot{r}) | C_a\omega = 0, C_r\dot{r} = 0\}$. This result will be used in the subsequent development.

3.4 Asymptotic Stability Analysis

We now develop conditions on the damping matrices C_a and C_r and on the inertia matrix $M(r)$ for (local) asymptotic stability of the equilibrium $(\omega, r, \dot{r}) = (0, r_e, 0)$. The reader should be cautioned that stability has been established for the equilibrium. Thus, throughout the remaining sections, the equilibrium is always stable in the sense of Lyapunov, even when it is not asymptotically stable. The conditions obtained in the subsequent sections are only for asymptotic stability of the equilibrium.

While general conditions follow from LaSalle's invariance principle [5], which is related to nonlinear observability [4, 7], it is nontrivial to obtain concrete and easily verified conditions, due to the nonlinear dynamics and the noncanonical form of the Hamiltonian. We first present a trivial sufficient condition and a necessary condition as follows.

Proposition 3.4.1. *Assume $C_a > 0$, $C_r > 0$. Then the equilibrium is asymptotically stable.*

Proof. The set where $\dot{H} = 0$ is equivalent to $(\omega, \dot{r}) = (0, 0)$. Substituting this result into the equations of motion, we obtain $\frac{\partial V_s(r)}{\partial r} = 0$, which holds only at r_e in any small neighborhood of r_e. Consequently, the equilibrium is asymptotically stable by the invariance principle. □

Proposition 3.4.2. *The equilibrium is asymptotically stable only if $C_a \neq 0$.*

Proof. The attitude equation in (3.2.2) can be written as $\dot{\Pi} = \Pi \times \omega - C_a\omega$, where Π is the conjugate angular momentum. If $C_a = 0$, then $||\Pi(t)||_2 = ||\Pi(0)||_2$ for all $t \geq 0$ since $\frac{d||\Pi(t)||_2^2}{dt} = 2\Pi^T\dot{\Pi} = 0$. Hence for any initial $(\omega, \dot{r})$ in any small neighborhood of the equilibrium such that $\Pi(0) \neq 0$, $||\Pi(t)||_2 \neq 0$ for all $t \geq 0$. Moreover, since the equilibrium is stable, $M(r)$ is bounded along the solutions. This implies that $(\omega, \dot{r})$ does not converge to zero as $t \to \infty$. □

It is a challenge to obtain more concrete conditions. For simplicity, we focus on two cases in the subsequent development: the first case assumes C_a to be positive-definite, and the second case assumes C_r to be positive-definite. The first case makes use of the observability rank condition to obtain results for asymptotic stability. In the second case, both linear and nonlinear observability conditions fail to verify observability; thus a new approach is developed. In both cases, we emphasize how the coupling between the attitude dynamics and the shape dynamics can lead to asymptotic stability without full damping.

3.4.1 Asymptotic stability analysis: $C_a > 0, C_r \geq 0$

Throughout this section, we assume C_a to be positive-definite. In this case, the set where $\dot{H} = 0$ is the set $S = \{(\omega, \dot{r}) | \omega = 0,\ C_r\dot{r} = 0\}$. Let $\mathcal{M}$ be the largest invariant set in S; solutions in $\mathcal{M}$ must satisfy the following equations:

$$\frac{d\Big(M_{12}(r)\dot{r}\Big)}{dt} = 0, \tag{3.4.1}$$

$$\frac{d\Big(M_{22}(r)\dot{r}\Big)}{dt} = \frac{1}{2}\left(\dot{r}^T\frac{\partial M_{22}(r)}{\partial r}\dot{r}\right) - \frac{\partial V_s(r)}{\partial r}, \tag{3.4.2}$$

$$C_r\dot{r} = 0, \tag{3.4.3}$$

where

$$\left(\dot{r}^T\frac{\partial M_{22}(r)}{\partial r}\dot{r}\right) = \left(\dot{r}^T\frac{\partial M_{22}(r)}{\partial r_1}\dot{r}, \; \cdots, \; \dot{r}^T\frac{\partial M_{22}(r)}{\partial r_n}\dot{r}\right)^T \in \mathbb{R}^n.$$

Clearly, $(r,\dot{r}) = (r_e, 0)$ is a trivial solution to these equations.

The general conditions for asymptotic stability may be obtained using the concept of "local distinguishability" in [6]. We present a sufficient condition using the concept of nonlinear observability [4, 7]. Note that (3.4.2) can be written as

$$\begin{bmatrix}\dot{r}\\ \ddot{r}\end{bmatrix} = \underbrace{\begin{bmatrix}\dot{r}\\ M_{22}^{-1}(r)\Big\{-\dot{M}_{22}(r)\dot{r} + \frac{1}{2}\Big(\dot{r}^T\frac{\partial M_{22}(r)}{\partial r}\dot{r}\Big) - \frac{\partial V_s(r)}{\partial r}\Big\}\end{bmatrix}}_{f(r,\dot{r})}, \tag{3.4.4}$$

and we define an output function according to (3.4.1) and (3.4.3)

$$y = \underbrace{\begin{bmatrix}\dot{M}_{12}(r)\dot{r} + M_{12}(r)M_{22}^{-1}(r)\Big\{-\dot{M}_{22}(r)\dot{r} + \frac{1}{2}\Big(\dot{r}^T\frac{\partial M_{22}(r)}{\partial r}\dot{r}\Big) - \frac{\partial V_s(r)}{\partial r}\Big\}\\ C_r\dot{r}\end{bmatrix}}_{h(r,\dot{r})}, \tag{3.4.5}$$

where $h = (h_1, \cdots, h_{2n})$ and each h_i is a scalar function of $(r,\dot{r})$. Denote the observability co-distribution [4, 7] by dO defined as

$$dO(r,\dot{r}) = \mathrm{span}\Big\{dL_f^k h_i(r,\dot{r}), \;\; i = 1,\cdots,2n, \;\; k = 0,1,\cdots\Big\}.$$

Proposition 3.4.3. *Assume $C_a > 0$. The equilibrium is asymptotically stable if* $\mathrm{rank}\{dO\} = 2n$ *at* $(r,\dot{r}) = (r_e, 0)$.

Proof. If $\mathrm{rank}\{dO\} = 2n$ holds at $(r,\dot{r}) = (r_e, 0)$, then the system described by (3.4.4)–(3.4.5) is locally observable at $(r_e, 0)$ [4, 7]. This means that there exists a neighborhood $\mathcal{V}$ of $(r_e, 0)$ such that for any initial condition in $\mathcal{V}$, the output $y(t) = 0$, $t \geq 0$ if and only if the initial condition is equal to $(r_e, 0)$. This shows that the only solution in $\mathcal{V}$ that lies in the invariant set $\mathcal{M}$ is the trivial solution $r(t) = r_e$, $t \geq 0$. Hence, the equilibrium is asymptotically stable. □

Checking the observability rank condition using the observability co-distribution may require complicated computations; a restrictive but computationally tractable condition is based on linearization of (3.4.4)–(3.4.5) at the equilibrium. Because observability of the linearized system implies observability of the nonlinear system, we have the following result.

Corollary 3.4.1. *Assume* $C_a > 0$. *Let*

$$K_s = \frac{\partial^2 V_s(r_e)}{\partial r^2}.$$

Then the equilibrium is asymptotically stable if the following pair is observable

$$\left(\begin{bmatrix} -M_{12}(r_e)M_{22}^{-1}(r_e)K_s, & C_r \end{bmatrix}, \begin{bmatrix} 0 & I_n \\ -M_{22}^{-1}(r_e)K_s & 0 \end{bmatrix} \right).$$

Remark 3.4.1. A restrictive but easily verified sufficient condition is that the intersection of the kernel of $M_{12}(r_e)$ and the kernel of C_r only contains the zero element. Note that if $M_{12}(r_e)$ has full column rank (which implies that the number of shape degrees of freedom cannot be more than three), then the condition

$$\text{Ker}(M_{12}(r_e)) \cap \text{Ker}(C_r) = \{0\}$$

holds regardless of C_r, even when $C_r = 0$. This observation shows that asymptotic stability can be achieved via coupling between the attitude dynamics and the shape dynamics without full damping in the shape dynamics.

Remark 3.4.2. Suppose M_{12} and M_{22} are constant for all $r \in Q_s$ and the elastic potential energy

$$V_s(r) = \frac{1}{2}(r - r_e)^T K_s (r - r_e),$$

where K_s is a positive-definite constant matrix. Then the linear observability condition in Corollary 3.4.1 is also a necessary condition for asymptotic stability.

3.4.2 Asymptotic stability analysis: $C_r > 0, C_a \geq 0$

Throughout this section, C_r is assumed to be positive-definite, i.e., the shape dynamics are fully damped. In this case, the set where $\dot{H} = 0$ is the set $S = \{(\omega, \dot{r}) | C_a\omega = 0,\ \dot{r} = 0\}$, i.e., $\omega(t)$ is in the kernel of C_a while $r(t) = r_c$ for some constant r_c in a small neighborhood of r_e. Let $\mathcal{M}$ be the largest invariant set of solutions of (3.2.2) that lie in S. The solutions $\omega(t)$ that lie in $\mathcal{M}$ must satisfy the following equations:

$$M_{11}(r_c)\dot{\omega} = \left[M_{11}(r_c)\omega\right] \times \omega, \tag{3.4.6}$$

$$M_{21}(r_c)\dot{\omega} = \frac{1}{2}\left(\omega^T \frac{\partial M_{11}(r_c)}{\partial r}\omega\right) - \frac{\partial V_s(r_c)}{\partial r}, \tag{3.4.7}$$

$$C_a\omega = 0, \tag{3.4.8}$$

where

$$\left(\omega^T \frac{\partial M_{11}(r_c)}{\partial r}\omega\right) = \left(\omega^T \frac{\partial M_{11}(r_c)}{\partial r_1}\omega, \cdots, \omega^T \frac{\partial M_{11}(r_c)}{\partial r_n}\omega\right)^T \in \mathbb{R}^n.$$

Using a similar argument as in the proof of Proposition 3.4.2, we see that the equilibrium is asymptotically stable if and only if the solution $\omega(t)$ in $\mathcal{M}$ is identically zero. To see this, note that the magnitude of

$$\Pi(t) = M_{11}(r_c)\omega(t)$$

is conserved along the solutions in $\mathcal{M}$. Thus if $\omega(t) \neq 0$ at some $t \geq 0$, then $||\Pi(t)||_2$ cannot approach zero, which contradicts asymptotic stability. The sufficiency follows from the fact that if $\omega(t) = 0, t \geq 0$ lies in $\mathcal{M}$, then r_c must equal r_e.

The observability rank condition [4, 7] fails to show (nonlinear) observability at the equilibrium when C_a is strictly positive semidefinite. In fact, let (3.4.6) describe the solution ω, and define an output function according to (3.4.7)–(3.4.8). It can be shown that the rank of the observability co-distribution evaluated at the equilibrium $\omega = 0$ is equal to rank(C_a). Thus if C_a is strictly positive semidefinite, the observability rank condition, which is only sufficient, does not guarantee local observability or asymptotic stability. In the following, we provide an alternative approach that leads to necessary and sufficient conditions for asymptotic stability when C_a is strictly positive semidefinite.

A. Properties of solutions in the invariant set

We first study solutions of (3.4.6)–(3.4.8) by representing it by the following time-series expansion

$$\omega(t) = b_0 + \sum_{k=1}^{\infty} b_k \frac{t^k}{k!},$$

where $b_k \in \mathbb{R}^3, k = 0, 1, \cdots$, are constant and each b_k can be viewed as the kth-order time derivative of $\omega(t)$ at $t = 0$. Substituting this series expansion into (3.4.6)–(3.4.7) and equating the orders of time, we obtain the following (nonlinear) algebraic equations

$$\frac{M_{11}(r_c)}{k!} b_{k+1} = \sum_{j=0}^{k} \frac{1}{j!(k-j)!} M_{11}(r_c) b_j \times b_{k-j}, \quad k = 0, 1, 2, \cdots \tag{3.4.9}$$

and

$$M_{21}(r_c) b_1 = \frac{1}{2}\left(b_0^T \frac{\partial M_{11}(r_c)}{\partial r} b_0\right) - \frac{\partial V_s(r_c)}{\partial r}; \tag{3.4.10}$$

$$\frac{M_{21}(r_c)}{k!} b_{k+1} = \frac{1}{2}\sum_{j=0}^{k} \frac{1}{j!(k-j)!}\left(b_j^T \frac{\partial M_{11}(r_c)}{\partial r} b_{k-j}\right), \quad k \geq 1. \tag{3.4.11}$$

To be more specific, we list these algebraic equations up to order three as follows:

$$\begin{aligned} M_{11}(r_c)b_1 &= M_{11}(r_c)b_0 \times b_0, \\ M_{11}(r_c)b_2 &= M_{11}(r_c)b_0 \times b_1 + M_{11}(r_c)b_1 \times b_0, \\ \frac{1}{2!}M_{11}(r_c)b_3 &= \frac{1}{2!}M_{11}(r_c)b_0 \times b_2 + M_{11}(r_c)b_1 \times b_1 + \frac{1}{2!}M_{11}(r_c)b_2 \times b_0, \end{aligned}$$

and

$$\begin{aligned} M_{21}(r_c)b_1 &= \frac{1}{2}\left(b_0^T \frac{\partial M_{11}(r_c)}{\partial r} b_0\right) - \frac{\partial V_s(r_c)}{\partial r}, \\ M_{21}(r_c)b_2 &= \left(b_0^T \frac{\partial M_{11}(r_c)}{\partial r} b_1\right), \\ \frac{1}{2!}M_{21}(r_c)b_3 &= \frac{1}{2}\left(2b_0^T \frac{\partial M_{11}(r_c)}{\partial r} b_2 + b_1^T \frac{\partial M_{11}(r_c)}{\partial r} b_1\right). \end{aligned}$$

In addition, it is clear that (3.4.8) holds, i.e.,

$$C_a \omega(t) = 0 \text{ for } t \geq 0,$$

if and only if

$$C_a b_k = 0, \quad k = 0, 1, 2, \cdots. \tag{3.4.12}$$

We conclude that $\omega(t)$ is a solution of (3.4.6)–(3.4.8) if and only if b_k, $k \geq 1$, generated from (3.4.9) for a given b_0 satisfy (3.4.10)–(3.4.12) at some r_c.

One can solve $b_k, k \geq 1$, recursively via the algebraic equation (3.4.9) for a given b_0; this defines a sequence $\{b_k, k \geq 0\}$. Thus (3.4.9) can be viewed as a generating equation, while (3.4.10)–(3.4.12) can be viewed as constraint equations for $b_k, k \geq 0$. Solving the algebraic equation (3.4.9) for the solution $\omega(t)$ is much easier than solving the original nonlinear differential (3.4.6). Moreover, this makes it possible that asymptotic stability conditions are described by simple algebraic equations, which are easy to check via computational tools.

The following properties can be easily verified for b_k generated from (3.4.9):

P1. If $b_0 = 0$, then $b_k = 0$ for all $k \geq 1$ and $\omega(t) = 0,\ t \geq 0$;

P2. If $b_1 = 0$, then $b_k = 0$ for all $k \geq 2$ for any b_0, and thus $\omega(t) = b_0$ for all $t \geq 0$;

P3. For any $j \geq 1$, if $b_j = b_{j+1} = \cdots = b_{2j-1} = 0$, then $b_k = 0$ for all $k \geq 2j$ regardless what $b_0, b_1, \cdots, b_{j-1}$ are; in this case $\omega(t) = \sum_{i=0}^{j-1} b_i \frac{t^i}{i!},\ t \geq 0$.

We show two results that will be used for the subsequent analysis.

Lemma 3.4.1. *Let b_k, $k \geq 1$, be generated via (3.4.9) from some b_0. Then the identity*

$$\sum_{j=0}^{k} C_j^k b_j^T M_{11}(r_c) b_{k-j} = 0$$

holds for $k = 1, 2, \cdots$, *where*

$$C_j^k = \frac{k!}{j!(k-j)!}.$$

Proof. Note that

$$\omega^T(t)M_{11}(r_c)\dot{\omega}(t) = \omega^T(t)\Big(M_{11}(r_c)\omega(t) \times \omega(t)\Big) = 0 \ \text{ for all } \ t \geq 0$$

and

$$\frac{d\big(\omega^T(t)M_{11}(r_c)\omega(t)\big)}{dt} = 2\omega^T(t)M_{11}(r_c)\dot{\omega}(t) = 0, \ \text{ for all } \ t \geq 0.$$

Hence, $\omega^T(t)M_{11}(r_c)\omega(t)$ is constant for all $t \geq 0$. Moreover because

$$\begin{aligned}\omega^T(t)M_{11}(r_c)\omega(t) &= \left(\sum_{j=0}^{\infty} b_j \frac{t^j}{j!}\right)^T M_{11}(r_c)\left(\sum_{i=0}^{\infty} b_i \frac{t^i}{i!}\right), \\ &= \sum_{k=0}^{\infty}\left(\sum_{j=0}^{k} \frac{1}{j!(k-j)!} b_j^T M_{11}(r_c) b_{k-j}\right) t^k = \text{constant},\end{aligned}$$

for all $t \geq 0$, we have

$$\sum_{j=0}^{k} \frac{1}{j!(k-j)!} b_j^T M_{11}(r_c) b_{k-j} = 0 \ \text{ for all } \ k \geq 1.$$

As a result,

$$\sum_{j=0}^{k} C_j^k b_j^T M_{11}(r_c) b_{k-j} = \sum_{j=0}^{k} \frac{k!}{j!(k-j)!} b_j^T M_{11}(r_c) b_{k-j} = 0, \ k \geq 1. \qquad \square$$

Lemma 3.4.2. *Let* b_k, $k \geq 1$, *be generated via (3.4.9) from some* b_0. *Suppose* $b_2 \in \text{span}(b_0, b_1)$, *then* $b_k \in \text{span}(b_0, b_1)$ *for all* $k \geq 0$.

Proof. We show this by induction. Clearly, $b_0, b_1, b_2 \in \text{span}(b_0, b_1)$, following from the given conditions. Assume $b_j \in \text{span}(b_0, b_1)$ for $j = 0, 1, 2, \cdots, k$, where $k \geq 2$. Therefore, b_j can be expressed as $b_j = e_j^0 b_0 + e_j^1 b_1$ for real numbers e_j^0, e_j^1. Recalling

$$C_j^k = \frac{k!}{j!(k-j)!}$$

and using

$$\sum_{j=0}^{k} C_j^k e_j^0 e_{k-j}^1 = \sum_{j=0}^{k} C_j^k e_j^1 e_{k-j}^0,$$

we have

$$\begin{aligned}
& M_{11}(r_c)b_{k+1} \\
&= \sum_{j=0}^{k} C_j^k M_{11}(r_c)b_j \times b_{k-j} \\
&= \sum_{j=0}^{k} C_j^k M_{11}(r_c)\Big(e_j^0 b_0 + e_j^1 b_1\Big) \times \Big(e_{k-j}^0 b_0 + e_{k-j}^1 b_1\Big) \\
&= \sum_{j=0}^{k} C_j^k \bigg[e_j^0 e_{k-j}^0 M_{11}(r_c)b_0 \times b_0 + \frac{e_j^0 e_{k-j}^1 + e_j^1 e_{k-j}^0}{2}\Big(M_{11}(r_c)b_0 \times b_1 \\
&\qquad + M_{11}(r_c)b_1 \times b_0\Big) + e_j^1 e_{k-j}^1 M_{11}(r_c)b_1 \times b_1\bigg] \\
&= \sum_{j=0}^{k} C_j^k \bigg[e_j^0 e_{k-j}^0 M_{11}(r_c)b_1 + \frac{e_j^0 e_{k-j}^1 + e_j^1 e_{k-j}^0}{2} M_{11}(r_c)b_2 \\
&\qquad - (b_1^T b_0)e_j^1 e_{k-j}^1 M_{11}(r_c)b_0\bigg],
\end{aligned}$$

where we use the identities for b_1, b_2 and the formula

$$\begin{aligned}
M_{11}(r_c)b_1 \times b_1 &= \Big(M_{11}(r_c)b_0 \times b_0\Big) \times b_1 \\
&= (b_1^T M_{11}(r_c)b_0)b_0 - (b_1^T b_0)M_{11}(r_c)b_0 \\
&= -(b_1^T b_0)M_{11}(r_c)b_0.
\end{aligned}$$

Consequently, we obtain

$$b_{k+1} = \sum_{j=0}^{k} C_j^k \left[e_j^0 e_{k-j}^0 b_1 + \frac{e_j^0 e_{k-j}^1 + e_j^1 e_{k-j}^0}{2} b_2 - (b_1^T b_0)e_j^1 e_{k-j}^1 b_0\right]. \quad (3.4.13)$$

Hence, $b_{k+1} \in \text{span}(b_0, b_1)$ because $b_2 \in \text{span}(b_0, b_1)$. □

The following result expresses asymptotic stability conditions in terms of the sequence b_k.

Proposition 3.4.4. *Assume $C_r > 0$. In any small neighborhood $\mathcal{U}$ of $(b, r) = (0, r_e) \in \mathbb{R}^{3+n}$, if for any $(b_0, r_c) \in \mathcal{U}$ with $b_0 \neq 0$, there exists some b_k generated via (3.4.9) from b_0 that violates one of the constraint (3.4.10)–(3.4.12), then the equilibrium is asymptotically stable.*

Proof. Suppose the given conditions hold. Then in any small neighborhood of $b = 0 \in \mathbb{R}^3$, no sequence $\{b_k\}$ generated from nonzero b_0 is a solution in the invariant set $\mathcal{M}$. Consider a sequence $\{b_k\}$ with $b_0 = 0$. By P1, all $b_k = 0, k \geq 1$ and they satisfy the constraint (3.4.11)–(3.4.12) trivially. Thus $b_k = 0, k \geq 0$, which implies $\omega(t) = 0,\ t \geq 0$, is the only solution in $\mathcal{M}$. This also implies that r_c must equal r_e using the properties of $V_s(r)$. Hence, the equilibrium is asymptotically stable. □

B. Conditions for asymptotic stability

According to Proposition 3.4.2, the dimension of Ker(C_a), the kernel of C_a (or the null space of C_a), must be either one or two for asymptotic stability. We consider these cases individually.

We refer to the following conditions as A1: in any small neighborhood $\mathcal{U}$ of $(b, r) = (0, r_e) \in \mathbb{R}^{3+n}$, there exists $(b_0, r_c) \in \mathcal{U}$ with $b_0 \neq 0$ satisfying

A1.1 $C_a b_0 = 0$,

A1.2 $M_{11}(r_c)b_0 \times b_0 = 0$,

A1.3 $\frac{1}{2}\left(b_0^T \frac{\partial M_{11}(r_c)}{\partial r} b_0\right) = \frac{\partial V_s(r_c)}{\partial r}$.

Before proceeding to the case $\dim\{\text{Ker}(C_a)\} = 1$, we point out a necessary condition for asymptotic stability when C_a is strictly positive semidefinite.

Lemma 3.4.3. *Assume $C_r > 0$ and A1 holds. Then the equilibrium is* not *asymptotically stable.*

Proof. It follows from P2: the conditions in A1 imply that in any small neighborhood of the equilibrium, there is $\omega(t) = b_0 \neq 0$ for all $t \geq 0$ in the invariant set $\mathcal{M}$. This contradicts asymptotic stability. □

Proposition 3.4.5. *Assume $C_r > 0$ and* $\dim\{\text{Ker}(C_a)\} = 1$. *Then the equilibrium is asymptotically stable if and only if A1 fails.*

Proof. It is obvious that the necessity follows from Lemma 3.4.3. Before proving the sufficiency, we first show for $b_0 \neq 0$, (b_0, b_1) are linearly independent if $b_1 \neq 0$. Suppose it is not. Thus b_1 must be in the form of $b_1 = \alpha b_0$ for a nonzero real α. Moreover, recall that b_1 satisfies

$$M_{11}(r_c)b_1 = M_{11}(r_c)b_0 \times b_0.$$

Therefore,

$$b_0^T M_{11}(r_c)b_1 = \alpha b_0^T M_{11}(r_c)b_0 = b_0^T \left(M_{11}(r_c)b_0 \times b_0\right) = 0.$$

Since $M_{11}(r)$ is always positive-definite, we have $\alpha = 0$, which is a contradiction.

Now we show the sufficiency using the above result and Proposition 3.4.4. It is clear that if A1.1 fails, then Proposition 3.4.4 holds at $k = 0$, which means the equilibrium is asymptotically stable. Suppose A1.1 holds. Thus $\text{Ker}(C_a) = \text{span}\{b_0\}$ by the dimensional assumption for Ker(C_a). This implies $b_1 \in \text{Ker}(C_a) = \text{span}\{b_0\}$, which holds only if $b_1 = 0$ or equivalently $M_{11}(r_c)b_0 \times b_0 = 0$. This means the failure of A1.2 implies asymptotic stability. Using the constraint (3.4.10), we also see the failure of A1.3 implies asymptotic stability even if A1.1 and A1.2 hold. This completes the proof. □

Next we study the case where $\dim\{\text{Ker}(C_a)\} = 2$. We refer to the following conditions as A2: in any small neighborhood $\mathcal{U}$ of $(b, r) = (0, r_e) \in \mathbb{R}^{3+n}$, there exists $(b_0, r_c) \in \mathcal{U}$ with $b_0 \neq 0$ satisfying

A2.1 $b_0, b_1, b_2 \in \text{Ker}(C_a)$, i.e., $C_a b_0 = C_a b_1 = C_a b_2 = 0$;

A2.2 $M_{21}(r_c)b_1 = \frac{1}{2}\left(b_0^T \frac{\partial M_{11}(r_c)}{\partial r} b_0\right) - \frac{\partial V_s(r_c)}{\partial r}$, $M_{21}(r_c)b_2 = \left(b_0^T \frac{\partial M_{11}(r_c)}{\partial r} b_1\right)$, and $M_{21}(r_c)b_3 = \left(2b_0^T \frac{\partial M_{11}(r_c)}{\partial r} b_2 + b_1^T \frac{\partial M_{11}(r_c)}{\partial r} b_1\right)$;

A2.3 $\alpha_0\alpha_1 M_{21}(r_c)b_0 + (2\alpha_0 + \alpha_1^2)M_{21}(r_c)b_1 + 2\alpha_0 \frac{\partial V_s(r_c)}{\partial r} = 0$ when $b_1 \neq 0$, where

$$\alpha_0 = -\frac{b_1^T M_{11}(r_c) b_1}{b_0^T M_{11}(r_c) b_0} \text{ and } \alpha_1 = -\frac{b_0^T b_1}{\alpha_0},$$

where b_1, b_2, b_3 are generated via (3.4.9) from the nonzero b_0.

Proposition 3.4.6. *Assume $C_r > 0$ and $\dim\{\text{Ker}(C_a)\} = 2$. Then the equilibrium is asymptotically stable if and only if A2 fails.*

Proof. We first show the necessity. Suppose A2 holds. It is clear that if b_1 that satisfies A1 is equal to zero, then the equilibrium is not asymptotically stable by Lemma 3.4.3. We now focus on the case where $b_1 \neq 0$. The basic idea for the proof in this case is as follows: we show that all $b_k, k \geq 1$ generated via (3.4.9) from the nonzero b_0 satisfy the constraint (3.4.10)–(3.4.12) under the given assumptions; and this implies that in any small neighborhood of the equilibrium, there exists a nonzero solution ω that lies in the invariant set $\mathcal{M}$.

As shown in the proof of Proposition 3.4.5, b_0 and b_1 are linearly independent if $b_1 \neq 0$ ($b_0 \neq 0$ follows from the assumption). Thus $\text{span}(b_0, b_1) = \text{Ker}(C_a)$ since both lie in the kernel of C_a. Note that $C_a b_2 = 0$ implies $b_2 \in \text{span}(b_0, b_1)$. Therefore, using Lemma 3.4.2, all $b_k \in \text{span}(b_0, b_1) = \text{Ker}(C_a)$ for $k = 0, 1, 2, \cdots$. This shows that all b_k satisfy (3.4.12).

We now show that all b_k satisfy (3.4.10)–(3.4.11). The conditions in A2 have shown this holds up to $k = 2$; we show it for the remaining ks. In the following, we express each b_k as $b_k = e_k^0 b_0 + e_k^1 b_1$ for real e_k^0 and e_k^1 as in the proof of Lemma 3.4.2. To simplify the notation, we denote e_2^0 and e_2^1 by α_0 and α_1, respectively (i.e., $b_2 = \alpha_0 b_0 + \alpha_1 b_1$). It is easy to verify via the equations for b_2 and b_3 that

$$\alpha_0 = -\frac{b_1^T M_{11} b_1}{b_0^T M_{11} b_0} < 0,$$

$$-b_1^T b_0 = \alpha_0 \alpha_1,$$

and

$$b_3 = 3\alpha_0\alpha_1 b_0 + (2\alpha_0 + \alpha_1^2)b_1.$$

Thus using (3.4.13), we obtain the coefficients e_{k+1}^0 and e_{k+1}^1 for $b_{k+1}, k \geq 0$ as

$$e_{k+1}^0 = \alpha_0 \sum_{j=0}^{k} C_j^k \left[\left(e_j^0 + \alpha_1 e_j^1\right) e_{k-j}^1 \right], \tag{3.4.14}$$

$$e_{k+1}^1 = \sum_{j=0}^{k} C_j^k \left[\left(e_j^0 + \alpha_1 e_j^1\right) e_{k-j}^0 \right]. \tag{3.4.15}$$

Using these results, we obtain the following identity from A2:

$$\left[M_{21}(r_c)b_0, \quad M_{21}(r_c)b_1, \quad \frac{\partial V_s(r_c)}{\partial r}\right]\begin{bmatrix}0 & \alpha_0 & 3\alpha_0\alpha_1\\ 2 & \alpha_1 & (2\alpha_0+\alpha_1^2)\\ 2 & 0 & 0\end{bmatrix}$$

$$= \left[\left(b_0^T\frac{\partial M_{11}(r_c)}{\partial r}b_0\right), \quad \left(b_0^T\frac{\partial M_{11}(r_c)}{\partial r}b_1\right), \quad \left(b_1^T\frac{\partial M_{11}(r_c)}{\partial r}b_1\right)\right]\begin{bmatrix}1 & 0 & 2\alpha_0\\ 0 & 1 & 2\alpha_1\\ 0 & 0 & 1\end{bmatrix}.$$

Therefore,

$$\left(b_1^T\frac{\partial M_{11}(r_c)}{\partial r}b_1\right) = \alpha_0\alpha_1M_{21}(r_c)b_0 - (2\alpha_0+\alpha_1^2)M_{21}(r_c)b_1 - 4\alpha_0\frac{\partial V_s(r_c)}{\partial r}.$$

Moreover, for all $k \geq 1$,

$$\begin{aligned}\sum_{j=0}^{k} C_j^k b_j^T M_{11}(r_c)b_{k-j} &= \sum_{j=0}^{k} C_j^k\left(e_j^0b_0+e_j^1b_1\right)^T M_{11}(r_c)\left(e_{k-j}^0b_0+e_{k-j}^1b_1\right)\\ &= \sum_{j=0}^{k} C_j^k\left(e_j^0e_{k-j}^0b_0^TM_{11}(r_c)b_0 + e_j^1e_{k-j}^1b_1^TM_{11}(r_c)b_1\right).\end{aligned}$$

Using Lemma 3.4.1 and the identity

$$\alpha_0 = -\frac{b_1^TM_{11}(r_c)b_1}{b_0^TM_{11}(r_c)b_0},$$

it can be further shown

$$\sum_{j=0}^{k} C_j^k\left[e_j^0e_{k-j}^0 - \alpha_0e_j^1e_{k-j}^1\right] = 0, \quad k \geq 1. \tag{3.4.16}$$

With these results and A2.3, we have for $k \geq 3$,

$$\begin{aligned}\sum_{j=0}^{k} C_j^k\left[b_j^T\frac{\partial M_{11}(r_c)}{\partial r}b_{k-j}\right] &= \sum_{j=0}^{k} C_j^k\left[(e_j^0b_0+e_j^1b_1)^T\frac{\partial M_{11}(r_c)}{\partial r}(e_{k-j}^0b_0+e_{k-j}^1b_1)\right]\\ &= \sum_{j=0}^{k} C_j^k\left[e_j^0e_{k-j}^0\left(b_0^T\frac{\partial M_{11}(r_c)}{\partial r}b_0\right) + (e_j^0e_{k-j}^1+e_j^1e_{k-j}^0)\left(b_0^T\frac{\partial M_{11}(r_c)}{\partial r}b_1\right)\right.\\ &\qquad\left.+e_j^1e_{k-j}^1\left(b_1^T\frac{\partial M_{11}(r_c)}{\partial r}b_1\right)\right]\\ &= 2\sum_{j=0}^{k} C_j^k\left[e_j^0e_{k-j}^0M_{21}(r_c)b_1 + \frac{e_j^0e_{k-j}^1+e_j^1e_{k-j}^0}{2}M_{21}(r_c)b_2\right.\\ &\qquad\left.+e_j^1e_{k-j}^1\alpha_0\alpha_1M_{21}(r_c)b_0\right] + 2\sum_{j=0}^{k} C_j^k\left[e_j^0e_{k-j}^0 - \alpha_0e_j^1e_{k-j}^1\right]\frac{\partial V_s(r_c)}{\partial r}\end{aligned}$$

$$-\sum_{j=0}^{k} C_j^k e_j^1 e_{k-j}^1 \left[\alpha_0\alpha_1 M_{21}(r_c)b_0 + (2\alpha_0+\alpha_1^2)M_{21}(r_c)b_1 + 2\alpha_0\frac{\partial V_s(r_c)}{\partial r}\right]$$

$$= 2M_{21}(r_c)b_{k+1} - \sum_{j=0}^{k} C_j^k e_j^1 e_{k-j}^1 \left[\alpha_0\alpha_1 M_{21}(r_c)b_0 + (2\alpha_0+\alpha_1^2)M_{21}(r_c)b_1 \right.$$

$$\left. +2\alpha_0\frac{\partial V_s(r_c)}{\partial r}\right] = 2M_{21}(r_c)b_{k+1},$$

following Lemma 3.4.2, (3.4.14)–(3.4.16), and A2.3. Thus b_k, $k \geq 3$, satisfy (3.4.11). Finally, we conclude that in any small neighborhood of the equilibrium, there exists a nonzero solution ω that lies in the invariant set $\mathcal{M}$. This contradicts the assumption of asymptotic stability.

We now show the sufficiency. As in the proof of Proposition 3.4.5, the failure of A2.1 or A2.2 implies asymptotic stability, following from Proposition 3.4.4 at $k = 2$. Now suppose A2.1 and A2.2 hold but A2.3 fails, i.e., $\alpha_0\alpha_1 M_{21}(r_c)b_0 + (2\alpha_0 + \alpha_1^2)M_{21}(r_c)b_1 + 2\alpha_0\frac{\partial V_s(r_c)}{\partial r} \neq 0$ for some nonzero b_0 and b_1 at some r_c, where $\alpha_0 = -\frac{b_1^T M_{11}(r_c)b_1}{b_0^T M_{11}(r_c)b_0}$ and $\alpha_0\alpha_1 = -b_0^T b_1$. This implies $\alpha_0 < 0$. Following the proof for the necessity and using $\sum_{j=0}^{3} C_j^k e_j^1 e_{k-j}^1 = 6\alpha_1$, $\sum_{j=0}^{4} C_j^k e_j^1 e_{k-j}^1 = 2(8\alpha_0 + 7\alpha_1^2)$, we see that the identities $M_{21}(r_c)b_4 = \frac{1}{2}\sum_{j=0}^{3} C_j^k \left[b_j^T \frac{\partial M_{11}(r_c)}{\partial r} b_{k-j}\right]$, $M_{21}(r_c)b_5 = \frac{1}{2}\sum_{j=0}^{4} C_j^k \left[b_j^T \frac{\partial M_{11}(r_c)}{\partial r} b_{k-j}\right]$ cannot both hold. Thus by Proposition 3.4.4, asymptotic stability follows. □

We now collect some conditions that are useful in the subsequent examples based on the above results. As shown in Lemma 3.4.3 and in Propositions 3.4.5 and 3.4.6, a key step for asymptotic stability is to check if there exists a nonzero $b_0 \in \mathrm{Ker}(C_a)$ such that $M_{11}(r_c)b_0 \times b_0 = 0$ at any r_c in a small neighborhood of r_e. Recall that this condition has been referred to as A1.2. If A1.2 fails, asymptotic stability is immediately established; otherwise, we need to check other conditions. In the following, we give simplified steps to check this condition.

Proposition 3.4.7. *Assume $C_r > 0$. For each r_c in a neighborhood of r_e, let $U = (u_1^T, u_2^T, u_3^T)^T$, $u_i^T \in \mathbb{R}^3$ be a unitary matrix that diagonalizes $M_{11}(r_c)$. Then the following statements are true:*

(1) *Assume all eigenvalues of $M_{11}(r_c)$ are equal. Then A1.2 never fails at r_c: it holds for any $b_0 \in \mathbb{R}^3$;*

(2) *Assume exactly two eigenvalues of $M_{11}(r_c)$ are equal. Then*

  (2.1) *In the case $\dim\{\mathrm{Ker}(C_a)\} = 1$, A1.2 fails at r_c if and only if $\mathrm{Ker}(C_a) \cap \{\mathrm{span}(u_3^T) \cup \mathrm{span}(u_1^T, u_2^T)\} = \{0\}$, where u_3^T corresponds to the distinct eigenvalue;*

  (2.2) *In the case $\dim\{\mathrm{Ker}(C_a)\} = 2$, A1.2 never fails at r_c, i.e., there always exists a nonzero $b_0 \in \mathrm{Ker}(C_a)$ such that A1.2 holds at r_c;*

(3) *Assume all eigenvalues of $M_{11}(r_c)$ are distinct. Then A1.2 fails at r_c if and only if $u_i^T \notin \mathrm{Ker}(C_a)$, $i = 1, 2, 3$.*

Proof. Since $M_{11}(r)$ is symmetric and positive-definite for all r, it can be diagonalized by a unitary matrix U at any fixed r, i.e., $M_{11} = U^T \Lambda U$ at r_c, where $\Lambda = \text{diag}(\lambda_1, \lambda_2, \lambda_3)$ is diagonal. Moreover, we can always assume $\det(U) = +1$ (since if $\det(U) = -1$, we can choose the unitary matrix that diagonalizes $M_{11}(r_c)$ as $-U$). This implies U can be regarded as a rotation matrix in $SO(3)$. With these results, we express

$$M_{11}(r_c)b_0 \times b_0 = \left(U^T \Lambda U b_0\right) \times b_0 = U^T\left[\left(\Lambda U b_0\right) \times \left(U b_0\right)\right],$$

where we use $U(v \times w) = Uv \times Uw$ for any $v, w \in \mathbb{R}^3$ and $U \in SO(3)$. Let $h = (h_1, h_2, h_3)^T = U b_0 \in \mathbb{R}^3$, we further have

$$M_{11}(r_c)b_0 \times b_0 = U^T \begin{bmatrix} (\lambda_2 - \lambda_3) & & \\ & (\lambda_3 - \lambda_1) & \\ & & (\lambda_1 - \lambda_2) \end{bmatrix} \begin{bmatrix} h_2 h_3 \\ h_3 h_1 \\ h_1 h_2 \end{bmatrix}. \quad (3.4.17)$$

With this result, we see that Statement 1 holds. Now we show Statement 2. Without loss of generality, we assume $\lambda_1 = \lambda_2 \neq \lambda_3$. We see from (3.4.17) that A1.2 holds for a nonzero b_0 if and only if $h_3 = 0$ or $h_1 = h_2 = 0$, but not both. Since $b_0 = U^T h$, this condition is equivalent to $b_0 \in \text{span}(u_1^T, u_2^T)$ or $b_0 \in \text{span}(u_3^T)$. Hence Statement 2.1 holds. When $\dim\{\text{Ker}(C_a)\} = 2$, we claim that there must be some nonzero $b_0 \in \text{Ker}(C_a)$ that lies in $\text{span}(u_1^T, u_2^T)$. To see this, note that if the claim was not true, then the plane of $\text{Ker}(C_a)$ must be parallel to the plane of $\text{span}(u_1^T, u_2^T)$ but this contradicts the fact that the two planes intersect at the origin. Hence, A1.2 always holds in this case. This shows that Statement 2.2 is true. Finally, we prove Statement 3. In this case, using (3.4.17), we see that A1.2 holds for a nonzero b_0 if and only if two of (h_1, h_2, h_3) are zero, which is equivalent to $b_0 \in \text{span}(u_1^T) \cup \text{span}(u_2^T) \cup \text{span}(u_3^T)$. Thus Statement 3 holds. □

The following result, which only requires knowledge of $M_{11}(r)$ at r_e, follows from Statement 3, Proposition 3.4.5, and the continuity of $M(r)$.

Corollary 3.4.2. *Assume $C_r > 0$. If all eigenvalues of $M_{11}(r_e)$ are distinct, then there exists a positive semidefinite C_a with* $\dim\{\text{Ker}(C_a)\} = 1$ *such that the equilibrium is asymptotically stable.*

Another result is a direct consequence of Statement 1.

Corollary 3.4.3. *Assume $C_r > 0$, and in any small neighborhood $\mathcal{U}$ of $(b, r) = (0, r_e)$, there exists $(b_0, r_c) \in \mathcal{U}$ with $b_0 \neq 0$ and $b_0 \in$* $\text{Ker}(C_a)$ *such that all the eigenvalues of $M_{11}(r_c)$ are equal and that*

$$\frac{1}{2}\left(b_0^T \frac{\partial M_{11}(r_c)}{\partial r} b_0\right) = \frac{\partial V_s(r_c)}{\partial r}$$

then the equilibrium is not *asymptotically stable.*

Corollary 3.4.4. *Assume $C_r > 0$ and in any small neighborhood $\mathcal{V}$ of r_e, there exists a $r_c \in \mathcal{V}$ such that $M_{11}(r_c)$ and C_a are simultaneously diagonalizable by a unitary matrix and that*

$$\frac{1}{2}\left(b^T \frac{\partial M_{11}(r_c)}{\partial r} b\right) = \frac{\partial V_s(r_c)}{\partial r}$$

holds for any $b \in \mathrm{Ker}(C_a) \cap \mathcal{W}$, where $\mathcal{W}$ is any small neighborhood of $0 \in \mathbb{R}^3$. Then the equilibrium is asymptotically stable if and only if C_a is positive-definite.

Proof. The sufficiency is obvious; we show the necessity. Let U be the unitary matrix that simultaneously diagonalizes $M_{11}(r_c)$ and C_a, i.e., $M_{11}(r_c) = U^T \Lambda U$ and $C_a = U^T \Delta U$, where Λ and Δ are diagonal matrices and $U \in SO(3)$ as shown in the previous proposition. Moreover, let $h = (h_1, h_2, h_3)^T = U b_0$, $M_{11}(r_c) b_0 \times b_0$ can be expressed in the form in (3.4.17). Suppose C_a is not positive-definite. Then at least one of the diagonal elements of Δ is zero; without loss of generality, we assume the first element to be zero. Choose $b_0 = \epsilon U^T e_1$, where $\epsilon \neq 0$ and $e_1 = (1,0,0)^T$, thus $C_a b_0 = \epsilon U^T \Delta e_1 = 0$ and $h = (h_1, h_2, h_3)^T = U b_0 = \epsilon e_1$, thus $h_2 = h_3 = 0$. This implies $M_{11}(r_c) b_0 \times b_0 = 0$. The second condition means that $\frac{1}{2}\left(b_0^T \frac{\partial M_{11}(r_c)}{\partial r} b_0\right) = \frac{\partial V_s(r_c)}{\partial r}$ holds for the b_0 defined above in any small neighborhood of $(b, r) = (0, r_e)$ by choosing $|\epsilon|$ sufficiently small. Hence, in any small neighborhood of $(b, r) = (0, r_e)$, all the conditions in A1 are satisfied. Therefore by Lemma 3.4.3, we claim the equilibrium is not asymptotically stable. □

3.5 Examples

In this section, we apply the general results in the previous sections to two classes of multibody attitude systems: one class includes an elastic rotational degree of freedom, the other class includes an elastic translational degree of freedom.

3.5.1 A system with elastic rotational degrees of freedom

A. System description and equations of motion

The elastic subsystem of a multibody attitude system consists of n elastic rotational components that are attached to the base body through rotational linear elastic springs. We further assume the center of mass of the system, which is constant and independent of shape, is always at the pivot point. Thus gravity has no influence on the dynamics.

All the n components are assumed to have an axial symmetric mass distribution with respect to their rotation axes which are fixed with respect to the base body. Moreover, they are assumed to have identical physical properties. Let m_r be the identical mass of the rotational components. Denote the constant moments of inertia of each rotational component by J_s along its spin axis and by J_r along its transverse axes. Define a body fixed orthogonal coordinate frame for each component with origin at the center of the component so that its first axis is along the spin axis of the component. Let the rotation matrix from the ith component frame to the base body frame be R_i. It can be shown that the inertia matrix J_i of the ith rotational component, expressed in the base body coordinate frame, is given by

$$J_i = R_i \mathrm{diag}(J_s, J_r, J_r) R_i^T .$$

Choose a base body coordinate frame whose origin is located at the pivot point. Let ρ_i be the constant position vector of the center of mass of the ith component, and J_B be the inertia matrix of the base body, both expressed in the base body coordinates. The shape coordinates are $r = (\phi_1, \cdots, \phi_n)$, the rotation angles of the components.

Let

$$V_s(r) = \frac{1}{2} r^T K_s r$$

be the elastic potential energy, where K_s is symmetric and positive-definite, and without loss of generality, we assume that zero shape corresponds to zero elastic potential energy. The reduced Lagrangian is

$$L(\omega, r, \dot{r}) = \frac{1}{2} \begin{pmatrix} \omega \\ \dot{r} \end{pmatrix}^T \begin{bmatrix} M_{11} & M_{12} \\ M_{12}^T & M_{22} \end{bmatrix} \begin{pmatrix} \omega \\ \dot{r} \end{pmatrix} - \frac{1}{2} r^T K_s r,$$

where

$$M_{11} = J_B + \sum_{i=1}^{n} \left(m_r \widehat{\rho}_i^T \widehat{\rho}_i + J_i \right), \quad M_{12} = J_s \left[R_1 e_1, \ \cdots, \ R_n e_1 \right], \quad M_{22} = J_s I_n .$$

Note that all the inertia matrix components are constant and are independent of the shape. Moreover, $R_i e_1$ in M_{12} denotes the direction of the spin axis of the ith component expressed in the base body frame. We call the rotational components "independent" if $R_i e_1$, $i = 1, \cdots, n$, are linearly independent. The equations of motion are given by

$$\begin{bmatrix} M_{11} & M_{12} \\ M_{12}^T & M_{22} \end{bmatrix} \begin{bmatrix} \dot{\omega} \\ \ddot{r} \end{bmatrix} = \begin{bmatrix} \Pi \times \omega \\ -K_s r \end{bmatrix} - \begin{bmatrix} C_a \omega \\ C_r \dot{r} \end{bmatrix} . \tag{3.5.1}$$

Clearly, $(\omega, r, \dot{r}) = (0, 0, 0)$ is an isolated stable equilibrium of (3.5.1). Furthermore, because the equilibrium is globally stable, we have

Proposition 3.5.1. *The equilibrium $(\omega, r, \dot{r}) = (0, 0, 0)$ is globally asymptotically stable if and only if it is locally asymptotically stable.*

B. Asymptotic stability analysis: $C_a > 0$, $C_r \geq 0$

It is easy to see that this case satisfies all the conditions in Remark 3.4.2. Thus we obtain

Corollary 3.5.1. *Assume $C_a > 0$. The equilibrium is asymptotically stable if and only if the following pair is observable*

$$\left(\left[-M_{12} M_{22}^{-1} K_s, \ \ C_r \right], \ \begin{bmatrix} 0 & I_n \\ -M_{22}^{-1} K_s & 0 \end{bmatrix} \right) .$$

Corollary 3.5.2. *Assume $C_a > 0$. The following statements hold.*

(1) *Suppose the number of rotational degrees of freedom is no more than three, i.e., $1 \leq n \leq 3$, and they are independent. Then the equilibrium is asymptotically stable regardless of the shape damping C_r.*

(2) *Suppose there are more than three rotational degrees of freedom, of which three are independent. Then the equilibrium is asymptotically stable if the remaining $(n-3)$ degrees of freedom are fully damped.* □

C. Asymptotic stability analysis: $C_r > 0$, $C_a \geq 0$

In this case, the shape is fully damped. The equations that characterize a solution $\omega(t)$ in the invariant set $\mathcal{M}$ are given by $M_{11}\dot{\omega} = M_{11}\omega \times \omega$, $M_{21}\dot{\omega} = -K_s r_c$, and $C_a\omega = 0$, where r_c is some constant. The results in Section 3.4.2 can be greatly simplified and refined using the fact that M is constant. The following proposition describes asymptotic stability conditions for this case. Particularly, we obtain concrete necessary and sufficient conditions for the case $\dim\{\text{Ker}(C_a)\} = 1$.

Proposition 3.5.2. *Assume $C_r > 0$. Let $U = (u_1^T, u_2^T, u_3^T)^T$ be a unitary matrix that diagonalizes M_{11}, where $u_i^T \in \mathbb{R}^3$. The following statements hold.*

(1) *Assume all eigenvalues of M_{11} are equal. Then the equilibrium is asymptotically stable if and only if C_a is positive-definite.*

(2) *Assume exactly two eigenvalues of M_{11} are equal.*

  (2.1) *Suppose $\dim\{\text{Ker}(C_a)\} = 1$. Then the equilibrium is asymptotically stable if and only if $\text{Ker}(C_a) \cap \{\text{span}(u_3^T) \cup \text{span}(u_1^T, u_2^T)\} = \{0\}$, where u_3^T corresponds to the distinct eigenvalue.*

  (2.2) *Suppose $\dim\{\text{Ker}(C_a)\} = 2$. Then the equilibrium is* not *asymptotically stable.*

(3) *Assume all eigenvalues of M_{11} are distinct.*

  (3.1) *Suppose $\dim\{\text{Ker}(C_a)\} = 1$. Then the equilibrium is asymptotically stable if and only if $u_i^T \notin \text{Ker}(C_a)$ for $i = 1, 2, 3$.*

  (3.2) *Suppose $\dim\{\text{Ker}(C_a)\} = 2$. Then the equilibrium is asymptotically stable only if $u_i^T \notin \text{Ker}(C_a)$ for $i = 1, 2, 3$.*

Proof. We first simplify conditions A1 and A2 used in Lemma 3.4.3 and in Propositions 3.4.5 and 3.4.6. Since M_{11} is constant, A1.3 is satisfied if and only if $r_c = 0$ and for all $b_0 \in \mathbb{R}^3$. Thus we remove this condition from consideration. Therefore, satisfaction of A1 is equivalent to existence of a nonzero $b_0 \in \text{Ker}(C_a)$ satisfying $M_{11}b_0 \times b_0 = 0$, or equivalently the condition A1.2. Similarly, using the fact that M_{11} is constant, we see that satisfaction of A2 is equivalent to existence of a nonzero b_0 and r_c satisfying $(b_0, b_1, b_2) \in \text{Ker}(C_a)$ and $M_{21}b_1 = -K_s r_c$, $M_{21}b_2 = 0$, $M_{21}b_3 = 0$, $\alpha_0\alpha_1 M_{21}b_0 + (2\alpha_0 + \alpha_1^2)M_{21}b_1 + 2\alpha_0 K_s r_c = 0$ when $b_1 \neq 0$, where α_0, α_1 are given in A2.3.

We use these simplified conditions and Proposition 3.4.7 to prove the statements. If all the eigenvalues of M_{11} are equal, then A1.2 holds for any $b_0 \neq 0$. This means

A1 is satisfied for any $b_0 \neq 0$. By Lemma 3.4.3, the equilibrium is asymptotically stable if and only if C_a is positive-definite. Statement 2.1 follows from Statement 2.1 in Proposition 3.4.7, the simplified condition A1, and constant M_{11}. We now prove Statement 2.2. By Statement 2.2 in Proposition 3.4.7, we see that there always exists a nonzero $b_0 \in Ker(C_a)$ such that $b_1 = 0$ which implies $b_2 = b_3 = 0$ and $r_c = 0$. Thus A2 holds for such b_0. By Proposition 3.4.6, the equilibrium is *not* asymptotically stable. Statement 3.1 is due to Statement 3 in Proposition 3.4.7 and the fact that A1 is equivalent to A1.2. Statement 3.2 is a direct consequence of Statement 3 of Proposition 3.4.7: if one of u_i^T lies in $\text{Ker}(C_a)$, then we can always find a nonzero $b_0 \in \text{Ker}(C_a)$ such that $b_1 = 0$, which leads to $b_2 = b_3 = 0$ and $r_c = 0$. Hence A2 is satisfied and the equilibrium is not asymptotically stable. □

It can be seen from Proposition 3.4.7 that it is more difficult to achieve asymptotic stability when $\dim\{\text{Ker}(C_a)\} = 2$. This agrees with physical intuition. Note that Statement 3.2 only gives a necessary condition for asymptotic stability; we now present a sufficient condition that makes use of coupling effects between the attitude dynamics and the shape dynamics.

Corollary 3.5.3. *Assume $C_r > 0$ and assume that C_a satisfies $\dim\{\text{Ker}(C_a)\} = 2$ and*

$$u_i^T \notin \text{Ker}(C_a) \text{ for } i = 1, 2, 3.$$

Moreover, suppose $\text{rank}\{M_{21}\} = 3$. Then the equilibrium is asymptotically stable.

Proof. We show in the following that $M_{21} b_2 \neq 0$ for any $b_0 \neq 0$ under the given conditions. This thus implies the failure of A2 and leads to asymptotic stability. Note that if $u_i^T \notin \text{Ker}(C_a)$ for $i = 1, 2, 3$ holds, then for any nonzero $b_0 \in \text{Ker}(C_a)$, b_1 cannot be zero. Let b_1 and b_2 be generated from some nonzero $b_0 \in \text{Ker}(C_a)$ and suppose both lie in $\text{Ker}(C_a)$. (If one of them does not, we have asymptotic stability.) As shown in Proposition 3.4.6, $b_2 = \alpha_0 b_0 + \alpha_1 b_1$, where $\alpha_0 < 0$. Recall that (b_0, b_1) are linearly independent as shown in Proposition 3.4.5. Therefore, $b_2 \neq 0$. Consequently, if $\text{rank}(M_{21}) = 3$, then $M_{21} b_2 \neq 0$ as desired. □

Corollary 3.5.4. *Assume $C_r > 0$ and suppose M_{11} and C_a are simultaneously diagonalizable by a unitary matrix. Then the equilibrium is asymptotically stable if and only if C_a is positive-definite.* □

This result is a consequence of Proposition 3.4.6 and condition A1 and has some interesting implications. It shows that if both M_{11} and C_a are diagonal or all the eigenvalues of M_{11} are equal, both special cases of Corollary 3.5.4, then C_a must be positive-definite for asymptotic stability.

We present an example to illustrate these observations.

Example. Consider a multibody attitude system with the following inertia and damping assumptions:

$$M_{11} = \begin{bmatrix} J_{11} & J_{12} & 0 \\ J_{12} & J_{22} & 0 \\ 0 & 0 & J_{33} \end{bmatrix} > 0, \quad J_{12} \neq 0, \quad C_a = \text{diag}(c_1,\ c_2,\ c_3) \geq 0.$$

We consider two cases:

(1) Assume $\dim\{\mathrm{Ker}(C_a)\} = 1$. If $c_1 = 0, c_2 > 0, c_3 > 0$ or if $c_1 > 0, c_2 = 0, c_3 > 0$, then the equilibrium is asymptotically stable. If $c_1 > 0, c_2 > 0, c_3 = 0$, then the equilibrium is *not* asymptotically stable. These conclusions follow from Proposition 3.5.2. In fact, consider the first case where $C_a = (0, c_2, c_3)$. Choose $b_0 = e_1 \in \mathrm{Ker}(C_a)$; we obtain $M_{11}b_0 \times b_0 = -J_{12}e_3 \neq 0$. This shows asymptotic stability.

(2) Assume $\dim\{\mathrm{Ker}(C_a)\} = 2$. If $c_1 = c_2 = 0, c_3 > 0$, then the equilibrium is not asymptotically stable. To see this, let $V \in \mathbb{R}^{2\times 2}$ be a unitary matrix that diagonalizes the upper 2×2 block of M_{11}. Thus the following unitary matrix

$$U = \begin{bmatrix} V & 0 \\ 0 & 1 \end{bmatrix}$$

simultaneously diagonalizes M_{11} and C_a. Therefore, the equilibrium is not asymptotically stable. If $c_1 > 0, c_2 = c_3 = 0$ or $c_2 > 0, c_1 = c_3 = 0$, then the equilibrium is not asymptotically stable because we can choose $b_0 = e_3 \in \mathrm{Ker}(C_a)$ such that $M_{11}b_0 \times b_0 = 0$.

3.5.2 A system with elastic translational degrees of freedom

A. System description and equations of motion

The elastic subsystem of a multibody attitude system consists of n idealized mass particles that are attached to the base body through linear elastic springs. We assume that the particles are constrained to translate in a symmetric mode such that their motions do not change the center of mass of the system, which is assumed to be at the pivot point. Thus gravity does not influence the dynamics.

We choose a base body coordinate frame with its origin at the pivot point. Let J_B denote the inertia matrix of the base body expressed in this frame. Each mass particle is assumed to translate along a particular direction fixed with respect to the base body that is denoted by the unit vector ν_i. The shape coordinate r_i denotes the position along this direction with respect to the base body frame. Thus the shape coordinates are $r = (r_1, \cdots, r_n)$. Let ρ_{i0} denote the constant offset position vector from the origin to zero shape (i.e., $r_i = 0$) of ith mass particle, assuming no elastic deformation at the zero shape; this is assumed to correspond to zero elastic potential energy. The position vectors of the mass particles expressed in the body coordinate frame are $\rho_i(r) = \rho_{i0} + r_i\nu_i$, $i = 1, \cdots, n$ and the mass of the ith mass particle is denoted by m_i.

Let the elastic potential energy be $V_s(r) = \frac{1}{2}r^T K_s r$, where K_s defines the stiffness matrix of the elastic subsystem and K_s is assumed to be symmetric and positive-definite. The reduced Lagrangian is

$$L(\omega, r, \dot{r}) = \frac{1}{2}\begin{pmatrix} \omega \\ \dot{r} \end{pmatrix}^T \begin{bmatrix} M_{11}(r) & M_{12} \\ M_{12}^T & M_{22} \end{bmatrix} \begin{pmatrix} \omega \\ \dot{r} \end{pmatrix} - \frac{1}{2}r^T K_s r$$

where $M_{11}(r) = J_B + \sum_{i=1}^n m_i \widehat{\rho}_i^T(r)\widehat{\rho}_i(r)$, $M_{12} = \left[m_1(\rho_{10} \times \nu_1), \cdots, m_n(\rho_{n0} \times \nu_n)\right]$, and $M_{22} = \mathrm{diag}(m_1, \cdots, m_n)$. Note that the matrices M_{12} and M_{22} are constant and do not depend on the shape. The equations of motion for the damped multibody attitude system are given by

$$\begin{bmatrix} M_{11}(r) & M_{12} \\ M_{12}^T & M_{22} \end{bmatrix} \begin{bmatrix} \dot{\omega} \\ \ddot{r} \end{bmatrix} = \begin{bmatrix} -\dot{M}_{11}(r)\omega + \Pi \times \omega \\ \frac{1}{2}\left(\omega^T \frac{\partial M_{11}(r)}{\partial r}\omega\right) - K_s r \end{bmatrix} - \begin{bmatrix} C_a\omega \\ C_r\dot{r} \end{bmatrix}, \quad (3.5.2)$$

where $\Pi = M_{11}(r)\omega + M_{12}\dot{r}$ is the conjugate angular momentum. Clearly $(\omega, r, \dot{r}) = (0, 0, 0)$ is an isolated stable equilibrium of (3.5.2).

B. Asymptotic stability analysis: $C_a > 0,\ C_r \geq 0$

It is easy to see that all the conditions in Remark 3.4.2 are satisfied in this case since M_{12} and M_{22} are constant and $V_s(r) = \frac{1}{2}r^T K_s r$. Hence, Corollaries 3.5.1 and 3.5.2 are applicable to this case.

An interesting special case is that all the offset position vectors $\rho_{i0} = 0$. In this case, M_{12} is identically zero. By Corollary 3.5.1, we conclude that the equilibrium is asymptotically stable if and only if the pair $(C_r, M_{22}^{-1}K_s)$ is observable. Note that this condition is much more restrictive than the general condition for $M_{12} \neq 0$. Since M_{12} characterizes the coupling between the attitude dynamics and the shape dynamics, this observation demonstrates that this coupling can provide an important mechanism for asymptotic stability.

C. Asymptotic stability analysis: $C_r > 0,\ C_a \geq 0$

In this case, the shape dynamics are fully damped. The equations that characterize a solution in the invariant set are given by

$$M_{11}(r_c)\dot{\omega} = M_{11}(r_c)\omega \times \omega, \quad M_{12}^T\dot{\omega} = -K_s r_c, \quad C_a\omega = 0,$$

for some constant r_c.

We now obtain asymptotic stability conditions using the structure of $M_{11}(r)$. Note that $M_{11}(r)$ can be written as

$$M_{11}(r) = J_B + \sum_{i=1}^{n} m_i \widehat{\rho}_{i0}^T \widehat{\rho}_{i0} + \sum_{i=1}^{n} m_i \left(\widehat{\rho}_{i0}^T \widehat{\nu}_i + \widehat{\nu}_i^T \widehat{\rho}_{i0}\right) r_i + \sum_{i=1}^{n} m_i \widehat{\nu}_i^T \widehat{\nu}_i r_i^2,$$

and we can write

$$\left(b^T \frac{\partial M_{11}(r)}{\partial r} b\right)$$

$$= 2 \underbrace{\begin{bmatrix} m_1 (\rho_{10} \times b)^T (\nu_1 \times b) \\ \vdots \\ m_n (\rho_{n0} \times b)^T (\nu_n \times b) \end{bmatrix}}_{E(b)} \underbrace{+2\,\mathrm{diag}\left(m_1 \|\nu_1 \times b\|_2^2, \cdots, m_n \|\nu_n \times b\|_2^2\right)}_{G(b)} r.$$

Thus condition A1.3 is equivalent to $E(b) = -[K_s - G(b)]r$. Since K_s is positive-definite and invertible, for sufficiently small $\|b\|$, $G(b)$ is sufficiently small so that

$K_s - G(b)$ is invertible. Thus we see that by choosing $r_c = -[K_s - G(b_0)]^{-1}E(b_0)$, A1.3 is satisfied for any sufficiently small b_0. Using this observation, we obtain the following result similar to Corollary 3.4.2.

Corollary 3.5.5. *Assume $C_r > 0$ and suppose all eigenvalues of $M_{11}(0) = J_B + \sum_{i=1}^n m_i \widehat{\rho}_{i0}^T \widehat{\rho}_{i0}$ are distinct. Let $U = (u_1^T, u_2^T, u_3^T)^T$ be a unitary matrix that diagonalizes $M_{11}(0)$, where $u_i^T \in \mathbb{R}^3$. Then the equilibrium is asymptotically stable only if $u_i^T \notin \text{Ker}(C_a)$, $i = 1, 2, 3$. Moreover, if $\dim\{\text{Ker}(C_a)\} = 1$, then this condition is also sufficient and as such C_a always exists.*

We consider a special case using Propositions 3.4.7 and 3.5.2.

Corollary 3.5.6. *Assume $C_r > 0$. Suppose the offset position vectors $\rho_{i0} = 0$, $i = 1, \cdots, n$. Then the following statements hold.*

(1) *Assume all eigenvalues of J_B are equal. Then the equilibrium is asymptotically stable if and only if C_a is positive-definite.*

(2) *Assume exactly two eigenvalues of J_B are equal. Then there exist a C_a with $\dim\{\text{Ker}(C_a)\} = 1$ such that the equilibrium is asymptotically stable; the equilibrium is* not *asymptotically stable if $\dim\{\text{Ker}(C_a)\} = 2$.*

(3) *Assume all eigenvalues of J_B are distinct. Then there exists a C_a satisfying $\dim\{\text{Ker}(C_a)\} = 1$ such that the equilibrium is asymptotically stable.*

Proof. Note that $\rho_{i0} = 0$, $i = 1, \cdots, n$, implies $M_{21} = 0$, $M_{11}(0) = J_B$, and $\left(b^T \frac{\partial M_{11}(r)}{\partial r} b\right) = 2G(b)r$. Thus A1.3 and the first condition in A2.2 are satisfied if and only if $r_c = 0$ for any sufficiently small b_0. Hence, the satisfaction of A1 is equivalent to satisfaction of A1.2. The statements thus follow from Proposition 3.4.7.

3.6 Conclusions

In this chapter we have studied asymptotic stability of a class of multibody attitude systems using passive damping. Emphases have been given to the nonlinear dynamics in a noncanonical Hamiltonian form and to the coupling effects between the attitude dynamics and the shape dynamics that can enable asymptotic stability. Under the stated assumptions, the equilibrium is always stable in the sense of Lyapunov. A number of results that guarantee that the equilibrium is asymptotically stable have been obtained. These results have been presented in both a general form, and specific results have been presented for typical multibody attitude examples. Linear and nonlinear observability rank conditions lead to asymptotic stability results when the attitude dynamics are fully damped, but they fail for the case where the attitude dynamics are partially damped. Our results also suggest the importance of coupling effects in achieving asymptotic stability.

A key assumption made in this chapter is that there are no gravity effects in the multibody attitude system. If the center of mass of the system does not remain at the pivot, then gravity effects must be considered. This fundamentally changes the equilibrium structure of the multibody attitude system and its stability properties.

Bibliography

[1] D. S. Bernstein, N. H. McClamroch, and A. M. Bloch, "Development of air spindle and triaxial air bearing testbed for spacecraft dynamics and control experiments," *Proc. 2001 American Control Conference*, pp. 3967–3972, Arlington, VA, 2001.

[2] S. Cho, J. Shen, and N. H. McClamroch, "Mathematical models for the triaxial attitude control testbed," to appear in *Mathematical and Computer Modeling of Dynamical Systems*, 2003.

[3] S. Cho, J. Shen, N. H. McClamroch, and D. S. Bernstein, "Dynamics of the triaxial attitude control testbed," *Proc. 40th IEEE Conference on Decision and Control*, pp .3429–3434, Orlando, FL, 2001.

[4] A. Isidori, *Nonlinear Control Systems*, 3rd ed., Springer-Verlag, New York, 1995.

[5] H. K. Khalil, *Nonlinear Systems*, 2nd ed., Prentice-Hall, Upper Saddle River, NJ, 1996.

[6] R. K. Miller and A. N. Michel, "Asymptotic stability of systems: Results involving the system topology," *SIAM J. Control Optim.*, vol. 18, no. 2, pp. 181–190, 1980.

[7] H. Nijmeijer and A. van der Schaft, *Nonlinear Dynamical Control Systems*, Springer-Verlag, New York, 1990.

[8] J. Shen, *Nonlinear Control of Multibody Systems with Symmetries via Shape Change*, Ph.D. Dissertation, The University of Michigan, Department of Aerospace Engineering, Ann Arbor, MI, 2002.

[9] R. E. Skelton, *Dynamic Systems Control*, John Wiley, New York, 1988.

Chapter 4

Robust Regulation of Polytopic Uncertain Linear Hybrid Systems with Networked Control System Applications*

Hai Lin and Panos J. Antsaklis

Abstract: In this chapter, a class of discrete-time uncertain linear hybrid systems, affected by both parameter variations and exterior disturbances, is considered. The main question is whether there exists a controller such that the closed loop system exhibits desired behavior under dynamic uncertainties and exterior disturbances. The notion of attainability is introduced to refer to the specified behavior that can be forced to the plant by a control mechanism. We give a method for attainability checking that employs the predecessor operator and backward reachability analysis, and we introduce a procedure for controller design that uses finite automata and linear programming techniques. Finally, networked control systems (NCS) are proposed as a promising application area of the results and tools developed here, and the ultimate boundedness control problem for the NCS with uncertain delay, package-dropout, and quantization effects is formulated as a regulation problem for an uncertain hybrid system.

*This work was supported in part by the National Science Foundation (ECS99-12458 and CCR01-13131), and by the DARPA/ITO-NEST Program (AF-F30602-01-2-0526).

4.1 Introduction

Hybrid systems are heterogeneous dynamical systems whose behavior is determined by interacting continuous-variable and discrete-event dynamics [2, 26]. The last decade has seen considerable research activities in the modeling, analysis, and synthesis of hybrid systems involving researchers from a number of traditionally distinct fields. On the one hand, computer scientists extend their computational models and verification methods from discrete systems to hybrid systems by embedding the continuous dynamics into their discrete models. Typically these approaches are able to deal with complex discrete dynamics described by finite automata and emphasize analysis results (verification) and simulation methodologies. From this perspective, the safety or invariance properties have gained the most attention [3, 23]. Other properties investigated include the qualitative temporal notions of liveness, nonblocking, fairness along infinite trajectories, and qualitative ordering of events along trajectories [24]. One of the main formal methods is symbolic model checking, which is based on the computation of reachable sets for hybrid systems [1]. As a result, a good deal of research effort has been focused on developing sophisticated techniques drawn from optimal control, game theory, and computational geometry to calculate or approximate the reachable sets for various classes of hybrid systems [9, 3]. However, the reachability (hence verification) problem is undecidable for most interesting classes of systems [1]. Working in parallel, researchers from the areas of dynamical systems and control theory have viewed hybrid systems as collections of differential/difference equations with discontinuous or multivalued right-hand sides [7, 19, 4]. In these approaches, the models and methodologies for continuous-valued variables described by ordinary differential/difference equations were extended to include discrete variables that exhibit jumps or extend results to switching systems. Typically these approaches are able to deal with complex continuous dynamics and mainly concern stability [19, 10], robustness [15, 10], and synthesis issues [7, 19, 10]. However, there has been little work done on integrating these concerns within a framework for formal methods, perhaps because formal methods traditionally lie in the realm of discrete mathematics, while these concerns from control theory lie separately in the realm of continuous mathematics. In this chapter, we will attempt to integrate these concerns within a framework of formal methods.

The model uncertainty and robust control of hybrid systems is an underexplored and highly promising field [15, 24]. Reachability analysis for uncertain hybrid systems has appeared in [17, 21], and there is also some work on analyzing the induced gain of switched systems [14, 29, 30]. In [29], the $\mathcal{L}_2$ gain of continuous-time linear switched systems is studied by an average dwell time approach incorporated with a piecewise quadratic Lyapunov function, and the results are extended to the discrete-time case in [30]. In [14], the root-mean-square (RMS) gain of a continuous-time switched linear system is computed in terms of the solution to a differential Riccati equation when the interval between consecutive switchings is large. In [24], the authors give an abstract algorithm, based on modal logic formalism, to design the switching mechanism among a finite number of continuous systems, and the closed-loop system forms a hybrid automata and satisfies the specifications. In [13], the problem of controlling a poorly modeled continuous-time linear system is addressed. The proposed approach is to employ logic-based switching among a family of candidate controllers. Note that most of the existing methods for synthesizing hybrid control systems either manually decouple the synthesis of the continuous control law and the design of the discrete event control signal, or only design the switching mechanism. Considering that the continuous dynamics and discrete dynamics are interacting (coupling) tightly in hybrid systems, we believe that the synthesis problem of hybrid systems has not been solved in a satisfactory way. In this chapter we attempt to present an integrated framework that directly addresses synthesis issues for both the continuous and discrete parts of the hybrid control systems.

In this chapter, we concentrate on a class of uncertain hybrid systems with polytopic uncertain continuous dynamics, called "discrete-time polytopic uncertain linear hybrid systems." The motivation for introducing uncertainty into the hybrid dynamical systems model can be described as follows. First, uncertainty of the plant and environment is one of the main challenges to control theory and engineering. Therefore, it is very important for the controller design stage to ensure that the desired performances are preserved even under the effect of uncertainties. The system parameters are often subject to unknown, possibly time-varying, perturbations. Moreover, the real processes are often affected by disturbances and it is necessary to consider them in control design. Another challenge to control theory and engineering is the nonlinearity of the real world dynamics, since no general methodologies that deal effectively with nonlinear systems exist as yet. In order to avoid dealing directly with a set of nonlinear equations, one may choose to work with sets of simpler equations (e.g., linear) and switch among these simpler models. This is a rather common approach in modeling physical phenomena. In control, switching among simpler dynamical systems has been used successfully in practice for many decades. Recent efforts in hybrid systems research along these lines typically concentrate on the analysis of the dynamic behaviors and aim to design controllers with guaranteed stability and performance, see, for example, [28, 16, 5, 18] and the references therein.

Uncertain systems with strong nonlinearities are often of interest. If we use ordinary piecewise linear systems to approximate and study such nonlinear systems, we have to shrink the operating region of the linearization. This results in a large number of linearized models, which makes the subsequent analysis and synthesis computationally expensive or even intractable. So we propose to introduce a bundle of linearization, whose convex hull covers the original (may be uncertain) nonlinear dynamics, instead of approximating with just a single linearization. In this way, we may keep the operating region from shrinking and we may study uncertain nonlinear systems in a systematic way with less computational burden (see Figure 4.1.1).

Our control objective is for the closed-loop system to exhibit certain desired behavior despite the uncertainty and disturbance. Specifically, given finite number of regions $\{\Omega_0, \Omega_1, \cdots, \Omega_M\}$ in the state space, our goal is for the closed-loop system trajectories, starting from the given initial region Ω_0, to go through the sequence of finite number of regions $\Omega_1, \Omega_2, \cdots, \Omega_M$ in the desired order and finally to reach the final region Ω_M and then remain in Ω_M. This kind of specification is analogous to the ordinary tracking and regulation problem in pure continuous-variable

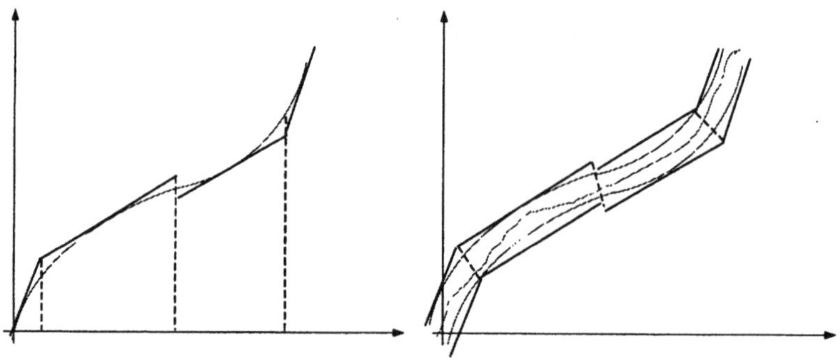

Figure 4.1.1: Piecewise linear approximation and uncertain piecewise linear coverage

dynamical control systems. In addition, it also reflects the qualitative ordering of event requirements along trajectories. One of the main questions is to determine whether there exists admissible control law such that the region sequence can be followed. If there exists such admissible control law, the region sequence specification $\{\Omega_0, \Omega_1, \cdots, \Omega_M\}$ is called attainable. The attainability checking is based on the backward reachability analysis and the symbolic model checking method. The next question is how to design an admissible control law to satisfy the closed-loop specification. An optimization-based method is given in this chapter to design such admissible control law.[†]

The organization of the chapter is as follows. Section 4.2 defines polytopic uncertain linear hybrid systems and formulates the tracking and regulation problems. Then in Section 4.3, a robust one-step predecessor operator for the uncertain linear hybrid systems is studied, which serves as the basic tool for the analysis that follows. In Section 4.4, the necessary and sufficient conditions for checking the safety, reachability, and attainability are given. Then, the robust tracking and regulation controller synthesis problem for the polytopic uncertain linear hybrid systems is formulated and solved in Section 4.5, which is based on linear programming techniques. In addition, Networked Control System (NCS) is proposed as a possible application field of the theoretic results and tools developed here, and the ultimate boundedness control for NCS is formulated as a regulation problem for the uncertain hybrid systems studied in this chapter. Finally, concluding remarks are made.

4.2 Problem Formulation

We are interested in the following discrete-time uncertain hybrid dynamical systems.

Definition 4.2.1. The discrete-time polytopic uncertain linear hybrid systems are de-

[†]This chapter is an extension of our group's previous work [18] to uncertain systems and to more general cases. Earlier work appeared in [21, 22].

fined by

$$x(t+1) = \tilde{A}_{q(t)}x(t) + \tilde{B}_{q(t)}u(t) + E_{q(t)}d(t) \tag{4.2.1}$$

$$q(t) = \delta(q(t-1), \pi(x(t)), \sigma_c(t), \sigma_u(t)) \tag{4.2.2}$$

where $q \in Q = \{q_1, q_2, \cdots, q_s\}$ and Q is the collection of discrete states (modes); $x \in X \subseteq \mathbb{R}^n$ and X stands for the continuous state space. For mode q, the continuous control $u \in \mathcal{U}_q \subset \mathbb{R}^m$, and the continuous disturbance $d \in \mathcal{D}_q \subset \mathbb{R}^p$, where $\mathcal{U}_q$, $\mathcal{D}_q$ are bounded convex polyhedral sets. Denote

$$\mathcal{U} = \bigcup_{q \in Q} \mathcal{U}_q, \quad \mathcal{D} = \bigcup_{q \in Q} \mathcal{D}_q.$$

- $\tilde{A}_q \in \mathbb{R}^{n \times n}$, $\tilde{B}_q \in \mathbb{R}^{n \times m}$, and $E_q \in \mathbb{R}^{n \times p}$ are the system matrices for the discrete state q. The entries in $\tilde{A}_q$ and $\tilde{B}_q$ are unknown, and may be time-variant, but $[\tilde{A}_q, \tilde{B}_q]$ are contained in a convex hull in $\mathbb{R}^{n \times n} \times \mathbb{R}^{n \times m}$, that is

$$[\tilde{A}_q, \tilde{B}_q] = \sum_{i=1}^{N_q} \lambda_i [A_q^i, B_q^i], \quad \lambda_i \geq 0, \sum_{i=1}^{N_q} \lambda_i = 1.$$

- $\pi : X \to X/E_\pi$ partitions the continuous state space $X \subset \mathbb{R}^n$ into polyhedral equivalence classes.
- $q(t) \in \text{act}(\pi(x(t)))$, where $\text{act} : X/E_\pi \to 2^Q$ defines the active mode set.
- $\delta : Q \times X/E_\pi \times \Sigma_c \times \Sigma_u \to Q$ is the discrete state transition function. Here $\sigma_c \in \Sigma_c$ denotes a controllable event and Σ_u the collection of uncontrollable events.
- The guard $G(q, q')$ of the transition (q, q') is defined as the set of all continuous states x such that $q' \in \text{act}(\pi(x(t)))$ and there exist controllable event $\sigma_c \in \Sigma_c$ such that $q' = \delta(q, \pi(x), \sigma_c, \sigma_u)$ for every uncontrollable event $\sigma_u \in \Sigma_u$. The guard of the transition describes the region of the continuous state space where the transition can be forced to take place independently of the disturbances generated by the environment.

Remark 4.2.1. Note that in the above definition, we do not consider "state jumps" (reset) for continuous state x explicitly. However, the reset function can be easily included in our model by adding some auxiliary modes.

In the following we assume the existence of the solution for such uncertain hybrid systems under given initial conditions. And we assume that exact state measurement (q, x) is available. An admissible control input (or law) is one that satisfies the input constraints $(\Sigma_c, \mathcal{U}_q)$. The elements of an allowable disturbance sequence are contained in $(\Sigma_u, \mathcal{D}_q)$.

We consider specifications that are described with respect to regions of the hybrid state space. Consider a finite number of regions $\{\Omega_0, \Omega_1, \cdots, \Omega_M\} \subset Q \times X$, where $\Omega_i = (\mathbf{q}_i, P_i)$ are regions in the hybrid state space. Note that the continuous part P_i does not necessarily coincide with the partitions of π in Definition 4.2.1, and this gives us more flexibility. However, it is required that the following consistency condition holds:

$$P_i \subset \bigcap_{q_i \in \mathbf{q}_i} \mathrm{Inv}(q_i) \tag{4.2.3}$$

where $\mathrm{Inv}(q_i) = \{x \in X : q_i \in \mathrm{act}(\pi(x))\}$. $\mathrm{Inv}(q_i)$ is similar to the concept of invariant set of mode q_i in hybrid automata.

Our control objective is for the closed-loop system trajectories to follow a given sequence of regions $\{\Omega_0, \Omega_1, \cdots, \Omega_M\} \subset Q \times X$, despite the uncertainty and disturbance. One of the main questions is to check whether there exists an admissible control law, $\sigma_c[q(t), x(t)] \in \Sigma_c$ and $u[q(t), x(t)] \in \mathcal{U}_{q(t)}$, such that the hybrid state trajectory $(q(t), x(t))$ goes through the regions, $\Omega_0, \Omega_1, \Omega_2, \cdots$, in the specified order and the closed-loop system satisfies some desired requirements. This solution involves sequencing of events and eventual execution of actions. If there exist admissible control laws such that the region sequence starting from Ω_0 can be followed, we call the sequence of regions specification $\{\Omega_0, \Omega_1, \cdots, \Omega_M\} \subset Q \times X$ attainable. To check the attainability of a sequence of regions specification, two different kinds of properties should be checked: the direct reachability from region Ω_i to Ω_{i+1} for $0 \le i < M$ and the safety (or controlled invariance) for region Ω_M. The analysis problems for safety and direct reachability are formulated as follows.

- Safety: Given a region $\Omega \subset Q \times X$, determine whether there exist admissible control laws such that the evolution of the system starting from Ω will remain inside the region for all time, despite the presence of dynamic uncertainties and disturbances.
- Reachability: Given two regions $\Omega_1, \Omega_2 \subset Q \times X$, determine whether there exist admissible control laws such that all the states in Ω_1 can be driven into Ω_2 in finite number steps without entering a third region.

The safety, reachability, and attainability checking are all based on the backward reachability analysis and the symbolic model checking method. In the next section, we will briefly discuss the backward reachability analysis, which serves as one of the basic tools for the analysis that follows. After answering how to check the attainability of a specification, we will design the admissible control law, $\sigma_c[q(t), x(t)]$ and $u[q(t), x(t)]$, such that the hybrid state trajectory $(q(t), x(t))$ goes through the regions, $\Omega_0, \Omega_1, \cdots$, in the specified order and such that the closed-loop system satisfies some desired requirements.

4.3 Robust One-Step Predecessor Set

The basic building block to be used for backward reachability analysis is the robust predecessor operator, which is defined below.

Definition 4.3.1. The robust one-step predecessor set, $\text{pre}(\Omega)$, is the set of states in $Q \times X$, for which, despite disturbances and dynamic uncertainties, admissible control inputs exist and guarantee that the system will be driven to Ω in one step, i.e.,

$$\text{pre}(\Omega) = \{(q(t), x(t)) \in Q \times X | \forall \sigma_u \in \Sigma_u, d(t) \in \mathcal{D}_{q(t)}, \exists \sigma_c \in \Sigma_c, u(t) \in \mathcal{U}_{q(t)}, \\ s.t. \left(q(t+1), \tilde{A}_{q(t)} x(t) + \tilde{B}_{q(t)} u(t) + E_{q(t)} d(t)\right) \subseteq \Omega\}$$

where

$$[\tilde{A}_{q(t)}, \tilde{B}_{q(t)}] \in \text{Conv}_{i=1}^{N_q} [A^i_{q(t)}, B^i_{q(t)}].$$

The predecessor operator has the following properties.

Proposition 4.3.1. *For all Ω_1 and Ω_2, $\Omega_1 \subseteq \Omega_2 \Rightarrow \text{pre}(\Omega_1) \subseteq \text{pre}(\Omega_2)$. If Ω is given by the union, $\Omega = \bigcup_i \Omega_i$, then* $\text{pre}(\Omega) = \bigcup_i \text{pre}(\Omega_i)$.

Next, we assume that a region of the state space is defined as $\Omega = (\mathbf{q}, P) \subset Q \times X$, where P is a piecewise linear set. Without loss of generality, we assume that P is convex and can be represented by $P = \{x \in X | Gx \leq w\}$, where $G \in \mathbb{R}^{v \times n}$, $w \in \mathbb{R}^v$. Here $a \leq b$ means that all entries in the vector $(a - b)$ are all nonpositive. If P is nonconvex, then it is known that nonconvex piecewise linear set P can be written as finite union of convex piecewise linear set P_i, that is, $P = \cup_{i=1}^m P_i$ [27]. And $\Omega = \bigcup_i \Omega_i = \bigcup_i (\mathbf{q}, P_i)$. Because of the above proposition, we have $\text{pre}(\Omega) = \bigcup_i \text{pre}(\Omega_i)$. Similarly, without loss of generality we assume that the discrete part of Ω contains only one mode, that is $|\mathbf{q}| = 1$ or $\mathbf{q} = \{q\}$.

We are interested in computing the set of all the states that can be driven to $\Omega = (q, \{Gx \leq w\})$ by both continuous and discrete transitions despite the presence of dynamic uncertainties and disturbances. To calculate $\text{pre}(\Omega)$, we first calculate the predecessor set for Ω either purely by discrete transition, $\text{pre}_d(\Omega)$, or purely by continuous transition at mode q, $\text{pre}_c^q(P)$. Then an algorithm is given for $\text{pre}(\Omega)$ by considering the coupling between $\text{pre}_d(\Omega)$ and $\text{pre}_c^q(P)$.

4.3.1 Discrete transitions

The predecessor operator for discrete transitions is denoted by $\text{pre}_d : 2^{Q \times X} \rightarrow 2^{Q \times X}$, and it is used to compute the set of states that can be driven to the region Ω by a discrete instantaneous transition $q' \rightarrow q$ which can be forced by the controller for any uncontrollable event. The predecessor operator in this case is defined as follows:

$$\text{pre}_d(\Omega) = \{(q', x) \in Q \times P | \forall \sigma_u \in \Sigma_u, \exists \sigma_c \in \Sigma_c, \ q = \delta(q', x, \sigma_c, \sigma_u)\}.$$

For every discrete transition that can be forced by a controllable event we have that

$$\text{pre}_d(\Omega) = \bigcup_{q' \in \text{act}(P)} \{q'\} \times (G(q', q) \cap P)$$

where $G(q', q)$ is the guard set of transition $q' \rightarrow q$.

4.3.2 Continuous transitions

In the case of continuous transitions, we define the continuous predecessor operator under mode q as

$$\text{pre}_c^q : 2^X \to 2^X .$$

It computes the set of states for which there exists a control input so that the continuous state will be driven into the set P for every disturbance and uncertainty, while the system is at the discrete mode q. The action of the operator is described by

$$\text{pre}_c^q(P) = \{x \in X | \forall d \in \mathcal{D}_q, \forall [\tilde{A}_q, \tilde{B}_q] \in \text{Conv}_{i=1}^{N_q}([A_q^i, B_q^i]), \exists u \in \mathcal{U}_q, \quad s.t.\ \tilde{A}_q x + \tilde{B}_q u + E_q d \in P\}.$$

4.3.3 Computation of the predecessor operator

As explained above, the predecessor operator for discrete transitions is given by the union of the guards of those transitions that are feasible and can be forced by a control mechanism. Since the guards are regions of the state space that are included in the description of the model, we concentrate on the predecessor operator for the continuous transitions.

Let us denote $\text{pre}_{c,i}^q(P)$ the continuous predecessor set of the ith vertex $[A_q^i, B_q^i]$ for $1 \leq i \leq N_q$. That is,

$$\text{pre}_{c,i}^q(P) = \{x \in X | \forall d \in \mathcal{D}_q, \exists u \in \mathcal{U}_q,\ s.t.\ A_q^i x + B_q^i u + E_q d \in P\}$$

Because of linearity and convexity, we can derive the relationship between $\text{pre}_{c,i}^q(P)$ and $\text{pre}_c^q(P)$ as the following proposition.

Proposition 4.3.2.

$$\text{pre}_c^q(P) = \bigcap_{i=1}^{N_q} \text{pre}_{c,i}^q(P).$$

Remark 4.3.1. The significance of the proposition is that the calculation for the continuous predecessor for the polytopic uncertain linear hybrid systems can be boiled down to the finite intersection of continuous predecessor sets corresponding to the dynamic matrix polytope vertices, which have deterministic continuous dynamics. The predecessor set under deterministic continuous dynamics, $\text{pre}_{c,i}^q(P)$, can be computed by Fourier-Motzkin elimination [25] and linear programming techniques, given in [18].

In the following, we describe an algorithm for calculating the robust predecessor set $\text{pre}(\Omega)$ under both discrete and continuous transitions. Consider the uncertain hybrid systems of Definition 4.2.1 and a region $\Omega = (\mathbf{q}, P)$. We denote $\{P_i^\pi\}$ $(i = 1, \cdots, N)$ the partition of the continuous state space X by the map π as given in Definition 4.2.1. The following algorithm computes all the states of the hybrid system that can be driven to Ω in one time step.

Algorithm 4.3.1. Predecessor Operator

INPUT: $\Omega = (\mathbf{q}, P)$, $S = \emptyset$;
for $i = 1, \cdots, N$,
  $Q_i = P \cap P_i^{\pi}$
  if $Q_i \neq \emptyset$
    for $q' \in \text{act}(P_i^{\pi})$
      $S_i^{q'} = G(q', q) \cap Q_i$
      if $S_i^{q'} \neq \emptyset$
        $V = X$
        for $j = 1, \cdots, N_{q'}$
          $V = V \cap \text{pre}_{c,j}^{q'}(S_i^{q'})$
        end
        if $V \neq \emptyset$
          $S = S \cup (\{q'\} \times V)$
        end if
      end if
    end
  end if
end
OUTPUT: $\text{pre}(\Omega) = S$

Remark 4.3.2. Note that $\text{pre}_{c,i}^{q}(P)$ is a piecewise linear and piecewise linear sets are closed under finite intersections, so

$$\bigcap_{i=1}^{N_q} \text{pre}_{c,i}^{q}(P)$$

is also a piecewise linear set. Piecewise linear sets are relatively easy to calculate and be efficiently represented in computers.

Let us see an numerical example for the predecessor operator.

Example 4.3.1 (Predecessor Operator). Consider an uncertain hybrid system with two modes, q_0 and q_1. The continuous dynamics are described by

$$x(t+1) = \begin{cases} \tilde{A}_0 x(t) + \tilde{B}_0 u(t) + E_0 d(t), & q = q_0 \\ \tilde{A}_1 x(t) + \tilde{B}_1 u(t) + E_1 d(t), & q = q_1 \end{cases}$$

where

$$A_0^1 = \begin{pmatrix} 0.9568 & 0.1730 & 0.2523 \\ 0.5226 & 0.9797 & 0.8757 \\ 0.8801 & 0.2714 & 0.7373 \end{pmatrix}, \quad A_0^2 = \begin{pmatrix} 1.0934 & 0.3721 & 0.5367 \\ 0.5343 & 1.2785 & 1.3450 \\ 1.7740 & 0.9329 & 0.8021 \end{pmatrix}$$

$$B_0^1 = \begin{pmatrix} 0.9883 \\ 0.5828 \\ 0.4235 \end{pmatrix}, \quad B_0^2 = \begin{pmatrix} 1.5038 \\ 0.9167 \\ 0.8564 \end{pmatrix}, \quad E_0 = \begin{pmatrix} 0.2259 \\ 0.5798 \\ 0.7604 \end{pmatrix}$$

$$A_1^1 = \begin{pmatrix} 0.1509 & 0.8600 & 0.4966 \\ 0.6979 & 0.8537 & 0.8998 \\ 0.3784 & 0.5936 & 0.8216 \end{pmatrix}, \quad A_1^2 = \begin{pmatrix} 0.2154 & 0.8942 & 0.5500 \\ 0.7797 & 0.8826 & 0.9725 \\ 0.4444 & 0.6277 & 0.8526 \end{pmatrix}$$

$$B_1^1 = \begin{pmatrix} 0.8385 \\ 0.5681 \\ 0.3704 \end{pmatrix}, \quad B_1^2 = \begin{pmatrix} 0.9088 \\ 0.6227 \\ 0.4149 \end{pmatrix}, \quad E_1 = \begin{pmatrix} 0.6946 \\ 0.6213 \\ 0.7948 \end{pmatrix}.$$

The partition of the state space is obtained by considering the following hyperplane $h_1(x) = x_1 - 5$, $h_2(x) = x_2 - 5$, $h_3(x) = x_3 - 5$, $h_4(x) = x_1$, $h_5(x) = x_2$, and $h_6(x) = x_3$. Assume $u \in \mathcal{U} = [-1, 1]$, $d \in \mathcal{D} = [-0.1, 0.1]$. Here we assume that $X = \mathbb{R}^3$, and $G_{q_0}^{q_1} = G_{q_1}^{q_0} = X$. Consider region $\Omega = (\{q_0, q_1\}, P)$, where P is the tube with edge 5. In order to calculate $\text{pre}(\Omega)$, we first calculate the continuous predecessor sets $\text{pre}_c^{q_0}(P)$, $\text{pre}_c^{q_0}(P \cap G_{q_0}^{q_1})$, $\text{pre}_c^{q_1}(P)$, and $\text{pre}_c^{q_1}(P \cap G_{q_1}^{q_0})$. It turns out that $\text{pre}_c^{q_0}(P \cap G_{q_0}^{q_1}) = \text{pre}_c^{q_0}(P)$, which is shown in Figure 4.3.1, and $\text{pre}_c^{q_1}(P \cap G_{q_1}^{q_0}) = \text{pre}_c^{q_1}(P)$ as shown in Figure 4.3.2. The predecessor operator is given by

$$\text{pre}(\Omega) = (q_0, \text{pre}_c^{q_0}(P)) \cup (q_1, \text{pre}_c^{q_1}(P)).$$

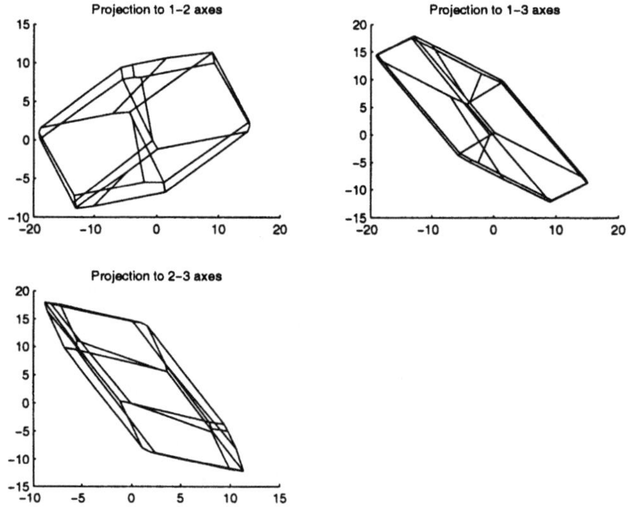

Figure 4.3.1: Illustration for the predecessor sets

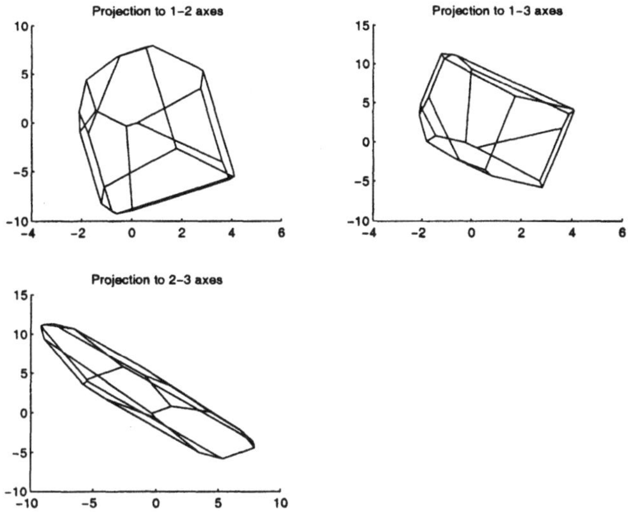

Figure 4.3.2: Illustration for the predecessor sets

4.4 Safety, Reachability, and Attainability

In this section, we first present necessary and sufficient conditions for checking the safety for a given region $\Omega \subset Q \times X$ and the direct reachability between two given regions Ω_1 and Ω_2. Then a necessary and sufficient condition for checking the attainability of a given specification is presented.

4.4.1 Safety

The following is an important, well-known geometric condition for a set to be safe (controlled invariant).

Theorem 4.4.1. *The set Ω is safe if and only if $\Omega \subseteq \mathrm{pre}(\Omega)$.*

The proof follows immediately from the definition of the predecessor set $\mathrm{pre}(\Omega)$. Testing for safety include the following: compute $\mathrm{pre}(\Omega)$, which can be efficiently done by the predecessor operator algorithm described in the former section; test whether $\Omega \subseteq \mathrm{pre}(\Omega)$, which can be checked by the feasibility of a linear programming problem. So this condition can be efficiently tested by solving a finite number of linear programming problems that depend on the number of regions and discrete states of the system.

4.4.2 Reachability

Consider the reachability problem for uncertain linear hybrid systems in Definition 4.2.1. It should be emphasized that we are interested only in the case when reachability between two regions Ω_1 and Ω_2 is defined so that the state is driven to Ω_2 directly from the region Ω_1 in finite steps without entering a third region. This is a problem of practical importance in hybrid systems since it is often desirable to drive the state to a target region of the state space while satisfying constraints on the state and input during the operation of the system.

The problem of deciding whether a region Ω_2 is directly reachable from Ω_1 can be solved by recursively computing all the states that can be driven to Ω_2 from Ω_1 using the predecessor operator. Given Ω_1 and Ω_2, define an N-step directly reachable set from Ω_1 to Ω_2,

$$\mathrm{PRE}_{\Omega_1}^N(\Omega_2) = \underbrace{\mathrm{pre}(\cdots \mathrm{pre}(\mathrm{pre}(\Omega_2) \cap \Omega_1) \cap \Omega_1 \cdots)}_{N \text{ times}}$$

assume $\mathrm{PRE}_{\Omega_1}^0(\Omega_2) = \Omega_1 \cap \Omega_2$. With the introduction of the directly reachable set from Ω_1 to Ω_2 by taking union of all the finite step directly reachable set from Ω_1 to Ω_2, that is $CR_{\Omega_1}(\Omega_2) = \bigcup_{N=0}^{\infty} \mathrm{PRE}_{\Omega_1}^N(\Omega_2)$, the geometric condition to check the direct reachability can be given as follows.

Theorem 4.4.2. *Consider an uncertain hybrid system described by Definition 4.2.1 and the regions Ω_1 and Ω_2. The region Ω_2 is directly reachable from Ω_1 if and only if $\Omega_1 \subseteq CR_{\Omega_1}(\Omega_2)$.*

In general, the reachability problem for hybrid systems is undecidable. So the above procedure is semidecidable [1] because the termination of the procedure is not guaranteed. To formulate a constructive procedure for reachability, two approaches may be employed. First, we consider an upper bound on the time horizon and we examine the reachability only for the predetermined finite horizon. Second, we formulate a termination condition for the reachability algorithm based on a grid-based approximation of the piecewise linear regions of the state space [18].

4.4.3 Attainability

Given a finite number of regions $\{\Omega_0, \Omega_1, \cdots, \Omega_M\} \subset Q \times X$, the attainability for this sequence of region specification is equivalent to the following two different kinds of properties, that is, the direct reachability from region Ω_i to Ω_{i+1} for $0 \leq i < M$ and the safety (or controlled invariance) for region Ω_M. Therefore the attainability checking can be expressed as follows.

Theorem 4.4.3. *The specification $\{\Omega_0, \Omega_1, \cdots, \Omega_M\} \subset Q \times X$ is attainable if and only if the following conditions hold: First, Ω_M is safe, and second, the region Ω_{i+1} is directly reachable from Ω_i, for $i = 0, 1, \cdots, M-1$.*

4.5 Hybrid Regulation

The hybrid regulation problem considered in this section is to design the admissible control law, $\sigma_c[q(t), x(t)]$ and $u[q(t), x(t)]$, such that the hybrid state trajectory $(q(t), x(t))$ goes through the regions, $\Omega_0, \Omega_1, \Omega_2, \cdots$, in the specified order and so that the closed-loop system satisfies some desired requirements. The requirements include sequencing of events and eventual execution of actions. In Section 4.4, we have specified the conditions for the existence of such control laws so that the closed-loop system satisfies the specifications, that is, safety, reachability, and attainability. A regulator is designed as a dynamical system to implement the desired control policy. Here we design such control laws based on optimization techniques. From the discussion in the previous section, the attainability regulation problem can be divided into two basic problems, that is, safety regulation and direct reachability regulation. The two basic regulation problems are formulated as follows.

- Safety Regulation: Given a safe set $\Omega \subset Q \times X$, determine the admissible control laws such that the evolution of the system starting from Ω will remain inside the set for all time, despite the presence of dynamic uncertainties and disturbances.
- Reachability Regulation: Given two regions Ω_1, $\Omega_2 \subset Q \times X$, where Ω_2 is directly reachable from Ω_1, determine the admissible control laws such that all the states in Ω_1 can be driven into Ω_2 in finite number steps without entering a third region.

In the following we present a systematic procedure for the regulator design for these two basic cases based on optimization techniques. Then a procedure for attainability regulation is given.

4.5.1 Safety regulator

First we consider terminating the safety regulation for the terminating region, $\Omega_M = (q_M, P_M)$, of the state space. We assume that $P_M = \{x\colon G_M x \le w_M\}$. We define the cost functional $J_M\colon Q \times [0,1]^{N_q} \times \mathcal{U}_q \to \mathbb{R}$

$$J_M(q,\lambda,u) = \left\| G_M \sum_{i=1}^{N_q} [\lambda_i A_q^i, \lambda_i B_q^i] \begin{pmatrix} x(t) \\ u(t) \end{pmatrix} \right\|_\infty$$

where $\|\cdot\|_\infty$ stands for the infinite norm. The control signal is selected as the solution to the following minmax optimization problem:

$$\min_{u\in\mathcal{U}_q} \max_{\lambda\in[0,1]^{N_q}} J_M(q,\lambda,u)$$
$$s.t. \begin{cases} \tilde{A}_q x(t) + \tilde{B}_q u(t) + E_q d(t) \in P_M \\ u \in \mathcal{U}_q, \ \ d \in \mathcal{D}_q \end{cases}$$

Because of the linearity and convexity, the constraints can be equivalently transformed into the following form.

$$\min_{u\in\mathcal{U}_q} \max_{\lambda\in[0,1]^{N_q}} J_M(q,\lambda,u)$$
$$s.t. \begin{cases} G_M[A_q^1 x(t) + B_q^1 u(t)] \le w_M - \delta \\ G_M[A_q^2 x(t) + B_q^2 u(t)] \le w_M - \delta \\ \cdots\cdots \\ G_M[A_q^{N_q} x(t) + B_q^{N_q} u(t)] \le w_M - \delta \\ u \in \mathcal{U}_q \end{cases}$$

where δ is a vector whose components are given by $\delta_j = \max_{d \in \mathcal{D}_q} g_j^T E_q d$, and g_j^T is the jth row of matrix G_M. The optimal action of the controller is one that tries to minimize the maximum cost, and also tries to counteract the worst disturbance and the worst model uncertainty. This kind of solution is referred to as a Stackelberg solution. The above minmax optimization problem can be equivalently transformed to the following semi-infinite programming problem [6],

$$\min_{u \in \mathcal{U}_q} z$$

$$s.t. \begin{cases} J_M(q, \lambda, u) \leq z \\ G_M[A_q^1 x(t) + B_q^1 u(t)] \leq w_M - \delta \\ G_M[A_q^2 x(t) + B_q^2 u(t)] \leq w_M - \delta \\ \cdots\cdots \\ G_M[A_q^{N_q} x(t) + B_q^{N_q} u(t)] \leq w_M - \delta \\ u \in \mathcal{U}_q. \end{cases}$$

In addition, because of the special form of J_M, the above optimization problems can be reduced to the following form:

$$\min_{u \in \mathcal{U}_q} z$$

$$s.t. \begin{cases} G_M[A_q^1 x(t) + B_q^1 u(t)] \leq z \\ G_M[A_q^2 x(t) + B_q^2 u(t)] \leq z \\ \cdots\cdots \\ G_M[A_q^{N_q} x(t) + B_q^{N_q} u(t)] \leq z \\ G_M[A_q^1 x(t) + B_q^1 u(t)] \leq w_M - \delta \\ G_M[A_q^2 x(t) + B_q^2 u(t)] \leq w_M - \delta \\ \cdots\cdots \\ G_M[A_q^{N_q} x(t) + B_q^{N_q} u(t)] \leq w_M - \delta \\ u \in \mathcal{U}_q. \end{cases}$$

The above problem can be solved very efficiently by solving a linear programming problem for each possible discrete mode. The following algorithm describes the procedure for the synthesis of the safety regulator for an given initial condition (q_0, x_0) containing in a specified region $\Omega_M = (\mathbf{q}_M, P_M)$.

Algorithm 4.5.1. Safety Regulator

INPUT: $\Omega_M = (\mathbf{q}_M, P_M)$, (q_0, x_0);
if $\min_u \max_\lambda J_M(q_0, x_0, \lambda, u)$ feasible
  $u^* = \arg\min_u J_M(q_0, x_0, \lambda, u)$
  $q^* = q_0$
else
  for $i = 1, \cdots, |\mathbf{q}_M|$,
    $q_i = \mathbf{q}_M(i)$
    if $x_0 \in G_{q_0}^{q_i}$
      $J_M^{q_i} = \min_u \max_\lambda J_M(q_i, x_0, \lambda, u)$

    end
  end
  $q^* = \arg\min_{q_i \in \mathbf{q}_M} J_M^{q_i}$
  $u^* = \arg\min_u J_M^{q_i}$
end
OUTPUT: u^*, q^*

In the procedure, we first try to retain the mode and avoid switching, simply because switching may be costly. However, sticking to mode q_0 may not be a good choice, and there may not exist a feasible control signal. So the procedure tries to take possible mode switchings into consideration and choose the mode that can make the next continuous state farthest from the boundary. It is claimed that there must exist at least one mode $q \in \mathbf{q}_M$ such that the above optimization problem is feasible. Otherwise, it leads to a contradiction to the safety of the region $\Omega_M = (\mathbf{q}_M, P_M)$. Here q^* stands for the mode that corresponds to the minimum cost value J_M, then the candidate control input is selected as $(\sigma_c(t), u^*(t))$ where $q^* = \delta(q(t), \pi(x(t)), \sigma_c(t), \Sigma_u)$ and u^* is the solution of the above optimization procedure. In a formal way we can express it as the following proposition.

Proposition 4.5.1. If the region $\Omega_M = (\mathbf{q}_M, P_M)$ is safe, then the procedure described in Algorithm 4.5.1 can solve the safety problem.

4.5.2 Reachability regulator

Next we consider the reachability specification between the regions $\Omega_k = (\mathbf{q}_k, P_k)$ and $\Omega_{k+1} = (\mathbf{q}_{k+1}, P_{k+1})$. The control objective is to drive every state in Ω_k to Ω_{k+1}. Let the convex polyhedral set $P_k = \{x\colon Gx \leq w\}$. For a pair of modes $q_k \in \mathbf{q}_k$ and $q_k' \in \mathbf{q}_{k+1}$, assume that the intersection of the guard set for (q_k, q_k'), $G_{q_k}^{q_k'}$, with the common region of P_k and P_{k+1}, is not empty. Denote this polytope as $P_C^{(q_k,q_k')} = P_k \cap P_{k+1} \cap G_{q_k}^{q_k'} = \{x\colon G_C x \leq w_C\}$. Because of the direct reachability between Ω_k and Ω_{k+1}, the existence of nonempty $P_C^{(q_k,q_k')}$ can be shown. We define the cost functional, $J_C\colon Q \times Q \times [0,1]^{N_{q_k}} \times \mathcal{U}_{q_k} \to \mathbb{R}$

$$J_C(q_k, q_k', \lambda, u) = \left\| G_C \sum_{i=1}^{N_{q_k}} [\lambda_i A_{q_k}^i, \lambda_i B_{q_k}^i] \begin{pmatrix} x(t) \\ u(t) \end{pmatrix} \right\|_\infty .$$

The control signal is selected as the solution to the following minmax optimization problem:

$$\min_{u \in \mathcal{U}_{q_k}} \max_{\lambda \in [0,1]^{N_{q_k}}} J_C(q_k, q_k', \lambda, u)$$

$$s.t. \begin{cases} G[A_{q_k}^1 x(t) + B_{q_k}^1 u(t)] \leq w - \delta \\ G[A_{q_k}^2 x(t) + B_{q_k}^2 u(t)] \leq w - \delta \\ \cdots\cdots \\ G[A_{q_k}^{N_{q_k}} x(t) + B_{q_k}^{N_{q_k}} u(t)] \leq w - \delta \\ u \in \mathcal{U}_{q_k}. \end{cases}$$

Similarly, the above minmax optimization problem can be equivalently transformed to the following linear programming problem:

$$\min_{u \in \mathcal{U}_{q_k}} z$$

$$s.t. \begin{cases} G_C[A_{q_k}^1 x(t) + B_{q_k}^1 u(t)] \leq z \\ G_C[A_{q_k}^2 x(t) + B_{q_k}^2 u(t)] \leq z \\ \cdots\cdots \\ G_C[A_{q_k}^{N_{q_k}} x(t) + B_{q_k}^{N_{q_k}} u(t)] \leq z \\ G[A_{q_k}^1 x(t) + B_{q_k}^1 u(t)] \leq w - \delta \\ G[A_{q_k}^2 x(t) + B_{q_k}^2 u(t)] \leq w - \delta \\ \cdots\cdots \\ G[A_{q_k}^{N_{q_k}} x(t) + B_{q_k}^{N_{q_k}} u(t)] \leq w - \delta \\ u \in \mathcal{U}_{q_k}. \end{cases}$$

The following algorithm designs the regulator to guarantee the direct reachability.

Algorithm 4.5.2. Reachability Regulator

INPUT: $\Omega_k = (\mathbf{q}_k, P_k)$, $\Omega_{k+1} = (\mathbf{q}_{k+1}, P_{k+1})$, (q_0, x_0), feasibility $= 0$;
for $j = 1, \cdots, |\mathbf{q}_{k+1}|$,
  $q'_j = \mathbf{q}_{k+1}(j)$
  if $\min_u \max_\lambda J_C(q_0, q'_j, x_0, \lambda, u)$ feasible
    $J_C^{(q_0, q'_i)} = \min_u \max_\lambda J_C(q_0, q'_j, x_0, \lambda, u)$
    feasibility $= 1$
  end
end
if feasibility $== 1$
  $ind = \arg\min_{q'_j \in \mathbf{q}_{k+1}} J_C^{(q_0, q'_j)}$
  $u^* = \arg\min_u J_C^{(q_0, ind)}$
  $q^* = q_0$
else
  for $i = 1, \cdots, |\mathbf{q}_k|$,
    $q_i = \mathbf{q}_k(i)$
    if $x_0 \in G_{q_0}^{q_i}$
      for $j = 1, \cdots, |\mathbf{q}_{k+1}|$,
        $q'_j = \mathbf{q}_{k+1}(j)$
        $J_C^{(q_i, q'_j)} = \min_u \max_\lambda J_C(q_i, q'_j, x_0, \lambda, u)$
      end
    end
  end
  $[q^*, q'] = \arg\min_{q_i \in \mathbf{q}_k; q'_j \in \mathbf{q}_{k+1}} J_M^{(q_i, q'_j)}$
  $u^* = \arg\min_u J_M^{q_i}$
end
OUTPUT: u^*, q^*

Similarly, we have the following proposition.

Proposition 4.5.2. *If the region* $\Omega_k = (\mathbf{q}_k, P_k)$ *is directly reachable to* $\Omega_{k+1} = (\mathbf{q}_{k+1}, P_{k+1})$, *then the procedure described in Algorithm 4.5.2 can solve the reachability problem.*

4.5.3 Attainability regulator

The following algorithm designs the regulator to guarantee the direct attainability for the specification described by $\{\Omega_0, \Omega_1, \cdots, \Omega_M\}$.

Algorithm 4.5.3. Attainability Regulator
INPUT: $\{\Omega_1, \cdots, \Omega_M\}$, (q_0, x_0);
for *n =1,· · · ,M-1,*
  while $x_0 \in \Omega_n$ and $x_0 \notin \Omega_{n+1}$
  Design Reachability Regulator from Ω_n to Ω_{n+1}
  end
end
Design Safety Regulator for Ω_M
OUTPUT: u^*, q^*

From Theorem 4.4.3 and the previous two propositions on safety and direct reachability regulation, we can conclude the following proposition.

Proposition 4.5.3. *If the region sequence* $\{\Omega_0, \Omega_1, \cdots, \Omega_M\}$ *is attainable, then the procedure described in Algorithm 4.5.3 can solve the attainability problem.*

Let us turn to an example to illustrate the regulation method.

Example 4.5.1 (Temperature Control System). The system consists of a furnace that can be switched on and off. The control objective is to control the temperature at a point of the system by applying the heat input at a different point. So the discrete mode contains only two states, which are the furnace is "off," q_0, and the furnace is "on," q_1. The continuous dynamics are described by ‡

$$x(t+1) = \begin{cases} \tilde{A}_0 x(t) + \tilde{B}_0 u(t) + E_0 d(t), & q = q_0 \\ \tilde{A}_1 x(t) + \tilde{B}_1 u(t) + E_1 d(t), & q = q_1 \end{cases}$$

where

$$\begin{aligned} A_0^1 &= \begin{pmatrix} 0.825 & 0.135 \\ 0.68 & 1 \end{pmatrix}, \; A_0^2 = \begin{pmatrix} 1 & 0.35 \\ 0.068 & 0.555 \end{pmatrix} \\ B_0^1 &= \begin{pmatrix} 1.7 \\ 0.06 \end{pmatrix}, \; B_0^2 = \begin{pmatrix} 1.9 \\ 0.08 \end{pmatrix}, \; E_0 = \begin{pmatrix} 0.0387 \\ 0.3772 \end{pmatrix} \\ A_1^1 &= \begin{pmatrix} -0.664 & 0.199 \\ 0.199 & 0.264 \end{pmatrix}, \; A_1^2 = \begin{pmatrix} -0.7 & 0.32 \\ 0.32 & 0.44 \end{pmatrix} \\ B_1^1 &= \begin{pmatrix} 0.8 \\ 0.1 \end{pmatrix}, \; B_1^2 = \begin{pmatrix} 0.9 \\ 0.2 \end{pmatrix}, \; E_1 = \begin{pmatrix} 0.1369 \\ 0.5363 \end{pmatrix}. \end{aligned}$$

‡using zero-order hold sampling with $T = 1$ sec.

The state space is partitioned by the following hyperplane $h_1(x) = x_1 - 20$, $h_2(x) = x_2 - 5$, $h_3(x) = x_2$, and $h_4(x) = x_1$. Assume $u \in \mathcal{U} = [-1, 1]$, $d \in \mathcal{D} = [-0.1, 0.1]$. Consider region $\Omega_1 = (\{q_0, q_1\}, P_1)$ and $\Omega_2 = (\{q_0, q_1\}, P_2)$, where $P_1 = \{x \in \mathbb{R}^2 | (0 \leq x_1 \leq 20) \wedge (-20 \leq x_2 \leq 0)\}$, and $P_2 = \{x \in \mathbb{R}^2 | (0 \leq x_1 \leq 20) \wedge (0 \leq x_2 \leq 5)\}$. Our control objective is that for every initial state (q_0, x_0) within region Ω_1 there exist control $u \in \mathcal{U}$ and $\sigma_c \in \Sigma_c$ so that from (q_0, x_0) the state can be driven to Ω_2 without entering a third region. Then the state will stay inside Ω_2, no matter what the dynamic uncertainty and continuous and discrete disturbances. Let us check the attainability. We first calculate pre(Ω_2), which covers the region Ω_2, so Ω_2 is safe. By recursively using pre($\cdot$), we find that Ω_1 can be driven to Ω_2 in three steps, i.e., Ω_2 reachable from Ω_1. So the attainability of the specification is satisfied. Then we design the regulator and plot the simulation result for nominal plant (here we choose the epicenter of the state matrix , i.e., $\frac{1}{2}(A_q^1 + A_q^2)$) in Figures4.5.1 and 4.5.2. Also the control signal output (σ_c, u) of the regulator is plotted in Figures 4.5.1 and 4.5.2.

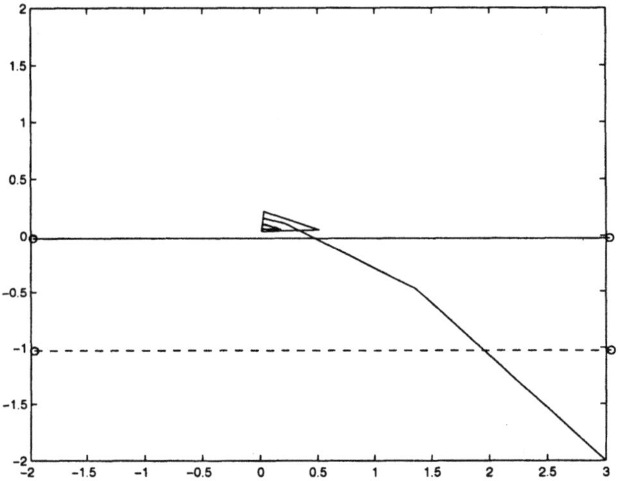

Figure 4.5.1: Simulation for closed-loop nominal plant (assuming $d = 0$)

4.6 Networked Control Systems (NCS)

By NCS, we mean feedback control systems where networks, typically digital band-limited serial communication channels, are used for the connections between spatially distributed system components like sensors and actuators to controllers. These channels may be shared by other feedback control loops. In traditional feedback control systems these connections are through point-to-point cables. Compared with the point-to-point cables, there are many attractive advantages of introducing serial communication networks, like high system testability and resource utilization, as well as

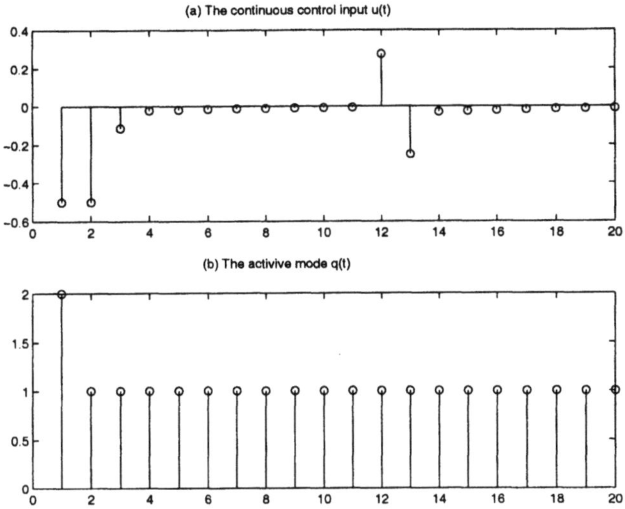

Figure 4.5.2: The control signals output (u, σ_c) of the regulator

low weight, space, power, and wiring requirements. These advantages make the use of networks in control systems connecting sensors/actuators to controllers more and more popular in many applications, including traffic control, satellite clusters, and mobile robotics. Modeling and control of networked control systems with limited communication capability has recently emerged as a topic of significant interest to the control community.

In NCS, the real-time requirement is critical, and time delay usually has negative effects on the NCS stability and performance. There are several situations where time delay may arise. First, transmission delay is caused by the limited bit rate of the communication channels. Second, the channel in NCS is usually shared by multiple sources of data, and the channel is usually multiplexed by the time-division method. Therefore, there are delays caused by a node waiting to send out a message through a busy channel, which is usually called "accessing delay" and serves as the main source of delays in NCS. There are also some delays caused by processing and propagation, which are usually negligible for NCS. Another interesting problem in NCS is the package-dropout issue. Because of the uncertainties and noise in the communication channels, there may exist unavoidable errors in the transmitted package or even loss. If this happens, the corrupted package is dropped and the receiver (controller or actuator) uses the package that it received most recently. In addition, package-dropout may occur when one package, say sampled values from the sensor, reaches the destination later than its successors. In such a situation, the old package is dropped, and its successive package is used instead. Finally, according to the finite bit rate constraint, only quantized signals can be transmitted through the network. So the quantization scheme and its effects have to be considered in real NCS. The primary objective of NCS is to efficiently use the finite channel capacity while maintaining good closed-loop control system performance, including stability, disturbance attenuation, rising time, overshoot, and other design criteria. Therefore, quantization and limited bit rate issues have attracted many researchers' attention, see for example [8, 12, 11]. In this section, we will formulate an NCS with uncertain time delay, package-dropout, and quantization effects into the framework of polytopic uncertain hybrid (switched) systems. Then the methods developed here and existing methods for hybrid (switched) systems can be employed to study NCS.

4.6.1 The delay and package-dropout

The NCS model discussed in this section is shown in Figure 4.6.1. For simplicity, but without loss of generality, we may combine all the time-delay and package-dropout effects into the sensor to controller path and assume the controller-actuator communicates ideally.

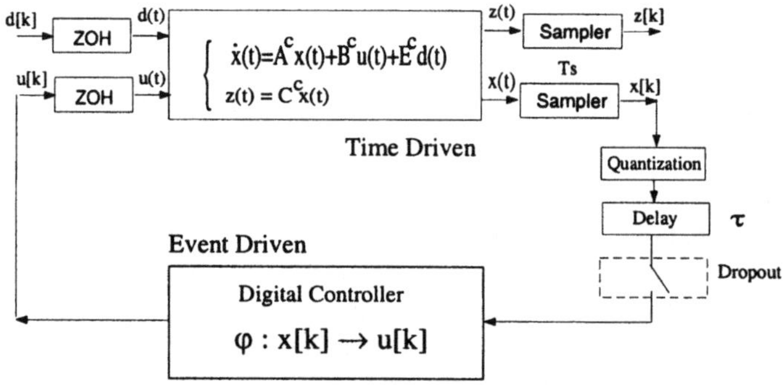

Figure 4.6.1: The networked control systems model

We assume that the plant can be modeled as a continuous-time linear time-invariant system described by

$$\begin{cases} \dot{x}(t) = A^c x(t) + B^c u(t) + E^c d(t) \\ z(t) = C^c x(t) \end{cases}$$

where $x(t) \in \mathbb{R}^n$ is the state variable, and $z(t) \in \mathbb{R}^p$ is the controlled output. $u(t) \in \mathbb{R}^m$ is the control input. The disturbance input $d(t)$ is contained in $\mathcal{D} \subset \mathbb{R}^r$. $A^c \in \mathbb{R}^{n\times n}$, $B^c \in \mathbb{R}^{n\times m}$, and $E^c \in \mathbb{R}^{n\times r}$ are state matrices, and $C^c \in \mathbb{R}^{p\times n}$ is the output matrix.

It is assumed that the sensors work in a time-driven fashion. After each clock cycle (sampling time T_s), the output node (sensor) attempts to send a package containing the most recent state (output) samples. If the communication bus is idle, then the package will be transmitted to the controller. Otherwise, if the bus is busy, then the output node will wait for, say $\varpi < T_s$, and try again. After several attempts or time elapses, if the transmission still cannot be completed, then the package is discarded. The controller and actuator are event driven and work in a simpler way. The controller, as a receiver, has a receiving buffer that contains the most recently received data package from the sensors (the overflow of the buffer may be dealt with as package-dropout). The controller reads the buffer periodically at a higher frequency than the sampling frequency, say every $\frac{T_s}{N}$ for some integer N large enough. Whenever there is new data in the buffer, the controller will calculate the new control signal and transmit it to the actuators instantly. Upon the arrival of the new control signal, the actuators update the output of the zero-order hold to the new value.

4.6.2 Models for NCS

In this section, we will consider the sampled-data model of the plant. Because we do not assume synchronization between the sampler and the digital controller, the control signal is no longer of constant value within a sampling period. Therefore the control signal within a sampling period has to be divided into subintervals corresponding to the controller's reading buffer period, $T = \frac{T_s}{N}$. Within each subinterval, the control signal is constant in view of the assumptions in the previous section. Hence the continuous-time plant may be discretized into the following sampled-data system using the lifting method:

$$x[k+1] = Ax[k] + \underbrace{[B\ B\ \cdots B]}_{N} \begin{bmatrix} u^1[k] \\ u^2[k] \\ \vdots \\ u^N[k] \end{bmatrix} + Ed[k] \tag{4.6.1}$$

where $A = e^{A^c T_s}$, $B = \int_0^{\frac{T_s}{N}} e^{A^c\eta} B^c d\eta$, and $E = \int_0^{T_s} e^{A^c\eta} E^c d\eta$. Note that for linear time-invariant plant and constant-periodic sampling, the matrices A, B, and E are constant.

During each sampling period, several different cases may arise. They are listed below.

(1) Delay $\tau = h \times \frac{T_s}{N}$, where $h = 0, 1, 2, \cdots, D_{max}$. For this case $u^1[k] = u^2[k] = \cdots u^h[k] = u[k-1]$, $u^{h+1}[k] = u^{h+2}[k] = \cdots u^N[k] = u[k]$, and (4.6.1) can be written as

$$\begin{aligned} x[k+1] &= Ax[k] + [B\ B\ \cdots B] \begin{bmatrix} u[k-1] \\ \vdots \\ u[k-1] \\ u[k] \\ \vdots \\ u[k] \end{bmatrix} + Ed[k] \\ &= Ax[k] + h \cdot Bu[k-1] + (N-h) \cdot Bu[k] + Ed[k]. \end{aligned}$$

If we let $\hat{x}[k] = \begin{bmatrix} u[k-1] \\ x[k] \end{bmatrix}$, then the above equations can be written as

$$\hat{x}[k+1] = \begin{bmatrix} 0 & 0 \\ hB & A \end{bmatrix} \hat{x}[k] + \begin{bmatrix} I \\ (N-h)B \end{bmatrix} u[k] + \begin{bmatrix} 0 \\ E \end{bmatrix} d[k]$$

where $h = 0, 1, 2, \cdots, D_{max}$. Note that $h = 0$ implies $\tau = 0$, which corresponds to the "no delay" case. And the controlled output $z[k]$ is given by

$$z[k] = \begin{bmatrix} 0 & C \end{bmatrix} \hat{x}[k]$$

where $C = C^c$.

(2) In the case of package-dropout due to corrupted package or delay $\tau > D_{max} \times \frac{T_s}{N}$, the actuator will implement the previous control signal, i.e., $u^1[k] = u^2[k] = \cdots u^N[k] = u[k-1]$. The state transition equation (4.6.1) for this case can be written as follows.

$$\begin{aligned} x[k+1] &= Ax[k] + [B\, B\, \cdots B] \begin{bmatrix} u[k-1] \\ u[k-1] \\ \vdots \\ u[k-1] \end{bmatrix} + Ed[k] \\ &= Ax[k] + N \cdot Bu[k-1] + Ed[k]. \end{aligned}$$

By introducing new state variables,

$$\hat{x}[k+1] = \begin{bmatrix} 0 & 0 \\ NB & A \end{bmatrix} \hat{x}[k] + \begin{bmatrix} I \\ 0 \end{bmatrix} u[k] + \begin{bmatrix} 0 \\ E \end{bmatrix} d[k].$$

The controlled output $z[k]$ is given by

$$z[k] = \begin{bmatrix} 0 & C \end{bmatrix} \hat{x}[k]$$

where $C = C^c$.

4.6.3 NCS hybrid model

In this section, we reformulate the NCS problem of the previous section as a hybrid (switched) system with $D_{max} + 2$ different modes. In particular,

$$\begin{cases} \hat{x}[k+1] &= A_h \hat{x}[k] + B_h u[k] + E_h d[k] \\ z[k] &= C_h \hat{x}[k] \end{cases} \tag{4.6.2}$$

where $A_h = \begin{bmatrix} 0 & I \\ hB & A \end{bmatrix}$, $B_h = \begin{bmatrix} I \\ (N-h)B \end{bmatrix}$, $E_h = \begin{bmatrix} 0 \\ E \end{bmatrix}$, and $C_h = \begin{bmatrix} 0 & C \end{bmatrix}$ for $h = 0, 1, 2, \cdots D_{max}, N$. The set of modes Q is given by $Q = \{0, 1, 2, \cdots D_{max}, N\}$. Note that $h = 0$ implies $\tau = 0$, which corresponds to the "no delay" case, while $h = N$ corresponds to the "package-dropout" case. The discrete dynamics (switching signal) may be modeled as a finite state machine (FSM), whose discrete states correspond to the $D_{max} + 2$ modes of NCS. The discrete transition rule of the FSM should reflect the delay and package-dropout pattern of the NCS, and it may either be specified in the average dwell-time sense or the stochastic sense.

Another important issue, the quantization effect, is not considered in the above NCS model. With finite bit-rate constraints, quantization has to be taken into consideration in NCS. It has been known that an exponential data representation scheme is most efficient [8, 12]. Here we focus on the floating point representation. Floating point quantization can be viewed as a nonlinear operation described by a time-variant sector gain, i.e., $Q(x) = k(x)$, $k \in [1-\epsilon, 1]$, with ϵ depending on the mantissa length, and for this reason the quantization effect can be dealt with as a model parameter uncertainty. Now we can model the NCS with quantization effects as a switched system with parameter uncertainty, which is a specific subclass of the polytopic uncertain hybrid systems defined in Section 4.2. In particular,

$$\hat{x}[k+1] = \tilde{A}_h \hat{x}[k] + \tilde{B}_h u[k] + E_h d[k]$$

where the parameter uncertainties in $\tilde{A}_h$ and $\tilde{B}_h$ reflect the quantization effects in NCS. Next we consider the robust stabilization problem for the NCS based on such a hybrid model. Because of the parameter uncertainties in the NCS hybrid model and exteriors disturbances, the convergence of all the closed-loop trajectories to the origin (assumed to be the equilibrium) may not be achievable. Instead, we consider the convergence to a small region containing the origin, and it is required that the closed-loop trajectories of NCS be driven to a small region containing the origin for all bounded initial conditions. In the literature this is usually called "ultimate boundedness control" or "practical stability problem." In the following, we will show that the ultimate boundedness control problem for NCS with uncertain delay, package-dropout, and quantization effects may be formulated as a regulation problem for the uncertain hybrid (switched) systems.

Consider the semiglobal asymptotic practical stabilization problem by assuming bounded initial states. If we outer-approximate the bounded region containing all the initial conditions with a polytope Ω_0, and inner-approximate the small region containing the origin with another polytope Ω_1 (as illustrated in Figure 4.6.2), then the ultimate boundedness problem of NCS can be transformed into a regulation problem. The ultimate boundedness of NCS can be checked by checking the attainability of the appropriately chosen $\{\Omega_0, \Omega_1\}$. The ultimate boundedness control law may be designed by solving the optimization-based regulator synthesis problem developed in the previous section.

4.7 Conclusions

In this chapter, the regulation problem for the polytopic uncertain linear hybrid systems was formulated and solved. Using the optimization-based regulator introduced, the closed-loop system exhibits the desired behavior under dynamic uncertainties, continuous disturbances, and uncontrollable events. The existence of a controller such that the closed-loop system follows a desired sequence of regions under uncertainty and disturbance was studied first. Then, based on the novel notion of attainability for the desired behavior of piecewise linear hybrid systems, we presented a

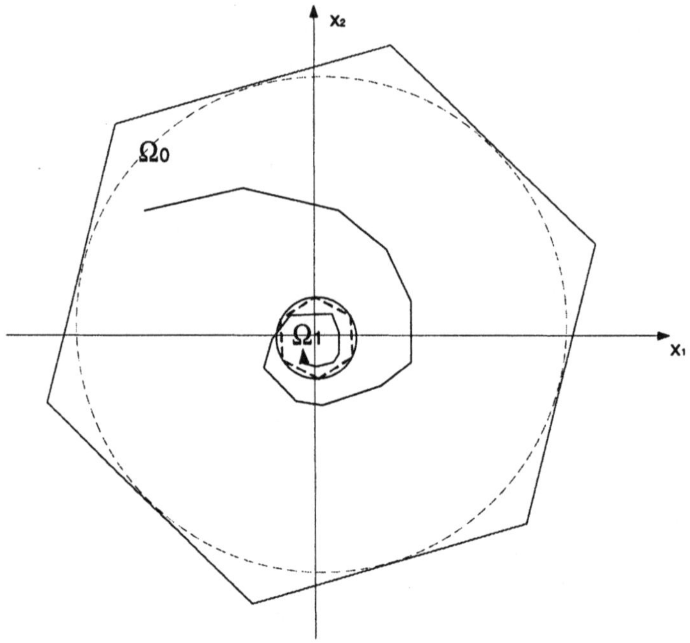

Figure 4.6.2: The networked control systems ultimate boundedness control as a regulation problem

systematic procedure for controller design by using finite automata and linear programming techniques. The design procedure may be seen as a one-step moving horizon optimal control. Our future work is intended to generalize the procedure into finite N-step moving horizon optimal control, and the main issue for this generalization is the feasibility of the optimization problems. The predecessor operator, attainability checking, and robust regulator design methods have been implemented in a Matlab toolbox called HySTAR [20].

The procedures given in this chapter for checking reachability and attainability, which were based on the backward reachability analysis, are semidecidable and their termination is not guaranteed. Future work includes specifying the class of polytopic uncertain linear hybrid systems that makes the procedure decidable. One way to obtain such decidable class is to simplify the continuous dynamics, see for example [1]. However, this approach may not be attractive to control applications, where simple continuous dynamics may not be adequate to capture the system's dynamics. Alternatively, one may simplify the discrete dynamics instead of the continuous dynamics, on which our current research effort is focused.

Finally, we proposed NCS as one of the promising application areas of the methods developed in this chapter. An NCS hybrid model was developed and the ultimate boundedness control problem for NCS was formulated as a hybrid regulation problem. The advantage from formulating the ultimate boundedness control problem of NCS with uncertain delay, package-dropout, and quantization effects as a regulation problem for uncertain hybrid switched systems comes from the systematic methods developed in this chapter and the existence of rich results in the field of hybrid (switched) systems, jump linear systems, etc. Our future work includes research into other promising application areas for the hybrid regulation methods developed here, such as network congestion control, chemical industrial process control, traffic management, manufacturing systems, and robotics.

Bibliography

[1] R. Alur, T. Henzinger, G. Lafferriere, and G. J. Pappas, "Discrete abstractions of hybrid systems," in [2], pp. 971–984, July 2000.

[2] P. Antsaklis, Ed., *Proc. IEEE*, Special Issue on Hybrid Systems: Theory and Applications, vol. 88, no. 7, July 2000.

[3] E. Asarin, O. Maler, and A. Pnueli, "Reachability analysis of dynamical systems having piecewise-constant derivatives," *Theoretical Computer Science*, vol. 138, pp. 35–65, 1995.

[4] A. Bemporad and M. Morari, "Control of systems integrating logic, dynamics, and constraints," *Automatica*, vol. 35, no. 3, pp. 407–427, 1999.

[5] A. Bemporad, G. Ferrari-Trecate, and M. Morari, "Observability and controllability of piecewise affine and hybrid systems," *IEEE Trans. Automatic Control*, vol. 45, no. 10, pp. 1864–1876, 2000.

[6] D. P. Bertsekas, *Nonlinear Programming*, 2nd ed., Athena Scientific, 1999.

[7] M. S. Branicky, V. S. Borkar, and S. K. Mitter, "A unified framework for hybrid control: Model and optimal control theory," *IEEE Trans. Automatic Control*, vol. 43, no. 1, pp. 31–45, 1998.

[8] R. W. Brockett and D. Liberzon, "Quantized feedback stabilization of linear systems," *IEEE Trans. Automatic Control*, vol. 45, no. 7, pp. 1279–1289, 2000.

[9] A. Chotinan and B. H. Krogh, "Verification of infinite-state dynamic systems using approximate quotient transition systems," *IEEE Trans. Automatic Control*, vol. 46, no. 9, pp. 1401–1410, 2001.

[10] R. A. Decarlo, M. S. Branicky, S. Pettersson, and B. Lennartson, "Perspectives and results on the stability and stabilizability of hybrid systems," in [2], pp. 1069–1082, July 2000.

[11] D. F. Delchamps, "Stabilizing a linear system with quantized state feedback," *IEEE Trans. Automatic Control*, vol. 35, no. 8, pp. 916–924, 1990.

[12] N. Elia and S. K. Mitter, "Stabilization of linear systems with limited information," *IEEE Trans. Automatic Control*, vol. 46, no. 9, pp. 1384–1400, 2001.

[13] J. Hespanha, D. Liberzon, A. S. Morse, B. D. O. Anderson, T. S. Brinsmead, and F. De Bruyne, "Multiple model adaptive control. Part 2: Switching," *Int. J. Robust and Nonlinear Control*, vol. 11, no. 5, pp. 479–496, 2001.

[14] J. P. Hespanha, "Computation of root-mean-square gains of switched linear systems," *Proc. 5th International Workshop on Hybrid Systems: Computation and Control*, pp. 239–252, Stanford, CA, Mar. 2002.

[15] C. Horn and P. J. Ramadge, "Robustness issues for hybrid systems," in *Proc. 34th IEEE Conf. Decision and Control*, pp. 1467–1472, New Orleans, LA, 1995.

[16] M. Johansson, *Piecewise Linear Control Systems*, Ph.D. Thesis, Lund Institute of Technology, Sweden, 1999.

[17] U. T. Jönsson, "On reachability analysis of uncertain hybrid systems," in *Proc. 41th IEEE Conf. Decision and Control*, pp. 2397–2402, Las Vegas, NV, 2002.

[18] X. D. Koutsoukos and P. J. Antsaklis, "Hierarchical control of piecewise linear hybrid dynamical systems based on discrete abstractions," *ISIS Technical Report*, ISIS-2001-001, Feb. 2001.

[19] D. Liberzon and A. S. Morse, "Basic problems in stability and design of switched systems," *IEEE Control Systems Magazine*, vol. 19, no. 15, pp. 59–70, 1999.

[20] H. Lin, X. D. Koutsoukos, and P. J. Antsaklis, "HySTAR: A toolbox for hierarchical control of piecewise linear hybrid dynamical systems," in *Proc. 2002 American Control Conf.*, Anchorage, AK, 2002.

[21] H. Lin, X. D. Koutsoukos, and P. J. Antsaklis, "Hierarchical control of a class uncertain piecewise linear hybrid dynamical systems," *Proc. IFAC World Congress*, Barcelona, Spain, July 2002.

[22] H. Lin, and P. J. Antsaklis, "Controller synthesis for a class of uncertain piecewise linear hybrid dynamical systems," in *Proc. 41th IEEE Conf. Decision and Control*, Las Vegas, NV, 2002.

[23] J. Lygeros, C. Tomlin, and S. Sastry, "Controllers for reachability specifications for hybrid systems," *Automatica*, vol. 35, no. 3, pp. 349–370, 1999.

[24] T. Moor and J. M. Davoren, "Robust controller synthesis for hybrid systems using modal logic," HSCC 2001, pp. 433–446, 2001.

[25] T. Motzkin, *The Theory of Linear Inequalities*, Rand Corp., Santa Monica, CA, 1952.

[26] A. van der Schaft and H. Schumacher, *An Introduction to Hybrid Dynamical Systems*, Springer, New York, 2000.

[27] E. Sontag, "Remarks on piecewise-linear algebra," *Pacific Journal of Mathematics*, vol. 92, no. 1, pp. 183–210, 1982.

[28] E. Sontag, "Interconnected automata and linear systems: A theoretical framework in discrete-time," *Hybrid Systems III*, vol. 1066 of Lecture Notes in Computer Science, pp. 436–448. Springer, New York, 1996.

[29] G. Zhai, B. Hu, K. Yasuda, and A. N. Michel, "Disturbance attenuation properties of time controlled switched systems," *J. Franklin Institute*, vol. 338, pp. 765–779, 2001.

[30] G. Zhai, B. Hu, K. Yasuda, and A. N. Michel, "Qualitative analysis of discrete-time switched systems," *Proc. 2002 American Control Conf.*, vol. 3, pp. 1880–1885, Anchorage, AK, 2002.

Chapter 5

Stability Analysis of Swarms in a Noisy Environment*

Kevin M. Passino

Abstract: Bees swarm when moving to their new nest site. Some birds flock during migration and some fish forage in schools. Groups of robots can work together cooperatively to perform tasks that a single robot could not (e.g., moving a large object). Groups of uninhabited autonomous air vehicles are currently being developed for military operations. Suppose that for simplicity we refer to groups of such "agents" as "swarms." One key to the success of swarms is their ability to maintain cohesive behavior. There has been a significant amount of recent research activity on proving that cohesive group level behaviors (e.g., threat evasion, effective foraging) emerge from simple individual agent actions. Here we overview some of the research in this area and show how swarm cohesion can be characterized as a stability property and analyzed in a Lyapunov framework.

5.1 Introduction

Swarming has been studied extensively in biology [1, 2] and there is a significant relevant literature in physics where the collective behavior of "self-propelled particles" is studied. Swarms have also been studied in the context of engineering applications, particularly in collective robotics where there are teams of robots working together by communicating over a communication network [3, 4]. For example, in the work

*The author obtained financial support from the DARPA MICA program (Contract No. F33615-01-C-3151) and AFRL/VA and AFOSR for support of the Collaborative Center of Control Science (Grant F33615-01-2-3154). The author gratefully acknowledges the helpful suggestions from Mr. Yanfei Liu.

in [5], their "social potential functions" are related to how we view attraction repulsion forces. Special types of swarms have been studied in "intelligent vehicle highway systems" [6] and in "formation control" for robots, aircraft, and in cooperative control for uninhabited autonomous (air) vehicles. Early work on swarm stability is in [7, 8]. Later work where there is asynchronism and time delays appears in [9–12]. Also relevant is the study in [13] where the authors use virtual leaders and artificial potentials.

In this chapter we continue some earlier work by studying stability properties of foraging swarms. The main difference with previous work is that here we give a simple example of how to consider the effect of sensor noise and noise in sensing the gradient of a "resource profile" (e.g., a nutrient profile). It is shown that even with very noisy measurements, the swarm can achieve cohesion and follow a nutrient profile in the proper direction. In essence, the agents can forage in noisy environments more efficiently as a group than individually, a principle that has been found for some organisms [14]. So, in particular, the work here builds on the work in (i) [15, 16], where the authors provide a class of attraction/repulsion functions and provide conditions for swarm stability (size of ultimate swarm size and ultimate behavior), and (ii) [17–19] that represents progress in the direction of combining the study of aggregating swarms and how during this process decisions about foraging or threat avoidance can affect the collective/individual motion of the swarm and swarm members (i.e., typical characteristics influencing social foraging). Additional work on gradient climbing by swarms, including work on climbing noisy gradients, is in [20, 21]. There, similar to [17, 18], the authors study climbing gradients but also consider noise effects and coordination strategies for climbing. Here we study noise effects but do not constrain the agents to be in a particular formation, nor do we control aspects of that formation.

5.2 Swarm and Environment Models

Here we describe the agent, communications, and environment that the agents move in.

5.2.1 Agent dynamics and communications

Rather than focusing on the particular characteristics of one type of animal or autonomous vehicle, we consider a swarm to be composed of an interconnection of N "agents" each of which has point mass dynamics given by

$$\dot{x}^i = v^i, \quad \dot{v}^i = \frac{1}{M_i} u^i \tag{5.2.1}$$

where $x^i \in \Re^n$ is the position, $v^i \in \Re^n$ is the velocity, M_i is the mass, and $u^i \in \Re^n$ is the (force) control input for the ith agent. We use this simple linear model for an agent to illustrate the basic features of swarming and stability analysis of cohesion. Other nonlinear and stochastic models for agents in swarms have been considered (see the literature referenced in the Introduction). Here we will use $n = 3$ for swarms moving in a three-dimensional space. We will assume that each agent can sense information about the position and velocity of other agents, but only with some noise that we will define below.

The agents interact to form groups, and in some situations groups will split. Some think of having local interactions between agents, which "emerge" into a global behavior for the group. One way to represent which agents can interact with each other is via a directed graph (G, A) where $G = \{1, 2, \cdots, N\}$ is a set of nodes (the agents) and $A = \{(i, j) \colon i, j \in G, i \neq j\}$ represents a "sensing/communication topology" (in general, then, each "link" (i, j) could be a dynamical system). For example, if $(i, j) \in A$, then it could be assumed that agent i can sense the position and velocity of agent j. In some vehicular systems it may be possible that A is fixed and independent of vehicle positions and velocities. Clearly, however, for biological systems it is often the case that A is not fixed, but changes dynamically based on the positions of the agents (e.g., so that only agents within a line-of-sight and close enough can be sensed). Here, for simplicity we will assume that A does not change based on the positions and velocities of the agents, and that A is fully connected (e.g., that for all $i \in G$, $(i, j) \in A$). We also assume that there are no delays in communicating, communications are not range constrained, and that there is infinite bandwidth. More general formulations may also include a "communications topology" that specifies which agents can send and receive messages to other agents. Such messages could represent a wide variety of communication capabilities of the agents (e.g., bee scout leadership in a swarm, sounds).

5.2.2 Agent to agent attraction and repulsion

Agent to agent interactions considered here are of the "attract-repel" type where each agent seeks to be in a position that is "comfortable" relative to its neighbors (and for us all other agents are its neighbors). Attraction indicates that each agent wants to be close to every other agent and provides the mechanism for achieving grouping and cohesion of the group of agents. Repulsion provides the mechanism where each agent does not want to be too close to every other agent (e.g., for animals to avoid collisions and excessive competition for resources). There are many ways to define attraction and repulsion, each of which can be represented by characteristics of how we define u^i for each agent, and we list a few of these below:

- *Attraction:* There can be linear attraction and in this case we have terms in u^i of, for example, the form $-k_a \left(x^i - x^j\right)$, where $k_a > 0$ is a scalar that represents the strength of attraction. If the agents are far apart, then there is a large attraction between them, and if they are close there is a small attraction. In other cases there can be nonlinear attraction terms that can be expressed in terms of nonlinear functions of $\left(x^i - x^j\right)$. Some attraction mechanisms are "local" (i.e., for range-constrained sensing where the agent only tries to move to other agents that are close to it) and others that are "global" (i.e., where agents can be attracted to move near other agents no matter how far away they are). Attraction terms can be specified in terms of a variety of agent variables. For example, the above term is for positions, but we could have similar terms for velocity so that the agents will try to match the velocities of other agents. Moreover, the above term is "static" but we would have a local "dynamic" controller that tries to match agent variables.

- *Repulsion:* As with attraction there are many types of repulsion terms, some local, global, static, or dynamic, each of which can be expressed in terms of a variety of agent variables. Sometimes repulsion is defined in terms that also quantify attraction, other times they quantify only a repulsion.

  - *Seek a "comfortable distance":* For example, a term in u^i may take the form
$$\left[-k\left(||x^i - x^j|| - d\right)\right]\left(x^i - x^j\right),$$
where
$$||x^i - x^j|| = \sqrt{\left(x^i - x^j\right)^\top \left(x^i - x^j\right)},$$
$k > 0$ is the magnitude of the repulsion, and d can be thought of as a comfortable distance between the ith and jth agents. Here the quantity in the brackets sets the size of the repulsion. When $||x^i - x^j||$ is small (relative to d) the term in the bracket is positive so that agents i and j try to move away from each other (there is repulsion). When $||x^i - x^j||$ is big (relative to d) then the term in the brackets is negative so that the agents are attracted to each other. Balance between attraction and repulsion (a basic concept in swarm dynamics that is sometimes referred to as an "equilibrium" even though it may not be one in the stability-theoretic sense) may be achieved when $||x^i - x^j|| = d$ so that the term above is zero.

  - *Repel when close:* Another type of repulsion term in u^i that may be used with, for example, a linear attraction term, may take the form
$$k_r \exp\left(\frac{-\frac{1}{2}||x^i - x^j||^2}{r_s^2}\right)\left(x^i - x^j\right) \qquad (5.2.2)$$
where $k_r > 0$ is the magnitude of the repulsion, and $r_s > 0$ quantifies the region size around the agent from which it will repel its neighbors. When $||x^i - x^j||$ is big relative to r_s the whole term approaches zero.

  - *Hard repulsion for collision avoidance:* The above repulsion terms are "soft" in the sense that when the two agents are at the same location the repulsion force is finite. In some cases it is appropriate to use a repulsion term that becomes increasingly large as two agents approach each other. One such term, which does not have an attraction component, is in the form of
$$\left[\max\left\{\left(\frac{a}{b||x^i - x^j|| - w} - \epsilon\right), 0\right\}\right]\left(x^i - x^j\right). \qquad (5.2.3)$$
Here, $w > 0$ affects the radius of repulsion of the agents, $a > 0$ is a gain on the magnitude of the repulsion, $b > 0$ can change the shape of the

repulsion gain, and $\epsilon > 0$ can be used to define the term so that it only has a local influence. For instance, the radius of the repulsion is

$$R' = \left(\frac{a}{\epsilon} + w\right) / b.$$

For agent positions such that $||x^i - x^j|| \geq R'$ the term in the brackets is zero. For given values of w, a, and b you can choose ϵ to get any value of $R' > 0$. Note that a key feature of this repulsion term is that as $||x^i - x^j||$ goes from a large value to w/b, the value in the brackets goes to infinity to provide a "hard" repelling action and hence avoid the possibility that two agents ever end up at the same position (i.e., to avoid collisions). Another way to define a hard repulsion is to consider agents to be solid "balls" where if they collide they do not deform.

For more ideas on how to define attraction and repulsion terms see the literature cited in the Introduction.

5.2.3 Environment and foraging

Next we will define the environment the agents move in. While there are many possibilities, here we will simply consider the case where they move over what we will call the "resource profile" (e.g., nutrient profile) $J(x)$, where $x \in \Re^n$. We will, however, think of this profile as being something where the agents want to be in certain regions of the profile and avoid other regions (e.g., where there are noxious substances). We will assume that $J(x)$ is continuous with finite slope at all points. Agents move in the direction of the negative gradient of $J(x)$

$$-\nabla J(x) = -\frac{\partial J}{\partial x}$$

in order to move away from bad areas and into good areas of the environment (e.g., to avoid noxious substances and find nutrients). Hence, they will use a term in their u^i that holds the negative gradient of $J(x)$.

Clearly there are many possible shapes for $J(x)$, including ones with many peaks and valleys. Here, we list two simple forms for $J(x)$ as follows:

- *Plane:* In this case we have $J(x) = J_p(x)$ where

$$J_p(x) = R^\top x + r_o,$$

  and where $R \in \Re^n$ and r_o is a scalar. Here, $\nabla J_p(x) = R$.

- *Quadratic:* In this case we have $J(x) = J_q(x)$,

$$J_q(x) = \frac{r_m}{2}||x - R_c||^2 + r_o,$$

  where r_m and r_o are scalars, $R_c \in \Re^n$, and $\nabla J_q(x) = r_m\,(x - R_c)$.

We will assume below that each agent can sense the gradient of the resource profile, but only with some noise.

It could be that the environment has many different types of agents in it, or the same types of agents with different objectives. In this case there may be different resource profiles for each agent, or the agents may switch the profiles they follow, or strategies for following them. Different agents may have different capabilities to sense the profile (e.g., sensing only at a point, or sensing a range-constrained region) and move over it. If the agents are consuming food this may change the shape of the profile, and of course an agent may pollute its environment so it may affect the profile in that manner also. This would create a time-varying profile that is dependent on agent position, and possibly many other inputs.

5.3 Stability Analysis of Swarm Cohesion Properties

Cohesion and swarm dynamics can be quantified and analyzed using stability analysis (e.g., via Lyapunov's method). You can pick agent dynamics, interactions, sensing capabilities, attraction/repulsion characteristics, and foraging environment characteristics, then quantify and analyze cohesion properties. Here we will do this for a few simple cases to give a flavor of the type of analysis that is possible and to provide insight into swarm properties and dynamics during social foraging.

5.3.1 Sensing, noise, and error dynamics

First, let

$$\bar{x} = \frac{1}{N}\sum_{i=1}^{N} x^i$$

be the center of the swarm and

$$\bar{v} = \frac{1}{N}\sum_{i=1}^{N} v^i$$

be the average velocity (vector), which we view as the velocity of the group of agents. We assume that each agent can sense the distance from itself to $\bar{x}$, and the difference between its own velocity and $\bar{v}$. We assume that each agent knows its own velocity, but not its own position. Note that for some animals, senses and sensory processing may naturally provide the distance to $\bar{x}$ and $\bar{v}$, but not all the individual positions and velocities of all agents that could be used to compute these. Also, we will consider below the case where there is noise in sensing these quantities.

The objective of each agent is to move so as to end up at or near $\bar{x}$ and have its velocity equal to $\bar{v}$. In this way an emergent behavior of the group is produced where they aggregate dynamically and end up near each other and ultimately move in the same direction (i.e., they achieve cohesion). The problem is that since all the agents are moving at the same time, $\bar{x}$ and $\bar{v}$ are generally time varying; hence, in order to study the stability of swarm cohesion we study the dynamics of an error system with $e_p^i = x^i - \bar{x}$ and $e_v^i = v^i - \bar{v}$. Other choices for error systems are also possible and have been used in some studies of swarm stability. For instance, you could use

$$\tilde{e}_p^i = \sum_{j=1}^{N} \left(x^i - x^j\right).$$

This corresponds to computing the errors to each other agent and then trying to get all those errors to go to zero. Note, however, that

$$\tilde{e}_p^i = N\left(x^i - \frac{1}{N}\sum_{j=1}^{N} x^j\right) = N\left(x^i - \bar{x}\right) = N e_p^i,$$

a scaled version of what we consider here. The same relationship holds for velocity.

Given the above choices, the error dynamics are given by

$$\dot{e}_p^i = e_v^i, \quad \dot{e}_v^i = \frac{1}{M_i}u^i - \frac{1}{N}\sum_{j=1}^{N}\frac{1}{M_j}u^j. \tag{5.3.1}$$

The challenge is to specify the u^i so that we get good cohesion properties and successful social foraging.

Assume that each agent can sense its position and velocity relative to $\bar{x}$ and $\bar{v}$, but with some bounded errors. In particular, let $d_p^i(t) \in \Re^n$, $d_v^i(t) \in \Re^n$ be these errors for agent i, respectively. We assume that $d_p^i(t)$ and $d_v^i(t)$ are sufficiently smooth and are independent of the state of the system. Each agent will try to follow the resource profile J_p defined earlier (we use the plane profile for the sake of illustration as it will show how swarm dynamics operate over a simple but representative surface). We assume that each agent senses the gradient of J_p, but with some sufficiently smooth error $d_f^i(t) \in \Re^n$, so it senses $\nabla J_p\left(x^i\right) - d_f^i$. You may think of this as either a sensing error or as variations (e.g., high-frequency ripples) on the resource profile. Below we will refer to the signals $d_p^i(t)$, $d_v^i(t)$, and $d_f^i(t)$ as "noise" signals, but clearly there is no underlying probability space and all signals in this section are deterministic. You may think of these signals as being generated by, for example, a chaotic dynamic system.

We assume that all the sensing errors are bounded such that

$$\|d_p^i\| \le D_p, \quad \|d_v^i\| \le D_v, \quad \|d_f^i\| \le D_f,$$

where $D_p > 0$, $D_v > 0$, and $D_f > 0$ are known constants. Related results to what we find below can be found for the more general case where we have

$$\|d_p^i\| \le D_{p_1}\left\|E^i\right\| + D_{p_2}$$

$$\|d_v^i\| \le D_{v_1}\left\|E^i\right\| + D_{v_2}$$

and $\|d_f^i\| \le D_f$ where D_{p_1}, D_{p_2}, D_{v_1} and D_{v_2} are known positive constants.

Thus each agent can sense noise-corrupted versions of e_p^i and e_v^i as

$$\hat{e}_p^i = e_p^i - d_p^i$$
$$\hat{e}_v^i = e_v^i - d_v^i.$$

Also, each agent can sense $\nabla J_p\left(x^i\right) - d_f^i$ at the location x^i where the agent is located.

Suppose that in order to steer itself each agent uses

$$\begin{aligned} u^i &= -M_i k_a \hat{e}_p^i - M_i k_a \hat{e}_v^i - M_i k_v v^i \\ &\quad + M_i k_r \sum_{j=1, j \neq i}^{N} \exp\left(\frac{-\frac{1}{2}\|\hat{e}_p^i - \hat{e}_p^j\|^2}{r_s^2}\right)\left(\hat{e}_p^i - \hat{e}_p^j\right) \\ &\quad - M_i k_f\left(\nabla J_p\left(x^i\right) - d_f^i\right). \end{aligned} \tag{5.3.2}$$

Here we assume that each agent knows its own mass M_i and velocity v^i. We think of the scalar $k_a > 0$ as the "attraction gain" that indicates how aggressive the agents are in aggregating. The gain k_r is a "repulsion gain" which sets how much the agents want to be away from each other. Note that use of the repulsion term assumes that the ith agent knows, within some errors, the relative distance of all other agents from the swarm center. Also, since

$$\begin{aligned} \hat{e}_p^i - \hat{e}_p^j &= \left(\left(x^i - \bar{x}\right) - d_p^i\right) - \left(\left(x^j - \bar{x}\right) - d_p^j\right) \\ &= \left(x^i - x^j\right) - \left(d_p^i - d_p^j\right), \end{aligned}$$

we are assuming that the ith agent knows its position (and velocity) relative to each other agent within some bounded errors. Also note that if $D_p = D_v = 0$, there is no sensing error on attraction and repulsion, thus,

$$\hat{e}_p^i = e_p^i, \; \hat{e}_v^i = e_v^i,$$

and

$$e_p^i - e_p^j = x^i - x^j,$$

and we get a repulsion term of the form explained in (5.2.2). The sensing errors create the possibility that agents will try to move away from each other when they may not really need to, and they may move toward each other when they should not. Clearly this complicates the ability of the agents to avoid collisions with their neighbors. The last term in (5.3.2) indicates that each agent wants to move along the negative gradient of the resource profile with the gain k_f proportional to the agents' desire to follow the profile.

5.3.2 Groups can increase foraging effectiveness

Next we will substitute this choice for u^i into the error dynamics described in (5.3.1) and study their stability properties. First, however, we will study how the group can follow the resource profile in the presence of noise. To do this, consider $\dot{e}_v^i = \dot{v}^i - \dot{\bar{v}}$. First note that

$$\begin{aligned}
\dot{\bar{v}} &= \frac{1}{N}\sum_{i=1}^{N}\frac{1}{M_i}u^i \\
&= -\frac{k_a}{N}\sum_{i=1}^{N}(\hat{e}_p^i + \hat{e}_v^i) - \frac{k_v}{N}\sum_{i=1}^{N}v^i \\
&\quad + \frac{k_r}{N}\sum_{i=1}^{N}\sum_{j=1, j\neq i}^{N}\exp\left(\frac{-\frac{1}{2}\|\hat{e}_p^i - \hat{e}_p^j\|^2}{r_s^2}\right)(\hat{e}_p^i - \hat{e}_p^j) \\
&\quad - \frac{k_f}{N}\sum_{i=1}^{N}(R - d_f^i).
\end{aligned} \tag{5.3.3}$$

Notice that

$$\frac{1}{N}\sum_{i=1}^{N}\hat{e}_p^i = \frac{1}{N}\sum_{i=1}^{N}((x^i - \bar{x}) - d_p^i) = \bar{x} - \frac{1}{N}N\bar{x} - \frac{1}{N}\sum_{i=1}^{N}d_p^i = -\frac{1}{N}\sum_{i=1}^{N}d_p^i.$$

Also, the term due to repulsion in (5.3.3) is zero as we show next. Note that

$$\begin{aligned}
&\sum_{i=1}^{N}\sum_{j=1, j\neq i}^{N}\exp\left(\frac{-\frac{1}{2}\|\hat{e}_p^i - \hat{e}_p^j\|^2}{r_s^2}\right)(\hat{e}_p^i - \hat{e}_p^j) = \\
&\qquad \left[\sum_{i=1}^{N}\hat{e}_p^i\sum_{j=1, j\neq i}^{N}\exp\left(\frac{-\frac{1}{2}\|\hat{e}_p^i - \hat{e}_p^j\|^2}{r_s^2}\right)\right] \\
&\qquad - \left[\sum_{i=1}^{N}\sum_{j=1, j\neq i}^{N}\exp\left(\frac{-\frac{1}{2}\|\hat{e}_p^i - \hat{e}_p^j\|^2}{r_s^2}\right)\hat{e}_p^j\right].
\end{aligned} \tag{5.3.4}$$

The last term in (5.3.4)

$$\sum_{i=1}^{N}\sum_{j=1, j\neq i}^{N}\exp\left(\frac{-\frac{1}{2}\|\hat{e}_p^i - \hat{e}_p^j\|^2}{r_s^2}\right)\hat{e}_p^j = \sum_{j=1}^{N}\sum_{i=1, i\neq j}^{N}\exp\left(\frac{-\frac{1}{2}\|\hat{e}_p^i - \hat{e}_p^j\|^2}{r_s^2}\right)\hat{e}_p^j$$

and since

$$\exp\left(\frac{-\frac{1}{2}\|\hat{e}_p^i - \hat{e}_p^j\|^2}{r_s^2}\right) = \exp\left(\frac{-\frac{1}{2}\|\hat{e}_p^j - \hat{e}_p^i\|^2}{r_s^2}\right)$$

we have

$$\sum_{j=1}^{N}\sum_{i=1, i\neq j}^{N}\exp\left(\frac{-\frac{1}{2}\|\hat{e}_p^i - \hat{e}_p^j\|^2}{r_s^2}\right)\hat{e}_p^j = \sum_{j=1}^{N}\hat{e}_p^j\sum_{i=1, i\neq j}^{N}\exp\left(\frac{-\frac{1}{2}\|\hat{e}_p^j - \hat{e}_p^i\|^2}{r_s^2}\right)$$

but this last value is the same as the first term in (5.3.4). So overall its value is zero. This gives us

$$\dot{\bar{v}} = \frac{k_a}{N}\sum_{i=1}^{N} d_p^i + \frac{k_a}{N}\sum_{i=1}^{N} d_v^i + \frac{k_f}{N}\sum_{i=1}^{N} d_f^i - k_v\bar{v} - k_f R.$$

Letting $\bar{d_p}(t) = \frac{1}{N}\sum_{i=1}^{N} d_p^i(t)$ and similarly for $\bar{d_v}(t)$ and $\bar{d_f}(t)$ we get

$$\dot{\bar{v}} = -k_v\bar{v} + \underbrace{k_a\bar{d_p} + k_a\bar{d_v} + k_f\bar{d_f} - k_f R}_{z(t)}. \tag{5.3.5}$$

This is an exponentially stable system with a time-varying but bounded input $z(t)$ so we know that $\bar{v}(t)$ is bounded. To see this, choose a Lyapunov function $V_{\bar{v}} = \frac{1}{2}\bar{v}^\top\bar{v}$ defined on $D = \{\bar{v} \in \Re^n \mid ||\bar{v}|| < r_v\}$ for some $r_v > 0$, and we have

$$\dot{V}_{\bar{v}} = \bar{v}^\top\dot{\bar{v}} = -k_v\bar{v}^\top\bar{v} + z(t)^\top\bar{v}$$

with

$$\left\|\frac{\partial V_{\bar{v}}}{\partial \bar{v}}\right\| = ||\bar{v}||.$$

Note that $||z(t)|| \le \left\|k_a\bar{d_p}\right\| + \left\|k_a\bar{d_v}\right\| + \left\|k_f\bar{d_f}\right\| + ||k_f R|| \le \delta$, where $\delta = k_a D_p + k_a D_v + k_f D_f + k_f||R||$. If $\delta < k_v\theta r_v$ for all $t \ge 0$, for some positive constant $\theta < 1$, and all $\bar{v} \in D$, then it can be proven that for all $||\bar{v}(0)|| < r_v$, and some finite T we have

$$||\bar{v}(t)|| \le \exp\left[-(1-\theta)k_v t\right]||\bar{v}(0)||, \ \forall\, 0 \le t < T$$

and

$$||\bar{v}(t)|| \le \frac{\delta}{k_v\theta}, \ \forall\, t \ge T.$$

Since this holds globally we can take $r_v \to \infty$ so these inequalities hold for all $\bar{v}(0)$. If δ and θ are fixed, with increasing k_v we get that $||\bar{v}(t)||$ decreases faster for $0 \le t < T$ and smaller bound on $||\bar{v}(t)||$ for $t \ge T$. If δ gets larger with k_v and θ fixed, $||\bar{v}(t)||$ has larger bound for $t \ge T$; hence if the magnitude of the noise increases, this increases δ and hence there can be larger magnitude changes in the ultimate average velocity of the swarm (e.g., the average velocity could oscillate). Note that if in (5.3.5) $z(t) \approx 0$ (e.g., due to noise that destroys the directionality of the resource profile R), then the above bound may be reduced but the swarm could be going in the wrong direction.

Regardless of the size of the bound, it is interesting to note that while the noise can destroy the ability of an individual agent to follow a gradient accurately, the average sensing errors of the group are what changes the direction of the group's movement relative to the direction of the gradient of $J_p(x)$. In some cases when the swarm is large (N big) it can be that $\bar{d}_p \approx \bar{d}_v \approx \bar{d}_f \approx 0$ to give a zero average sensing error and the group will perfectly follow the proper direction for foraging (this may be a reason why for some organisms, large group size is favorable). In the case when $N = 1$ (i.e., single agent), there is no opportunity for a cancellation of the sensor errors. Hence an individual may not be able to climb a noisy gradient as easily as a group, and in some cases a group may be able to follow a profile where an individual cannot. This characteristic has been found in biological swarms [14]. From an optimization perspective you should think of an individual trying to execute a gradient optimization method that we know can result in it getting stuck in local minima. The group is producing a type of approximation to the gradient by a larger spatial sampling and attraction/repulsion terms. Intuitively, it filters out the noise and moves in the proper direction. Of course the group itself can get stuck in a local minimum if the basin of attraction of that minimum is large.

It is also important to note that there is an intimate relationship between sensor noise and observations of biological swarms (e.g., in bee swarms [22]) where there is a type of "inertia" of a swarm. Note that for large swarms (high N) there can be regions where the average sensor noise is small so that agents in that region move in the right direction. In other regions there may be alignments of the errors and hence the agents may not be all moving in the right direction so they may get close to each other and impede each other's motion, having the effect of slowing down the whole group. With no noise, the group inertia effect is not found since each agent is moving in the right direction. The presence of sensor noise generally can make it more difficult to get the group moving in the right direction (e.g., for foraging, migration, or movement to a nest site). Large swarms can help move the group in the right direction, but at the expense of possibly slowing their movement initially in a transient period.

5.3.3 Cohesive social foraging in noise

Next we return to the problem of finding the error dynamics and then stability analysis by considering the $\dot{v}^i$ term of $\dot{e}_v^i = \dot{v}^i - \dot{\bar{v}}$ in the error dynamics of (5.3.1). Note that

$$\begin{aligned}
\dot{v}^i &= \frac{1}{M_i} u^i = -k_a \hat{e}_p^i - k_a \hat{e}_v^i - k_v v^i \\
&\qquad + k_r \sum_{j=1, j\neq i}^{N} \exp\left(\frac{-\frac{1}{2}\|\hat{e}_p^i - \hat{e}_p^j\|^2}{r_s^2} \right) (\hat{e}_p^i - \hat{e}_p^j) \\
&\qquad - k_f \left(\nabla J_p (x^i) - d_f^i \right) \\
&= -k_a e_p^i + k_a d_p^i - k_a e_v^i + k_a d_v^i - k_v v^i \\
&\quad + k_r \sum_{j=1, j\neq i}^{N} \exp\left(\frac{-\frac{1}{2}\|\hat{e}_p^i - \hat{e}_p^j\|^2}{r_s^2} \right) (\hat{e}_p^i - \hat{e}_p^j) - k_f \left(R - d_f^i \right).
\end{aligned}$$

Hence, we have

$$\begin{aligned}\dot{e}_v^i = \dot{v}^i - \dot{\bar{v}} &= -k_a e_p^i - k_a e_v^i - k_v e_v^i + k_a\left(d_p^i - \bar{d}_p\right) + k_a\left(d_v^i - \bar{d}_v\right)\\ &+k_r \sum_{j=1, j\neq i}^{N} \exp\left(\frac{-\frac{1}{2}\|\left(x^i - x^j\right) - \left(d_p^i - d_p^j\right)\|^2}{r_s^2}\right)\left(\left(x^i - x^j\right)\right.\\ &\left.- \left(d_p^i - d_p^j\right)\right) + k_f\left(d_f^i - \bar{d}_f\right).\end{aligned}$$

To study the stability of the error dynamics, and hence swarm xyz cohesiveness, define $E^i = \left[{e_p^i}^\top, {e_v^i}^\top\right]^\top$ and $E = \left[{E^1}^\top, {E^2}^\top, \cdots, {E^N}^\top\right]^\top$, and choose a Lyapunov function $V(E) = \sum_{i=1}^{N} V_i\left(E^i\right)$, where $V_i\left(E^i\right) = {E^i}^\top P E^i$ with $P = P^\top$ and $P > 0$ (a positive-definite matrix). We know that

$$\lambda_{\min}(P){E^i}^\top E^i \leq {E^i}^\top P E^i \leq \lambda_{\max}(P){E^i}^\top E^i.$$

Notice that with I being an $n \times n$ identity matrix, we have

$$\dot{E}^i = \underbrace{\begin{bmatrix} 0 & I \\ -k_a I & -\left(k_a + k_v\right) I \end{bmatrix}}_{A} E^i + \underbrace{\begin{bmatrix} 0 \\ I \end{bmatrix}}_{B} g^i(E)$$

where

$$\begin{aligned}g^i(E) &= k_a\left(d_p^i - \bar{d}_p\right) + k_a\left(d_v^i - \bar{d}_v\right)\\ &+k_r \sum_{j=1, j\neq i}^{N} \exp\left(\frac{-\frac{1}{2}\|\hat{e}_p^i - \hat{e}_p^j\|^2}{r_s^2}\right)\left(\hat{e}_p^i - \hat{e}_p^j\right)\\ &+k_f\left(d_f^i - \bar{d}_f\right).\end{aligned} \tag{5.3.6}$$

Note that any matrix

$$\begin{bmatrix} 0 & I \\ -k_1 I & -k_2 I \end{bmatrix}$$

with $k_1 > 0$ and $k_2 > 0$ has eigenvalues given by the roots of $\left(s^2 + k_2 s + k_1\right)^n$, which are in the strict left half plane. Since $k_a > 0$ and $k_v > 0$, the matrix A above is Hurwitz (i.e., has eigenvalues all in the strict left half plane).

We have

$$\dot{V}_i = {E^i}^\top P \dot{E}^i + {\dot{E}^i}{}^\top P E^i = {E^i}^\top \underbrace{\left(PA + A^\top P\right)}_{-Q} E^i + 2{E^i}^\top P B g^i(E). \tag{5.3.7}$$

Note that if Q, defined in this manner, is such that $Q = Q^\top$ and $Q > 0$, then the unique solution P of $PA + A^\top P = -Q$ has $P = P^\top$ and $P > 0$ as needed.

Also, since $\|B\| = 1$, ${E^i}^\top Q E^i \geq \lambda_{\min}(Q){E^i}^\top E^i$, and $\|P\| = \lambda_{\max}(P)$ with $P = P^\top > 0$, we have

$$\begin{aligned}\dot{V}_i &\leq -\lambda_{\min}(Q)\left\|E^i\right\|^2 + 2\left\|E^i\right\|\lambda_{\max}(P)\|g^i(E)\| \\ &= -\lambda_{\min}(Q)\left(\left\|E^i\right\| - \frac{2\lambda_{\max}(P)}{\lambda_{\min}(Q)}\|g^i(E)\|\right)\left\|E^i\right\|. \qquad (5.3.8)\end{aligned}$$

Suppose for a moment that for each $i = 1, 2, \cdots, N$, $\|g^i(E)\| < \beta$ for some known β. Then, if

$$\left\|E^i\right\| > \frac{2\lambda_{\max}(P)}{\lambda_{\min}(Q)}\|g^i(E)\| \qquad (5.3.9)$$

we have that $\dot{V}_i < 0$. By choosing $Q = k_q I$, for any $k_q > 0$, we minimize [23]

$$\frac{\lambda_{\max}(P)}{\lambda_{\min}(Q)}.$$

Here we later consider some choices for k_q. The set

$$\Omega_b = \left\{E : \left\|E^i\right\| \leq 2\frac{\lambda_{\max}(P)}{\lambda_{\min}(Q)}\|g^i(E)\|,\ i = 1, 2, \cdots, N\right\} \qquad (5.3.10)$$

is attractive and compact. Also we know that within a finite amount of time, $E^i \to \Omega_b$. This means that we can guarantee that if the swarm is not cohesive, it will seek to be cohesive, but this can only be guaranteed if it is a certain distance from cohesiveness as indicated by (5.3.9).

It remains to show that for each i, $\|g^i(E)\| < \beta$ for some β. Note that

$$\begin{aligned}\|g^i(E)\| &\leq k_a\|d_p^i - \bar{d}_p\| + k_a\|d_v^i - \bar{d}_v\| + k_f\|d_f^i - \bar{d}_f\| \\ &+ k_r \sum_{j=1, j\neq i}^{N} \exp\left(\frac{-\frac{1}{2}\|\psi\|^2}{r_s^2}\right)\|\psi\| \qquad (5.3.11)\end{aligned}$$

where $\psi = \hat{e}_p^i - \hat{e}_p^j = \left(x^i - x^j\right) - \left(d_p^i - d_p^j\right)$. Notice that $\frac{1}{N}\sum_{j=1}^{N}\|d_p^j\| \leq D_p$ since $\|d_p^j\| \leq D_p$. Also

$$d_p^i - \frac{1}{N}\sum_{j=1}^{N} d_p^j \leq \|d_p^i\| + \frac{1}{N}\left\|\sum_{j=1}^{N} d_p^j\right\| \leq \|d_p^i\| + \frac{1}{N}\sum_{j=1}^{N}\|d_p^j\|$$

$\|d_p^i - \bar{d}_p\| \leq 2D_p$, $\|d_v^i - \bar{d}_v\| \leq 2D_v$, and $\|d_f^i - \bar{d}_f\| \leq 2D_f$.

For the last term in (5.3.11), note that as $\|x^i - x^j\|$ becomes large for all i and j, the agents are all far from each other and the repulsion term goes to zero. Also, the term due to the repulsion is bounded with a unique maximum point [15]. To find this point note that

$$\frac{\partial}{\partial\|\psi\|}\left(\|\psi\|\exp\left(\frac{-\frac{1}{2}\|\psi\|^2}{r_s^2}\right)\right) = \exp\left(\frac{-\frac{1}{2}\|\psi\|^2}{r_s^2}\right) - \frac{\|\psi\|^2}{r_s^2}\exp\left(\frac{-\frac{1}{2}\|\psi\|^2}{r_s^2}\right).$$

The maximum point occurs at a point such that

$$1 - \frac{\|\psi\|^2}{r_s^2} = 0$$

or when $\|\psi\| = r_s$. Hence, we have

$$\begin{aligned}\|g^i(E)\| &\leq 2k_a\left(D_p + D_v\right) + 2k_f D_f + k_r \sum_{j=1, j\neq i}^{N} \exp\left(-\frac{1}{2}\right) r_s \\ &= 2k_a\left(D_p + D_v\right) + 2k_f D_f + k_r r_s (N-1)\exp\left(-\frac{1}{2}\right) = \beta.\end{aligned}$$

If you substitute this value for β into (5.3.10) you get the set Ω_b which ultimately all the trajectories will end up in.

5.3.4 Cohesive social foraging with no noise: Optimization

When there is no noise, tighter bounds and stronger results can be obtained. First we can eliminate the effect of P via $\lambda_{\max}(P)$ on the bound for the no-noise case. Assume there is no sensor noise so $D_p = D_v = D_f = 0$. Choose

$$\begin{aligned}u^i &= -M_i k_a e_p^i - M_i k_a e_v^i - M_i k_v v^i \\ &+ M_i k_r\left(B^\top P^{-1} B\right) \sum_{j=1, j\neq i}^{N} \exp\left(\frac{-\frac{1}{2}\|e_p^i - e_p^j\|^2}{r_s^2}\right)\left(e_p^i - e_p^j\right) \\ &- M_i k_f R \end{aligned} \tag{5.3.12}$$

where $P = P^\top$, $P > 0$ was defined earlier, so P^{-1} exists. Also

$$\begin{aligned}\dot{V}_i &\leq -\lambda_{\min}(Q)\left\|E^i\right\|^2 \\ &+ 2{E^i}^\top PB\left(k_r B^\top P^{-1} B \sum_{j=1, j\neq i}^{N} \exp\left(\frac{-\frac{1}{2}\|e_p^i - e_p^j\|^2}{r_s^2}\right)\left(e_p^i - e_p^j\right)\right) \\ &= -\lambda_{\min}(Q)\left\|E^i\right\|^2 + 2k_r {E^i}^\top B \sum_{j=1, j\neq i}^{N} \exp\left(\frac{-\frac{1}{2}\|e_p^i - e_p^j\|^2}{r_s^2}\right)\left(e_p^i - e_p^j\right) \\ &\leq -\lambda_{\min}(Q)\left\|E^i\right\|^2 + 2k_r\left\|E^i\right\|(N-1)\exp\left(-\frac{1}{2}\right) r_s.\end{aligned}$$

So $\dot{V}_i < 0$ if $\left\|E^i\right\| > \frac{2k_r(N-1)r_s}{\lambda_{\min}(Q)}\exp\left(-\frac{1}{2}\right)$. Let

$$\Omega_b' = \left\{E : \left\|E^i\right\| \leq \frac{2k_r r_s(N-1)}{\lambda_{\min}(Q)}\exp\left(-\frac{1}{2}\right),\ i = 1, 2, \cdots, N\right\}.$$

Next note that in the set Ω_b, we have bounded e_p^i and e_v^i but we are not guaranteed that $e_v^i \to 0$ for any i. Achieving $e_v^i \to 0$ for all i would be a desirable property since this represents that $v^i = \bar{v}$ for all i so that the group will move cohesively in the same direction. To study this, consider Ω_b', and consider a Lyapunov function $V^o(E) = \sum_{i=1}^{N} V_i^o\left(E^i\right)$ with

$$V_i^o\left(E^i\right) = \frac{1}{2}k_a {e_p^i}^\top e_p^i + \frac{1}{2}k_a {e_v^i}^\top e_v^i + k_r r_s^2 \sum_{j=1,j\neq i}^{N} \exp\left(\frac{-\frac{1}{2}\|e_p^i - e_p^j\|^2}{r_s^2}\right).$$

Note that this Lyapunov function satisfies $V_i^o\left(E^i\right) \geq 0$. You should view the objective of the agents as being that of *minimizing* this Lyapunov function. The agents try to minimize the distance to the center of the swarm, match the average velocity of the group, and minimize the repulsion effect (to do that the agents move away from each other). We have

$$\begin{aligned} \nabla_{e_p^i} V_i^o &= k_a e_p^i - k_r \sum_{j=1,j\neq i}^{N} \exp\left(\frac{-\frac{1}{2}\|e_p^i - e_p^j\|^2}{r_s^2}\right)\left(e_p^i - e_p^j\right) \\ \nabla_{e_v^i} V_i^o &= e_v^i \end{aligned}$$

so $\dot{V}_i^o = \left(\nabla J\left(E^i\right)\right)^\top \dot{E}_i = -\left(k_a + k_v\right) {e_v^i}^\top e_v^i$ and

$$\dot{V}^o = -\left(k_a + k_v\right) \sum_{i=1}^{N} \|e_v^i\|^2 \leq 0$$

on $E \in \Omega$ for a compact set Ω. Choose Ω so it is positively invariant, which is clearly possible, and so $\Omega_e \in \Omega$ where

$$\Omega_e = \{E : \dot{V}^o(E) = 0\} = \{E : e_v^i = 0, i = 1, 2, \cdots, N\}.$$

From LaSalle's Invariance Principle we know that if $E(0) \in \Omega$ then $E(t)$ will converge to the largest invariant subset of Ω_e. Hence $e_v^i(t) \to 0$, as $t \to \infty$. When $R = 0$ (no resource profile effect), $\bar{v}(t) \to 0$ and hence $v^i(t) \to 0$ as $t \to \infty$ for all i (i.e., ultimately there will be no oscillations in the average velocity). If $R \neq 0$, then $\dot{\bar{v}} = -k_v\bar{v} - k_f R$ and $\bar{v}(t) \to -\frac{k_f}{k_v}R$ as $t \to \infty$, and thus, $v^i(t) \to -\frac{k_f}{k_v}R$ for all i as $t \to \infty$, i.e, the group follows the profile. These results help to highlight the effects of the noise. The noise makes it so that the swarm may not follow the profile as well (but makes following it possible when it may not be possible for a single individual), and it destroys tight cohesion characterized by getting $e_v^i(t) \to 0$. Next, we will study additional characteristics of swarms by analyzing the results of this and the previous sections in more detail.

5.4 Cohesion Characteristics and Swarm Dynamics

Here we will study the effects of various parameters on cohesion characteristics and then provide a simulation to provide insight into swarm dynamics, especially transient behavior.

5.4.1 Effects of parameters on swarm size

The size of Ω_b in (5.3.10), which we denote by $|\Omega_b|$, is directly a function of several known parameters. Consider the following cases.

- *No sensing errors:* If there are no sensing errors, i.e., $D_p = D_v = D_f = 0$, and if $Q = k_a I$, we obtain

$$\Omega_b = \left\{ E : \|E^i\| \le \frac{2k_r r_s (N-1)}{k_a} \lambda_{\max}(P) \exp\left(-\frac{1}{2}\right), \ i = 1, \cdots, N \right\}.$$

  If N, k_r, and r_s are fixed, then if k_a increases from zero to infinity we get $\frac{\lambda_{\max}(P)}{k_a} \to 1$ from above and a decrease in $|\Omega_b|$, but only up to a certain point.

- *Sensing errors:* There are several characteristics of interest.
  - *Noise cancellation:* In the special situation when $d_p^i = d_p^j$, $d_v^i = d_v^j$, and $d_f^i = d_f^j$ for all i and j, then $d_p^i - \bar{d}_p = d_v^i - \bar{d}_v = d_f^i - \bar{d}_f$ for all i and it is as if there were no error and $|\Omega_b|$ is smaller.
  - *Repel effects:* For fixed values of N, k_a, and k_r if we increase r_s each agent has a larger region from which it will repel its neighbors so $|\Omega_b|$ is larger. For fixed k_r, k_a, and r_s if we let $N \to \infty$, then $|\Omega_b| \to \infty$ as we expect due to the repulsion (the bound is conservative since it depends on the special case of all agents being aligned on a line so there are $N - 1$ interagent distances that sum to make the bound large).
  - *Attraction can amplify noise:* Let $D_s = D_p + D_v$ and J quantify the size of the set Ω_b. Next we study the special case of choosing $Q = k_a I$. Fix all values of the parameters except k_a and D_s. A plot of J versus k_a and D_s is shown in Figure 5.4.1, where the locus of points are those values of k_a that minimize J for each given value of D_s. This plot shows that if there is a set magnitude of the noise, then to get the best cohesiveness (smallest Ω_b) k_a should not be too small (or it would not hold the group together), but also not too large since then the noise is also amplified by the attraction gain and poor cohesion results. Could you interpret the plot as a type of fitness function, and then make any conclusions about the evolution of the agent parameters?

- *Swarm size N:* In some situations, when N is very large, $\bar{d}_p \approx \bar{d}_v \approx \bar{d}_f \approx 0$ when there is no biasing of sensing errors so that on average they are zero and this reduces the above bound on $\|g^i(E)\|$.

5.4.2 Swarm dynamics: Individual and group

Here we will simply simulate a swarm for the no-noise case to provide some insights into the dynamics, especially the transient behavior. We will in particular seek to study the individual motions and how they collectively move as a group to achieve cohesion, and the dynamics of the motion of the group. We use linear attraction,

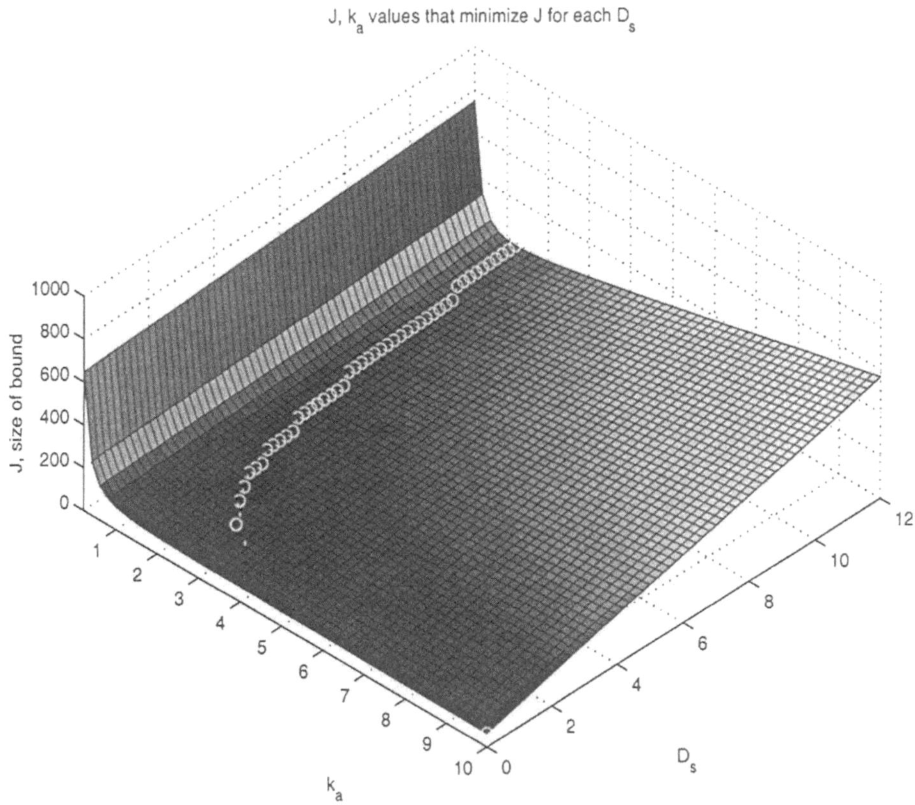

Figure 5.4.1: Values of k_a that minimize J for given values of noise magnitude D_s

velocity damping, the Gaussian form for the repulsion term, and the resource profile with the shape of a plane. The parameters for the simulation are $N = 50$, $k_a = 1$, $k_r = 10$, $r_s^2 = 0.1$, $k_v = k_f = 0.1$, $R = [1, 2, 3]^\top$, and $r_o = 0$. Simulating for 10 sec. we get the agent trajectories in Figure 5.4.2.

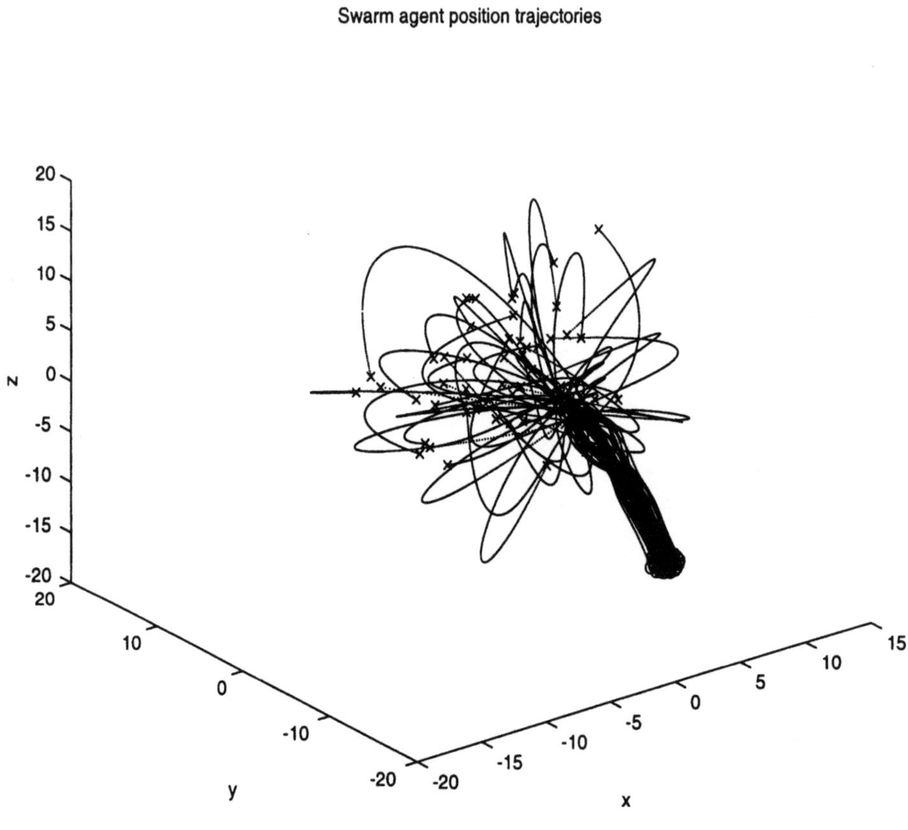

Figure 5.4.2: Agent trajectories in a swarm

Notice that initially the agents move to achieve cohesion. By the end of the simulation the agents are moving at the same velocity and as a group, with some constant interagent spacing between each pair of agents. Moreover, due to the choice of initial conditions (actually random), the group achieves a certain level of aggregation relatively quickly, then the *group* moves to follow the foraging profile. Of course some orientation toward following the profile is achieved during the initial aggregation period, but in this simulation some reorientation toward following the resource profile occurs *after* there is relatively tight aggregation.

5.5 Conclusions

We gave a relatively simple example of how to model and analyze cohesion for a multiagent system that operates as a swarm. This is currently a very active research area and a number of other topics are being studied, including (i) modeling and analysis of particular swarms of agents (e.g., robots, bees, fish), (ii) formation control where the position of each agent relative to other agents matters (especially important in some military and robotic applications), (iii) consideration of alternative local rules to maintain spacing and more complicated dynamics for agents, (iv) effects of noise and topology of the interconnecting communication network, and (v) installation of "intelligent" local behaviors in agents and its impact on emergent behaviors and performance.

Bibliography

[1] J. Parrish and W. Hamner, Eds., *Animal Groups in Three Dimensions.* Cambridge Univ. Press, Cambridge, England, 1997.

[2] L. Edelstein-Keshet, *Mathematical Models in Biology.* Birkhäuser Mathematics Series, Random House, New York, 1989.

[3] T. Balch and R. C. Arkin, "Behavior-based formation control for multirobot teams," *IEEE Trans. Robotics and Automation*, vol. 14, pp. 926–939, Dec. 1998.

[4] I. Suzuki and M. Yamashita, "Distributed anonymous mobile robots: Formation of geometric patterns," *SIAM J. Computing*, vol. 28, no. 4, pp. 1347–1363, 1999.

[5] J. H. Reif and H. Wang, "Social potential fields: A distributed behavioral control for autonomous robots," *Robotics and Autonomous Systems*, vol. 27, pp. 171–194, 1999.

[6] D. Swaroop, *String Stability of Interconnected Systems: An Application to Platooning in Automated Highway Systems*, Ph.D. Thesis, Department of Mechanical Engineering, University of California, Berkeley, CA, 1995.

[7] K. Jin, P. Liang, and G. Beni, "Stability of synchronized distributed control of discrete swarm structures," *Proc. IEEE International Conference on Robotics and Automation*, pp. 1033–1038, San Diego, CA, May 1994.

[8] G. Beni and P. Liang, "Pattern reconfiguration in swarms–Convergence of a distributed asynchronous and bounded iterative algorithm," *IEEE Trans. Robotics and Automation*, vol. 12, pp. 485–490, June 1996.

[9] Y. Liu, K. M. Passino, and M. Polycarpou, "Stability analysis of one-dimensional asynchronous swarms," *Proc. American Control Confrence*, pp. 716–721, Arlington, VA, June 2001.

[10] Y. Liu, K. M. Passino, and M. Polycarpou, "Stability analysis of one-dimensional asynchronous mobile swarms," in *Proc. IEEE Conference on Decision and Control*, pp. 1077–1082, Orlando, FL, Dec. 2001.

[11] Y. Liu, K. M. Passino, and M. M. Polycarpou, "Stability analysis of m-dimensional asynchronous swarms with a fixed communication topology," *Proc. American Control Confrence*, pp. 1278–1283, Anchorage, AK, May 2002.

[12] V. Gazi and K. M. Passino, "Stability of a one-dimensional discrete-time asynchronous swarm," *Proc. Joint IEEE Int. Symp. Intelligent Control/IEEE Confrence on Control Applications*, pp. 19–24, Mexico City, Mexico, Sept. 2001.

[13] N. E. Leonard and E. Fiorelli, "Virtual leaders, artificial potentials and coordinated control of groups," *Proc. IEEE Conference on Decision and Control*, pp. 2968–2973, Orlando, FL, Dec. 2001.

[14] D. Grunbaum, "Schooling as a strategy for taxis in a noisy environment," *Evolutionary Ecology*, vol. 12, pp. 503–522, 1998.

[15] V. Gazi and K. M. Passino, "Stability analysis of swarms," *Proc. American Control Confrence*, pp. 1813–1818, Anchorage, AK, May 2002.

[16] V. Gazi and K. M. Passino, "A class of attraction/repulsion functions for stable swarm aggregations," *Proc. IEEE Conference on Decision and Control*, pp. 2842–2847, Las Vegas, NV, Dec. 2002.

[17] V. Gazi and K. M. Passino, "Stability analysis of swarms in an environment with an attractant/repellent profile," in *Proc. American Control Confrence*, pp. 1819–1824, Anchorage, AK, May 2002.

[18] V. Gazi and K. M. Passino, "Stability analysis of social foraging swarms: Combined effects of attractant/repellent profiles," *Proc. IEEE Confrence on Decision and Control*, pp. 2848–2853, Las Vegas, Nevada, Dec. 2002.

[19] V. Gazi and K. Passino, "Modeling and analysis of the aggregation and cohesiveness of honey bee clusters and in-transit swarms," submitted to *J. Theoretical Biology*, 2002.

[20] R. Bachmayer and N. E. Leonard, "Vehicle networks for gradient descent in a sampled environment," *Proc. IEEE Confrence on Decision and Control*, pp. 113–117, Las Vegas, Nevada, Dec. 2002.

[21] P. Ogren, E. Fiorelli, and N. E. Leonard, "Formations with a mission: Stable coordination of vehicle group maneuvers," *Proc. Symposium on Mathematical Theory of Networks and Systems*, Notre Dame, IN, Aug. 2002.

[22] T. D. Seeley, R. A. Morse, and P. K. Visscher, "The natural history of the flight of honey bee swarms," *Psyche*, vol. 86, pp. 103–113, 1979.

[23] H. K. Khalil, *Nonlinear Systems*, 3rd ed., Prentice-Hall, Englewood Cliffs, NJ, 1999.

Chapter 6

Stability of Discrete Time-Varying Linear Delay Systems and Applications to Network Control

Mihail L. Sichitiu and Peter H. Bauer

Abstract: This chapter presents a stability condition for time-varying uncertain discrete delay systems, where the uncertainty set is finite and constraints on the allowable sequences of system matrices are imposed. The proposed stability condition is necessary and sufficient and can be viewed as a modified combination of two existing conditions for time-varying systems with polytopic uncertainties. The introduced stability test is especially useful in the area of network embedded control systems and network congestion control where time-varying delays with constraints occur.

6.1 Introduction

The problem of robust stability of uncertain time-varying discrete-time systems with polytopic uncertainties has been the subject of several publications (e.g., see [1–3]). This type of system finds applications in fields such as fault tolerant, adaptive, switched, and network embedded systems. The typical system investigated has a zero input state-space representation, where the time-varying system matrix $A(n)$ is taken from a matrix polytope. In some problem settings, additional constraints on the rate of change of the system parameters (system matrix) were imposed [4, 5]. Asymptotic/exponential stability was shown using a number of conditions, three of which were shown to be equivalent in complexity as well as necessary and sufficient in nature [1–3, 6].

Motivated by the applications in network embedded control and congestion control systems, this chapter investigates the problem of asymptotic stability of time-varying uncertain systems where the uncertainty set is finite and constraints on permissible matrix sequences exist. The derived stability condition is necessary and sufficient and can be viewed as a modified combination of two existing necessary and sufficient stability conditions on polytopic uncertain systems. The developed approach drastically reduces the computational complexity of the test. The gains over existing methods will be illustrated by several examples from the area of network control.

6.2 The Stability Test

Consider the linear, time-varying, and uncertain zero input systems of the form:

$$x(n+1) = A(n)x(n) \tag{6.2.1}$$

$$A(n) \in \mathcal{S}, \quad \mathcal{S} = \{A_1, A_2, \cdots, A_N\} \subset \mathcal{R}^{L\times L} \tag{6.2.2}$$

$$A(n) = A_i \Rightarrow A(n+1) \in \mathcal{S}_{A_i} \subset \mathcal{S} \tag{6.2.3}$$

where the $\mathcal{S}_{A_i}$ are nonempty subsets of $\mathcal{S}$ dependent only on the choice of the matrix $A(n)$ at time instant n, i.e., on A_i. Let there be a nonempty subset* $\mathcal{S}_{A_i}$ of $\mathcal{S}$ for each A_i, $i = 1, \cdots, N$. (This problem reduces to the typical robust stability problem for time-varying systems with polytopic uncertainties if $\mathcal{S}$ is replaced by the convex hull of the matrices in $\mathcal{S}$ and condition (6.2.3) is dropped.)

The problem addressed in this chapter is to establish asymptotic stability conditions for system (6.2.1) under the constraints (6.2.2), (6.2.3). It is well known [6] that this problem is NP hard even if the set $\mathcal{S}$ contains only two elements and condition (6.2.3) is dropped.

Two special cases deserve mentioning.

(a) If each $\mathcal{S}_{A_i}$, $i = 1, \cdots, N$, is of cardinality 1, then the system is time varying with a known coefficient trajectory. This system (except possibly for some initial states) is periodic and its stability is easily analyzed using the eigenvalues of the full period transition matrix.

(b) If $\mathcal{S}_{A_i} = \mathcal{S}$, $i = 1, \cdots, N$, we have no additional constraints and the result obtained for this system also guarantees stability of the entire convex hull defined by the vertex matrices $A_1, \cdots, A_N$ [1, 2].

The following theorem is a generalization of the results in [1] and provides a necessary and sufficient condition for time-varying systems with polytopic uncertainties.

*The exact definition for each of the $\mathcal{S}_{A_i}$ is application dependent and will be illustrated in Section 6.4.

Theorem 6.2.1. *The system*

$$x(n+1) = A(n)x(n), \quad A(n) \in \text{conv}(A_1, \cdots, A_N)$$

is exponentially stable iff $\exists$ *a sufficiently large integer* k *such that*

$$\|A_{i_1} \cdot \cdots \cdot A_{i_k}\| \leq \gamma < 1 \;\; \forall (i_1, \cdots, i_k) \in \{1, \cdots, N\}^k$$

where by $\text{conv}(A_1, \cdots, A_N)$ *we denote the convex hull (convex matrix polyhedron) of the set of constant matrices* $\{A_1, \cdots, A_N\}$ *and* $\|\cdot\|$ *is any vector induced matrix norm.*

Comments.

- Theorem 6.2.1 is more general than the main result in [1] due to generalization of the norm type [6] and of the polytope form. Exponential stability was explicitly shown in [6].
- The problem is NP hard and the test is often inconclusive because the required k is very large.
- The theorem is only theoretically necessary and sufficient since instability cannot be determined in practice.

The practical implementation of the stability test implies testing the norms of all the combinations $A_{i_1} \cdot \cdots \cdot A_{i_k}$. The number of combinations is N^k, which limits the test usability for large values of k. We may be able to use a lower product length k if instead of applying Theorem 6.2.1 to the original system we first transform it and then apply the theorem.

Corollary 6.2.1. *The system*

$$x(n+1) = A(n)x(n), \quad A(n) \in \text{conv}(A_1, \cdots, A_N)$$

is exponentially stable iff $\exists$ *a sufficiently large integer* k *such that*

$$\|P^{-1}A_{i_1} \cdot \cdots \cdot A_{i_k}P\| \leq \gamma < 1 \;\; \forall (i_1, \cdots, i_k) \in \{1, \cdots, N\}^k$$

where P *is some invertible matrix.*

Proof. The proof is straightforward and follows from a similarity transformation on the polytope. Consider the following transformation:

$$x(n) = P\tilde{x}(n). \tag{6.2.4}$$

If P is invertible, (6.2.1) becomes with (6.2.4):

$$\tilde{x}(n+1) = P^{-1}A(n)P\tilde{x}(n). \tag{6.2.5}$$

Defining

$$P^{-1}A(n)P = \tilde{A}(n) \tag{6.2.6}$$

we now have an equivalent description of the system in (6.2.1):

$$\tilde{x}(n+1) = \tilde{A}(n)\tilde{x}(n), \quad \tilde{A}(n) \in \tilde{\mathcal{S}} \tag{6.2.7}$$

where

$$\tilde{\mathcal{S}} = \mathrm{conv}(\tilde{A}_1, \cdots, \tilde{A}_N).$$

Now applying the condition in Theorem 6.2.1 to the new vertex matrices

$$\tilde{A}_i = P^{-1} A_i P$$

yields the desired result. □

If one could find a similarity transformation that will *simultaneously* make all the norms $\|\tilde{A}_i\|$ smaller than 1, then Corollary 6.2.1 would be satisfied for $k = 1$. However, for a square matrix P this is not always possible [2].

Theorem 6.2.2 ([2]). *The system*

$$x(n+1) = A(n)x(n),$$

$$A(n) \in \mathrm{conv}(A_1, \cdots, A_N),$$

$$\{A_1, \cdots, A_N\} \subset \mathcal{R}^{L \times L}$$

is exponentially stable iff $\exists$ *a matrix* $P = (p_1, \cdots, p_M)$ *where* $p_i,\ i = 1, \cdots, M$ *are column vectors with* $\mathrm{rank}(P) = \dim(A(n)) = L \leq M$ *such that the matrices* $\tilde{A}_i,\ i = 1, \cdots, N,\ \tilde{A}_i \in \mathcal{R}^{M \times M}$ *defined by*

$$A_i^T P = P \tilde{A}_i^T, \quad i = 1, \cdots, N$$

satisfy the condition

$$\|\tilde{A}_i\|_\infty < 1, \quad i = 1, \cdots, N.$$

As we already mentioned Theorem 6.2.1 (and Corollary 6.2.1) cannot be used to prove instability. We will introduce a sufficient condition for instability in the next lemma.

Lemma 6.2.1. *The system (6.2.1)–(6.2.3) is unstable if there exists* $(i_1, i_2, \cdots, i_k) \in \{1, \cdots, N\}^k$ *such that the matrices* $A_{i_1}, \cdots, A_{i_k}$ *satisfy condition (6.2.3) and*

$$\rho\left(A_{i_k} \cdot \cdots \cdot A_{i_1}\right) > 1$$

where by $\rho(\cdot)$ *we denote the spectral radius of a matrix.*

Proof. The result follows immediately if one considers the periodic sequence

$$A_{i_1}, A_{i_2}, \cdots, A_{i_k}$$

which results in an unstable full period transition matrix $A_k = A_{i_k} \cdot \cdots \cdot A_{i_1}$ if the condition of Lemma 6.2.1 is satisfied. □

The system (6.2.1)–(6.2.3) is different from the one considered in Theorem 6.2.1: it has a finite number of possible system matrices $A_1, \cdots, A_N$ and it additionally imposes condition (6.2.3) on permissible matrix sequences. We will next present a lemma that tailors Corollary 6.2.1 to the system (6.2.1)–(6.2.3).

Lemma 6.2.2. *The system (6.2.1)–(6.2.3) is exponentially stable iff* $\exists$ *a sufficiently large integer* k *such that*

$$\|P^{-1}A_{i_k}\cdot\cdots\cdot A_{i_1}P\| \leq \gamma < 1 \;\; \forall(i_1,\cdots,i_k) \;\; \textit{such that} \;\; A_{i_{j+1}} \in \mathcal{S}_{A_{i_j}} \tag{6.2.8}$$

where P *is any invertible square matrix of appropriate dimension.*

Proof. Sufficiency: Since P is invertible we can use a similarity transformation as in (6.2.4). Applying (6.2.5) recursively we can write

$$x(k+1) = P^{-1}A(k)A(k-1)\cdots A(1)Px(1). \tag{6.2.9}$$

If there is an integer k such that (6.2.8) is satisfied then (6.2.9) is a contraction mapping, which assures that the system (6.2.1)–(6.2.3) is asymptotically stable. Additionally, since

$$\|x(k\,n)\| \leq \gamma^n \|x(0)\|$$

it can be easily shown that it is exponentially stable.

Necessity: Assume that there exists no k such that (6.2.8) is satisfied. Therefore, given an arbitrary value for k, we may always satisfy

$$\|P^{-1}A_{i_k}\cdot\cdots\cdot A_{i_1}P\| \geq 1 \tag{6.2.10}$$

for some choice of a sequence $A(1),\cdots,A(k)$. Then there exists in initial vector $x(1)$ such that

$$\|\tilde{x}(k+1)\| \geq \|\tilde{x}(1)\|, \;\; \text{independent of the choice of } k,$$

which contradicts the condition for globally asymptotic stability for the transformed system which in turn contradicts the stability condition for the original system. □

Comments.

- In [1, 2] it was shown that the stability of the vertex matrices $A_1, A_2, \cdots, A_N$ is necessary and sufficient for the stability of the entire convex hull conv(A_1, $A_2, \cdots, A_N$).
- The cost of performing the test in Theorem 6.2.1 (or Corollary 6.2.1) is $O(N^k)$, i.e., it grows exponentially with k.
- The cost of performing the test in Lemma 6.2.2 depends on the cardinality of the sets $\mathcal{S}_{A_i}$. For example, if card$(\mathcal{S}_{A_i}) = Q \;\forall i = 1,\cdots,N$, then the complexity of the test is $O(Q^k)$.
- The stability conditions in Theorem 6.2.1, Corollary 6.2.1 and Lemma 6.2.2 are highly dependent on the chosen norm. Moreover, for some norms one might find a quadratic transformation matrix P which satisfies Lemma 6.2.2 for $k = 1$, while for others it may not be possible [2].
- A well-chosen similarity transformation P in Corollary 6.2.1 or Lemma 6.2.2 can significantly decrease the necessary number of matrix products k, and since the complexity of the stability test increases exponentially with k, a good algorithm for computing P is warranted.

6.3 The Algorithm

It was shown in [7] that computing the generalized transformation P in Theorem 6.2.2 is itself an NP-hard problem. We present an algorithm that is simple to implement but does not guarantee that it will reduce the necessary product length k to $k = 1$.

By assumption, all matrices $A_i, \quad i = 1, \cdots, N$, have the eigenvalues in the unit circle. For any matrix (without repeated eigenvalues) we can find a transformation P_i which will diagonalize the matrix, such that $P_i^{-1} A_i P_i = D_i$ where

$$D_i = \text{diag}(\lambda_{i,1}, \cdots, \lambda_{i,N}).$$

Obviously $||D_i||_1 = ||D_i||_\infty < 1$. Unfortunately the matrix P_i that results in

$$||P_i^{-1} A_i P_i||_1 < 1$$

does not necessarily result in $||P_i^{-1} A_j P_i||_1 < 1$ for $j \neq i$. Therefore the following scheme is proposed.

Compute the weighted average A_0 of the matrices $A_i, \quad i = 1, \cdots, N$,

$$\mu_i = ||A_i|| \quad \forall i = 1, \cdots, N \tag{6.3.1}$$

$$A_0 = \frac{\sum_{i=1}^{N} \mu_i A_i}{\sum_{i=1}^{N} \mu_i}, \tag{6.3.2}$$

then compute the matrix P that will diagonalize the weighted average A_0 and use it as a similarity transformation to obtain the transformed system (6.2.7).

We can repeat the algorithm for the transformed system until the results do not further improve. In order to measure the "improvement," the approach described as follows is taken.

Obviously, the best measure is the minimum product length k that will satisfy the condition in (6.2.8). This k needs to be minimized. But finding the minimum product length k is equivalent to performing the proposed stability test itself, which is NP hard. Hence one needs to find different, less ideal, but easily computable measures.

One candidate for such a measure is the lower bound on the necessary product length. This is computed as the minimum positive integer γ such that $||A_i^\gamma|| < 1, \quad \forall i = 1, \cdots, N$. Of course, one needs to minimize the lower bound γ on the necessary product length k in the hope that this will result in a lower necessary product length. Note that $\gamma \leq k$.

Another measure candidate is the number of the matrices $\tilde{A}_i$ with the norm smaller than 1:

$$\delta = \text{card}(\{\tilde{A}_i \mid ||\tilde{A}_i|| < 1\}).$$

δ should be as high as possible, primarily since it may enable us to avoid the test in (6.2.8): one can compute the matrix norms $\tilde{\mu}_i = ||\tilde{A}_i||$ beforehand and if

$$\prod_{i \in \{i_1, \cdots, i_k\}} \tilde{\mu}_i < 1 \tag{6.3.3}$$

then there is no need to compute the matrix multiplications and the norm in (6.2.8) since

$$\|PA_{i_1} \cdot A_{i_2} \cdot \dots \cdot A_{i_k} P^{-1}\| \leq \prod_{i \in \{i_1, \cdots, i_k\}} \tilde{\mu}_i < 1.$$

The third measure candidate is the maximum of all the norm of the matrices $\tilde{A}_i$: $\nu = \max_i\{\|\tilde{A}_i\|\}$. As this maximum decreases the chance that the replacement test (6.3.3) will hold increases.

It has been observed that the iterative application of the algorithm converges quickly (a few iterations) and that it does not always converge to the optimal solution. Therefore the choice of the optimum number of iterations is divided in two phases. In the first phase, a fixed number of iterations (I_N) is performed and the measures for each iteration are recorded. In the second phase, after the recursion is stopped the optimum number of iterations o is chosen based on the performance measures recorded during the iteration phase. Notice that I_N is usually not the optimum number of iterations. Since we have more than one performance measure we will give priorities as follows: the highest priority is given to the lower bound γ on the product length k, the second highest priority is given to the number of matrices δ with norm smaller than 1, and finally, the maximum norm ν receives the lowest priority.

Therefore, we will use the following procedure to choose the optimum number of iterations.

- First, find the set I_1 of the number of iterations that result in the minimum lower bound γ. I_1 is a set because there might be more than one iteration that results in the minimum γ.
- Second, find the set I_2, subset of I_1, of the number of iterations such that the number of matrices with the norm smaller than 1 is maximized.
- Third, find the set I_3, subset of I_2, of the number of iterations such that the maximum norm ν is minimized.
- Finally, choose the minimum element of the set I_3 to be the optimum number of iterations o.

$$I_1 = \left\{ j | \gamma_j = \min_{i=1\cdots I_N} \{\gamma_i\} \right\} \quad (6.3.4)$$

$$I_2 = \left\{ j | \delta_j = \max_{i \in I_1} \{\delta_i\} \right\} \quad (6.3.5)$$

$$I_3 = \left\{ j | \nu_j = \min_{i \in I_2} \{\nu_i\} \right\} \quad (6.3.6)$$

$$o = \min_{j \in I_3} \{j\} \quad (6.3.7)$$

Example. Assume that during the iteration phase the measures presented in Table 6.3.1 have been recorded. We denoted with γ_j, δ_j, and ν_j the measures γ, δ, and γ respectively, measured after the jth iteration. Then choosing the optimum number

Table 6.3.1: Example of recorded measures during the iteration phase

j	0	1	2	3	4	5	6	7	8	9	10
γ_j	237	9	5	4	4	4	4	5	5	5	5
δ_j	0	1	4	4	5	5	4	4	4	4	4
ν_j	4.45	1.52	1.43	1.37	1.40	1.54	1.48	1.45	1.46	1.46	1.46

of iterations according to (6.3.4)–(6.3.7) will result in $I_1 = \{3,4,5,6\}, I_2 = \{4,5\}$, $I_3 = \{4\}$ and $o = 4$, i.e., one should use the transformation matrix P produced at the fourth iteration.

In the rare case that the weighted average A_0 cannot be diagonalized, one can use a different norm (which changes the coefficients μ_i), a different type of weighted average, or use a different system representation.

To conclude the description of the algorithm we present the pseudocode in the appendix of this chapter.

6.4 Applications in Communication Network Control

To illustrate our results we present two examples from the field of network congestion control. The objective of a congestion control system is to regulate the data flow through the network such that there are no packet losses (due to buffer overflows in the intermediate switches) and the network is fully utilized, i.e., there is always data to be sent in the buffers. One way to achieve these objectives is to design a set point control system for the buffers of the intermediate switches. Such a control system is shown in Figure 6.4.1 for the particular case of the available bit rate (ABR) option in asynchronous transfer mode (ATM) networks [8].

The congested switch controls the rates of the sources by providing explicit rate requests to the sources through the return path. The feedback data transmitted by the switch encounters time-varying delays as it propagates through the network. The sources adjust their transmission rate to the one specified in the most recently received rate request and continue to transmit at that rate until another request arrives. We assume that the time-varying delays on the return paths are bounded:

$$0 \leq \tau_{1,i}(n) \leq \bar{\tau}_{1,i} \quad i = 1, \cdots, M_c. \tag{6.4.1}$$

Figure 6.4.2 depicts the model for a communication link with time-varying delays. The time-varying coefficients $\alpha_j(n)$ satisfy $\alpha_j(n) \in \{0,1\}$ $\forall j = 1, \cdots, \bar{\tau}$ and at each time instant exactly one of the coefficients is not-zero, resulting in

$$\sum_{j=0}^{\bar{\tau}} \alpha_j(n) = 1. \tag{6.4.2}$$

Since the source "holds" the same rate until it receives "fresh" information from the switch, we model the delays on the return path $\tau_{1,i}(n)$ as time-varying delays

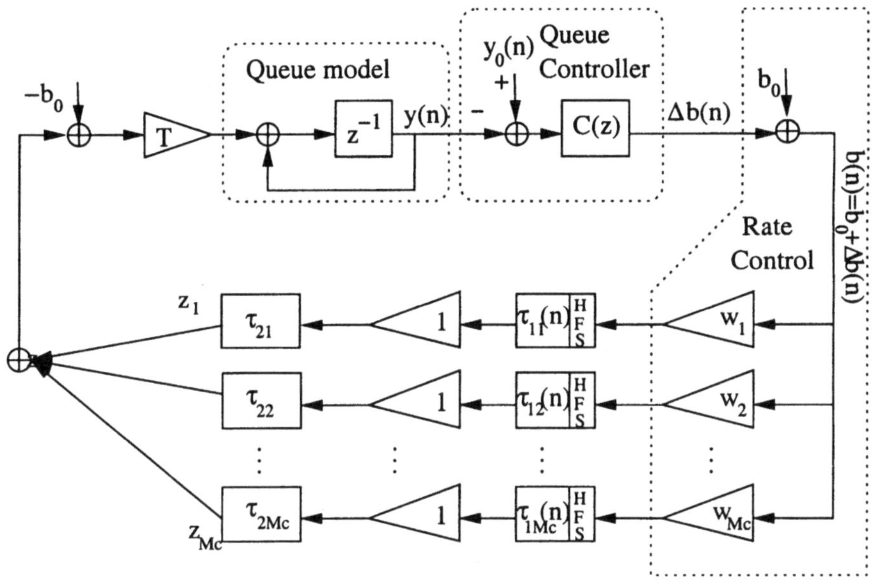

Figure 6.4.1: System model for an ATM congestion control system with time-varying delays: $C(z)$ is the rate controller, w_i are constant weights, M_c is the number of active connections, T is the controller sampling rate, $y(n)$ is the buffer occupancy level, y_0 is the set point for the buffer occupancy, b_0 is the bandwidth of the outgoing link, and $\Delta b(n)$ is the total rate change as computed by the controller.

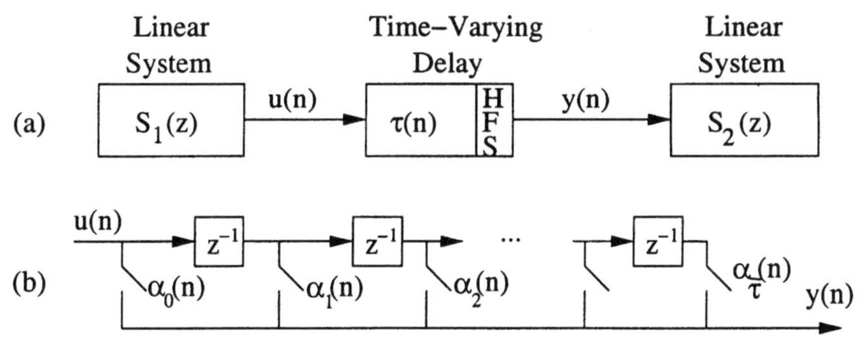

Figure 6.4.2: A network signal propagation/interface delay model. (a) Signal propagation delay model interfaced with a linear system. (b) Detailed block diagram of a signal propagation delay/interface.

with a hold freshest sample (HFS) interface (also called input variable delay in [8]). For an HFS interface the coefficients $\alpha_j(n)$ cannot vary arbitrarily from one time instant to another: the integer delay $\tau(n)$ is restricted by $\tau(n+1) \leq \tau(n) + 1$, and hence we have

$$\alpha_j(n) = 1 \Rightarrow \alpha_k(n+1) = 0 \ \ \forall k > j+1. \tag{6.4.3}$$

In other words the used sample from the time-varying delay output maximally ages with time, but not faster.

Corresponding to each combination of the delays $\tau_{1,i}, \ i = 1, \cdots, M_c$, we have a different system matrix $A(n)$. We thus have

$$N = \prod_{i=1}^{M_c} (\bar{\tau}_{1,i} + 1)$$

matrices in the polytope $\mathcal{S}$. Since the delays on an HFS interface cannot vary arbitrarily, the system matrix $A(n)$ cannot vary arbitrarily inside the polytope. Therefore the sets $\mathcal{S}_i$ in (6.2.3) are often smaller than the entire polytope $\mathcal{S}$.

The delays $\tau_{2,i}, \ i = 1, \cdots, M_c$, correspond to the delays encountered by the transmitted data on the forward path. For simplicity we assume they are constants.

6.4.1 The case of two sources

We consider the case of a single congested switch with two sources ($M_c = 2$). We will use the following parameters for our system:

- $b_0 = 10^6$ cells/s.
- $T = 10^{-3}$s.
- $\bar{\tau}_{1,1} = \bar{\tau}_{1,2} = 2 \cdot 10^{-3}$ s, i.e., there are three possible delays: 0, 1, and 2.
- $\tau_{2,1}(n) = \tau_{2,2}(n) = 10^{-3}$s.
- $w_1(n) = w_2(n) = 0.5$.
- We used a proportional controller with a gain G.

A time-invariant analysis of the system (i.e., analyzing stability of all the frozen time systems) yields the following (necessary) stability condition:

$$G \in (0, \ 445). \tag{6.4.4}$$

The sufficient stability condition for the time-varying case introduced in [9] yields the following stability condition:

$$G \in (0, \ 77.06). \tag{6.4.5}$$

Using Theorem 6.2.1 directly to determine stability is impractical due to the large product length k. In Table 6.4.1 the minimum product length that needs to be checked for $G = 1$ is shown for several matrix norms.

Using Lemma 6.2.2 results in a drastic reduction of the necessary product length. By choosing the transformation P appropriately we managed to have the 2 norm of

Table 6.4.1: Minimum product lengths in Theorem 6.2.1 for various matrix induced norms

Used norm	1	2	∞
Minimum product length $k \geq$	1387	696	1387

Table 6.4.2: Provable stabilizing gain ranges for G as a function of k in Lemma 6.2.2 (two source case)

Product length k	1	2	3	4	5	6
Controller gain $0 < G \leq$	66.8	118.3	254.3	279.1	322.7	366.4

all matrices smaller than 1 for G up to 66.8, which corresponds to $k = 1$. In Table 6.4.2 the resulting bounds for the controller gain G are shown. It can be seen that the results are significantly less conservative than the sufficient condition presented in (6.4.5).

It should be mentioned that only the upper limit of the interval for G is obtained using the algorithm based on Lemma 6.2.2. The lower limit is known to be zero from [9].

Corresponding to each combination of the delays $\tau_{1,1}(n)$ and $\tau_{1,2}(n)$ we have a corresponding system matrix $A(n)$. Let us denote by $A_{i,j}(n)$ the system matrix corresponding to the delays $\tau_{1,1}(n) = i,\ \tau_{1,2}(n) = j;\ i,j \in \{0,1,2\}$. The entire matrix polytope $\mathcal{S}$ is given by $\mathcal{S} = \{A_{i,j} | 0 \leq i \leq 2,\ 0 \leq j \leq 2\}$. The matrix sets $\mathcal{S}_{A_{i,j}}$ are given by the following equation:

$$\mathcal{S}_{A_{i,j}} = \{A_{k,l} | 0 \leq k \leq \min\{2, i+1\},\ 0 \leq l \leq \min\{2, j+1\}\}\,.$$

For example, $\mathcal{S}_{A_{0,1}} = \{A_{0,0}, A_{0,1}, A_{0,2}, A_{1,0}, A_{1,1}, A_{1,2}\}$. For the results presented in Table 6.4.2, we used the phase-variable representation for the system matrices $A(n)$.

The transformation matrix P corresponding to the product length of 6 and the gain $G = 366.4$ is, for example, given by

$$P = \begin{pmatrix} 0.3921-0.0065i & 0.3784+0.1031i & -0.0668-0.0231i & 0.0590-0.0388i \\ 0.3973-0.2136i & 0.3322+0.3051i & 0.1366-0.0866i & -0.1537-0.0502i \\ 0.2931-0.4280i & 0.1784+0.4872i & -0.0400+0.3679i & 0.1296+0.3467i \\ 0.0735-0.5921i & -0.0749+0.5919i & -0.6013-0.5966i & 0.4354-0.7266i \end{pmatrix}.$$

For this transformation matrix P, 7 of the 9 $\tilde{A}_i$ matrices had a norm smaller than 1. This enabled us to use the replacement test in (6.3.3) in 58.8% of the cases. Moreover, only 26% of all the possible combinations are "HFS compliant," i.e., satisfy the HFS constraints in (6.4.3). Combining the two improvements, we performed the norm test in (6.2.8) only for 10% of the cases.

6.4.2 The case of a single source

We present here the case of a single source $M_c = 1$ with the following parameters:

- $b_0 = 1500$ cells/s.
- $T = 10^{-3}$ s.
- $\bar{\tau}_{1,1} = 10^{-2}$s.
- $\tau_{2,1}(n) = 10^{-3}$s.

In this case the time-invariant analysis results in the following stability condition:

$$G \in (0,\ 136.48). \tag{6.4.6}$$

The sufficient stability condition for the time-varying case introduced in [9] yields the following stability condition:

$$G \in (0,\ 6.57). \tag{6.4.7}$$

Let us denote with $A_i(n)$ the system matrix corresponding to the delay $\tau(n) = i$ with $i = 0, \cdots, 10$. Then the matrix sets $\mathcal{S}_{A_i}$ are given by

$$\mathcal{S}_{A_i} = \{A_j | 0 \le j \le \min\{10, i+1\}\}.$$

In Table 6.4.3 the results of the stability test discussed in Lemma 6.2.2 are presented for the linear time-varying case.

In this case (product length $k = 9$) five matrices $\tilde{A}_i$ (of 11) have a norm smaller than 1. The number of "HFS compliant" combinations is only 0.127% of the total number of combinations, which greatly reduced the computational time.

Table 6.4.3: The case of a single source for the linear case: Stable proportional controller gains as a function of k

k	2	3	4	5	6	7	8	9
$0 < G \le$	3.92	21.67	31.23	34.42	36.66	37.58	38.31	38.68

6.5 Conclusions

This chapter presented a new stability test for time-varying uncertain systems, where the uncertainty set is finite and additional constraints on the time variance of the system exist. The main result is based on two previous results on stability of discrete time-varying systems with polytopic uncertainties. It is shown that the introduced necessary and sufficient stability condition is especially useful for systems with time-varying communication delays. These systems presently occur in network congestion control, networked control systems, teleoperation, and sensor/actuator networks.

Acknowledgments

We would like to thank Dr. Kamal Premaratne, Mr. Ernest Kulasekere, and Mr. Jinsong Zhang for several fruitful discussions and insights.

6.6 Appendix: Pseudocode for the Stability Testing Algorithm

In this appendix we provide the pseudocode of the stability testing algorithm based on Lemma 6.2.2 and described in Section 6.3. In steps 2–8 the transformation matrix P is computed in constant time, while in the steps 9–13 the stability test presented in Lemma 6.2.2 is implemented. The pseudocode tests the condition in Lemma 6.2.2 for a given gain of the controller G and for a given product length k.

(1) Generate all vertex matrices $A_i^{(0)}(G)\ \forall\, i = 1, \cdots, N$;

(2) For $j = 1$ to I_N,

(3) $\mu_i^{(j)} \leftarrow \|A_i^{(j)}\|\ \forall\, i = 1, \cdots, N$;

(4) $A_0^{(j)} \leftarrow \frac{\sum_{i=1}^{N} A_i^{(j)} \mu_i^{(j)}}{\sum_{i=1}^{N} \mu_i^{(j)}}$;

(5) Compute P to diagonalize $A_0^{(j)}$ [i.e., $P^{-1} A_0^{(j)} P = \mathrm{diag}(\lambda_{A_0^{(j)}})$];

(6) $A_i^{(j+1)} \leftarrow P^{-1} A_i^{(j)} P\ \ \forall i = 1, \cdots, N$;

(7) Next j.

(8) Choose the optimum iteration o as described in (6.3.4)–(6.3.7).

(9) Generate the first compliant index combination of length k: $(i_1,\ i_2, \cdots, i_k)$;

(10) If $\prod_{l=1}^{k} \|A_{i_l}^{(o)}\| < 1$ then goto step 12 else goto step 11;

(11) If $\|\prod_{l=1}^{k} A_{i_l}^{(o)}\| < 1$ then goto step 12 else the test is inconclusive;

(12) Generate the next compliant index combination of length k: $(i_1,\ i_2, \cdots, i_k)$;

(13) If all compliant index combinations have been tested conclude that the system is stable, else goto step 10

By compliant index combination we understand a index combination that satisfies the matrix sequence constraint from (6.2.3).

Bibliography

[1] P. H. Bauer, K. Premaratne, and J. Durán, "A necessary and sufficient condition for robust asymptotic stability of time-variant discrete systems," *IEEE Trans. Automatic Control*, vol. 38, pp. 1427–1430, Sept. 1993.

[2] A. P. Molchanov, "Lyapunov functions for nonlinear discrete-time control systems," *Automation and Remote Control*, vol. 48, no. 6, pp. 728–736, 1987.

[3] R. Brayton and C. Tong, "Constructive stability and asymptotic stability of dynamical systems," *IEEE Trans. on Circuits and Systems*, vol. 27, pp. 1121–1130, 1980.

[4] K. Premaratne and M. Mansour, "Robust stability of time-variant discrete-time systems with bounded parameter perturbations," *IEEE Trans. Circuits and Systems I: Fundamental Theory and Applications*, vol. 42, pp. 40–45, Jan. 1995.

[5] S. Dasgupta, G. Chockalingam, B. D. O. Anderson, and M. Fu, "Lyapunov functions for uncertain systems with applications to the stability of time varying systems," *IEEE Trans. Circuits and Systems I*, vol. 41, pp. 93–106, Feb. 1994.

[6] A. Bhaya and F. Mota, "Equivalence of stability concepts for discrete time-varying systems," *Int. J. Robust and Nonlinear Control*, vol. 4, pp. 725–740, 1994.

[7] J. N. Tsitsiklis and V. D. Blondel, "Lyapunov exponent and joint spectral radius of pairs of matrices are hard - when not impossible - to compute and to approximate," *Mathematics of Control, Signals, and Systems*, vol. 10, no. 1, pp. 31–40, 1997.

[8] M. M. Ekanayake, *Robust Stability of Discrete Time Nonlinear Systems*. Ph.D. Thesis, University of Miami, 1999.

[9] P. H. Bauer, M. L. Sichitiu, and K. Premaratne, "Closing the loop through communication networks: The case of an integrator plant and multiple controllers," *Proc. 38th IEEE Conference on Decision and Control*, pp. 2180–2185, Phoenix, AZ, Dec. 1999.

Chapter 7

Stability and $\mathcal{L}_2$ Gain Analysis of Switched Symmetric Systems

Guisheng Zhai

Abstract: We study stability and $\mathcal{L}_2$ gain properties for a class of switched systems that are composed of a finite number of linear time-invariant symmetric subsystems. We consider both continuous-time and discrete-time switched systems. We show that when all the subsystems are (Hurwitz or Schur) stable, the switched system is exponentially stable under arbitrary switching. Furthermore, we show that when all the subsystems have $\mathcal{L}_2$ gains less than a positive scalar γ, the switched system has $\mathcal{L}_2$ gain less than the same γ under arbitrary switching. We also extend the results to switched symmetric systems with time delay in system state for continuous-time switched systems. The key idea for both stability and $\mathcal{L}_2$ gain analysis in this chapter is to establish a common Lyapunov function for all the subsystems in the switched system.

7.1 Introduction and Problem Formulation

By a switched system, we mean a hybrid dynamical system that is composed of a family of continuous-time or discrete-time subsystems and a rule orchestrating the switching among the subsystems. In the last two decades, there has been increasing interest in the stability analysis and controller design for such switched systems; see, e.g., [1–21] and the references cited therein. The motivation for studying switched systems is that many practical systems are inherently multimodal in the sense that several dynamical subsystems are required to describe their behavior, which may depend on various environmental factors [1], and that the methods of intelligent control design are based on the idea of switching among different controllers [2–4]. For recent progress and perspectives in the field of switched systems, see survey papers [3, 5] and the references cited therein.

As also pointed out in [3, 5], there are three basic problems in the study of switched systems: (i) finding conditions for stabilizability under arbitrary switching; (ii) identifying the class of stabilizing switching signals; and (iii) constructing a stabilizing switching signal. There are many existing works on (ii) and (iii). For example, [6–9] considered (ii) using piecewise Lyapunov functions, and [10–12] considered (ii) for switched systems with pairwise commutation or Lie-algebraic properties. [13] considered (iii) by dividing the state space associated with appropriate switching depending on state, and [14, 15] considered quadratic stabilization, which belongs to (iii), for switched systems composed of a pair of unstable linear subsystems by using a linear stable combination of unstable subsystems. However, we see very few references dealing with the first problem, although it is desirable to permit arbitrary switching in many real applications. [16] showed that when all subsystems are stable and commutative pairwise, the switched linear system is stable under arbitrary switching. There are some other results concerning common Lyapunov functions for the subsystems in a switched system, but we do not find any explicit answer to (i) except [16]. In this chapter, rather than considering the condition of stabilizability under arbitrary switching for a given switched system, we are interested in the following question: What kind of switched systems are stable under arbitrary switching? Specifically, is there a switched system whose subsystems are stable but not commutative pairwise, yet it is stable under arbitrary switching?

For switched systems, there are a few results concerning $\mathcal{L}_2$ gain analysis. Hespanha considered such a problem in his Ph.D. dissertation [17], by using a piecewise Lyapunov function approach. In [18], a modified approach was proposed for more general switched systems and more exact results have been obtained. In that context, it has been shown that when all subsystems are stable and have $\mathcal{L}_2$ gains less than a positive scalar γ_0, the switched system under an average dwell time scheme [6] has a weighted $\mathcal{L}_2$ gain γ_0, and the weighted $\mathcal{L}_2$ gain approaches normal $\mathcal{L}_2$ gain if the average dwell time is chosen sufficiently large. However, the results obtained in [17] and [18] are conservative, and it is supposed that the main reason is the use of piecewise Lyapunov functions. Recently, [19] considered the computation of $\mathcal{L}_2$ gain for switched linear systems with large dwell time and gave an algorithm by considering the separation between the stabilizing and antistabilizing solutions to a set of algebraic Riccati equations. Noticing that these papers deal with the class of switching signals with large (average) dwell time, we are motivated to ask the following question: Is there a switched system that preserves its subsystems' $\mathcal{L}_2$ gain properties under arbitrary switching?

For the above questions concerning stability and $\mathcal{L}_2$ gain, we give a clear (although not complete) answer here. More exactly, we will show that a class of switched systems, which are composed of a finite number of linear time-invariant symmetric subsystems and called shortly "switched symmetric systems," will preserve its subsystems' stability and $\mathcal{L}_2$ gain properties under arbitrary switching. We take symmetric systems into consideration since such systems appear quite often in many engineering disciplines (for example, electrical and power networks, structural systems, viscoelastic materials) and thus belong to an important class in control engineering.

In this chapter, we will discuss not only continuous-time but also discrete-time switched systems. We include discrete-time switched systems for several reasons. First, discrete-time switched systems appear frequently in real applications. For example, a multimodal dynamical system may be composed of several discrete-time dynamical subsystems due to its physical structure. Even when all subsystems are continuous time, the case of considering sampled-data control for the entire system can be dealt with in the framework of discrete-time switched systems [20, 21]. Second, we find that the extension from continuous-time switched systems to discrete-time ones is not obvious in most cases. In our recent paper [15], we pointed out that we can easily find a stable convex combination of subsystem matrices [14] for quadratic stabilizability of continuous-time switched systems, but for discrete-time switched systems we cannot derive such a combination without involving a Lyapunov matrix. In our opinion, some of the reason stems from the fact that fast switching is not available theoretically for discrete-time switched systems.

The continuous-time switched systems under consideration are described as

$$\begin{cases} \dot{x}(t) &= A_{\sigma(t)}x(t) + B_{\sigma(t)}w(t)\,, \qquad x(0) = x_0 \\ z(t) &= C_{\sigma(t)}x(t) + D_{\sigma(t)}w(t)\,, \end{cases} \tag{7.1.1}$$

and the discrete-time switched systems are

$$\begin{cases} x[k+1] &= A_{\sigma(k)}x[k] + B_{\sigma(k)}w[k]\,, \qquad x[0] = x_0 \\ z[k] &= C_{\sigma(k)}x[k] + D_{\sigma(k)}w[k]\,, \end{cases} \tag{7.1.2}$$

where $x(t)(x[k]) \in \Re^n$ is the state, $w(t)(w[k]) \in \Re^m$ is the input, $z(t)$ $(z[k]) \in \Re^p$ is the output, x_0 is the initial state. $\sigma(t)(\sigma(k)) : \Re^+(\mathcal{I}^+) \to \mathcal{I}_N = \{1, 2, \cdots, N\}$ is a piecewise constant function, called a "switching signal," which is assumed to be arbitrary. Here, $\Re^+(\mathcal{I}^+)$ denotes the set of all nonnegative real numbers (integers), A_i, B_i, C_i, D_i $(i \in \mathcal{I}_N)$ are constant matrices of appropriate dimensions denoting the subsystems, and $N > 1$ is the number of subsystems. Throughout this chapter, we assume that all the subsystems in (7.1.1) and (7.1.2) are symmetric, i.e.,

$$A_i = A_i^T\,, \qquad B_i = C_i^T\,, \qquad D_i = D_i^T\,, \qquad \forall i \in \mathcal{I}_N\,. \tag{7.1.3}$$

It should be noted here that (7.1.3) does not cover all symmetric subsystems, and the more general definition is that $T_iA_i = A_i^TT_i$, $T_iB_i = C_i^T$, $D_i = D_i^T$ holds for some nonsingular symmetric matrix T_i [22, 23]. However, since (7.1.3) represents an interesting class of symmetric systems [24], we focus our attention on this kind of symmetric system. We will give some remarks in Section 7.4 for the extension to more general cases.

Furthermore, in the case of continuous-time switched systems, we will extend the discussion to an important class of switched symmetric time-delay systems described by equations of the form

$$\begin{cases} \dot{x}(t) = A_{\sigma(t)}x(t) + \bar{A}_{\sigma(t)}x(t-\tau)) + B_{\sigma(t)}w(t) \\ x(t) = \phi(t)\,, \quad \forall t \in [-\tau, 0] \\ z(t) = C_{\sigma(t)}x(t) + D_{\sigma(t)}w(t)\,, \end{cases} \tag{7.1.4}$$

where τ is the time delay in the state, $\phi(\cdot)$ is the initial condition, $\bar{A}_i = \bar{A}_i^T$ is a constant matrix, and all the other notation are the same as in (7.1.1). The motivation of including time delay in (7.1.4) is that it exists in most real control systems. For example, in a networked control systems [25] where all information (reference input, plant output, control input, etc.) are exchanged through a network among control system components (sensors, controller, actuators, etc.), there exist various time delays inevitably due to busy and/or unreliable transmission paths. There have been extensive publications concerning time-delay systems. For good references on time-delay systems and discussions of the importance of such systems, the reader may want to refer to, e.g., [26, 27].

We now give two definitions concerning stability and $\mathcal{L}_2$ gain properties of the above switched systems.

Definition 7.1.1. We say the continuous-time switched system (7.1.1) is exponentially stable if $\|x(t)\| \leq ce^{-\alpha t}\|x_0\|$ with $c > 0, \alpha > 0$ holds for any $t > 0$ and any initial state x_0, and say the switched system (7.1.1) has $\mathcal{L}_2$ gain less than γ if $\int_0^t z^T(s)z(s)ds \leq \gamma^2 \int_0^t w^T(s)w(s)ds$ holds for any $t > 0$ when $x_0 = 0$ and $w(t) \not\equiv 0$. These definitions are also valid for the system (7.1.4) and all the subsystems in (7.1.1) and (7.1.4).

Definition 7.1.2. We say the discrete-time switched system (7.1.2) is exponentially stable if $\|x[k]\| \leq c\mu^k\|x_0\|$ with $c > 0, 0 < \mu < 1$ holds for any $k > 0$ and any initial state x_0, and say the switched system (7.1.2) has $\mathcal{L}_2$ gain less than γ if $\sum_{j=0}^k z^T[j]z[j] \leq \gamma^2 \sum_{j=0}^k w^T[j]w[j]$ holds for any integer $k > 0$ when $x_0 = 0$ and $w[k] \not\equiv 0$. These definitions are also valid for all the subsystems in (7.1.2).

This chapter is organized as follows. In Section 7.2, we analyze stability properties of the switched systems (7.1.1) and (7.1.2). Assuming that all the subsystems are (Hurwitz or Schur) stable, we show that there exists a common Lyapunov function for all the subsystems, and that the switched system is exponentially stable under arbitrary switching. We then extend the results to a class of continuous-time switched linear systems with time delay in the state and obtain parallel results. In Section 7.3, we consider $\mathcal{L}_2$ gain properties for the switched systems (7.1.1) and (7.1.2). Assuming that all the subsystems have $\mathcal{L}_2$ gains less than a positive scalar γ, we prove that there exists a common Lyapunov function for all the subsystems in the sense of $\mathcal{L}_2$ gain, and that the switched system has $\mathcal{L}_2$ gain less than the same γ under arbitrary switching. We also extend the discussions concerning $\mathcal{L}_2$ gain to the case of continuous-time switched systems with time delay and obtain parallel results. Finally, we conclude the chapter in Section 7.4.

7.2 Stability Analysis

In this section, we set $w \equiv 0$ in the switched systems (7.1.1), (7.1.2), and (7.1.4) to consider stability properties of the systems under arbitrary switching.

7.2.1 Switched symmetric systems without time delay

We first state and prove the stability result for the continuous-time switched system (7.1.1).

Theorem 7.2.1. *If all the subsystems in (7.1.1) are Hurwitz stable, then the switched symmetric system (7.1.1) is exponentially stable under arbitrary switching.*

Proof. Since all the subsystems are Hurwitz stable, we obtain $A_i < 0\,, \forall i \in \mathcal{I}_N$. Then it is always possible to find a positive scalar α such that

$$A_i < -\alpha I\,, \qquad \forall i \in \mathcal{I}_N\,. \tag{7.2.1}$$

We consider the following Lyapunov function candidate

$$V(x) = x^T x\,, \tag{7.2.2}$$

and obtain easily for any $t > 0$ that

$$\frac{dV(x)}{dt} = 2x^T(t)A_{\sigma(t)}x(t) \le -2\alpha V(x) \tag{7.2.3}$$

from (7.2.1). Since this inequality holds for any $t > 0$ and there is no state jump in the switched systems, we obtain $V(x) \le e^{-2\alpha t}V(x_0)$, and thus $\|x(t)\| \le e^{-\alpha t}\|x_0\|$. This implies that the switched system is exponentially stable under arbitrary switching. □

Remark 7.2.1. According to Theorem 7.2.1 and its proof, if all A_is in (7.1.1) are Hurwitz stable, there exists a common Lyapunov matrix $P = I$ for all A_is, satisfying the linear matrix inequality (LMI)

$$A_i^T P + PA_i < 0\,, \qquad \forall i \in \mathcal{I}_N\,. \tag{7.2.4}$$

Hence, $V(x) = x^T x$ serves as a common Lyapunov function for all the subsystems in (7.1.1).

To consider stability for the discrete-time switched system (7.1.2), we need the following preliminary result.

Lemma 7.2.1. *Consider the linear time-invariant symmetric system*

$$x[k+1] = Ax[k]\,, \tag{7.2.5}$$

where $x[k] \in \Re^n$ is the state and $A = A^T$ is a constant matrix. The system (7.2.5) is Schur stable if and only if

$$A^2 < I\,. \tag{7.2.6}$$

Proof. Since A is a symmetric matrix, there exists a nonsingular matrix Q such that

$$Q^T A Q = \mathrm{diag}\{\lambda_1,\ \lambda_2,\ \cdots,\ \lambda_n\},\quad Q^T = Q^{-1}, \tag{7.2.7}$$

where λ_i, $i = 1, 2, \cdots, n$, are A's real eigenvalues (noticing that A has only real eigenvalues), and thus $Q^T A^2 Q = \mathrm{diag}\{\lambda_1^2,\ \lambda_2^2,\ \cdots,\ \lambda_n^2\}$. The system (7.2.5) is Schur stable if and only if $|\lambda_i| < 1$, $i = 1, 2, \cdots, n$, which is equivalent to $Q^T A^2 Q < I$. This is true if and only if (7.2.6) is satisfied. This completes the proof.

Remark 7.2.2. Lemma 7.2.1 implies that if all A_is in (7.1.2) are Schur stable, there exists a common Lyapunov matrix $P = I$ for all A_is, satisfying the LMI

$$A_i^T P A_i - P < 0, \qquad \forall i \in \mathcal{I}_N. \tag{7.2.8}$$

Hence, $V(x) = x^T x$ serves as a common Lyapunov function for all subsystems in (7.1.2).

Theorem 7.2.2. *When all the subsystems in (7.1.2) are Schur stable, the switched symmetric system (7.1.2) is exponentially stable under arbitrary switching.*

Proof. Since all the subsystems in (7.1.2) are Schur stable, from Lemma 7.2.1, the matrix inequality $A_i^2 < I$ holds for all $i \in \mathcal{I}_N$, and thus there exists a scalar $\alpha \in (0, 1)$ such that $A_i^2 < \alpha I$, $\forall i \in \mathcal{I}_N$. Now, for the Lyapunov function candidate (7.2.2), we obtain for any integer $k > 0$ that

$$V(x[k]) = x^T[k]x[k] \le \alpha x^T[k-1]x[k-1] = \alpha V(x[k-1]) \tag{7.2.9}$$

holds under arbitrary switching, and thus we have $V(x[k]) \le \alpha^k V(x_0)$, which means $\|x[k]\|| \le (\sqrt{\alpha})^k \|x_0\|$. Therefore, the switched symmetric system (7.1.2) is exponentially stable under arbitrary switching. □

7.2.2 Switched symmetric systems with time delay

We first recall an existing result.

Lemma 7.2.2 ([29]). *Consider the linear time-delay system*

$$\dot{x}(t) = Ax(t) + \bar{A}x(t-\tau), \tag{7.2.10}$$

where $x(t) \in \Re^n$ is the state, and $\tau > 0$ is the time delay. If there exist $P > 0$ and $\bar{P} > 0$ satisfying the LMI

$$\begin{bmatrix} A^T P + PA + \bar{P} & P\bar{A} \\ \bar{A}^T P & -\bar{P} \end{bmatrix} < 0, \tag{7.2.11}$$

then the system (7.2.10) is asymptotically stable, and one of the Lyapunov functions is given by

$$V(x,t) = x^T(t)Px(t) + \int_0^{\tau} x^T(t-s)\bar{P}x(t-s)ds. \tag{7.2.12}$$

Next we state and prove a preliminary result, which will be used to establish our main result in this subsection. It should be noted that the idea of this result and its proof are motivated by a lemma in [24], which is extended to time-delay systems here.

Lemma 7.2.3. *Assume that the linear time-delay system (7.2.10) is symmetric with* $A = A^T$ *and* $\bar{A} = \delta I_n$*, where* δ *is a known scalar. If there exists* $P > 0$ *satisfying the LMI*

$$\begin{bmatrix} AP + PA + P & \delta P \\ \delta P & -P \end{bmatrix} < 0, \tag{7.2.13}$$

then the system (7.2.10) is asymptotically stable, and

$$\begin{bmatrix} 2A + I & \delta I \\ \delta I & -I \end{bmatrix} < 0. \tag{7.2.14}$$

Proof. Suppose that $P_0 > 0$ satisfies (7.2.13). Then, it is obvious that (7.2.11) holds with $P = \bar{P} = P_0$. Therefore, according to Lemma 7.2.2, the system (7.2.10) is asymptotically stable. From now, we prove (7.2.14).

Since $P_0 > 0$, there always exists a nonsingular matrix U satisfying $U^T = U^{-1}$ such that

$$U^T P_0 U = \Sigma_0 = \mathrm{diag}\{\sigma_1, \sigma_2, \cdots, \sigma_n\}, \tag{7.2.15}$$

where $\sigma_i > 0, i = 1, 2, \cdots, n$. From (7.2.13) and (7.2.15), we obtain

$$\begin{aligned} &\begin{bmatrix} U^T & 0 \\ 0 & U^T \end{bmatrix} \begin{bmatrix} AP_0 + P_0 A + P_0 & \delta P_0 \\ \delta P_0 & -P_0 \end{bmatrix} \begin{bmatrix} U & 0 \\ 0 & U \end{bmatrix} \\ &= \begin{bmatrix} A_U \Sigma_0 + \Sigma_0 A_U + \Sigma_0 & \delta \Sigma_0 \\ \delta \Sigma_0 & -\Sigma_0 \end{bmatrix} < 0, \end{aligned} \tag{7.2.16}$$

where $A_U = U^T A U$. Pre- and postmultiplying the first and second rows and columns in (7.2.16) by Σ_0^{-1} leads to

$$\begin{bmatrix} A_U \Sigma_0^{-1} + \Sigma_0^{-1} A_U + \Sigma_0^{-1} & \delta \Sigma_0^{-1} \\ \delta \Sigma_0^{-1} & -\Sigma_0^{-1} \end{bmatrix} < 0. \tag{7.2.17}$$

Since $\sigma_1 > 0$, there always exists a scalar λ_1 such that

$$0 < \lambda_1 < 1, \qquad \lambda_1 \sigma_1 + (1 - \lambda_1) \sigma_1^{-1} = 1. \tag{7.2.18}$$

Then, by computing $\lambda_1 \times$(7.2.16) + $(1 - \lambda_1) \times$(7.2.17), we obtain

$$\begin{bmatrix} A_U \Sigma_1 + \Sigma_1 A_U + \Sigma_1 & \delta \Sigma_1 \\ \delta \Sigma_1 & -\Sigma_1 \end{bmatrix} < 0, \tag{7.2.19}$$

where

$$\begin{aligned}\Sigma_1 &= \text{diag}\{\lambda_1\sigma_1 + (1-\lambda_1)\sigma_1^{-1}, \lambda_1\sigma_2 + (1-\lambda_1)\sigma_2^{-1}, \cdots, \\ &\qquad \lambda_1\sigma_n + (1-\lambda_1)\sigma_n^{-1}\} \\ &\triangleq \text{diag}\{1, \bar{\sigma}_2, \cdots, \bar{\sigma}_n\} > 0. \end{aligned} \tag{7.2.20}$$

Similar to (7.2.17), we can obtain

$$\begin{bmatrix} A_U\Sigma_1^{-1} + \Sigma_1^{-1}A_U + \Sigma_1^{-1} & \delta\Sigma_1^{-1} \\ \delta\Sigma_1^{-1} & -\Sigma_1^{-1} \end{bmatrix} < 0 \tag{7.2.21}$$

from (7.2.19). Since $\bar{\sigma}_2 > 0$, there exists a scalar λ_2 such that

$$0 < \lambda_2 < 1, \qquad \lambda_2\bar{\sigma}_2 + (1-\lambda_2)\bar{\sigma}_2^{-1} = 1. \tag{7.2.22}$$

Then, the linear combination of (7.2.19) and (7.2.21) results in

$$\begin{bmatrix} A_U\Sigma_2 + \Sigma_2 A_U + \Sigma_2 & \delta\Sigma_2 \\ \delta\Sigma_2 & -\Sigma_2 \end{bmatrix} < 0, \tag{7.2.23}$$

where

$$\begin{aligned}\Sigma_2 &= \text{diag}\left\{1, \lambda_2\bar{\sigma}_2 + (1-\lambda_2)\bar{\sigma}_2^{-1}, \cdots, \lambda_2\bar{\sigma}_n + (1-\lambda_2)\bar{\sigma}_n^{-1}\right\} \\ &\triangleq \text{diag}\{1, 1, \tilde{\sigma}_3, \cdots, \tilde{\sigma}_n\} > 0. \end{aligned} \tag{7.2.24}$$

By repeating this process, we see that $\Sigma_n = I$ also satisfies (7.2.16), i.e.,

$$\begin{bmatrix} 2A_U + I & \delta I \\ \delta I & -I \end{bmatrix} < 0. \tag{7.2.25}$$

Pre- and postmultiplying this inequality by $\text{diag}\{U, U\}$ and $\text{diag}\{U^T, U^T\}$, respectively, we obtain (7.2.14). This completes the proof.

Remark 7.2.3. Lemma 7.2.3 implies that if for the linear symmetric time-delay system (7.2.10) with $A = A^T$ and $\bar{A} = \delta I_n$ there is a Lyapunov function (7.2.12) where $P = \bar{P}$ satisfies (7.2.13), then

$$V_I(x,t) = x^T(t)x(t) + \int_0^{\tau} x^T(t-s)x(t-s)ds \tag{7.2.26}$$

serves also as a Lyapunov function.

We now state and prove the main result of this subsection.

Theorem 7.2.3. *Assume that in (7.1.4) $\bar{A}_i = \delta_i I$, $\forall i \in \mathcal{I}_N$. If all the subsystems are asymptotically stable in the sense that there exists $P_i > 0$ satisfying the LMI*

$$\begin{bmatrix} A_iP_i + P_iA_i + P_i & \delta_iP_i \\ \delta_iP_i & -P_i \end{bmatrix} < 0, \tag{7.2.27}$$

then (7.2.26) is a common Lyapunov function for all the subsystems in (7.1.4), and thus the switched system (7.1.4) is asymptotically stable under arbitrary switching.

Proof. If (7.2.27) is true, we know from Lemma 7.2.3 that

$$\begin{bmatrix} 2A_i + I & \delta_i I \\ \delta_i I & -I \end{bmatrix} < 0, \qquad \forall i \in \mathcal{I}_N, \tag{7.2.28}$$

and thus there exists a scalar $\epsilon > 0$ such that

$$\begin{bmatrix} 2A_i + I & \delta_i I \\ \delta_i I & -I \end{bmatrix} < -\epsilon I, \ \forall i \in \mathcal{I}_N. \tag{7.2.29}$$

We then obtain easily that for any $t > 0$,

$$\begin{aligned} \frac{d}{dt}V_I(x,t) &= \begin{bmatrix} x(t) \\ x(t-\tau) \end{bmatrix}^T \begin{bmatrix} 2A_i + I & \delta_i I \\ \delta_i I & -I \end{bmatrix} \begin{bmatrix} x(t) \\ x(t-\tau) \end{bmatrix} \\ &\leq -\epsilon \left[x^T(t)x(t) + x^T(t-\tau)x(t-\tau) \right] \end{aligned} \tag{7.2.30}$$

holds for every subsystem. Therefore, (7.2.26) is a common Lyapunov function for all the subsystems in (7.1.4). Under any switching signal, since $V_I(x,t)$ always decreases quadratically and there is no state jump in our switching, the switched system (7.1.4) is asymptotically stable under arbitrary switching. □

Remark 7.2.4. Lemma 7.2.3 and Theorem 7.2.3 can be extended easily to multidelay systems. Suppose that the subsystems are described as

$$\dot{x}(t) = A_i x(t) + \sum_{j=1}^{L} \bar{A}_{ij} x(t - \tau_j), \tag{7.2.31}$$

where $A_i = A_i^T$, $\bar{A}_{ij} = \delta_{ij} I$, $j = 1, \cdots, L$, and τ_js are the time delays. If there exists $P_i > 0$ satisfying the LMI

$$\begin{bmatrix} A_i P_i + P_i A_i + L P_i & \delta_{i1} P_i & \cdots & \delta_{iL} P_i \\ \delta_{i1} P_i & -P_i & \cdots & 0 \\ \vdots & \vdots & \ddots & \vdots \\ \delta_{iL} P_i & 0 & \cdots & -P_i \end{bmatrix} < 0, \qquad \forall i \in \mathcal{I}_N, \tag{7.2.32}$$

then all the subsystems in (7.2.31) are asymptotically stable, and

$$\begin{bmatrix} 2A_i + LI & \delta_{i1} I & \cdots & \delta_{iL} I \\ \delta_{i1} I & -I & \cdots & 0 \\ \vdots & \vdots & \vdots & \vdots \\ \delta_{iL} I & 0 & \cdots & -I \end{bmatrix} < 0, \qquad \forall i \in \mathcal{I}_N. \tag{7.2.33}$$

Furthermore,

$$V_I^L(x,t) = x^T(t)x(t) + \sum_{j=1}^{L} \int_0^{\tau_j} x^T(t-s)x(t-s)ds \tag{7.2.34}$$

is a common Lyapunov function for all the subsystems, and the switched system composed of (7.2.31) is asymptotically stable under arbitrary switching.

7.3 $\mathcal{L}_2$ Gain Analysis

In this section, we study $\mathcal{L}_2$ gain properties of the switched symmetric systems (7.1.1), (7.1.2), and (7.1.4) under arbitrary switching.

7.3.1 Switched symmetric systems without time delay

We first recall the well-known Bounded Real Lemma concerning $\mathcal{L}_2$ gain of continuous-time linear time-invariant (LTI) systems.

Lemma 7.3.1 ([28, 29]). *Consider the continuous-time LTI system*

$$\dot{x}(t) = Ax(t) + Bw(t), \quad z(t) = Cx(t) + Dw(t) \tag{7.3.1}$$

where $x(t)$, $w(t)$, and $z(t)$ are the same as in (7.1.1), and A, B, C, D are constant matrices of appropriate dimensions. The system (7.3.1) has $\mathcal{L}_2$ gain less than γ if and only if there exists $P > 0$ satisfying the LMI

$$\begin{bmatrix} A^TP + PA & PB & C^T \\ B^TP & -\gamma I & D^T \\ C & D & -\gamma I \end{bmatrix} \leq 0. \qquad \square$$

The following result can be proved easily by using the proof technique in [24] with small modifications.

Lemma 7.3.2. *Assume that the continuous-time LTI system (7.3.1) is symmetric in the sense of satisfying $A = A^T$, $B = C^T$, $D = D^T$. Then the system (7.3.1) has $\mathcal{L}_2$ gain less than γ if and only if*

$$\begin{bmatrix} 2A & B & C^T \\ B^T & -\gamma I & D \\ C & D & -\gamma I \end{bmatrix} \leq 0.$$

Based on Lemmas 7.3.1 and 7.3.2, we now state and prove the $\mathcal{L}_2$ gain analysis result for the switched symmetric system (7.1.1).

Theorem 7.3.1. *When all the subsystems in (7.1.1) have $\mathcal{L}_2$ gains less than γ, the switched symmetric system (7.1.1) has $\mathcal{L}_2$ gain less than the same γ under arbitrary switching.*

Proof. According to Lemma 7.3.2, the matrix inequality

$$\begin{bmatrix} 2A_i & B_i & C_i^T \\ B_i^T & -\gamma I & D_i \\ C_i & D_i & -\gamma I \end{bmatrix} \leq 0 \tag{7.3.2}$$

or equivalently,

$$\begin{bmatrix} 2A_i + \frac{1}{\gamma}C_i^T C_i & B_i + \frac{1}{\gamma}C_i^T D_i \\ B_i^T + \frac{1}{\gamma}D_i C_i & -\gamma I + \frac{1}{\gamma}D_i^2 \end{bmatrix} \leq 0 \tag{7.3.3}$$

holds for $\forall i \in \mathcal{I}_N$. Comparing the matrix inequality (7.3.2) with Lemma 7.3.1, we get a hint that $V(x) = x^T x$ should play a role of Lyapunov function in the sense of $\mathcal{L}_2$ gain. We compute the derivative of $V(x) = x^T x$ along the trajectory of any subsystem to obtain

$$\begin{aligned} \frac{dV(x)}{dt} &= x^T(t)(A_i x(t) + B_i w(t)) + (A_i x(t) + B_i w(t))^T x(t) \\ &= \begin{bmatrix} x^T(t) & w^T(t) \end{bmatrix} \begin{bmatrix} 2A_i & B_i \\ B_i^T & 0 \end{bmatrix} \begin{bmatrix} x(t) \\ w(t) \end{bmatrix} \\ &\leq -\begin{bmatrix} x^T(t) & w^T(t) \end{bmatrix} \begin{bmatrix} \frac{1}{\gamma}C_i^T C_i & \frac{1}{\gamma}C_i^T D_i \\ \frac{1}{\gamma}D_i C_i & \frac{1}{\gamma}D_i^2 - \gamma I \end{bmatrix} \begin{bmatrix} x(t) \\ w(t) \end{bmatrix} \\ &= -\frac{1}{\gamma}\Gamma(t), \end{aligned} \tag{7.3.4}$$

where $\Gamma(t) \triangleq z^T(t)z(t) - \gamma^2 w^T(t)w(t)$ and (7.3.3) was used to obtain the inequality.

For an arbitrary piecewise constant switching signal and any given $t > 0$, we let $t_1, \cdots, t_r (r \geq 1)$ denote the switching points of $\sigma(\cdot)$ over the interval $[0, t)$. Then, using the inequality (7.3.4), we obtain

$$\begin{aligned} V(x(t)) - V(x(t_r)) &\leq -\frac{1}{\gamma}\int_{t_r}^{t} \Gamma(s)ds \\ V(x(t_r)) - V(x(t_{r-1})) &\leq -\frac{1}{\gamma}\int_{t_{r-1}}^{t_r} \Gamma(s)ds \\ \cdots \quad &\cdots \quad \cdots \\ V(x(t_1)) - V(x_0) &\leq -\frac{1}{\gamma}\int_{0}^{t_1} \Gamma(s)ds, \end{aligned} \tag{7.3.5}$$

and thus

$$V(x(t)) - V(x_0) \leq -\frac{1}{\gamma}\int_0^t \Gamma(s)ds. \tag{7.3.6}$$

In this inequality, we use the assumption that $x_0 = 0$ and the fact that $V(x(t)) \geq 0$ to obtain

$$\int_0^t z^T(s)z(s)ds \leq \gamma^2 \int_0^t w^T(s)w(s)ds. \tag{7.3.7}$$

Since the above inequality holds for any $t > 0$ including the case of $t \to \infty$ and we did not add any limitation on the switching signal, we declare that the switched symmetric system (7.1.1) has $\mathcal{L}_2$ gain less than γ under arbitrary switching. □

Next we will proceed to study the $\mathcal{L}_2$ gain property for the discrete-time switched symmetric system (7.1.2). To do that, we need the following key lemma.

Lemma 7.3.3. *Consider the discrete-time symmetric system*

$$\begin{cases} x[k+1] = Ax[k] + Bw[k] \\ \quad z[k] = Cx[k] + Dw[k] \, , \end{cases} \tag{7.3.8}$$

where $x[k] \in \Re^n$, $w[k] \in \Re^m$, $z[k] \in \Re^p$, $A = A^T$, $B = C^T$, $D = D^T$, *and* A, B, C, D *are constant matrices of appropriate dimensions. The system (7.3.8) has* $\mathcal{L}_2$ *gain less than* γ *if and only if*

$$\begin{bmatrix} -I & A & B & 0 \\ A & -I & 0 & C^T \\ B^T & 0 & -\gamma I & D \\ 0 & C & D & -\gamma I \end{bmatrix} \le 0 \, . \tag{7.3.9}$$

Proof. Sufficiency: The condition (7.3.9) means that the matrix inequality

$$\begin{bmatrix} -P & PA & PB & 0 \\ A^TP & -P & 0 & C^T \\ B^TP & 0 & -\gamma I & D^T \\ 0 & C & D & -\gamma I \end{bmatrix} \le 0 \tag{7.3.10}$$

is satisfied with $P = I$. Hence, according to the Bounded Real Lemma [28] for discrete-time LTI systems, the system (7.3.8) has $\mathcal{L}_2$ gain less than γ.

Necessity: Suppose that the system (7.3.8) has $\mathcal{L}_2$ gain less than γ. Then there exists a matrix $P_0 > 0$ such that

$$\begin{bmatrix} -P_0 & P_0A & P_0B & 0 \\ AP_0 & -P_0 & 0 & B \\ B^TP_0 & 0 & -\gamma I & D \\ 0 & B^T & D & -\gamma I \end{bmatrix} \le 0 \, . \tag{7.3.11}$$

Now we prove that $P = I$ is also a solution of the above matrix inequality (i.e., (7.3.11) holds when replacing P_0 with I). The proof technique is similar to that of Lemma 7.2.3. We use the same transform (7.2.15). Then pre- and postmultiplying (7.3.11) by $\mathrm{diag}\{U^T, U^T, I, I\}$ and $\mathrm{diag}\{U, U, I, I\}$, respectively, we obtain

$$\begin{bmatrix} -\Sigma_0 & \Sigma_0\bar{A} & \Sigma_0\bar{B} & 0 \\ \bar{A}\Sigma_0 & -\Sigma_0 & 0 & \bar{B} \\ \bar{B}^T\Sigma_0 & 0 & -\gamma I & D \\ 0 & \bar{B}^T & D & -\gamma I \end{bmatrix} \le 0 \, , \tag{7.3.12}$$

where $\bar{B} = U^T B$. Furthermore, pre- and postmultiplying the first and second rows and columns in (7.3.12) by Σ_0^{-1} leads to

$$\begin{bmatrix} -\Sigma_0^{-1} & \bar{A}\Sigma_0^{-1} & \bar{B} & 0 \\ \Sigma_0^{-1}\bar{A} & -\Sigma_0^{-1} & 0 & \Sigma_0^{-1}\bar{B} \\ \bar{B} & 0 & -\gamma I & D \\ 0 & \bar{B}\Sigma_0^{-1} & D & -\gamma I \end{bmatrix} \leq 0 . \tag{7.3.13}$$

In (7.3.13), we exchange the first and second rows and columns, and then exchange the third and fourth rows and columns, to obtain

$$\begin{bmatrix} -\Sigma_0^{-1} & \Sigma_0^{-1}\bar{A} & \Sigma_0^{-1}\bar{B} & 0 \\ \bar{A}\Sigma_0^{-1} & -\Sigma_0^{-1} & 0 & \bar{B} \\ \bar{B}\Sigma_0^{-1} & 0 & -\gamma I & D \\ 0 & \bar{B} & D & -\gamma I \end{bmatrix} \leq 0 . \tag{7.3.14}$$

Using the scalar λ_1 satisfying (7.2.18) to compute $\lambda_1 \times (7.3.12) + (1 - \lambda_1) \times (7.3.14)$, we obtain

$$\begin{bmatrix} -\Sigma_1 & \Sigma_1\bar{A} & \Sigma_1\bar{B} & 0 \\ \bar{A}\Sigma_1 & -\Sigma_1 & 0 & \bar{B} \\ \bar{B}^T\Sigma_1 & 0 & -\gamma I & D \\ 0 & \bar{B}^T & D & -\gamma I \end{bmatrix} \leq 0 , \tag{7.3.15}$$

where Σ_1 was defined in (7.2.20).

Similar to (7.3.14), we can obtain

$$\begin{bmatrix} -\Sigma_1^{-1} & \Sigma_1^{-1}\bar{A} & \Sigma_1^{-1}\bar{B} & 0 \\ \bar{A}\Sigma_1^{-1} & -\Sigma_1^{-1} & 0 & \bar{B} \\ \bar{B}^T\Sigma_1^{-1} & 0 & -\gamma I & D \\ 0 & \bar{B}^T & D & -\gamma I \end{bmatrix} \leq 0 \tag{7.3.16}$$

from (7.3.15). Also, using the scalar λ_2 satisfying (7.2.22), we obtain the linear combination of (7.3.15) and (7.3.16) as

$$\begin{bmatrix} -\Sigma_2 & \Sigma_2\bar{A} & \Sigma_2\bar{B} & 0 \\ \bar{A}\Sigma_2 & -\Sigma_2 & 0 & \bar{B} \\ \bar{B}^T\Sigma_2 & 0 & -\gamma I & D \\ 0 & \bar{B}^T & D & -\gamma I \end{bmatrix} \leq 0 , \tag{7.3.17}$$

where Σ_2 was defined in (7.2.24).

By repeating this process, we see that $\Sigma_n = I$ also satisfies (7.3.12), i.e.,

$$\begin{bmatrix} -I & \bar{A} & \bar{B} & 0 \\ \bar{A} & -I & 0 & \bar{B} \\ \bar{B}^T & 0 & -\gamma I & D \\ 0 & \bar{B}^T & D & -\gamma I \end{bmatrix} \leq 0 . \tag{7.3.18}$$

Pre- and postmultiplying this inequality by $\mathrm{diag}\{U, U, I, I\}$ and $\mathrm{diag}\{U^T, U^T, I, I\}$, respectively, we obtain (7.3.9). This completes the proof.

We assume that all the subsystems in (7.1.2) have $\mathcal{L}_2$ gains less than γ. Then, according to Lemma 7.3.3, we have

$$\begin{bmatrix} -I & A_i & B_i & 0 \\ A_i & -I & 0 & C_i^T \\ B_i^T & 0 & -\gamma I & D_i \\ 0 & C_i & D_i & -\gamma I \end{bmatrix} \leq 0 \tag{7.3.19}$$

for all $i \in \mathcal{I}_N$, which is equivalent to

$$\begin{bmatrix} A_i^2 + \frac{1}{\gamma} C_i^T C_i - I & A_i B_i + \frac{1}{\gamma} C_i^T D_i \\ B_i^T A_i + \frac{1}{\gamma} D_i C_i & B_i^T B_i + \frac{1}{\gamma} D_i^2 - \gamma I \end{bmatrix} \leq 0 \,. \tag{7.3.20}$$

We now compute the difference of the Lyapunov function candidate $V(x) = x^T x$ along the trajectory of any subsystem to obtain

$$\begin{aligned} & V(x[k+1]) - V(x[k]) \\ &\quad = x^T[k+1]x[k+1] - x^T[k]x[k] \\ &\quad = (A_i x[k] + B_i w[k])^T (A_i x[k] + B_i w[k]) - x^T[k]x[k] \\ &\quad = \begin{bmatrix} x^T[k] & w^T[k] \end{bmatrix} \begin{bmatrix} A_i^2 - I & A_i^T B_i \\ B_i^T A_i & B_i^T B_i \end{bmatrix} \begin{bmatrix} x[k] \\ w[k] \end{bmatrix} \\ &\quad \leq - \begin{bmatrix} x^T[k] & w^T[k] \end{bmatrix} \begin{bmatrix} \frac{1}{\gamma} C_i^T C_i & \frac{1}{\gamma} C_i^T D_i \\ \frac{1}{\gamma} D_i C_i & \frac{1}{\gamma} D_i^2 - \gamma I \end{bmatrix} \begin{bmatrix} x[k] \\ w[k] \end{bmatrix} \\ &\quad = -\frac{1}{\gamma} \Gamma[k] \,, \end{aligned} \tag{7.3.21}$$

where $\Gamma[k] \triangleq z^T[k]z[k] - \gamma^2 w^T[k]w[k]$ and (7.3.20) was used to obtain the inequality.

For an arbitrary piecewise constant switching signal and any given integer $k > 0$, we let $k_1, \cdots, k_r (r \geq 1)$ denote the switching points of $\sigma(k)$ over the interval $[0, k)$. Then, using the difference inequality (7.3.21), we obtain

$$\begin{aligned} V(x[k]) - V(x[k_r]) &\leq -\frac{1}{\gamma} \sum_{j=k_r}^{k-1} \Gamma[j] \\ V(x[k_r]) - V(x[k_{r-1}]) &\leq -\frac{1}{\gamma} \sum_{j=k_{r-1}}^{k_r-1} \Gamma[j] \\ \cdots \quad & \cdots \quad \cdots \\ V(x[k_1]) - V(x[0]) &\leq -\frac{1}{\gamma} \sum_{j=0}^{k_1-1} \Gamma[j] \,, \end{aligned} \tag{7.3.22}$$

and thus

$$V(x[k]) - V(x[0]) \leq -\frac{1}{\gamma}\sum_{j=0}^{k-1}\Gamma[j]\,. \tag{7.3.23}$$

In this inequality, we use the assumption that $x[0] = 0$ and the fact that $V(x[k]) \geq 0$ to obtain

$$\sum_{j=0}^{k} z^T[j]z[j] \leq \gamma^2 \sum_{j=0}^{k} w^T[j]w[j]\,. \tag{7.3.24}$$

We note that the above inequality holds for any $k > 0$ including the case of $k \to \infty$, and that we did not add any limitation on the switching signal.

We summarize the above discussion in the following theorem.

Theorem 7.3.2. *When all the subsystems in (7.1.2) have $\mathcal{L}_2$ gains less than γ, the switched symmetric system (7.1.2) has $\mathcal{L}_2$ gain less than the same γ under arbitrary switching.*

Remark 7.3.1. From Lemma 7.3.3 and the proof of Theorem 7.3.2, we see that if all the subsystems in (7.1.2) have $\mathcal{L}_2$ gains less than γ, then there exists a common Lyapunov matrix $P = I$ for all subsystems, satisfying the LMI

$$\begin{bmatrix} -P & PA_i & PB_i & 0 \\ A_i^T P & -P & 0 & C_i^T \\ B_i^T P & 0 & -\gamma I & D_i^T \\ 0 & C_i & D_i & -\gamma I \end{bmatrix} \leq 0\,, \qquad \forall i \in \mathcal{I}_N\,. \tag{7.3.25}$$

Hence, $V(x) = x^T x$ serves as a common Lyapunov function for all the subsystems in the sense of $\mathcal{L}_2$ gain.

Remark 7.3.2. According to Theorems 7.3.1 and 7.3.2, the problem of finding a tight $\mathcal{L}_2$ gain for the switched symmetric system (7.1.1) or (7.1.2) is reduced to the one of minimizing $\gamma > 0$ subject to the constraint of

$$\begin{bmatrix} 2A_i & B_i & C_i^T \\ B_i^T & -\gamma I & D_i \\ C_i & D_i & -\gamma I \end{bmatrix} \leq 0\,, \qquad \forall i \in \mathcal{I}_N\,, \tag{7.3.26}$$

or (7.3.25).

7.3.2 Switched symmetric systems with time delay

In this subsection we assume $\phi(t) = 0,\ t \in [-\tau, 0]$ and study $\mathcal{L}_2$ gain property of the switched symmetric system (7.1.4) under arbitrary switching. To proceed, we need the following lemma.

Lemma 7.3.4. *Consider the linear time-delay system*

$$\begin{cases} \dot{x}(t) = Ax(t) + \bar{A}x(t-\tau) + Bw(t) \\ x(t) = 0, \quad \forall t \in [-\tau, 0] \\ z(t) = Cx(t) + Dw(t), \end{cases} \tag{7.3.27}$$

where $x(t)$, $w(t)$, and $z(t)$ are the same as before, and $\tau > 0$ is the time delay. If there exist $P > 0$ and $\bar{P} > 0$ satisfying the LMI

$$\begin{bmatrix} A^T P + PA + \bar{P} & P\bar{A} & PB & C^T \\ \bar{A}^T P & -\bar{P} & 0 & 0 \\ B^T P & 0 & -\gamma I & D^T \\ C & 0 & D & -\gamma I \end{bmatrix} \leq 0, \tag{7.3.28}$$

then the system (7.3.27) has $\mathcal{L}_2$ gain less than γ.

Proof. For the Lyapunov function candidate (7.2.12), we compute its derivative along the trajectories of (7.3.27) as

$$\frac{dV(x,t)}{dt} = \begin{bmatrix} x(t) \\ x(t-\tau) \\ w(t) \end{bmatrix}^T \begin{bmatrix} A^T P + PA + \bar{P} & P\bar{A} & PB \\ \bar{A}^T P & -\bar{P} & 0 \\ B^T P & 0 & 0 \end{bmatrix} \begin{bmatrix} x(t) \\ x(t-\tau) \\ w(t) \end{bmatrix}.$$

Since (7.3.28) is equivalent to

$$\begin{bmatrix} A^T P + PA + \bar{P} + \frac{1}{\gamma} C^T C & P\bar{A} & PB + \frac{1}{\gamma} C^T D \\ \bar{A}^T P & -\bar{P} & 0 \\ B^T P + \frac{1}{\gamma} D^T C & 0 & -\gamma I + \frac{1}{\gamma} D^T D \end{bmatrix} \leq 0, \tag{7.3.29}$$

we obtain that

$$\begin{aligned} \frac{dV(x,t)}{dt} &\leq -\begin{bmatrix} x(t) \\ x(t-\tau) \\ w(t) \end{bmatrix}^T \begin{bmatrix} \frac{1}{\gamma} C^T C & 0 & \frac{1}{\gamma} C^T D \\ 0 & 0 & 0 \\ \frac{1}{\gamma} D^T C & 0 & -\gamma I + \frac{1}{\gamma} D^T D \end{bmatrix} \begin{bmatrix} x(t) \\ x(t-\tau) \\ w(t) \end{bmatrix} \\ &= -\frac{1}{\gamma} \Gamma(t). \end{aligned} \tag{7.3.30}$$

Integrating both sides of (7.3.30) and using $V(x(t), t) \geq 0$, $V(x(0), 0) = 0$, we obtain

$$\int_0^t z^T(s) z(s) ds \leq \gamma^2 \int_0^t w^T(s) w(s) ds, \tag{7.3.31}$$

which implies the system (7.3.27) has $\mathcal{L}_2$ gain less than γ. □

Next we state and prove a lemma that plays an important role in the remaining discussion of this subsection.

Lemma 7.3.5. *Assume that the linear time-delay system (7.3.27) is symmetric in the sense of satisfying* $A = A^T$, $\bar{A} = \delta I_n$, $B = C^T$, $D = D^T$. *If there exists* $P > 0$ *satisfying the LMI*

$$\begin{bmatrix} AP + PA + P & \delta P & PB & B \\ \delta P & -P & 0 & 0 \\ B^T P & 0 & -\gamma I & D \\ B^T & 0 & D & -\gamma I \end{bmatrix} \leq 0, \tag{7.3.32}$$

then the system (7.3.27) has $\mathcal{L}_2$ *gain less than* γ. *Furthermore,*

$$\begin{bmatrix} 2A + I & \delta I & B & B \\ \delta I & -I & 0 & 0 \\ B^T & 0 & -\gamma I & D \\ B^T & 0 & D & -\gamma I \end{bmatrix} \leq 0. \tag{7.3.33}$$

Proof. Suppose that $P_0 > 0$ satisfies (7.3.32), i.e.,

$$\begin{bmatrix} AP_0 + P_0 A + P_0 & \delta P_0 & P_0 B & B \\ \delta P_0 & -P_0 & 0 & 0 \\ B^T P_0 & 0 & -\gamma I & D \\ B^T & 0 & D & -\gamma I \end{bmatrix} \leq 0. \tag{7.3.34}$$

Then, it is obvious that (7.3.28) is satisfied with $P = \bar{P} = P_0$, and thus the system (7.3.27) has $\mathcal{L}_2$ gain less than γ according to Lemma 7.3.4. We now proceed to prove (7.3.33).

We use the same idea as in the proofs of Lemmas 7.2.3 and 7.3.3. Since $P_0 > 0$, we do the same similarity transformation as in (7.2.15), and pre- and postmultiply (7.3.34) by $\text{diag}\{U^T, U^T, I, I\}$ and $\text{diag}\{U, U, I, I\}$, respectively, to obtain

$$\begin{bmatrix} A_U \Sigma_0 + \Sigma_0 A_U + \Sigma_0 & \delta \Sigma_0 & \Sigma_0 B_U & B_U \\ \delta \Sigma_0 & -\Sigma_0 & 0 & 0 \\ B_U^T \Sigma_0 & 0 & -\gamma I & D \\ B_U^T & 0 & D & -\gamma I \end{bmatrix} \leq 0. \tag{7.3.35}$$

Furthermore, pre- and postmultiplying the first and second rows and columns in (7.3.35) by Σ_0^{-1}, and then exchanging the third and the fourth rows and columns, leads to

$$\begin{bmatrix} A_U \Sigma_0^{-1} + \Sigma_0^{-1} A_U + \Sigma_0^{-1} & \delta \Sigma_0^{-1} & \Sigma_0^{-1} B_U & B_U \\ \delta \Sigma_0^{-1} & -\Sigma_0^{-1} & 0 & 0 \\ B_U^T \Sigma_0^{-1} & 0 & -\gamma I & D \\ B_U^T & 0 & D & -\gamma I \end{bmatrix} \leq 0. \tag{7.3.36}$$

Using λ_1 that satisfies (7.2.18) to compute $\lambda_1\times$(7.3.35) + $(1-\lambda_1)\times$(7.3.36), we obtain

$$\begin{bmatrix} A_U\Sigma_1+\Sigma_1 A_U+\Sigma_1 & \delta\Sigma_1 & \Sigma_1 B_U & B_U \\ \delta\Sigma_1 & -\Sigma_1 & 0 & 0 \\ B_U^T\Sigma_1 & 0 & -\gamma I & D \\ B_U^T & 0 & D & -\gamma I \end{bmatrix} \le 0\,, \tag{7.3.37}$$

where Σ_1 is defined in (7.2.20).

By repeating this process as is done in Lemmas 7.2.3 and 7.3.3, we see that $\Sigma_n = I$ also satisfies (7.3.35), i.e.,

$$\begin{bmatrix} 2A_U+I & \delta I & B_U & B_U \\ \delta I & -I & 0 & 0 \\ B_U^T & 0 & -\gamma I & D \\ B_U^T & 0 & D & -\gamma I \end{bmatrix} \le 0\,. \tag{7.3.38}$$

Pre- and postmultiplying this inequality by $\mathrm{diag}\{U,U,I,I\}$ and $\mathrm{diag}\{U^T,U^T,I,I\}$, respectively, we obtain (7.3.33). This completes the proof.

We now state and prove the main result of this subsection.

Theorem 7.3.3. *Assume that in (7.1.4) $\bar{A}_i = \delta_i I$, $\forall i \in \mathcal{I}_N$. If all the subsystems have $\mathcal{L}_2$ gains less than γ in the sense that there exists $P_i > 0$ satisfying the LMI*

$$\begin{bmatrix} A_iP_i+P_iA_i+P_i & \delta_iP_i & P_iB_i & B_i \\ \delta_iP_i & -P_i & 0 & 0 \\ B_i^TP_i & 0 & -\gamma I & D_i \\ B_i^T & 0 & D_i & -\gamma I \end{bmatrix} \le 0\,, \tag{7.3.39}$$

then (7.2.26) is a common Lyapunov function for all the subsystems in the sense of $\mathcal{L}_2$ gain, and that the switched system (7.1.4) has $\mathcal{L}_2$ gain less than the same γ under arbitrary switching.

Proof. If (7.3.39) is true, then according to Lemma 7.3.5,

$$\begin{bmatrix} 2A_i+I & \delta_iI & B_i & B_i \\ \delta_iI & -I & 0 & 0 \\ B_i^T & 0 & -\gamma I & D_i \\ B_i^T & 0 & D_i & -\gamma I \end{bmatrix} \le 0 \tag{7.3.40}$$

holds for all $i \in \mathcal{I}_N$. From Lemmas 7.3.4 and 7.3.5 and their proofs, we obtain easily that the derivative of (7.2.26) along the trajectories of any subsystem satisfies

$$\frac{d}{dt}V_I(x,t) \le -\frac{1}{\gamma}\Gamma(t)\,, \tag{7.3.41}$$

which implies that (7.2.26) is a common Lyapunov function for all the subsystems in (7.1.4) in the sense of $\mathcal{L}_2$ gain.

As before, for any piecewise constant switching signal $\sigma(t)$ and any given $t > 0$, we let $t_1, \cdots, t_r$ $(r \geq 1)$ denote the switching points of $\sigma(t)$ over the interval $[0, t)$. From (7.3.41), we obtain

$$\begin{aligned} V_I(t, x(t)) - V_I(t_r, x(t_r)) &\leq -\frac{1}{\gamma}\int_{t_r}^{t} \Gamma(s)ds \\ V_I(t_r, x(t_r)) - V_I(t_{r-1}, x(t_{r-1})) &\leq -\frac{1}{\gamma}\int_{t_{r-1}}^{t_r} \Gamma(s)ds \\ \cdots \quad & \cdots \\ V_I(t_1, x(t_1)) - V_I(0, x(0)) &\leq -\frac{1}{\gamma}\int_{0}^{t_1} \Gamma(s)ds\,, \end{aligned} \tag{7.3.42}$$

and thus

$$V_I(t, x(t)) - V_I(0, x(0)) \leq -\frac{1}{\gamma}\int_0^t \Gamma(s)ds\,. \tag{7.3.43}$$

We use the assumption that $\phi(t) = 0$ $(-\tau \leq t \leq 0)$ and the fact that $V_I(t, x(t)) \geq 0$ to obtain

$$\int_0^t z^T(s)z(s)ds \leq \gamma^2 \int_0^t w^T(s)w(s)ds\,. \tag{7.3.44}$$

We note that the above inequality holds for any $t > 0$ including the case of $t \to \infty$, and that we did not add any limitation on the switching signal. Therefore, the switched system (7.1.4) has $\mathcal{L}_2$ gain less than γ under arbitrary switching. □

Remark 7.3.3. Lemma 7.3.5 and Theorem 7.3.3 can be extended easily to multidelay systems. Suppose that the subsystems are described as

$$\begin{cases} \dot{x}(t) = A_i x(t) + \sum_{j=1}^{L} \bar{A}_{ij} x(t - \tau_j) + B_i w(t) \\ z(t) = C_i x(t) + D_i w(t)\,, \end{cases} \tag{7.3.45}$$

where $A_i = A_i^T$, $\bar{A}_{ij} = \delta_{ij} I$, $B_i = C_i^T$, $D_i = D_i^T$, $j = 1, \cdots, L$, and τ_js are the time delays. If there exists $P_i > 0$ satisfying the LMI

$$\begin{bmatrix} A_iP_i + P_iA_i + LP_i & \delta_{i1}P_i & \cdots & \delta_{iL}P_i & P_iB_i & B_i \\ \delta_{i1}P_i & -P_i & \cdots & 0 & 0 & 0 \\ \vdots & \vdots & \ddots & \vdots & 0 & 0 \\ \delta_{iL}P_i & 0 & \cdots & -P_i & 0 & 0 \\ B_i^TP_i & 0 & \cdots & 0 & -\gamma I & D_i \\ B_i^T & 0 & \cdots & 0 & D_i & -\gamma I \end{bmatrix} \leq 0\,, \tag{7.3.46}$$

then all the subsystems in (7.3.45) have $\mathcal{L}_2$ gains less than γ, and

$$\begin{bmatrix} 2A_i + LP_i & \delta_{i1}I & \cdots & \delta_{iL}I & B_i & B_i \\ \delta_{i1}I & -I & \cdots & 0 & 0 & 0 \\ \vdots & \vdots & \ddots & \vdots & 0 & 0 \\ \delta_{iL}I & 0 & \cdots & -I & 0 & 0 \\ B_i^T & 0 & \cdots & 0 & -\gamma I & D_i \\ B_i^T & 0 & \cdots & 0 & D_i & -\gamma I \end{bmatrix} \le 0 . \tag{7.3.47}$$

Furthermore, (7.2.34) is a common Lyapunov function for all the subsystems in the sense of $\mathcal{L}_2$ gain, and the switched system composed of (7.3.45) has $\mathcal{L}_2$ gain less than γ under arbitrary switching.

7.4 Conclusions

In this chapter we have studied stability and $\mathcal{L}_2$ gain properties for a class of switched systems that are composed of a finite number of LTI symmetric subsystems. Assuming that all subsystems are (Hurwitz or Schur) stable and have $\mathcal{L}_2$ gains less than a positive scalar γ, we have shown for both stability and $\mathcal{L}_2$ gain analysis that there exists a common Lyapunov function $V(x) = x^T x$ for all the subsystems, and that the switched system is exponentially stable and has $\mathcal{L}_2$ gain less than the same γ under arbitrary switching. The discussions have also been extended to switched continuous-time systems with time delay in the state.

We note finally that the results of the present chapter can be extended to the switched symmetric systems in a more general sense. More precisely, if the equations $TA_i = A_i^T T$, $\bar{A}_i = \delta_i I_n$, $TB_i = C_i^T$, $D_i = D_i^T$ are satisfied for a constant matrix $T > 0$, then we consider the similarity transformation $A_{\star i} = T^{\frac{1}{2}} A_i T^{-\frac{1}{2}}$, $B_{\star i} = T^{\frac{1}{2}} B_i$, $C_{\star i} = C_i T^{-\frac{1}{2}}$, $D_{\star i} = D_i$. Since the stability and $\mathcal{L}_2$ gain properties of the system in this transformation do not change and we can easily confirm that $A_{\star i} = A_{\star i}^T$, $B_{\star i} = C_{\star i}^T$ and $D_{\star i}^T = D_{\star i}$, we can apply the results we have obtained up to now for the systems represented by $(A_{\star i}, B_{\star i}, C_{\star i}, D_{\star i})$ and derive corresponding results for the original switched system under arbitrary switching.

Acknowledgments

The author would like to thank Dr. Anthony N. Michel and Dr. Bo Hu of the University of Notre Dame, Dr. Xinkai Chen of Kinki University, Dr. Masao Ikeda of Osaka University, Dr. Kazunori Yasuda of Wakayama University, and Dr. João Pedro Hespanha of the University of California, Santa Barbara, for their valuable discussions, which greatly contributed to this chapter.

Bibliography

[1] W. P. Dayawansa and C. F. Martin, "A converse Lyapunov theorem for a class of dynamical systems which undergo switching," *IEEE Trans. Automatic Control*,

vol. 44, no. 4, pp. 751–760, 1999.

[2] A. S. Morse, "Supervisory control of families of linear set-point controllers, Part 1: Exact matching," *IEEE Trans. Automatic Control*, vol. 41, no. 10, pp. 1413–1431, 1996.

[3] D. Liberzon and A. S. Morse, "Basic problems in stability and design of switched systems," *IEEE Control Systems Magazine*, vol. 19, no. 5, pp. 59–70, 1999.

[4] B. Hu, G. Zhai, and A. N. Michel, "Hybrid output feedback stabilization of two-dimensional linear control systems," *Proc. American Control Conference*, pp. 2184–2188, Chicago, IL, 2000; see also *Linear Algebra and its Applications*, vol. 351–352, pp. 475–485, 2002.

[5] R. DeCarlo, M. Branicky, S. Pettersson, and B. Lennartson, "Perspectives and results on the stability and stabilizability of hybrid systems," *Proc. IEEE*, vol. 88, no. 7, pp. 1069–1082, 2000.

[6] J. P. Hespanha and A. S. Morse, "Stability of switched systems with average dwell-time," *Proc. 38th IEEE Conference on Decision and Control*, pp. 2655–2660, Phoenix, AZ, 1999.

[7] G. Zhai, B. Hu, K. Yasuda, and A. N. Michel, "Stability analysis of switched systems with stable and unstable subsystems: An average dwell time approach," *Int. J. Systems Science*, vol. 32, no. 8, pp. 1055–1061, 2001.

[8] G. Zhai, B. Hu, K. Yasuda, and A. N. Michel, "Piecewise Lyapunov functions for switched systems with average dwell time," *Asian J. Control*, vol. 2, no. 3, pp. 192–197, 2000.

[9] M. A. Wicks, P. Peleties, and R. A. DeCarlo, "Construction of piecewise Lyapunov functions for stabilizing switched systems," *Proc. 33rd IEEE Conference on Decision and Control*, pp. 3492–3497, Orlando, FL, 1994.

[10] B. Hu, X. Xu, A. N. Michel, and P. J. Antsaklis, "Stability analysis for a class of nonlinear switched systems," *Proc. 38th IEEE Conference Decision and Control*, pp. 4374–4379, Phoenix, AZ, 1999.

[11] G. Zhai and K. Yasuda, "Stability analysis for a class of switched systems," *Trans. Society of Instrument and Control Engineers*, vol. 36, no. 5, pp. 409–415, 2000.

[12] D. Liberzon, J. P. Hespanha, and A. S. Morse, "Stability of switched systems: A Lie-algebraic condition," *Systems and Control Letters*, vol. 37, no. 3, pp. 117–122, 1999.

[13] S. Pettersson and B. Lennartson, "LMI for stability and robustness of hybrid systems," *Proc. American Control Conference*, pp. 1714–1718, Albuquerque, NM, 1997.

[14] M. A. Wicks, P. Peleties, and R. A. DeCarlo, "Switched controller design for the quadratic stabilization of a pair of unstable linear systems," *European J. Control*, vol. 4, pp. 140–147, 1998.

[15] G. Zhai, "Quadratic stabilizability of discrete-time switched systems via state and output feedback," *Proc. 40th IEEE Conference Decision and Control*, pp. 2165–2166, Orlando, FL, 2001.

[16] K. S. Narendra and V. Balakrishnan, "A common Lyapunov function for stable LTI systems with commuting A-matrices," *IEEE Trans. Automatic Control*, vol. 39, no. 12, pp. 2469–2471, 1994.

[17] J. P. Hespanha, *Logic-Based Switching Algorithms in Control*, Ph.D. Dissertation, Yale University, 1998.

[18] G. Zhai, B. Hu, K. Yasuda, and A. N. Michel, "Disturbance attenuation properties of time-controlled switched systems," *J. Franklin Institute*, vol. 338, no. 7, pp. 765–779, 2001.

[19] J. P. Hespanha, "Computation of $\mathcal{L}_2$-induced norms of switched linear systems," *Proc. 5th International Workshop of Hybrid Systems: Computation and Control*, pp. 238–252, Stanford University, Stanford, CA, 2002.

[20] B. Hu and A. N. Michel, "Stability analysis of digital feedback control systems with time-varying sampling periods," *Automatica*, vol. 36, pp. 897–905, 2000.

[21] M. Rubensson and B. Lennartson, "Stability and robustness of hybrid systems using discrete-time Lyapunov techniques," *Proc. American Control Conference*, pp. 210–214, Chicago, IL, 2000.

[22] M. Ikeda, "Symmetric controllers for symmetric plants," *Proc. 3rd European Control Conference*, pp. 988–993, Rome, Italy, 1995.

[23] M. Ikeda, K. Miki, and G. Zhai, "$\mathcal{H}_\infty$ controllers for symmetric systems: A theory for attitude control of large space structures," *Proc. 2001 International Conference on Control, Automation and Systems*, pp. 651–654, Cheju National University, Korea, 2001.

[24] K. Tan and K. M. Grigoriadis, "Stabilization and $\mathcal{H}^\infty$ control of symmetric systems: An explicit solution," *Systems and Control Letters*, vol. 44, pp. 57–72, 2001.

[25] W. Zhang, M. S. Branicky, and S. M. Phillips, "Stability of networked control systems," *IEEE Control Systems Magazine*, vol. 20, pp. 84–99, 2001.

[26] J. Hale, *Functional Differential Equations*, Springer-Verlag, New York, 1971.

[27] Y. Kuang, *Delay Differential Equations with Applications in Population Dynamics*, Academic Press, New York, 1993.

[28] T. Iwasaki, R. E. Skelton, and K. M. Grigoriadis, *A Unified Algebraic Approach to Linear Control Design*, Taylor & Francis, London, 1998.

[29] S. Boyd, L. El Ghaoui, E. Feron, and V. Balakrishnan, *Linear Matrix Inequalities in System and Control Theory*, SIAM, Philadelphia, 1994.

PART II

NEURAL NETWORKS AND SIGNAL PROCESSING

Chapter 8

Approximation of Input-Output Maps using Gaussian Radial Basis Functions

Irwin W. Sandberg

Abstract: Radial basis functions are of interest in connection with a variety of approximation problems in the neural networks area, and in other areas as well. Here we show that the members of some interesting families of shift-varying input-output maps, which take a function space into a function space, can be uniformly approximated over an infinite time or space domain in a certain special way using Gaussian radial basis functions.

8.1 Introduction

Radial basis functions are of interest in connection with a variety of approximation problems in the neural networks area, and in other areas as well. Much is understood about the properties of these functions (see, for instance, [1–4]). It is known [2], for example, that arbitrarily good approximation in $L_1(\mathbb{R}^n)$ of a general $f \in L_1(\mathbb{R}^n)$ is possible using uniform smoothing factors and radial basis functions generated in a certain natural way from a single g in $L_1(\mathbb{R}^n)$ if and only if g has a nonzero integral. As another example, in [3] it is proved that Gaussian radial basis functions can uniformly approximate arbitrarily well any continuous real functional defined on a compact convex subset of $\mathbb{R}^n$.

In [5] an approximation result involving the concept of locally compact metric spaces * is proved concerning Gaussian radial basis functions in a general inner product space, and two applications are given. In particular, it is shown that Gaussian radial basis functions defined on $\mathbb{R}^n$ can in fact uniformly approximate arbitrarily well over *all* of $\mathbb{R}^n$ any continuous real functional f on $\mathbb{R}^n$ that meets the condition that

$$\lim_{\|x\|\to\infty} |f(x)| = 0.$$

This generalizes the result in [3] because, by the Lebesgue-Urysohn extension theorem [6, p. 63] (sometimes attributed to Tietze), any continuous real functional defined on a bounded closed subset of $\mathbb{R}^n$ can be extended so that it is defined and continuous on all of $\mathbb{R}^n$ and meets the above condition.† The second application concerns the problem of classifying signals, and related results concerning the structure of reconfigurable classifiers are given in [7].

This chapter too is concerned with the capabilities of Gaussian radial basis functions and the concept of locally compact metric spaces. Here we show that the members of some interesting families of shift-varying input-output maps G, which take a function space into a function space, can be uniformly approximated in a certain special way. More specifically, we consider a setting in which A denotes, for example, the (noncompact) time set $[0,\infty)$ or the entire plane $\mathbb{R}^2$, and X_2 (we avoid the use of X or X_1 because they are used differently in the following section) is taken to be a certain subset of an inner-product function space S_2. Using a result in [7], we show in Section 8.3 (see Theorem 8.3.1 and its corollary) that the elements of important families of Gs from X_2 into the set of real-valued functions defined on A can be uniformly approximated arbitrarily well using Gaussian radial basis functions, in the sense that given $\epsilon > 0$ there are a positive integer q, real functions $c_1, \cdots, c_q$ on A, positive numbers $\beta_1, \cdots, \beta_q$, and elements $u_1, \cdots, u_q$ of a certain subset of S_2 such that

$$|(Gx)(a) - \sum_{k=1}^{q} c_k(a) \exp\{-\beta_k \|x - u_k\|^2\}| < \epsilon, \quad a \in A \tag{8.1.1}$$

for all $x \in X_2$, in which $\|\cdot\|$ is the norm in S_2. Assuming that $A = [0,\infty)$, we show also that the coefficients $c_k(a)$ can be taken to be given by

$$c_k(a) = \alpha_k \exp\{-\beta_k |a - a_k|^2\}$$

in which $\alpha_1, \cdots, \alpha_q$ are real numbers and $a_1, \cdots, a_q$ belong to A. Results of this kind are of interest in connection with, for example, system identification and adaptive systems.‡

---

* A metric space M is locally compact if for each point x of M there is an open subset O_x of M such that $x \in O_x$ and the closure of O_x is compact. For example, $\mathbb{R}^n$ is locally compact.

† For example, let a continuous f_0 defined on a bounded closed subset A of $\mathbb{R}^n$ be given, and let A be contained in an open ball B centered at the origin of $\mathbb{R}^n$. Then, since the complement C of B with respect to $\mathbb{R}^n$ is closed (and thus $A \cup C$ is closed), by the Lebesgue-Urysohn extension theorem there is a continuous extension f of f_0 defined on $\mathbb{R}^n$ such that $f(x) = 0, x \in C$.

‡ For material related in a general sense, but with the emphasis on shift-invariant systems, see, for instance, [8].

A key assumption under which (8.1.1) holds is that $(Gx)(a)$ vanishes at infinity uniformly in x, meaning that for each $\gamma > 0$ there is a compact subset A_γ of A for which

$$|(Gx)(a)| < \gamma \ \ (a \notin A_\gamma, x \in X_2). \tag{8.1.2}$$

This condition is often met in situations in which inputs and outputs vanish at infinity. A detailed example involving systems governed by integral equations is given in Section 8.4. Results related to our main result described above, concerning approximations using functions other than Gaussian functions, are discussed in Section 8.5.

8.2 Preliminaries

We first describe a theorem in [7] that provides the foundation for our main result. Let S be a real inner-product space [§] (i.e., a real pre-Hilbert space) with inner product $\langle \cdot , \cdot \rangle$ and norm $\| \cdot \|$ derived in the usual way from $\langle \cdot , \cdot \rangle$. Let X be a metric space whose points are a subset of the points of S, and denote the metric in X by d. With V any convex subset of S such that $\{ \| \cdot + v \| : v \in V \}$ separates the points of X (i.e., such that for x and y in X with $x \neq y$ there is a $v \in V$ for which $\| x + v \| \neq \| y + v \|$) and with P any nonempty subset of $(0, \infty)$ that is closed under addition, let $\mathcal{X}_0$ denote the set of functions g defined on X that have the representation

$$g(x) = \alpha \exp\{-\beta \|x - v\|^2\} \tag{8.2.1}$$

in which $\alpha \in I\!R$, $\beta \in P$, and $v \in V$.

Let $\mathcal{X}$ stand for the set of continuous functions f from X to the reals $I\!R$ with the property that for each $\epsilon > 0$ there is a compact subset $X_{f,\epsilon}$ of X such that

$$|f(x) - f(y)| < \epsilon$$

for $x, y \notin X_{f,\epsilon}$. And let $\mathcal{X}_\infty$ denote the family of fs in $\mathcal{X}$ such that for each f and each $\epsilon > 0$ there is a compact subset $X_{f,\epsilon}$ of X such that $|f(x)| < \epsilon$ for $x \notin X_{f,\epsilon}$.

Now let S_1 and S_2 be two additional real inner-product spaces, with inner products $\langle \cdot , \cdot \rangle_1$ and $\langle \cdot , \cdot \rangle_2$, and norms $\| \cdot \|_1$ and $\| \cdot \|_2$, respectively. The norms in S_1 and S_2 are derived from their respective inner products in the usual way. Take S to be the inner-product space $S_1 \times S_2$ with inner product $\langle \cdot , \cdot \rangle$ given by

$$\langle (a_1, b_1)\,(a_2, b_2) \rangle = \langle a_1, a_2 \rangle_1 + \langle b_1, b_2 \rangle_2 .$$

Let (X_1, d_1) and (X_2, d_2) be two metric spaces whose points are subsets of the points of S_1 and S_2, respectively. (The metrics d_1 and d_2 are not necessarily derived in the usual way from $\| \cdot \|_1$ and $\| \cdot \|_2$.) Take X to be the metric space $X_1 \times X_2$ with the metric given by

$$d[(a_1, b_1), (a_2, b_2)] = d_1(a_1, a_2) + d_2(b_1, b_2).$$

[§] Theorems 8.2.1 and 8.3.1, which follow, hold also if all inner-product spaces considered are complex inner-product spaces.

Recall that V is a convex subset of S such that $\{\|\cdot + v\| : v \in V\}$ separates the points of X (i.e., such that for x and y in X with $x \neq y$ there is a $v \in V$ for which $\|x + v\| \neq \|y + v\|$) and that P is any nonempty subset of $(0, \infty)$ that is closed under addition. For $v \in S$, we use v_1 and v_2 to denote the components of v belonging to S_1 and S_2, respectively.

The result in [7] is the following:

Theorem 8.2.1. *Assume that (X_1, d_1) and (X_2, d_2) are locally compact but that (X_1, d_1) or (X_2, d_2), or both, are not compact, and that X is such that $\mathcal{X}_0 \subset \mathcal{X}_\infty$. Then for each $f \in \mathcal{X}_\infty$ and each $\epsilon > 0$ there are a positive integer q, real numbers $\alpha_1, \cdots, \alpha_q$, numbers $\beta_1, \cdots, \beta_q$ belonging to P, and elements $v_1, \cdots, v_q$ of V such that*

$$\left| f(a,b) - \sum_{k=1}^{q} \alpha_k \exp\{-\beta_k \|a - v_{k_1}\|_1^2\} \exp\{-\beta_k \|b - v_{k_2}\|_2^2\} \right| < \epsilon \qquad (8.2.2)$$

for all $(a, b) \in X$.¶

Comments. Since the condition that $\|x - v\| \neq \|y - v\|$ is equivalent to the condition that $2\langle x - y, v\rangle \neq \|x\|^2 - \|y\|^2$, and assuming that $x \neq y$ in X implies that $x \neq y$ in S, we see that V can be taken to be, for instance, any convex subset of S that contains the points of an open ball in S. Of course, P can be taken to be $(0, \infty)$ or $\{1, 2, 3, \cdots\}$, etc.

Since $\exp\{-\beta \|x - v\|^2\} \to 0$ as $\|x\| \to \infty$ when $\beta > 0$, we have $\mathcal{X}_0 \subset \mathcal{X}_\infty$ when for each $\gamma > 0$ there is a compact subset X_γ of X such that $\|x\| \geq \gamma$ for $x \in X$ with $x \notin X_\gamma$.

8.3 Radial Basis Function Approximations of Input-Output Maps

The following is our main result.

Theorem 8.3.1. *Let A be an unbounded closed subinterval of $\mathbb{R}$. Suppose that (X_2, d_2) is compact, and let U be any convex subset of S_2 such that*

$$\{\|\cdot + u\|_2 : u \in U\}$$

separates the points of X_2. Let f be a continuous function from $(A, |\cdot|) \times (X_2, d_2)$‖ *to $\mathbb{R}$ such that for each $\gamma > 0$ there is a compact subset A_γ of $(A, |\cdot|)$ for which*

$$|f(a, x)| < \gamma \ \ (a \notin A_\gamma, x \in X_2). \qquad (8.3.1)$$

¶Any $f \in \mathcal{X}$ can be similarly approximated by adding a real constant (that depends on f) to the sum in (8.2.2).

‖Here, as suggested, $|\cdot|$ denotes the metric associated with the absolute-value norm. In accord with the material in Section 8.3, the metric on $(A, |\cdot|) \times (X_2, d_2)$ is the sum of the two metrics.

Then for each $\epsilon > 0$ there are a positive integer q, real numbers $\alpha_1, \cdots, \alpha_q$, numbers $\beta_1, \cdots, \beta_q$ belonging to P, elements $a_1, \cdots, a_q$ of A, and elements $u_1, \cdots, u_q$ of U such that

$$\left| f(a,x) - \sum_{k=1}^{q} \alpha_k \exp\{-\beta_k |a - a_k|^2\} \exp\{-\beta_k \|x - u_k\|_2^2\} \right| < \epsilon \tag{8.3.2}$$

for all $(a, x) \in A \times X_2$.

Proof. The theorem is a consequence of Theorem 8.2.1 for $S_1 = \mathbb{R}$ with the usual multiplication inner product, $X_1 = (A, |\cdot|)$, and the observation that $A \times U$ is a convex subset of $S = S_1 \times S_2$ for which $\{ \|\cdot + v\| : v \in A \times U\}$ separates the points of X.** In particular, the space $X_1 = (A, |\cdot|)$ is locally compact but not compact. We have $f \in \mathcal{X}_\infty$ because f is continuous; closed balls in $(A \times X_2, d)$ are compact (we show this below); and by condition (8.3.1) and the expression for d, in which $\max_{b_1, b_2 \in X_2} d_2(b_1, b_2) < \infty$, for each $\epsilon > 0$ there is a closed ball in $(A \times X_2, d)$ outside of which $|f(a, x)| < \epsilon$. Specifically, let ϵ be given, and select a compact subset A_ϵ of $(A, |\cdot|)$ so that $|f(a, x)| < \epsilon$ for all $a \notin A_\epsilon$ and all $x \in X_2$. Choose any $(a_0, x_0) \in A \times X_2$, and let r be the radius of a closed ball $B_0(r)$ in $(A, |\cdot|)$ centered at a_0 such that $B_0(r)$ contains the points of A_ϵ. Observe that for $(a, x) \in A \times X_2$, but outside the closed ball in $(A \times X_2, d)$ of radius $r + \max_{b_1, b_2 \in X_2} d_2(b_1, b_2)$ centered at (a_0, x_0), we have $a \notin B_0(r)$ and thus $a \notin A_\epsilon$.

Since closed balls in $(A \times X_2, d)$ are compact, and $(A, |\cdot|)$ and (X_2, d_2) are unbounded and bounded, respectively, it is clear that for each $\xi > 0$ there is a compact subset B of $(A \times X_2, d)$ such that for $(a, b) \in A \times X_2$ with $(a, b) \notin B$, we have

$$\|(a, b)\| = \left(\|a\|_1^2 + \|b\|_2^2 \right)^{\frac{1}{2}} \geq \|b\|_2 \geq \xi.$$

This shows that $\mathcal{X}_0 \subset \mathcal{X}_\infty$.

Finally, let C denote a closed ball in (or any closed bounded subset of) $(A \times X_2, d)$, and let $\{(a_k, b_k)\}_{k=0}^\infty$ be a sequence in C. By the compactness of (X_2, d_2), and by the compactness of closed bounded subsets of $(A, |\cdot|)$, there is a point $(a_\infty, b_\infty) \in A \times X_2$ and a subsequence $\{(a_{0k}, b_{0k})\}_{k=0}^\infty$ of $\{(a_k, b_k)\}_{k=0}^\infty$ such that $|a_{0k} - a_\infty| \to 0$ as $k \to \infty$ and $d_2(b_{0k}, b_\infty) \to 0$ as $k \to \infty$, and thus such that $d[(a_{0k}, b_{0k}), (a_\infty, b_\infty)] \to 0$ as $k \to \infty$. Since C is closed, $(a_\infty, b_\infty) \in C$. Thus, by the equivalence of compactness and sequential compactness in metric spaces, C is compact. This completes the proof.

Theorem 8.3.1 can be easily extended to cover the case in which A is instead an unbounded closed n-dimensional subinterval of $\mathbb{R}^n$, in which n is an arbitrary positive integer. In that case, $|\cdot|$ in the theorem denotes the Euclidean norm on $\mathbb{R}^n$, and $A = A_1 \times \cdots \times A_n$, in which the A_j are closed subintervals of $\mathbb{R}$, and at least one of the A_j is unbounded. The proof is essentially the same.

**Here we could have replaced $A \times U$ with $E \times U$, where E is any open subinterval of $\mathbb{R}$, and the proof would then show that the numbers $a_1, \cdots, a_q$ in (8.3.2) can be drawn from E instead of A.

In connection with the discussion in Section 8.1 concerning input-output maps G, of course $f(a,x)$ can be written in the form $(Gx)(a)$. Also, using the observation that we have

$$|f(a,x)| \leq |\sum_{k=1}^{q} \alpha_k \exp\{-\beta_k |a-a_k|^2\}| + \epsilon$$

whenever (8.3.2) is met, we see that condition (8.3.1), which is a key hypothesis of Theorem 8.3.1, is a *necessary* condition that (8.3.2) holds for each $\epsilon > 0$.

8.3.1 A useful corollary

Here we record a useful consequence of Theorem 8.3.1.

With $I\!R_+$ the set of nonnegative numbers, and with ℓ a positive number and $\Phi : I\!R_+ \to I\!R_+$ bounded and such that $\Phi(t) \to 0$ as $t \to \infty$, let Y stand for the set of $I\!R$-valued maps on $I\!R_+$, and let Y_Φ denote the metric space of all functions x from $I\!R_+$ to $I\!R$ such that x is Lipschitz continuous with Lipschitz constant ℓ, and such that

$$|x(t)| \leq \Phi(t), \ \ t \geq 0$$

with the metric in Y_Φ given by $d_\Phi(x_a, x_b) = \sup_t |x_a(t) - x_b(t)|$.

Corollary 8.3.1. *Let G map Y_Φ into Y such that f given by $f(t,x) := (Gx)(t)$ is continuous on $I\!R_+ \times Y_\Phi$, and such that for each $\gamma > 0$ there is a $T > 0$ for which $|(Gx)(t)| < \gamma$ for $t > T$ and all $x \in Y_\Phi$. Then for each $\epsilon > 0$ there are a positive integer q, real numbers $\alpha_1, \cdots, \alpha_q$, numbers $\beta_1, \cdots, \beta_q$ belonging to P, elements $a_1, \cdots, a_q$ of $I\!R_+$, and elements $x_1, \cdots, x_q$ of Y_Φ such that*

$$\left| (Gx)(t) - \sum_{k=1}^{q} \alpha_k \exp\{-\beta_k |t-a_k|^2\} \exp\{-\beta_k \|x - x_k\|_2^2\} \right| < \epsilon \qquad (8.3.3)$$

for all $(t,x) \in I\!R_+ \times Y_\Phi$, in which $\|\cdot\|_2$ is the usual norm on $L_2(0,\infty)$.

Proof. Referring to the theorem, set $A = I\!R_+$, $S_2 = L_2(0,\infty)$, $(X_2, d_2) = Y_\Phi$, $f(a,x) = f(t,x)$, and observe that (8.3.1) is met. The subset Y_Φ of S_2 is convex. Given x_a and x_b in Y_Φ with $x_a \neq x_b$, we see that $u := -x_a \in Y_\Phi$ satisfies $\|x_a + u\|_2 \neq \|x_b + u\|_2$, showing that $\{\ \|\cdot + u\|_2 : u \in Y_\Phi\}$ separates the points of X_2. To complete the proof we need to show that Y_Φ is compact. We do that as follows.

Given any $\gamma > 0$, select $t_0 > 0$ so that $\Phi(t) < \gamma/2, \ \ t \geq t_0$. Using the compactness of the restriction $Y_\Phi(0,t_0)$ of Y_Φ to $[0,t_0]$ ($Y_\Phi(0,t_0)$ is a closed bounded subset of an equicontinuous set), let $x_1, \cdots, x_p$ be a $\gamma/2$-net in $Y_\Phi(0,t_0)$. Let B_γ denote the subset $\{y_m : m = 1, \cdots, p\}$ of the set B of bounded continuous maps from $I\!R_+$ to $I\!R$ with the usual sup metric, where $y_m(t) = x_m(t), \ 0 \leq t \leq t_0$ and $y_m(t) = x_m(t_0), \ t > t_0$. We see that for each x in Y_Φ there is an $m \in \{1, \cdots, p\}$ for which $\sup_t |x(t) - y_m(t)| < \gamma$. Because B is complete and Y_Φ is a closed subset of B, it follows from a standard result [9, p. 201] that Y_Φ is compact.[††] □

[††]We have noted that Y_Φ is closed because "compactness" in [9] means what is often called "relative compactness."

8.4 An Example

Many systems of practical interest (e.g., feedback systems) are described by integral equations of the form

$$y(t) + \int_0^t k(t-\tau)\psi[y(\tau),\tau]d\tau = x(t), \;\; t \geq 0 \qquad (8.4.1)$$

where t denotes time, x is the input (or a modified input that takes into account initial conditions), and y is the output. It is known that under very reasonable conditions on k and ψ, which we assume are met, (8.4.1) has a unique solution y in the set $\mathcal{C}$ of continuous $I\!R$-valued functions on $I\!R_+$ for each $x \in \mathcal{C}$.

Here we suppose that x is drawn from Y_Φ with

$$\Phi(t) = M \exp\{-\beta t\}, \;\; t \geq 0 \qquad (8.4.2)$$

in which M and β are given positive numbers. Let Y_β denote this Y_Φ, and notice that Y_β is an interesting class of inputs x that go to zero at infinity. Let G from Y_β into Y be defined by the condition that G maps x into a continuous y via (8.4.1). With (8.4.2) assumed, we are going to show that under a familiar circle condition involving k and ψ, G satisfies the hypotheses of Corollary 8.3.1.

Let α and β denote positive constants such that $\alpha \leq \beta$. By $k \in \mathcal{K}(\alpha, \beta)$ we mean that k is real valued, $k \in L_1(0,\infty)$, and with

$$K(j\omega) = \int_0^\infty k(t)e^{-j\omega t}\,dt,$$

the locus of $K(j\omega)$ for $\omega \in I\!R$ lies outside the circle C_1 of radius $0.5(\alpha^{-1} - \beta^{-1})$ centered on the real axis of the complex plane at $[-0.5(\alpha^{-1} + \beta^{-1}), 0]$ and does not encircle C_1 (which is often referred to as the "critical disk").

We make the following assumptions which we refer to as A.1.

(i) $k \in \mathcal{K}(\alpha,\beta)$, with k the inverse Laplace transform of a rational function.

(ii) ψ is real valued and defined on $I\!R_+ \times I\!R$, $\psi(0,t) = 0$ for $t \geq 0$, h given by $h(t) = \psi[w(t),t]$ is Lebesgue measurable on $I\!R_+$ whenever w is Lebesgue measurable there.

(iii)

$$\alpha \leq \frac{\psi(\xi_a,t) - \psi(\xi_b,t)}{\xi_a - \xi_b} \leq \beta \qquad (8.4.3)$$

for $t \in I\!R_+$ and all $\xi_a \neq \xi_b$.

We will establish the following fact.

Fact. Let G be as described in this section, and suppose that A.1 is met. Then for each $\epsilon > 0$ there are a positive integer q, real numbers $\alpha_1, \cdots, \alpha_q$, numbers

$\beta_1, \cdots, \beta_q$ belonging to P, elements $a_1, \cdots, a_q$ of $\mathbb{R}_+$, and elements $x_1, \cdots, x_q$ of Y_β such that

$$\left| (Gx)(t) - \sum_{k=1}^{q} \alpha_k \exp\{-\beta_k |t - a_k|^2\} \exp\{-\beta_k \|x - x_k\|_2^2\} \right| < \epsilon \qquad (8.4.4)$$

for all $(t, x) \in \mathbb{R}_+ \times Y_\beta$, in which $\|\cdot\|_2$ is the usual norm on $L_2(0, \infty)$.

To do this, we use Corollary 8.3.1. To show that f defined by $f(t, x) = (Gx)(t)$ is continuous on $\mathbb{R}_+ \times Y_\beta$, we make use of the following observation [7].

Lemma 8.4.1. *Let (Y_1, d_1), (Y_2, d_2), and (Y_3, d_3) be three metric spaces, and with d defined on the product set $Y := (Y_1 \times Y_2)$ by $d[(a_1, b_1), (a_2, b_2)] = d_1(a_1, a_2) + d_2(b_1, b_2)$ (here d_1, d_2, and d are not necessarily the metrics of Section 8.2), let g be a map from the metric space (Y, d) into (Y_3, d_3). Suppose that for each $a \in Y_1$, the map $g(a, \cdot)\colon (Y_2, d_2) \to (Y_3, d_3)$ is continuous. Suppose also that g is continuous on (Y_1, d_1) locally uniformly with respect to (Y_2, d_2), in the sense that given $(a_0, b_0) \in Y_1 \times Y_2$ and $\epsilon > 0$ there are positive constants δ_1 and δ_2 such that*

$$d_3[g(a, b), g(a_0, b)] < \epsilon \text{ for } d_1(a, a_0) < \delta_1 \text{ and } d_2(b, b_0) < \delta_2.$$

Under these conditions, g is continuous on (Y, d).

Using A.1, by Corollary 3(a) of [10] there is a positive constant r that depends only on k, α, and β such that

$$\sup_t |(Gx_a)(t) - (Gx_b)(t)| \leq r \sup_t |x_a(t) - x_b(t)|, \quad x_a, x_b \in Y_\beta.$$

Since $|(Gx_a)(t) - (Gx_b)(t)| \leq \sup_t |(Gx_1)(t) - (Gx_b)(t)|$ for each t, we see that $f(t, x)$ is continuous in x for each t. Now let $(t_0, x_0) \in \mathbb{R}_+ \times Y_\beta$ and $\epsilon > 0$ be given.

Using (8.4.1),

$$(Gx)(t) + \int_0^t k(t - \tau)\psi[(Gx)(\tau), \tau] d\tau = x(t), \quad t \geq 0 \qquad (8.4.5)$$

for $x \in Y_\beta$. Thus, for $t \geq 0$,

$$\begin{aligned}(Gx)(t) - (Gx)(t_0) &= x(t) - x(t_0) \\ &- \int_0^t k(t - \tau)\psi[(Gx)(\tau), \tau] d\tau + \int_0^{t_0} k(t_0 - \tau)\psi[(Gx)(\tau), \tau] d\tau. \qquad (8.4.6)\end{aligned}$$

Select a positive δ_0 for which $|x(t) - x(t_0)| < \epsilon/2$ for all $x \in Y_\beta$ when $|t - t_0| < \delta_0$ (recall that the elements of Y_β are uniformly Lipschitz continuous), and define $\xi(x)(t) = \psi[(Gx)(t), t]$, $t \geq 0$. By Theorem 1 of [10] together with (8.4.3) and the observation that Y_β is a bounded subset of $L_2(0, \infty)$, there is a constant c such that

$\|\xi(x)\|_2 \le c,\ x \in Y_\beta$. With regard to the right side of (8.4.6), we have, using the Schwarz inequality,

$$\left| -\int_0^t k(t-\tau)\psi[(Gx)(\tau),\tau]d\tau + \int_0^{t_0} k(t_0-\tau)\psi[(Gx)(\tau),\tau]d\tau \right|$$

$$\le \int_{t_0}^t |k(t-\tau)\xi(x)(\tau)|d\tau + \int_0^{t_0} |k(t_0-\tau) - k(t-\tau)| \cdot |\xi(x)(\tau)|d\tau$$

$$\le c\left(\int_{t_0}^t |k(t-\tau)|^2 d\tau\right)^{\frac{1}{2}} + c\left(\int_0^{t_0} |k(t_0-\tau) - k(t-\tau)|^2 d\tau\right)^{\frac{1}{2}}. \tag{8.4.7}$$

Because k is equivalent to a continuous function (here we have used the hypothesis that k belongs to $L_1(0,\infty)$ and is the inverse Laplace transform of a rational function), it is not difficult to check that there is a constant $\delta_1 < \delta_0$ such that the extreme right side of (8.4.7) is less than $\epsilon/2$ for $|t-t_0| < \delta_0$. Using Lemma 8.4.1, and the continuity of $f(t,x)$ in x for each t, this shows that f is continuous on $I\!R_+ \times Y_\beta$.

By the proof of Theorem 2 of [10], using the inequality $\|\xi(x)\|_2 \le c,\ x \in Y_\beta$, there are positive constants c_1 and c_2 such that $c_1 < \beta$ and

$$|(Gx)(t)| \le |x(t)| + c_2 \exp\{-c_1 t\},\quad t \ge 0$$

for all $x \in Y_\beta$. Therefore, $|(Gx)(t)| \le (M + c_2)\exp\{-c_1 t\}$ for $t \ge 0$ and all x in Y_β. By Corollary 8.3.1 our proof is complete.

8.5 Related Results and Comments

With A a closed subinterval of $I\!R$, let $\mathcal{F}$ be a collection of maps F from A to $I\!R$ such that for each $\gamma > 0$, each $\beta_0 \in P$, and each $a_0 \in A$, there is an $F \in \mathcal{F}$ such that

$$|F(a) - \exp\{-\beta_0|a - a_0|^2\}| < \gamma,\quad a \in A.$$

Similarly, let $\mathcal{G}$ be a set of maps G from X_2 to $I\!R$ such that for each $\gamma > 0$, each $\beta_0 \in P$, and each $u_0 \in U$, there is an $G \in \mathcal{G}$ such that

$$|G(x) - \exp\{-\beta_0 \|x - u_0\|_2^2\}| < \gamma,\quad x \in X_2.$$

Corollary 8.5.1. *Suppose that the hypotheses of Theorem 8.3.1 are satisfied. Then for each $\epsilon > 0$ there are a positive integer q, elements $F_1, \dots, F_q$ of $\mathcal{F}$, and elements $G_1, \dots, G_q$ of $\mathcal{G}$ such that*

$$|f(a,x) - \sum_{k=1}^q F_k(a) G_k(x)| < \epsilon \tag{8.5.1}$$

for all $(a,x) \in A \times X_2$.

Proof. Since there are only a finite number of terms in the sum over k in (8.3.2), it suffices to check that given $\gamma > 0$, $\beta_0 \in P$, $a_0 \in A$, and $u_0 \in U$, there are $F \in \mathcal{F}$ and $G \in \mathcal{G}$ for which

$$|F(a)G(x) - \exp\{-\beta_0|a - a_0|^2\}\exp\{-\beta_0 \|x - u_0\|_2^2\}| < \gamma, \quad (a, x) \in A \times X_2.$$

To do that, select F and G so that

$$|F(a) - \exp\{-\beta_0|a - a_0|^2\}| < \gamma/2, \quad a \in A$$

and

$$|G(x) - \exp\{-\beta_0 \|x - u_0\|_2^2\}| < \delta, \quad x \in X_2$$

in which δ is a positive constant satisfying $\delta(\gamma/2 + 1) < \gamma/2$. Observe that

$$|F(a)| < (\gamma/2 + 1) \text{ for all } a,$$

and that

$$\begin{aligned} &|F(a)G(x) - \exp\{-\beta_0|a - a_0|^2\}\exp\{-\beta_0 \|x - u_0\|_2^2\}| \\ = &|F(a)G(x) - F(a)\exp\{-\beta_0 \|x - u_0\|_2^2\} + F(a)\exp\{-\beta_0 \|x - u_0\|_2^2\} \\ &- \exp\{-\beta_0|a - a_0|^2\}\exp\{-\beta_0 \|x - u_0\|_2^2\}| \\ &\leq |F(a)| \cdot |G(x) - \exp\{-\beta_0 \|x - u_0\|_2^2\}| \\ &+|F(a) - \exp\{-\beta_0|a - a_0|^2\}| \cdot |\exp\{-\beta_0 \|x - u_0\|_2^2\}| \\ &< (\gamma/2 + 1)\delta + < \gamma/2 < \gamma \end{aligned}$$

for $(a, x) \in A \times X_2$, which finishes the proof.

For example, if $A = I\!R_+$ and α is any positive number, $\mathcal{F}$ can be taken to be the family of Laguerre functions of the form $\exp\{-\alpha t\}p(t)$ where $p(t)$ is a polynomial [6, p. 74].

The following version of Theorem 8.2.1 (see [7]) concerns the often less interesting case in which (X_1, d_1) and (X_2, d_2) are compact.

Theorem 8.5.1. *Assume that (X_1, d_1) and (X_2, d_2) are compact. Then for each continuous $f: X \to I\!R$, and each $\epsilon > 0$, there are a positive integer q, real numbers $\alpha_1, \cdots, \alpha_q$, numbers $\beta_1, \cdots, \beta_q$ belonging to P, and elements $v_1, \cdots, v_q$ of V such that*

$$\left| f(a, b) - \sum_{k=1}^{q} \alpha_k \exp\{-\beta_k \|a - v_{k_1}\|_1^2\}\exp\{-\beta_k \|b - v_{k_2}\|_2^2\} \right| < \epsilon \qquad (8.5.2)$$

for all $(a, b) \in X$.

This leads directly to the following.

Corollary 8.5.2. *Suppose that* (X_1, d_1) *and* (X_2, d_2) *are compact, that* $f: X \to \mathbb{R}$ *is continuous, and that* A *in the definition of* $\mathcal{F}$ *is replaced with* X_2. *Then for each* $\epsilon > 0$ *there are a positive integer* q, *elements* $F_1, \ldots, F_q$ *of* $\mathcal{F}$, *and elements* $G_1, \ldots, G_q$ *of* $\mathcal{G}$ *such that*

$$\left| f(a, x) - \sum_{k=1}^{q} F_k(a) G_k(x) \right| < \epsilon \tag{8.5.3}$$

for all $(a, x) \in A \times X_2$.

Results along the lines of Theorem 8.5.1 and Corollary 8.5.2, can be proved using a theorem due to Dieudonné [6, p. 66] which asserts that if Y is the Cartesian product of compact topological spaces Y_γ, $\gamma \in \Gamma$, then any continuous real functional on Y can be uniformly approximated by finite sums of finite products of continuous functions of one variable on Y. It is not difficult to check that for Γ a finite set and for each γ, these continuous functions of one variable on Y_γ can be assumed to belong to any family of $\mathbb{R}$-valued functions on Y_γ that is dense, in the sense of uniform approximation, in the set of continuous $\mathbb{R}$-valued functions on Y_γ. However, this approach does not quite lead to Theorem 8.5.1 in which the same constant β_k appears in both exponential factors in the kth term in the sum in (8.5.2).

Bibliography

[1] J. Park and I. W. Sandberg, "Universal approximation using radial basis-function networks," *Neural Computation*, vol. 3, no. 2, pp. 246–257, 1991.

[2] J. Park and I. W. Sandberg, "Approximation and radial-basis function networks," *Neural Computation*, vol. 5, no. 2, pp. 305–316, Mar. 1993.

[3] E. J. Hartman, J. D. Keeler, and J. M. Kowalski, "Layered neural networks with Gaussian hidden units as universal approximators," *Neural Computation*, vol. 2, no. 2, pp. 210–215, 1990.

[4] E. Parzen, "On estimation of a probability density function and mode," *Annals of Mathematical Statistics*, vol. 33, pp. 1065–1076, 1962.

[5] I. W. Sandberg, "Gaussian radial basis functions and inner-product spaces," *Circuits, Systems and Signal Processing*, vol. 20, no. 6, pp. 635–642, 2001. (See also the Errata in vol. 21, no. 1, p. 123, 2002.)

[6] M. H. Stone, "A generalized Weierstrass approximation theorem," in *Studies in Modern Analysis*, R. C. Buck, Ed,. vol. 1 of MAA Studies in Mathematics, pp. 30–87, Prentice-Hall, Englewood Cliffs, NJ, 1962.

[7] I. W. Sandberg, "Indexed families of functionals, and Gaussian radial basis functions," *Neural Computation*, 2003 (to appear).

[8] I. W. Sandberg, J. T. Lo, C. Francourt, J. Principe, S. Katagiri, and S. Haykin, *Nonlinear Dynamical Systems: Feedforward Neural Network Perspectives*, John Wiley, New York, 2001.

[9] I. P. Natanson, *Theory of Functions of a Real Variable*, Vol. II, Frederick Ungar Publishing Co., New York, 1960.

[10] I. W. Sandberg, "Some results on the theory of physical systems governed by nonlinear functional equations," *The Bell System Technical J.*, vol. 44, no. 5, pp. 871–898, May 1965.

Chapter 9

Blind Source Recovery: A State-Space Formulation

Khurram Waheed and Fathi M. Salem

Abstract: Blind source recovery (BSR) denotes recovery of original sources or signals without any explicit identification of the environments which may include convolution, temporal variation, and even nonlinearity. This chapter provides an overview of a generalized (i.e., nonlinear and time-varying) state-space BSR formulation by the application of stochastic optimization principles to the Kullback-Lieblar divergence as an information-theoretic performance functional. The multivariable optimization technique is used to derive update laws for nonlinear time-varying dynamical systems, which are subsequently specialized to time-invariant and linear systems. Furthermore, the various possible state-space demixing network structures have been exploited to develop learning rules, capable of handling most filtering paradigms–which are conveniently extendible to nonlinear models. Distinct linear state-space algorithms are presented for the minimum phase and nonminimum phase mixing environment models. Illustrative simulation examples are then presented to demonstrate the on-line adaptation capabilities of the developed algorithms.

9.1 Introduction

Blind source recovery (BSR) is informally described as follows: several unknown but stochastically independent temporal signals propagate through a natural or synthetic dynamic mixing and filtering environment. By observing only the outputs of this environment, a system (e.g., a filter bank, a neural network, or a device) is constructed to counteract, to the extent possible, the effects of the environment and adaptively recover the best estimate of the original signals. The adaptive approach is a form of unsupervised or autonomous learning. In the context of source recovery, our prime interest is to obtain the best possible estimate of the actual source signals (up to a possible scaling and a permuted sequence), which may be achieved independent of, and even in the absence of, precise environment identifiability [37, 41, 44, 51, 57]. For example, consider the case of nonlinear mixing; there may be several demixing networks possible, which can result in the mutually independent estimated outputs [13, 24]. Similarly, even in the linear case, various possible demixing structures may be constructed to recover or estimate the sources. For the overcomplete case, where there are less observations than actual sources, there may not even exist an explicit recovering network. The focus of BSR, therefore, is to address the issue of recovery (or estimation) of the desired sources rather than to focus on the issue of accurate multi-input multi-output (MIMO) (environment, channel) identification.

For this unsupervised adaptive filtering task, the property of signal independence (or near-independence) is assumed. No additional a priori knowledge of the original signals/sources is required. BSR requires few assumptions and possesses the self-learning capability, which renders such networks attractive from the viewpoint of real world applications where on-line training is often desired. The challenges for BSR reside in the development of sound mathematical analyses and a framework capable of handling a variety of diverse problems. In the linear case, BSR includes the well-known adaptive filtering subproblems of blind source separation and blind source deconvolution. Blind source separation (BSS) is the process of recovering a number of original stochastically independent sources/signals when only their linear static mixtures are available. Blind source deconvolution (BSD) deals with deconvolving the effects of a temporal or a spatial mixing linear filter on signals without a priori information about the filtering medium/network or the original sources/signals [14, 15, 23].

As background, interest in the field of BSR has grown dramatically during recent years, motivated to a large extent by its similarity to the mixed signal separation capability of the human brain. The brain makes use of unknown parallel nonlinear and complex dynamic signal processing with auto-learning and self-organization ability to perform similar tasks. The peripheral nervous system integrates complex external stimuli and endogenous information into packets, which are transformed, filtered, and transmitted in a manner that is yet to be completely understood. This complex mixture of information is received by the central nervous system (brain), split again into original information, and relevant information relayed to various sections of cerebral cortex for further processing and action [35].

BSR is valuable in numerous applications, including telecommunication systems, sonar and radar systems, audio and acoustics, mining and oil drilling systems, imagery and feature analysis, and biomedical signal processing (EEG/MEG, EOG, EMG, ECG signals). Consider, for example, the audio and sonar applications where the original signals are sounds, and the mixed signals are the output of several microphones or sensors placed at different vantage points. A network would receive, via each sensor, a mixture of original sounds that usually undergo multipath delays. The network's role in this scenario is to dynamically reproduce, as closely as possible, the original signals. These separated signals can subsequently be channeled for further processing or transmission [16, 17]. Similar application scenarios can be described in situations involving measurement of neural, cardiac, or other vital biological parameters, communication in noisy environments, engine or plant diagnostics, and cellular mobile communications, to name a few [9, 14, 19, 23, 26].

Development of adequate models for the environment, which include time delays or filtering, multipath effects, time-varying parameters, and/or nonlinear sensor dynamics, is important as representatives of desired practical applications. The choice of inadequate models of the environment would result in highly sensitive and nonrobust processing by a network. Indeed, in order to render the network operable in real world scenarios, robust operations must account for the parameter variations, dynamic influences, and signal delays that often result in asynchronous signal propagation [20, 25, 32, 37, 41]. The environment needs to be modeled as a rich dynamic linear (or even nonlinear) system [13, 20, 24, 26, 28, 32, 41, 43, 49]. While other approaches to BSR have used more conventional signal processing structures, we propose a comprehensive BSR framework based on multivariable state-space representations, optimization theory, the calculus of variations, and higher-order statistics. This chapter presents a generalized framework based on the multivariable canonical state-space representation (see Appendix) of both the mixing environment and the demixing system. Various filtering paradigms have been consequently derived as special cases from the proposed framework. Simulation results verifying the theoretical developments have also been included in this chapter.

The state-space formulation provides a general framework capable of dealing with a variety of situations [8]. The multivariable state-space provides a compact and computationally attractive representation of MIMO filters. This rich state-space representation allows for the derivation of generalized iterative update laws for the BSR. The state notion abridges weighted past as well as filtered versions of input signals and can be easily extended to include nonlinear networks [8, 28]. There are several reasons for choosing this framework.

- State-space models give an efficient internal description of a system. Further, this choice allows various equivalent state-space realizations for a system, more importantly being the canonical observable and controllable forms. Transfer function models, although equivalent to linear state-space models when initial conditions are zero, do not exploit any internal features that may be present in the real dynamical systems.

- The inverse for a state-space representation is easily derived subject to the invertibility of the instantaneous relational mixing matrix between input-output, in case this matrix is not square; the condition reduces to the existence of the pseudo-inverse of this matrix. This feature quantifies and ensures recoverability of original sources provided the environment model is invertible.

- Parameterization methods are well known for specific classes of models. In particular, the state-space model allows much more general description than standard finite/infinite impulse response (FIR/IIR) convolutive filtering. All known (dynamic) filtering models, such as AR, MA, ARMA, ARMAX, and

Gamma filters, can be considered as special cases of flexible state-space models.

- The linear state-space formulation of a problem is conveniently extendible to include nonlinear models and specific component dynamics. The richness of state-space models enables them to represent generalized nonlinear models that include neural networks, genetic algorithms, and so on.

9.1.1 Adaptive BSR formulation

In the most general setting, the mixing/convolving environment may be represented by an unknown dynamic process $\overline{\mathbf{H}}$ with inputs being the n-d independent sources $\underline{s}$ and the outputs being the m-d measurements $\underline{m}$. In this extreme case, no structure is assumed about the model of the environment.

The environment can also be modeled as a dynamic system with a structure and fixed but unknown parameters. The processing network $\mathbf{H}$ must be constructed with the capability to recover or estimate the original sources. For the linear case, this is equivalent to computing the "inverse" (or the "closest to an inverse") of the environment model without assuming any knowledge of the mixing environment or the distribution structure of the unknown sources.

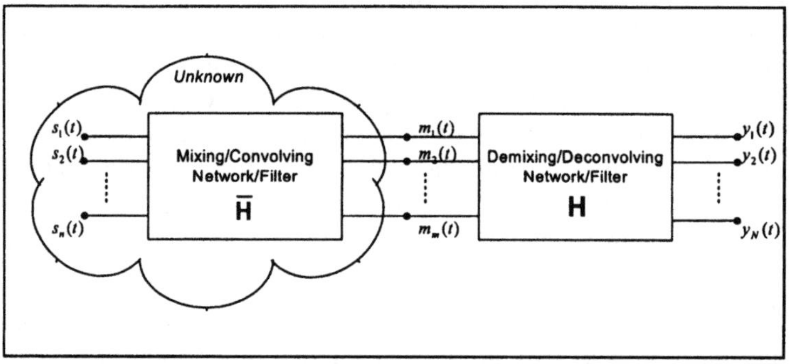

Figure 9.1.1: General framework for BSR

It is possible that an augmented network be constructed so that the inverse of the environment is merely a subsystem of the network with learning. In this case, even if the environment is unstable (e.g., due to existence of nonminimum phase zeros in a linear state-space model), the overall augmented network may represent a nonlinear adaptive dynamic system, which may converge to the desired parameters as a stable equilibrium point, thus achieving the global task of blind identification [18, 29, 30, 36, 40, 53, 56].

For the linear filtering environments [4, 9, 12, 18, 21, 25, 31, 36], this problem may be presented as

$$Y = \mathbf{H} * \underline{m} = \mathbf{H} * \bar{\mathbf{H}} * \underline{s} = \mathbf{P} * \mathbf{D} * \underline{s} \qquad (9.1.1)$$

where $\mathbf{P}$ is a generalized permutation matrix, $\mathbf{D}$ is a diagonal matrix of filters with each diagonal filter having only one nonzero tap, and $*$ represents ordinary matrix multiplication for the static mixing case, while it represents MIMO system convolution for the multitap deconvolution case.

For notational and mathematical convenience, we will derive all BSR algorithms for the case where $n = m = N$. This convenient choice allows for unambiguous mathematical manipulations in the derivation for all the algorithms. The undercomplete mixture case of BSR, where $m > n$, can be easily reduced to the aforementioned case by appropriately discarding some measurements using preprocessors such as principal component analysis (PCA) or eigendecomposition etc. [14, 22, 26]. On the other hand, the overcomplete mixture case for $m < n$, in general, is not tractable and cannot be handled by the proposed algorithms in any straightforward fashion. See [51] and the references cited therein for relevant work.

9.2 Optimization Framework for BSR

This section formally describes the BSR optimization framework. See [18, 39, 41] for more details. We first discuss the performance functional used for the derivation of the BSR update laws based on the stochastic gradient descent and constrained multivariable optimization. This performance functional is a well-known information-theoretic "distance" measure, appropriate for quantifying the mutual *dependence* of signals in a mixture. Further, we present our optimization framework based on this performance functional and derive an update algorithm for a generalized nonlinear dynamic structure. Subsequently, the algorithm is specialized for the case of a linear dynamical state-space model [8].

9.2.1 The performance functional

We employ the divergence of a random output vector y (Kullback-Lieblar divergence) as our performance measure. In the discrete case, assuming output signal properties to be ergodic [4, 6, 27, 39, 41], this relation is given by

$$L(y(k)) = \sum_{y \in Y} p_y(y(k)) \ln \left| \frac{p_y(y(k))}{\prod_{i=1}^{n} p_{y_i}(y_i(k))} \right| \tag{9.2.1}$$

where $p_y(y)$ is the probability density function of the random output vector y and $p_{y_i}(y_i)$ is the probability density function of the ith component of the output vector y.

This functional $L(y)$ is a "distance" measure with the following properties

- $L(y) \geq 0$
- $L(y) = 0$ iff $p_y(y) = \prod_{i=1}^{n} p_{y_i}(y_i)$.

This measure can provide an estimate of the degree of dependence among the various components of the recovered output signal vector and is an appropriate functional to be used in the optimization framework of the BSR problem. Further, invoking the independence assumption on the recovered outputs, we have

$$L(y(k)) = -H(y(k)) + \sum_{i=1}^{n} H_i(y_i(k)) \tag{9.2.2}$$

where $H(y(k))$ is the entropy of the signal vector $y(k)$, for the discrete time it is given by

$$H(y) = -E\left[\ln |p_y(y)|\right] = -\sum_{y \in Y} p_y(y) \ln |p_y(y)| \tag{9.2.3}$$

and $H_i(y_i(k))$ is the marginal entropy of a component signal $y_i(k)$.

9.2.2 Algorithms for the nonlinear dynamic case

Assume that the environment can be modeled as the following nonlinear discrete-time dynamic forward model [34]

$$X_e(k+1) = f_e(X_e(k), s(k), h_1) \tag{9.2.4}$$

$$m(k) = g_e(X_e(k), s(k), h_2) \tag{9.2.5}$$

where $s(k)$ is n-d vector of original source signals, $m(k)$ is m-d vector of measurements, $X_e(k)$ is N_e-d state vector for the environment, h_1 is constant parameter vector (or matrix) of dynamic state equation, h_2 is constant parameter vector (or matrix) of output equation, and $f_e(\cdot)$ and $g_e(\cdot)$ are differentiable nonlinear functions that specify the structure of the environment.

Further it is assumed that existence and uniqueness of solutions are satisfied for any given initial conditions $X_e(t_o)$ and sources $s(k)$ such that a Lipschitz condition on $f_e(\cdot)$ is satisfied [28].

The processing demixing network model may be represented by a dynamic feedforward or feedback network [34]. Focusing on a feedforward network model, we assume the network to be represented by

$$X(k+1) = f(X(k), m(k), w_1) \tag{9.2.6}$$

$$y(k) = g(X(k), m(k), w_2) \tag{9.2.7}$$

where $m(k)$ is m-d vector of measurements, $y(k)$ is N-d vector of network output, $X(k)$ is L-d state vector for the processing network, w_1 are parameters of the network state equation, w_2 are parameters of the network output equation, and $f(\cdot)$ and $g(\cdot)$ are differentiable nonlinear functions defining the structure of the demixing network.

The existence and uniqueness of solutions of the nonlinear difference equations is also assumed for the network model for any given initial conditions $X(t_o)$ and measurement vector $m(k)$ [28, 34].

In order to derive the update law, we abuse the notation for the sake of convenience so that $y(k)$ in (9.2.2) is represented as y_k and $L(y(k))$ is generalized to $L^k(y_k)$ so that the functional may also be represented as a function of the time index k. Thus, we formulate the following constrained optimization problem [39, 41] to be

$$\text{Minimize} \qquad J_o(w_1, w_2) = \sum_{k=n_o}^{N-1} L^k(y_k) \tag{9.2.8}$$

subject to $X_{k+1} = f^k(X_k, m_k, w_1), \ \ y_k = g^k(X_k, m_k, w_2)$ with the initial condition X_{k_o}.

The augmented cost functional to be optimized becomes

$$J(w_1, w_2) = \sum_{k=n_o}^{N-1} L^k(y_k) + \lambda_{k+1}^T (f^k(X_k, m_k, w_1) - X_{k+1}) \tag{9.2.9}$$

where λ_k is the Lagrange variable [33]. Define the Hamiltonian as

$$\mathrm{H}^k = L^k(y_k) + \lambda_{k+1}^T f^k(X_k, m_k, w_1). \tag{9.2.10}$$

Consequently, the necessary conditions for optimality are

$$X_{k+1} = \frac{\partial \mathrm{H}^k}{\partial \lambda_{k+1}} = f^k(X_k, m_k, w_1) \tag{9.2.11}$$

$$\lambda_k = \frac{\partial \mathrm{H}^k}{\partial X_k} = (f_{X_k}^k)^T \lambda_{k+1} + \frac{\partial L^k}{\partial X_k} \tag{9.2.12}$$

and the changes in "weight" parameters become

$$\Delta w_1 = -\eta_k \frac{\partial \mathrm{H}^k}{\partial w_1} = -\eta_k (f_{w_1}^k)^T \lambda_{k+1} \tag{9.2.13}$$

$$\Delta w_2 = -\eta_k \frac{\partial \mathrm{H}^k}{\partial w_2} = -\eta_k \frac{\partial L^k}{\partial w_2} \tag{9.2.14}$$

where $f_a^k \triangleq \frac{\partial f^k}{\partial a}$ represents the partial derivative w.r.t. the parameter a in the limit and η_k represents a positive learning rate that may be adaptive.

9.2.3 Algorithms for the linear dynamic case

In the linear dynamic case, the environment model is assumed to be in the state-space form [8]

$$X_e(k+1) = A_e X_e(k) + B_e s(k) \tag{9.2.15}$$

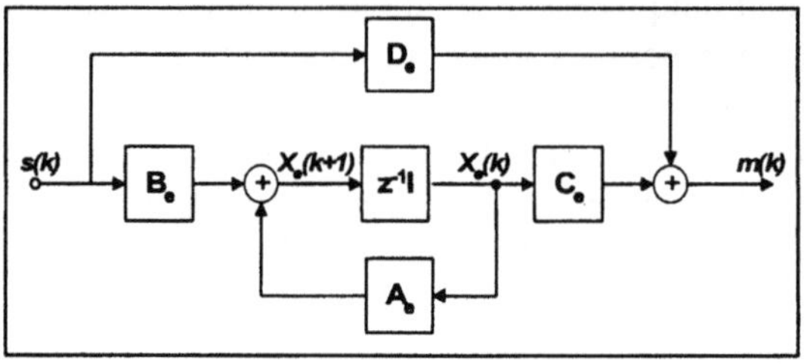

Figure 9.2.1: State-space feedback demixing structure

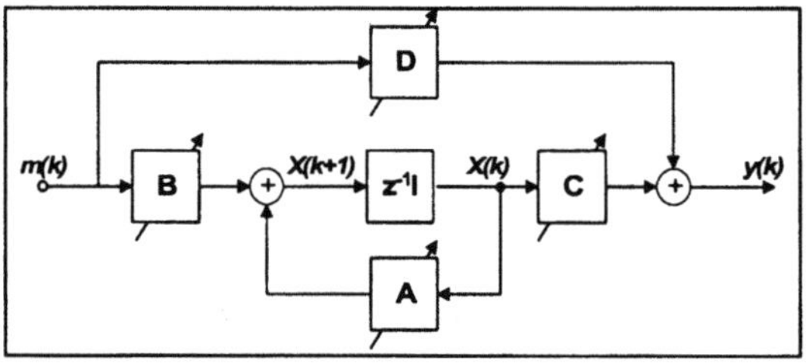

Figure 9.2.2: State-space demixing network model

$$m(k) = C_e X_e(k) + D_e s(k). \tag{9.2.16}$$

In this case the feedforward separating network will attain the state-space form [8]

$$X(k+1) = A\,X(k) + B\,m(k) \tag{9.2.17}$$

$$y(k) = C\,X(k) + D\,m(k). \tag{9.2.18}$$

The existence of an explicit solution in this case has been shown by [18, 36]. This existence of solution ensures that the network has the capacity to compensate for the environment and consequently recover the original signals (see Figures 9.2.1 and 9.2.2).

The derivation of the BSR algorithm is set up as an optimization problem subject to the constraints of a multivariable state-space representation. The Kullback-Lieblar divergence in the mutual information form is used as the performance ("distance") measure, see (9.2.2).

We minimize the performance functional

$$J_o(w_1, w_2) = \sum_{k=n_o}^{N-1} L^k(y_k) \tag{9.2.19}$$

subject to (9.2.17) and (9.2.18) with the initial conditions X_{k_o}. The augmented cost functional to be optimized becomes

$$J(w_1, w_2) = \sum_{k=n_o}^{N-1} L^k(y_k) + \lambda_{k+1}^T (A\, X_k + B\, m_k - X_{k+1}). \tag{9.2.20}$$

Again, define the Hamiltonian as

$$\mathrm{H}^k = L^k(y_k) + \lambda_{k+1}^T (A\, X_k + B\, m_k). \tag{9.2.21}$$

For the linear time-invariant case the ordinary stochastic gradient update laws are given by [39, 41]

$$X_{k+1} = \frac{\partial \mathrm{H}^k}{\partial \lambda_{k+1}} = A\, X_k + B\, m_k \tag{9.2.22}$$

$$\lambda_k = \frac{\partial \mathrm{H}^k}{\partial X_k} = A_k^T \lambda_{k+1} + C_k^T \frac{\partial L^k}{\partial y_k} \tag{9.2.23}$$

$$\Delta A = -\eta_k \frac{\partial \mathrm{H}^k}{\partial A} = -\eta_k \lambda_{k+1} X_k^T \tag{9.2.24}$$

$$\Delta B = -\eta_k \frac{\partial \mathrm{H}^k}{\partial B} = -\eta_k \lambda_{k+1} m_k^T \tag{9.2.25}$$

$$\Delta C = -\eta_k \frac{\partial \mathrm{H}^k}{\partial C} = -\eta_k \frac{\partial L^k}{\partial C} = -\eta_k \varphi(y) X^T \tag{9.2.26}$$

$$\Delta D = -\eta_k \frac{\partial \mathrm{H}^k}{\partial D} = -\eta_k \frac{\partial L^k}{\partial D} = \eta_k([D]^{-T} - \varphi(y) m^T). \tag{9.2.27}$$

The above derived update laws form a comprehensive theoretical algorithm and provides the update laws for the states X_k, the co-states λ_k, and all the parametric matrices in the state-space $\{A,\, B,\, C,\, D\}$. The invertibility of the state space, as discussed in [18, 36, 37], is guaranteed if the matrix D is invertible. In the above derived laws η_k is a positive learning rate of the algorithm which may be adaptive, $[D]^{-T}$ represents the transpose of the inverse of the matrix D if it is a square matrix or the transpose of its pseudo-inverse in case D is not a square matrix, and $\varphi(y)$ represents a vector of the usual nonlinearity or score function [4, 6, 7, 11, 30, 32, 38, 41, 45, 50] which acts individually on each component of the output vector y, with each component, say $\varphi(y_i)$, for exponential distributions is given as

$$\varphi(y_i) = -\frac{\partial \log p(y_i)}{\partial y_i} = -\frac{\partial p(y_i)/\partial y_i}{p(y_i)} \tag{9.2.28}$$

where $p(y_i)$ is the (estimate) of the probability density function of each source [46, 48, 52].

The update law in (9.2.26) and (9.2.27) is similar to the standard gradient descent results [4, 56], indicating its optimality for the Euclidean parametric structure. The update law provided above, although noncausal for the update of parametric matrices A and B, can be easily implemented using delay and memory storage as done in similar implementations of multichannel BSD problems [14]. A delay or latency in the recovered signal is acceptable in the BSR problem if the delay is fixed for every recovered component.

9.3 Extensions to the Natural Gradient

In this section, we extend the linear state-space algorithm for the problem of BSR derived in the previous section using the natural gradient [1, 2, 6, 7, 12]. We present specialized algorithms for the class of minimum phase mixing environments both in feedforward and feedback state-space configuration [39, 41, 42]. For the nonminimum phase mixing environment, the requisite demixing system becomes unstable due to the presence of poles outside the unit circle. These unstable poles are required to cancel the nonminimum phase transmission zeros of the environment. In order to avoid instability due to the existence of these poles outside the unit circle, the natural gradient algorithm may be derived with the constraint that the demixing system is a double-sided FIR filter, i.e., instead of trying to determine the IIR inverse of the environment, we will approximate the inverse using an all zero noncausal filter [4, 49]. The double-sided filters have been proven to adequately approximate IIR filters at least in the magnitude terms with a certain associated delay [10]. For the presented algorithms, only sketchy proofs are included; consult [39–42, 44–50] for in-depth discussions.

Theorem 9.3.1 (Feedforward Minimum Phase Demixing Network). *Assume the (mixing) environment is modeled by an MIMO minimum phase transfer function. Then the update laws for the zeros of the feedforward state-space demixing network using the natural gradient are given by*

$$\Delta C(k) = \eta\left((I_N - \varphi(y(k))y^T(k))C(k) - \varphi(y(k))X^T(k)\right) \tag{9.3.1}$$

$$\Delta D(k) = \eta(I_N - \varphi(y(k))y^T(k))D(k) \tag{9.3.2}$$

where $\varphi(y)$ is a nonlinear score function given by (9.2.28).

Proof. For a complete proof, please refer to [41, 42]. For the feedforward demixing network, its linear state-space representation is assumed to be as in (9.2.17)–(9.2.18). In order to achieve improved equivariant convergence properties [3, 4, 6, 7], the update laws for C and D matrices based on the Riemannian manifold may be derived using the output equation (9.2.18) and the natural gradient learning [1, 2, 6, 7]. Note that in the following derivation, instantaneous time index k has been dropped for convenience.

Define augmented vectors $\tilde{y}$ and $\tilde{x}$ and the matrix $\tilde{W}$ as

$$\tilde{y} = \begin{bmatrix} y \\ X \end{bmatrix}, \; \tilde{x} = \begin{bmatrix} m \\ X \end{bmatrix}, \; \tilde{W} = \begin{bmatrix} D & C \\ 0 & I \end{bmatrix},$$

so the augmented output equation becomes

$$\tilde{y} = \tilde{W}\, \tilde{x}. \tag{9.3.3}$$

Using the natural gradient [1, 2, 6, 7], the update laws for $\tilde{W}$ are

$$\Delta\tilde{W} = \eta \left[I - \varphi(\tilde{y})\tilde{y}^T\right] \tilde{W}. \tag{9.3.4}$$

Considering the update laws for matrices C and D only, we have

$$\Delta C(k) = \eta \left((I_N - \varphi(y(k))y^T(k))C(k) - \varphi(y(k))X^T(k)\right) \tag{9.3.5}$$

$$\Delta D(k) = \eta(I_N - \varphi(y(k))y^T(k))D(k). \tag{9.3.6}$$

The resulting update laws for the natural gradient update derived in [55] are in exact agreement with the update laws derived above [41]. The update law in (9.3.5) and (9.3.6) is related to the earlier derived update law (9.2.26) and (9.2.27) by the relation

$$\tilde{\nabla} l = \nabla l \begin{bmatrix} D^T D & D^T C \\ C^T D & I_M + C^T C \end{bmatrix} = \nabla l \begin{bmatrix} D & C \\ 0 & I_M \end{bmatrix}^T \begin{bmatrix} D & C \\ 0 & I_M \end{bmatrix} \tag{9.3.7}$$

where

$$\nabla l = \begin{bmatrix} \frac{\partial L^k}{\partial D} & \frac{\partial L^k}{\partial C} \end{bmatrix} \tag{9.3.8}$$

denotes the update according to the ordinary stochastic gradient. The conditioning matrix in (9.3.7) is symmetric and positive-definite. For a discussion on the local stability conditions of the algorithm, see [5, 6, 12, 14] and for other auxiliary conditions to be satisfied see [41, 42, 49]. Implementation issues for the above algorithm have been discussed in [41, 42, 44, 45, 56].

Theorem 9.3.2 (Feedback Minimum Phase Demixing). *Assume the MIMO (mixing) environment modeled by a minimum phase transfer function. Then the update laws for the zeros of the feedback state-space demixing network using the natural gradient are given by*

$$\Delta D = \eta \left[(I_N + D)(\varphi(y)y^T - I_N)\right] \tag{9.3.9}$$

$$\Delta C = \eta[(I_N + D)\varphi(y)X^T] \tag{9.3.10}$$

where $\varphi(y)$ is a nonlinear score function given by (9.2.28).

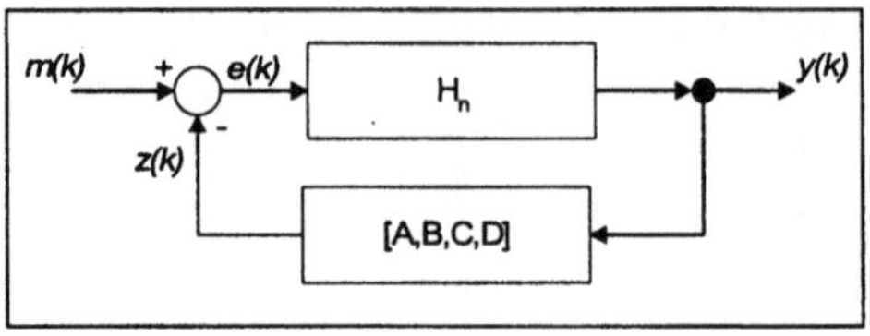

Figure 9.3.1: State-space feedback demixing structure

Proof. We propose an alternate structure [42] where we model the demixing network to be comprised of a feedback transfer function (see Figure 9.3.1). This feedback path is composed of tunable poles and zeros while the feedforward path is assumed to have fixed parameters.

The network equations for the feedback path are (see Figure 9.3.1)

$$z(k) = C\,X(k) + D\,y(k). \tag{9.3.11}$$

The output of the feedback structure is given by

$$y(k) = H_n * (m(k) - z(k)) \cong m(k) - z(k), \text{ assuming } H_n = I. \tag{9.3.12}$$

Rearranging terms, we get

$$(I_N + D)\,y(k) \; + \; C\,X(k) = m(k). \tag{9.3.13}$$

In matrix form, we can express (9.3.13) as

$$\begin{bmatrix} y \\ X \end{bmatrix} = \begin{bmatrix} I_N + D & C \\ 0 & I_M \end{bmatrix}^{-1} \begin{bmatrix} m \\ X \end{bmatrix} \tag{9.3.14}$$

Defining the augmented vectors $\tilde{y}$ and $\tilde{x}$ and the matrix $\tilde{W}$ as

$$\widetilde{X} = \begin{bmatrix} m \\ X \end{bmatrix}, \widetilde{Y} = \begin{bmatrix} y \\ X \end{bmatrix}, \widetilde{W} = \begin{bmatrix} I_N + D & C \\ 0 & I_M \end{bmatrix}$$

we have

$$\widetilde{Y} = \widetilde{W}^{-1}\widetilde{X} = \widetilde{\widetilde{W}}\widetilde{X} \text{ where } \widetilde{\widetilde{W}} = \widetilde{W}^{-1}. \tag{9.3.15}$$

Using the natural gradient, deriving the update law for $\widetilde{\widetilde{W}}$ [42] and expressing in terms of the state-space matrices C and D, we get

$$\Delta(I_N + D) = \Delta D = \eta\left[(I_N + D)(\varphi(y)y^T - I_N) + C\varphi(X)y^T\right] \tag{9.3.16}$$

$$\Delta C = \eta[(I_N + D)\varphi(y)X^T + C(\varphi(X)X^T - I_M)]. \tag{9.3.17}$$

Enforcing the following auxiliary conditions to the above update laws not only simplifies them computationally, but also improves the convergence properties of the proposed algorithm

$$0_{M\times N} = \varphi(X)y^T \tag{9.3.18}$$

$$\varphi(X)X^T = I_M. \tag{9.3.19}$$

The final update law for the feedback structure is given by

$$\Delta D = \eta\left[(I_N + D)(\varphi(y)y^T - I_N)\right] \tag{9.3.20}$$

$$\Delta C = \eta[(I_N + D)\varphi(y)X^T]. \tag{9.3.21}$$

Theorem 9.3.3 (Feedforward Nonminimum Phase Demixing). *Assume the MIMO (mixing) environment modeled by a nonminimum phase transfer function. Then the update law for the all-zero state-space FIR demixing network using the natural gradient is given by*

$$\Delta C_i = \eta(k)\left[C_i - \varphi(y(k))u(k-i)^T\right],\quad i = 1, 2, \cdots, M-1 \tag{9.3.22}$$

$$\Delta D = \eta(k)\left[D - \varphi(y(k))u(k)^T\right] \tag{9.3.23}$$

where the state-space matrix C is partitioned as

$$C = [C_1\ C_2\ \cdots\ C_{M-1}] \tag{9.3.24}$$

with C_i being the MIMO FIR filter coefficients corresponding to delay z^{-i}, and $u(k)$ represents an information back-propagation filter, given as

$$u(k) = \sum_{i=1}^{M-1} C_{M-i}^H(k)y(k-i+1) + D^H(k)y(k-M+1) \tag{9.3.25}$$

where H represents the matrix Hermitian operator, and $\varphi(y)$ is a nonlinear score function defined in (9.2.28).

Proof. Please refer to [49] for a comprehensive proof. For the nonminimum phase mixing case, in order to avoid stability issues we constrain the demixing network to be a double-sided FIR filter so as to approximate the intended unstable IIR inverse. Using the MIMO controller canonical form (see Appendix), the poles of the demixing network are all set to zero.

In order to use a more compact notation for the derivation we define a MIMO FIR filter as

$$\bar{W} \triangleq \left[\ W_0\quad W_1\quad \cdots\quad W_{N-1}\ \right] = [D\ \ C] \tag{9.3.26}$$

where the state-space matrices C and D are represented as

$$D = W_0,\ C = [\ W_1 \quad W_2 \quad \cdots \quad W_{N-1}\ ]. \tag{9.3.27}$$

Using a compact hybrid time-frequency form, we express the network output equation as

$$y_k = \bar{W}(z) m_k \tag{9.3.28}$$

where m_k represents an $N \times m$ matrix of observations at the kth iteration which includes the kth and previous $N-1$ observation vectors, and y_k represents the output vector at the kth iteration. Using the natural gradient in the systems space [6], the update law for $\bar{W}(z)$ is

$$\Delta \bar{W} = \eta_k \left[\bar{W}(z) - \varphi(y_k) u^T(z^{-1}) \right] \tag{9.3.29}$$

where

$$u_k = \bar{W}^T(z^{-1}) y_k \tag{9.3.30}$$

in which y_k represents an $N \times M$ matrix of outputs at the kth iteration, which includes the kth and previous $N-1$ output vectors, and u_k represents the information back-propagation (IBP) vector at the kth iteration. Note that the IBP filter is implemented by applying a transposed and time-reversed version of the forward filter on the network output [4, 45, 49].

Therefore the explicit update laws for the state-space matrices C and D, using the definition of (9.3.27), are

$$\Delta C_i = \eta(k) \left[C_i - \varphi(y(k)) u(k-i)^T \right], \quad i = 1, 2, \cdots, M-1 \tag{9.3.31}$$

$$\Delta D = \eta(k) \left[D - \varphi(y(k)) u(k)^T \right]. \tag{9.3.32}$$

In the state-space regime, the IBP filter assumes the form

$$\lambda(k) = A' \lambda(k+1) + C' y(k) \tag{9.3.33}$$

$$u(k) = B' \lambda(k) + D' y(k). \tag{9.3.34}$$

This algorithm (9.3.31)–(9.3.34) requires both forward and backward in time propagation by the FIR filter [44, 45, 49]. Although the derived computational structure is noncausal, this algorithm can be practically implemented using a delayed version of the algorithm. Further, the algorithmic delay and computational storage requirements can be minimized by operating both forward and backward in time filters on the same batch of observations and outputs [41, 49].

9.4 Simulation Results

MATLAB simulation results for the three proposed algorithms are presented. In all three cases the environment comprises a 3×3 IIR mixing/convolving filter with the following model [41, 44, 56]

$$\sum_{i=0}^{m-1} \mathbf{A}_i m(k-i) = \sum_{i=0}^{n-1} \mathbf{B}_i s(k-i) + v(k) \tag{9.4.1}$$

where $\mathbf{A}_i$, $\mathbf{B}_i$ are the coefficient matrices for the ith tap of the autoregressive and moving average sections of the filter respectively, $v(k)$ is the additive Gaussian measurement noise, and m, n are the lengths of the MIMO auto-regressive and moving average lags in the filter.

The demixing network for each simulation can be initialized in a number of ways, depending on the available environment information, which may be collected via auxiliary means [54, 56]. This may include complete, partial, or no knowledge of the poles of the mixing environment [44, 45]. The primary advantage of incorporating knowledge of poles in a BSR algorithm, if known, is to obtain a more compact demixing network representation. All three cases have been verified via simulation and the algorithms were able to converge satisfactorily using on-line update algorithms. However, due to limited space only typical results are being presented here, for more details see [41, 44–46, 49]. Unless otherwise specified, the original sources had sub-Gaussian, Gaussian, and super-Gaussian densities, respectively.

The convergence performance of the BSR algorithms, for the linear and convolutive mixing, is measured using the multichannel intersymbol interference (MISI) benchmark [4, 41, 56]. MISI is a measure of the global transfer function diagonalization and permutation as achieved by the demixing network and is defined as

$$\mathrm{MISI}_k = \sum_{i=1}^{N} \frac{\left| \sum_j \sum_p |G_{pij}| - \max_{p,j} |G_{pij}| \right|}{\max_{p,j} |G_{pij}|} + \sum_{j=1}^{N} \frac{\left| \sum_i \sum_p |G_{pij}| - \max_{p,i} |G_{pij}| \right|}{\max_{p,i} |G_{pij}|} \tag{9.4.2}$$

where $G(z) = H(z) * \bar{H}(z)$ is the global transfer function, $\bar{H}(z) = [A_e, B_e, C_e, D_e]$ is the transfer function of environment, and $H(z) = [A, B, C, D]$ is the transfer function of network.

As discussed earlier, the optimal score functions for the derived update laws depend on the density functions of the sources to be separated, which upon successful convergence of the algorithm are similar to the density of the separated outputs. We have used an adaptive score function [44–46, 50], which relies on the batch kurtosis of the output of the demixing system. This score function given by (9.4.3) converges to the optimal nonlinearity for the demixing system as the network's outputs approach stochastic independence:

$$\varphi(y) = \begin{cases} y - \alpha \tanh(\beta y) & \kappa_4(y) \le -\gamma \\ y & \text{for } |\kappa_4(y)| < \gamma \\ \alpha \tanh(\beta y) & \kappa_4(y) \ge \gamma \end{cases} \tag{9.4.3}$$

where $\kappa_4(y)$ represents batch Fisher kurtosis [27] of the output signals, and γ represents a positive density classification threshold.

9.4.1 BSR from minimum phase mixing environments

In this section, we present the BSR results for a minimum phase IIR mixing model of the form (9.4.1), where $m = n = 3$ and

$$A_0 = \begin{bmatrix} 1 & 1 & -1 \\ 1 & -1 & 1 \\ 1 & -1 & 1 \end{bmatrix}, A_1 = \begin{bmatrix} 0.5 & 0.8 & -0.7 \\ 0.8 & 0.3 & -0.2 \\ -0.1 & -0.5 & 0.4 \end{bmatrix}, A_2 = \begin{bmatrix} 0.06 & 0.4 & -0.5 \\ 0.16 & -0.1 & -0.4 \\ -0.3 & -0.06 & 0.3 \end{bmatrix}$$

$$B_0 = \begin{bmatrix} 1 & 0.6 & 0.8 \\ 0.3 & 1 & 0.1 \\ 0.6 & -0.8 & 1 \end{bmatrix}, B_1 = \begin{bmatrix} 0.5 & 0.5 & 0.6 \\ -0.3 & 0.2 & -0.3 \\ -0.2 & -0.43 & 0.6 \end{bmatrix}, B_2 = \begin{bmatrix} .125 & 0.06 & 0.2 \\ -0.1 & 0 & 0.4 \\ 0.08 & -0.13 & 0.3 \end{bmatrix}.$$

In the transfer function domain, this MIMO transfer function will constitute a 3×3 matrix with each element being an order 18 IIR filter (provided each element has been scaled to account for all the environment transmission poles). See Figure 9.4.1 for graphical representation of the simulated transfer function and the recovered demixing network.

Case I: Feedforward minimum phase mixing recovery

Using the presented feedforward structure (Theorem 9.3.1), three different simulation scenarios are presented, where the poles of the demixing network have been (a) fixed at the solution or (b) all set to zero. In all cases, the matrix C is initialized to have small random elements, the matrix D is initialized to be the identity matrix, while matrices A and B are correspondingly set to be in canonical form (see Appendix).

The number of states in the demixing state-space network was correspondingly chosen to be 18 and 60. For both cases, the pole-zero map of the final demixing network and the on-line MISI convergence characteristics are presented in Figure 9.4.2 after 30,000 on-line iterations. It can be observed that in both cases the global transfer function is diagonalized, although the demixing network have quite different pole-zero maps, indicating multiple possible solutions to the same problem. A quantitative comparison has been provided in Table 9.4.1.

Case II: Feedback minimum phase mixing recovery

For the presented feedback structure (Theorem 9.3.2), two different simulation scenarios are presented, where the demixing network poles have been (a) fixed at the solution or (b) all set to zero. The state-space matrices and the number of states for the feedback network were chosen similar to the feedforward case. Note that

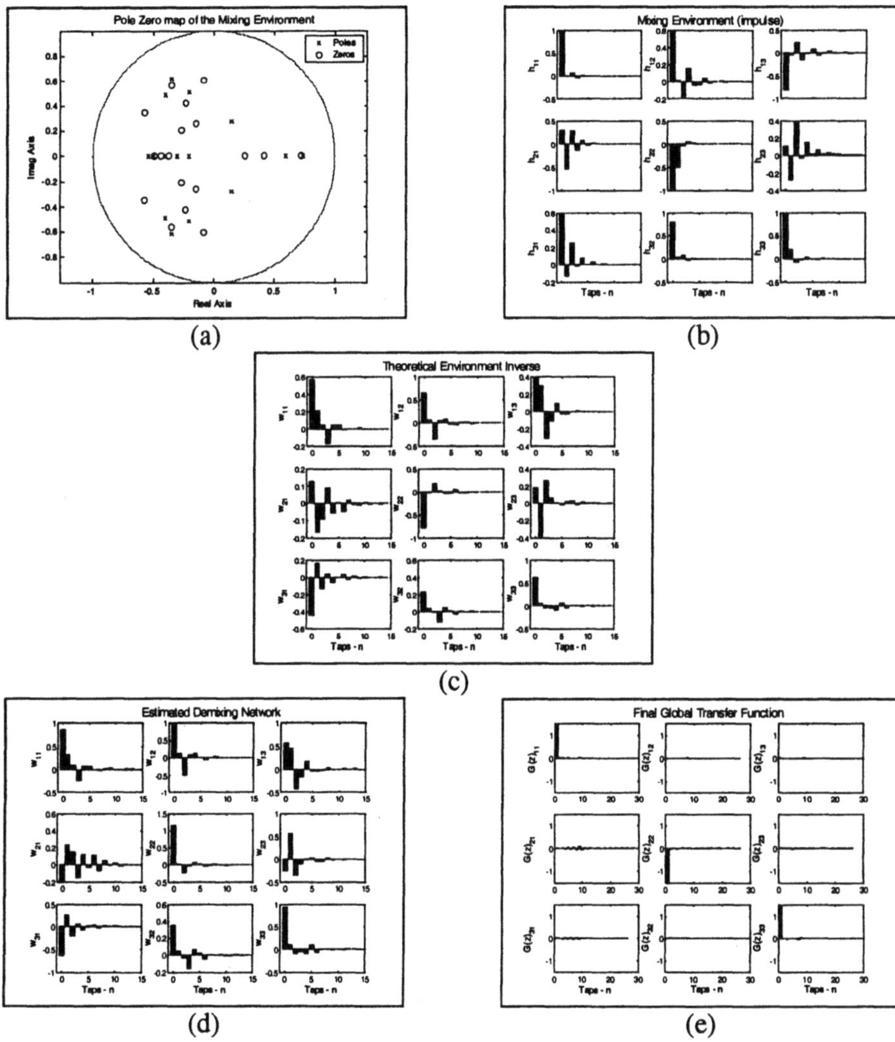

Figure 9.4.1: State-space BSR for a minimum phase mixing environment: (a) transmission pole-zero map of the environment model, (b) impulse response of the environment model, (c) theoretical environment inverse, (d) typical estimated demixing network using minimum phase BSR algorithms, (e) final global transfer function.

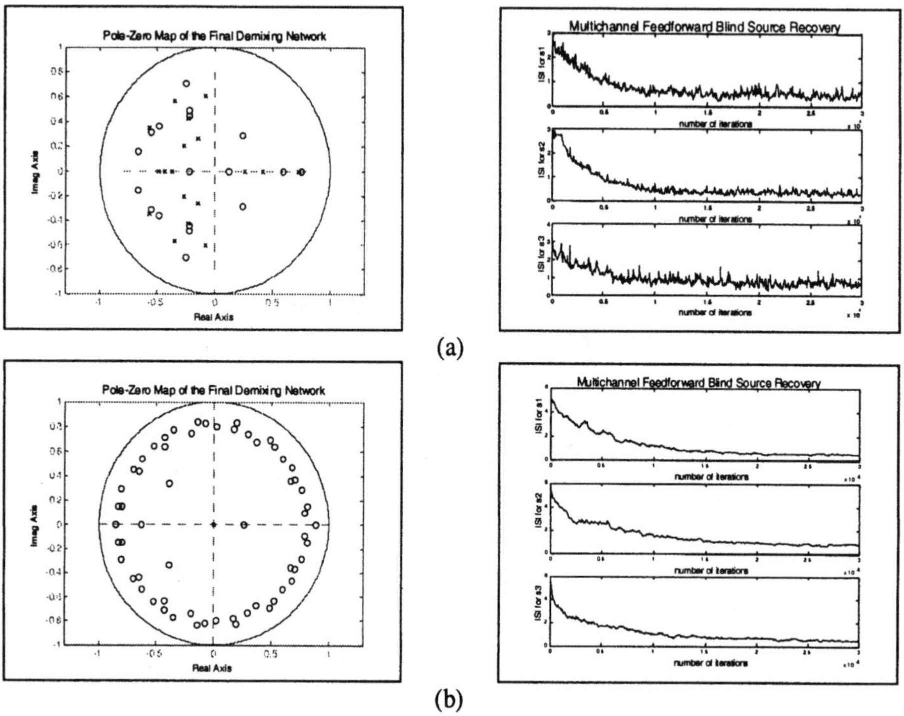

Figure 9.4.2: Transmission pole-zero map and convergence of MISI performance index for the minimum phase feedforward network, when (a) all poles are set to the theoretical solution, (b) all poles are fixed at zero.

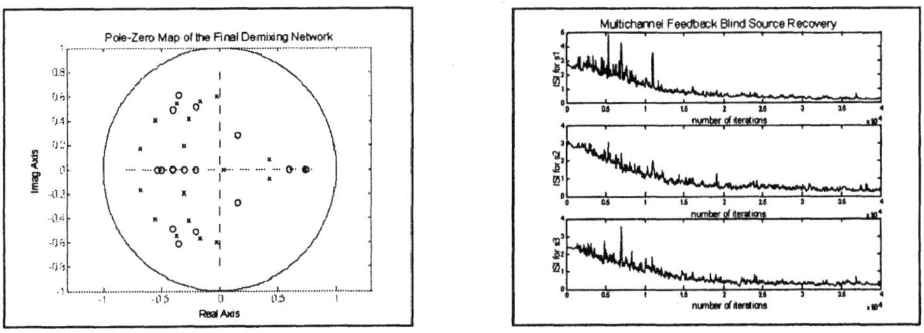

(a)

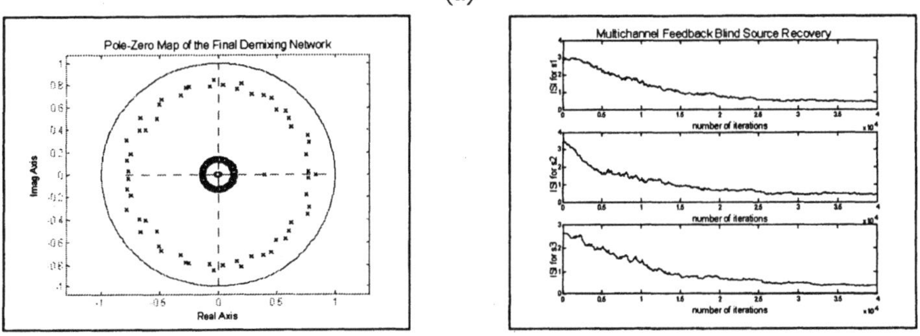

(b)

Figure 9.4.3: Feedback transmission pole-zero map and convergence of MISI performance index for the minimum phase feedback network, when (a) all poles are set to the theoretical solution, (b) all poles are fixed at zero.

again three different global diagonalizing demixing network pole-zero maps are obtained. The qualitative simulation result summary is presented in Figure 9.4.3. See Table 9.4.1 for a quantitative comparison summary.

9.4.2 BSR from nonminimum phase mixing environments

For the case of nonminimum phase IIR mixing environment model (9.4.1), we again choose $m = n = 3$ and the coefficient matrices are given by

$$A_0 = \begin{bmatrix} 1 & 1 & -1 \\ 1 & -1 & 1 \\ 1 & -1 & 1 \end{bmatrix}, A_1 = \begin{bmatrix} 0.5 & 0.8 & -0.7 \\ 0.8 & 0.3 & -0.2 \\ -0.1 & -0.5 & 0.4 \end{bmatrix}, A_2 = \begin{bmatrix} 0.06 & 0.4 & -0.5 \\ 0.16 & -0.1 & -0.4 \\ -0.3 & -0.06 & 0.3 \end{bmatrix}$$

$$B_0 = \begin{bmatrix} 1 & 0.6 & 0.8 \\ 0.5 & 1 & 0.7 \\ 0.6 & 0.8 & 1 \end{bmatrix}, B_1 = \begin{bmatrix} 0.5 & 0.7 & 0.16 \\ 0.7 & 0.2 & -0.3 \\ -0.2 & 0.53 & 0.6 \end{bmatrix}, B_2 = \begin{bmatrix} .425 & 0.3 & 0.7 \\ -0.1 & 0 & -0.4 \\ 0.08 & -0.13 & 0.3 \end{bmatrix}.$$

The theoretical inverse of this environment model is an unstable IIR filter with a minimum of 18 states per output. This instability stems from having two poles of the intended demixing network be outside the unit circle. However, the demixing network is set up to be composed of a doubly finite 3×3 FIR demixing filter with 31 taps per filter (i.e., a total of 90 states) and supplied with mixtures of multiple source distributions. The polynomial matrix $\bar{W} = [D\ C]$ is initialized to be full rank, while A and B are in canonical form.

Instantaneous update results after 40,000 iterations are presented in Figure 9.4.4. A close comparison of the theoretical environment inverse and the adaptively estimated inverse will reveal that the theoretical results form a part of the much larger (in taps), estimated inverse.

A quantitative performance comparison of presented BSR algorithms in a communication signals scenario is provided in Table 9.4.1.

Table 9.4.1: Quantitative signal recovery comparison of the proposed algorithms

		Minimum Phase Feedforward Demixing		Minimum Phase Feedback Demixing		Nonminimum Phase Demixing
		Poles known	*Poles at zero*	*Poles known*	*Poles at zero*	
ISI (in dB)	AM Signal	−22.55	−21.77	−27.67	−16.47	−17.67
	PSK Signal	−26.66	−18.48	−23.52	−18.85	−11.43
	White Noise	−21.09	−21.85	−26.03	−18.36	−11.39
SNR (in dB)	AM Signal	45.61	45.91	49.80	41.18	37.46
	PSK Signal	43.62	31.55	46.65	45.17	24.06

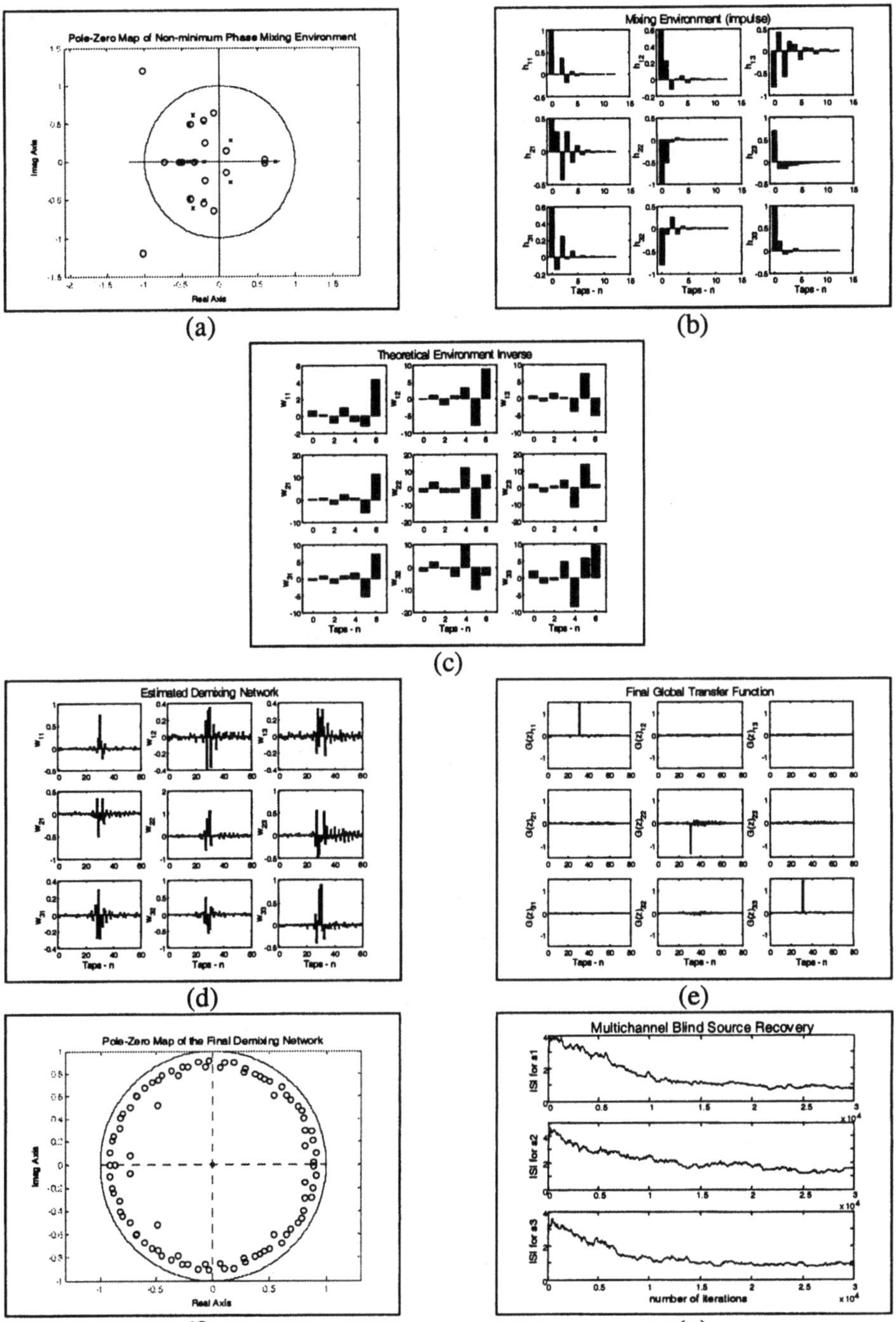

Figure 9.4.4: State-space BSR for a nonminimum phase system: (a) transmission pole-zero map for the mixing environment, (b) impulse response of the environment model, (c) theoretical environment inverse, (d) estimated demixing network, (e) global transfer function achieved, (f) pole-zero map of estimated demixing network, and (g) convergence of MISI performance index.

9.5 Conclusions

This chapter presents a generalized state-space BSR framework based on the theory of multivariable optimization and the Kullback-Lieblar divergence as the performance functional. The framework is then specialized to the state-space BSR for both static and dynamic environments, which may possess either a minimum phase or a nonminimum phase structure. For the case of minimum phase environments, two possible state-space demixing networks in feedforward and feedback configuration are proposed. For nonminimum phase mixing, to avoid any instability, the demixing network is chosen to be a noncausal all-zero feedforward state-space network. The update laws for all the cases have been derived using the natural (or the Riemannian contravariant) gradient.

The presented algorithms have been extensively simulated for both synthetic and physical data. The simulation results for all the cases have been provided using mixtures of multiple source densities, including one Gaussian source. The algorithms exhibit success in recovery of sources in all cases. Furthermore, for the case of minimum phase environments, various possible initial conditions have been explored. It can be observed from the final pole zero maps of the estimated demixing networks that a number of network structures are able to recover the original sources. A comparative table for a set of communication signals has also been provided as a guideline to the achievable separation/recovery results using the proposed on-line adaptive algorithms.

9.6 Appendix: MIMO Canonical Form

For all the BSR simulations and derivations, the network structure is assumed to be in MIMO controller form [8, 39]. In this form, the matrices A and B have the structure

$$A = \begin{bmatrix} -Q_1 & -Q_2 & \cdots & -Q_{M-1} & -Q_M \\ I & 0 & \cdots & 0 & 0 \\ 0 & I & \cdots & 0 & 0 \\ \vdots & \vdots & & \vdots & \vdots \\ 0 & 0 & \cdots & I & 0 \end{bmatrix}, \; B = \begin{bmatrix} I \\ 0 \\ 0 \\ 0 \\ 0 \end{bmatrix} \tag{9.6.1}$$

$$C = [\; P_1 \;\; P_2 \;\; \cdots \;\; P_{M-1} \;\; P_M \;],\; D = [P_0] \tag{9.6.2}$$

where the MIMO state-space filter has the transfer function representation

$$H(z) = [P(z)][Q(z)]^{-1} \tag{9.6.3}$$

and

$$P(z) = \sum_{i=0}^{n} P_i z^{-i} \tag{9.6.4}$$

$$Q(z) = \sum_{i=0}^{n} Q_i z^{-i} \text{ with } Q_0 = I_N.$$

Bibliography

[1] S. Amari, *Differential-Geometrical Methods in Statistics*, Lecture Notes in Statistics 28. Springer-Verlag, New York, 1985.

[2] S. Amari, "Neural learning in structured parameter spaces-Natural Riemannian gradient," in *Neural Information Processing Systems*, NIPS-96, pp. 127–133, MIT Press, Cambridge, MA, 1996.

[3] S. Amari and J. F. Cardoso, "Blind source separation: Semi-parametric statistical approach," *IEEE Trans. Signal Processing*, vol. 45, no. 11, pp. 2692–2700, 1997.

[4] S. Amari, S. C. Douglas, A. Cichocki, and H. H. Yang, "Multichannel blind deconvolution and equalization using the natural gradient," *Proc. IEEE Workshop on Signal Processing*, pp. 101–104, Paris, France, 1997.

[5] S. Amari, T.-P. Chen, and A. Cichocki, "Stability analysis of adaptive blind source separation," *Neural Networks*, vol. 10, no. 8, pp. 1345–1352, 1997.

[6] S. Amari, "Natural gradient works efficiently in learning," *Neural Computation*, vol. 10, pp. 251–276, 1998.

[7] S. Amari and H. Nagaoka, *Methods of Information Geometry*, AMS and Oxford University Press, London, 1999.

[8] P. J. Antsaklis and A. N. Michel, *Linear Systems*, McGraw-Hill, New York, 1997.

[9] J. Bell and T. J. Sejnowski, "An information-maximization approach to blind separation and blind deconvolution," *Neural Computation*, vol. 7, pp. 1129–1159, 1995.

[10] A. Benveniste, M. Goursat and G. Ruget, "Robust identification of a nonminimum phase system: Blind adjustment of a linear equalizer in data communications," *IEEE Trans. Automatic Control*, vol. 25, pp. 385–399, June 1980.

[11] J. F. Cardoso, "Blind signal processing: Statistical principles," *Proc. IEEE*, vol. 90, no. 8, pp. 2009–20026, Oct. 1998 (Special issue on Blind Identification and Estimation, R.-W. Liu and L. Tong, eds.).

[12] J. F. Cardoso and B. Laheld, "Equivariant adaptive source separation," *IEEE Trans. Signal Processing*, vol. 44, pp. 3017–3030, 1996.

[13] A. Cichocki, S. Amari, and J. Cao, "Neural network models for blind separation of time delayed and convolved signals," *Japanese IECE Trans. Fundamentals*, vol. E82-A, pp. 1595–1603, 1997.

[14] A. Cichocki and S. Amari, *Adaptive Blind Signal and Image Processing*, John Wiley, New York, 2002.

[15] S. C. Douglas and S. Haykin, "On the relationship between blind deconvolution and blind source separation," *Proc. 31st Asilomar Conference on Signals, Systems, and Computers*, vol. 2, pp. 1591–1595, Pacific Grove, CA, 1997.

[16] G. Erten and F. M. Salam, "Voice output extraction by signal separation," *Proc. 1998 IEEE International Symposium on Circuits and Systems*, vol. 3, pp. 5–8, Monterey, CA, 1998.

[17] G. Erten and F. M. Salam, "Voice extraction by on-line signal separation and recovery," *IEEE Trans. Circuits and Systems II: Analog and Digital Signal Processing*, vol. 46, no. 7, pp. 915–922, July 1999.

[18] A. B. A. Gharbi and F. M. Salam, "Separation of mixed signals in dynamic environments: Formulation and some implementation," *Proc. 37th Midwest Symposium on Circuits and Systems*, vol. 1, pp. 587–590, Lafayette, LA, 1994.

[19] A. B. A. Gharbi and F. M. Salam, "Implementation and test results of a chip for the separation of mixed signals," *IEEE Trans. Circuits and Systems*, vol. 42, no. 11, pp. 748–751, Nov. 1995.

[20] A. B. A. Gharbi and F. M. Salam, "Algorithms for blind signal separation and recovery in static and dynamic environments," *Proc. 1997 International Symposium on Circuits and Systems*, pp. 713–716, Hong Kong, June 1997.

[21] M. Girolami, *Advances in Independent Component Analysis*, Springer-Verlag, New York, 2000.

[22] S. Haykin, *Adaptive Filter Theory*, 3rd ed., Prentice-Hall, Upper Saddle River, NJ, 1996.

[23] S. Haykin, Ed., *Unsupervised Adaptive Filtering*, vol. I and II, John Wiley, New York, 2000.

[24] J. Herault and C. Jutten, "Space or time adaptive signal processing by neural network models," in *Neural Networks for Computing, AIP Conference Proceedings*, vol. 151, pp. 206–211, American Institute for Physics, New York, 1986.

[25] A. Hyvarinen and E. Oja, "A fast fixed-point algorithm for independent component analysis," *Neural Computation*, vol. 9, pp. 1483–1492, 1997.

[26] A. Hyvarinen, J. Karhunen, and E. Oja, *Independent Component Analysis*, John Wiley, New York, 2001.

[27] M. Kendall and A. Stuart, *The Advanced Theory of Statistics*, vol. I, Charles Griffin and Co. Ltd., London, 1977.

[28] H. K. Khalil, *Nonlinear Systems*, 3rd ed., Prentice-Hall, Upper Saddle River, NJ, 2002.

[29] R. Lambert, *Multichannel Blind Deconvolution: FIR Matrix Algebra and Separation of Multipath Mixtures*, Ph.D. Thesis, University of Southern California, Department of Electrical Engineering, 1996.

[30] T. W. Lee and T. Sejnowski, "Independent component analysis for sub-Gaussian and super-Gaussian mixtures," *Proc. 4th Joint Symposium on Neural Computation*, vol. 7, pp. 132–140, Institute for Neural Computation, 1997.

[31] T. W. Lee, A. J. Bell, and R. Orglmeister, "Blind source separation of real-world signals," *Proc. IEEE Conference on Neural Networks*, pp. 2129–2135, Houston, TX, 1997.

[32] T. W. Lee, M. Girolami, A. Bell, and T. J. Sejnowski, "A unifying information-theoretic framework for independent component analysis," *Int. J. Mathematical and Computer Modeling*, vol. 38, pp. 1–21, 2000.

[33] F. L. Lewis and V. L. Syrmos, *Optimal Control*, 2nd ed., Wiley, New York, 1995.

[34] A. N. Michel and D. Liu, *Qualitative Analysis and Synthesis of Recurrent Neural Networks*, Marcel Dekker, New York, 2002.

[35] J. Nicholls, P. A. Fuchs, A. R. Martin, and B. G. Wallace, *From Neuron to Brain*, 4th ed., Sinauer Associates Inc., Sunderland, MA, 2001.

[36] F. M. Salam, "An adaptive network for blind separation of independent signals," *Proc. IEEE International Symposium on Circuits and Systems*, vol. 1, pp. 431–434, Chicago, IL, May 1993.

[37] F. M. Salam and G. Erten, "Blind signal separation and recovery in dynamic environments," *Proc. 3rd IEEE Workshop on Nonlinear Signal and Image Processing*, Mackinac Island, MI, Sept. 1997 (available at http://www.ecn.purdue.edu/NSIP/ta34.ps).

[38] F. M. Salam and G. Erten, "Exact entropy series representation for blind source separation," *Proc. IEEE Conference on Systems, Man, and Cybernetics*, vol. 1, pp. 553–558, 1999.

[39] F. M. Salam and G. Erten, "The state-space framework for blind dynamic signal extraction and recovery," *Proc. 1999 IEEE International Symposium on Circuits and Systems*, vol. 5, pp. 66–69, Orlando, FL, 1999.

[40] F. M. Salam, A. B. A. Gharbi, and G. Erten, "Formulation and algorithms for blind signal recovery," *Proc. 40th Midwest Symposium on Circuits and Systems*, vol. 2, pp. 1233–1236, Sacramento, CA, 1998.

[41] F. M. Salam, G. Erten, and K. Waheed, "Blind source recovery: Algorithms for static and dynamic environments," *Proc. of INNS-IEEE International Joint Conference on Neural Networks*, vol. 2, pp. 902–907, Washington, DC, 2001.

[42] F. M. Salam and K. Waheed, "State-space feedforward and feedback structures for blind source recovery," *Proc. 3rd International Conference on Independent Component Analysis and Blind Signal Separation*, pp. 248–253, San Diego, CA, 2001.

[43] K. Torkkola, "Blind separation of convolved sources based on information maximization," *Proc. IEEE Workshop on Neural Networks for Signal Processing*, pp. 423–432, Kyoto, Japan, 1996.

[44] K. Waheed and F. M. Salam, "Blind source recovery: Some implementation and performance issues," *Proc. 44th IEEE Midwest Symposium on Circuits and Systems*, vol. 2, pp. 694–697, Dayton, OH, 2001.

[45] K. Waheed and F. M. Salam, "State-space blind source recovery for mixtures of multiple source distributions," *Proc. IEEE International Symposium on Circuits and Systems*, pp. 197–200, Scottsdale, AZ, May 2002,

[46] K. Waheed and F. M. Salam, "Blind source recovery using an adaptive generalized Gaussian score function," *Proc. 45th IEEE Midwest Symposium on Circuits and Systems*, vol. 2, pp. 418–421, Tulsa, OK, Aug. 2002.

[47] K. Waheed and F. M. Salam, "A data-derived quadratic independence measure for adaptive blind source recovery in practical applications," *Proc. 45th IEEE Midwest Symposium on Circuits and Systems*, vol. 3, pp. 473–476, Tulsa, OK, Aug. 2002.

[48] K. Waheed and F. M. Salam, "Cascaded structures for blind source recovery," *Proc. 45th IEEE Midwest Symposium on Circuits and Systems*, vol. 3, pp. 656–659, Tulsa, OK, Aug. 2002.

[49] K. Waheed and F. M. Salam, "State-space blind source recovery of non-minimum phase environments," *Proc. 45th IEEE Midwest Symposium on Circuits and Systems*, vol. 2, pp. 422–425, Tulsa, OK, Aug. 2002.

[50] K. Waheed and F. M. Salam, "New hyperbolic source density models for blind source recovery score functions," *IEEE International Symposium on Circuits and Systems*, Bangkok, Thailand, May 2003, forthcoming.

[51] K. Waheed and F. M. Salam, "Algebraic overcomplete independent component analysis," *4th International Symposium on Independent Component Analysis and Blind Source Separation*, Nara, Japan, Apr. 2003, forthcoming.

[52] H. H. Yang, S. Amari, and A. Cichocki, "Information back-propagation for blind separation of sources from non-linear mixtures," *Proc. IEEE Conference on Neural Networks*, pp. 2141–2146, Houston, TX, 1997.

[53] L. Q. Zhang, A. Cichocki, and S. Amari, "Multichannel blind deconvolution of nonminimum phase systems using information backpropagation," *Proc. 5th International Conference on Neural Information Processing*, pp. 210–216, Perth, Australia, Nov. 1999.

[54] L. Q. Zhang, A. Cichocki, and S. Amari, "Kalman filter and state-space approach to multichannel blind deconvolution," *Proc. IEEE Workshop on Neural Networks for Signal Processing*, pp. 425–434, Sydney, Australia, Dec. 2000.

[55] L. Q. Zhang, S. Amari, and A. Cichocki, "Semiparametric approach to multichannel blind deconvolution of non-minimum phase systems," *Advances in Neural Information Processing Systems*, vol. 12, pp. 363–369, MIT Press, Cambridge, MA, 2000.

[56] L. Q. Zhang, and A. Cichocki, "Blind deconvolution of dynamical systems: A state space approach," *J. Signal Processing*, vol. 4, no. 2 pp. 111–130, Mar. 2000.

[57] Website http://www.egr.msu.edu/bsr.

Chapter 10

Direct Neural Dynamic Programming

Lei Yang, Russell Enns, Yu-Tsung Wang, and Jennie Si

Abstract: This chapter is about approximate dynamic programming (ADP), which has been referred to by many different names, such as "reinforcement learning," "adaptive critics," "neuro-dynamic programming," and "adaptive dynamic programming." The fundamental issue under consideration is optimization over time by using learning and approximation to handle problems that severely challenge conventional methods due to their very large scale and/or lack of sufficient prior knowledge. In this chapter we discuss the relationships, results, and challenges of various approaches under the theme of ADP. We also introduce the fundamental principles of our direct neural dynamic programming (NDP). We demonstrate its application for a continuous state control problem using an industrial scale Apache helicopter model. This is probably one of the first studies where an ADP type of algorithm has been applied to a complex, realistic, continuous state problem, which is a major challenge in machine learning when dealing with scalability or generalization.

10.1 Introduction

Approximate dynamic programming (ADP) has been studied extensively in the past two decades, referred to by many different names, such as "reinforcement learning (RL)," "adaptive critics (AC)," and "neuro-dynamic programming (NDP)." These general-purpose learning mechanisms are expected to adapt to any dynamic environment and learn to maximize the expectation value of a utility function over time. The purpose is to develop design tools for doing optimization over time, by using learning and approximation to handle larger-scale and more difficult problems.

There is a long history that is closely related and attributed to the modern version of dynamic programming that dates back to 19th-century physics. The Bellman Equation represents a blend of ideas from physics, economics, mathematics, and dynamic system control. The dynamic programming approach to decision and control problems involving nonlinear dynamic systems provides the ultimate optimal solution in any stochastic or uncertain environment. Application areas of dynamic programming are broad and of substantial importance in fundamental science and engineering. These include industrial operations, medicine, energy, transportation and economics, to mention a few. However, dynamic programming suffers from the "curse of dimensionality" which has hindered its tremendous power and appeal as the way of providing the best solution to complex system design problems.

In the past two decades, there has been substantial progress, but not necessarily cohesively linked, made from different disciplines including neural networks, adaptive/optimal/robust control, computer science/machine learning, decision theory (especially the study of Markov decision processes), engineering, and operations research. The common goal is to address the question of how we can use approximation methods and/or learning to extend the power of the Bellman Equation, so that we can develop widely applicable and robust methods for obtaining approximately optimal strategies.

ADP refers to learning mechanisms in which processes or agents learn through trial-and-error interactions with a dynamic environment. An ADP algorithm is designed to optimize a (typically) scalar performance index called a "reinforcement signal." Compared to dynamic programming, ADP is an approach that focuses on how to approximate the value function through iterative learning and thus an approximate optimal solution instead of searching for the optimal value of the function.

In the following sections, we will give an overview of the ADP methodology, together with some well-known algorithms and applications. Then the chapter will focus on results developed and extensively tested by the authors. We introduce the basic framework of our direct NDP, the associated learning algorithms, and especially case studies to demonstrate the effectiveness of our direct NDP. We demonstrate the direct NDP in Apache helicopter stabilization, command tracking, and reconfiguration after actuator failures. This is probably one of the first times that an ADP type of algorithm has been applied to a complex real-life continuous state problem. Until now, reinforcement learning has been mostly successful in discrete state-space problems. On the other hand, prior ADP-based approaches to controlling continuous state-space systems have all been limited to smaller, linearized, or decoupled problems. Therefore the work presented here complements and advances the existing literature in the general area of learning approaches in ADP.

10.2 ADP Models and Algorithms

Figure 10.2.1 depicts a general reinforcement learning framework [1] which also applies to general ADP.

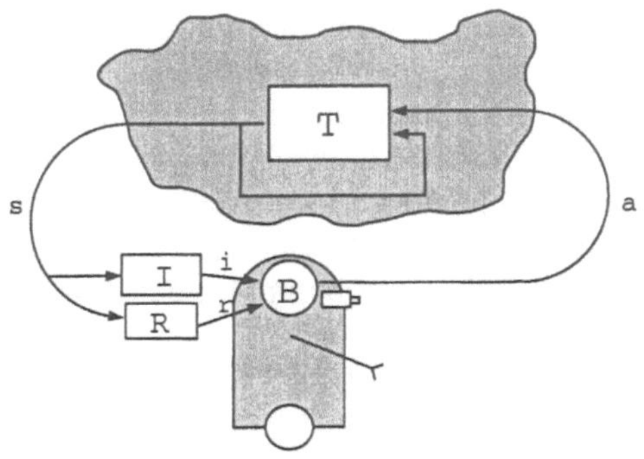

Figure 10.2.1: The standard ADP model

A reinforcement, and hence an ADP, model typically includes

- S, a discrete or discretized set of environment states,
- A, a discrete or discretized set of agent actions,
- R, a set of scalar reinforcement signals typically denoted by 0s and 1s or other real numbers.

Figure 10.2.1 also includes the agent's behavior B, or the agent's actions taken in consideration of optimizing a long-term sum of values of the reinforcement signal, I, which is an observable state from the system.

During each step of interaction the following interaction is repeated, imitating learning:

- The agent receives the input i, some indication of the current state.
- Then an action is chosen, a, to generate an output. The action is selected by B so that it tends to increase the long-run sum of values of the reinforcement signal.
- The action changes the state of the environment and the value of this state transition is communicated to the agent through a scalar reinforcement signal r.

This interaction will go on until an optimal policy π is founded, which is in the form of a mapping from system states to agent's actions so that an optimized long-term sum of reinforcement signals is achieved. The agent can learn to do this over time by a systematic trial-and-error process, guided by a wide variety of algorithms.

ADP algorithms can generally be divided into two main categories, model-based algorithms and model-free algorithms.

In model-based approaches, an explicit model, usually formulated as Markov decision process (MDP), is needed. The model consists of the knowledge of state transition probability function $T(s, a, s')$ and the reinforcement function $R(s, a)$. Since some a priori knowledge of the system is given, it tends to learn the optimal policy faster and with better convergence properties. Because an MDP model is explicitly used to find the optimal value of the value function, this class of algorithms can also be considered to be dynamic programming algorithms.

Model-free algorithms, on the other hand, do not require the explicit system model a priori. They can adaptively learn the optimal policy by interacting with the environment through a trial-and-error process. Model-free approaches are more challenging due to lack of system information prior to learning. However, they may offer a wider domain of applications since they can be implemented in an on-line or real-time fashion.

In the rest of this section, the ADP algorithms discussed mainly make use of an infinite-horizon discounted objective functional as the optimality model. The following notation is used in describing the algorithms. $V^*(s)$ is the optimal value of a state s, where in the following s and s' denote the current and the next state, respectively. It is written as

$$V^*(s) = \max_{\pi} E\left(\sum_{n=t}^{\infty} \gamma^n r_n\right). \tag{10.2.1}$$

$V^*(s)$ in (10.2.1) is unique and can be represented as a solution of the following simultaneous equation.

$$V^*(s) = \max_{a}\left(R(s,a) + \gamma \sum_{s' \in S} T(s,a,s')V^*(s')\right), \forall s \in S. \tag{10.2.2}$$

The optimal policy can also be specified as

$$\pi^*(s) = \arg\max_{a}\left(R(s,a) + \gamma \sum_{s' \in S} T(s,a,s')V^*(s')\right). \tag{10.2.3}$$

10.2.1 Value iteration

Value iteration is one of the earliest ADP methods for finding the optimal policies that maximize $V^*(s)$ [2, 3] and it is model based. The transition probability function T and the reinforcement function R are needed to search for the optimal policy.

For each iteration, a value is calculated and updated according to the following

$$V(s) := \max_{a}\left(R(s,a) + \gamma \sum_{s' \in S} T(s,a,s')V * (s')\right). \tag{10.2.4}$$

By updating the $V(s)$ infinitely for every state, it can be shown that $V(s)$ is guaranteed to converge to the optimal value $V^*(s)$ [3]. The order of which state to use in updating $V(s)$ is not critical so long as it is updated frequently enough in infinite runs [4].

The computational complexity of value-iteration-based algorithms is quadratic in the number of states and linear in the number of actions. The convergence rate slows down considerably as the discount factor γ approaches 1 [5].

10.2.2 Policy iteration

Policy iteration directly updates the policy π instead of the $V(s)$. A set of linear equations needs to be solved to update $V(s)$. The policy is then improved based on the updated state value. The updating rules of the state value and the policy are as follows:

$$V_\pi(s) := R(s, \pi(s)) + \gamma \sum_{s' \in S} T(s, a, s') V_\pi^*(s'), \tag{10.2.5}$$

$$\pi'(s) := \arg\max_a \left(R(s, a) + \gamma \sum_{s' \in S} T(s, a, s') V_\pi^*(s') \right). \tag{10.2.6}$$

First the algorithm tries to solve the state value for each state. Then policy is updated to improve the current state value. A new action is generated based on the new policy of that situation. When no improvement can be made, the policy is guaranteed to be optimal.

The algorithm terminates updating after at most an exponential number of iterations [6]. Compared to value iteration, it takes fewer iterations to converge to an optimal policy, but takes much longer time to complete one iteration.

10.2.3 Temporal difference methods

This class of algorithms is designed to address the critical issue of temporal credit assignment [7] in the context of reinforcement learning.

While an action is applied during training, we actually do not know whether it is a good action or a bad one at that time until the system reaches the "end of a trial." One solution is to remember all the states until the end. Then every previous action is rewarded based on the final result. But that takes too much memory and it is difficult to know when the "end" is. In another strategy, the temporal difference method, the state value of the total discounted reward is adjusted based on the difference between the estimation of the instantaneous reward and the estimated value of the next state, so that the state value $V(s)$ will be gradually updated toward the optimal state value $V^*(s)$.

10.2.3 Adaptive heuristic critic structure

The adaptive heuristic critic (AHC) structure is an abstraction of the structures in [8]. A function approximator is used to estimate the value function instead of calculating it by solving the set of linear equations in policy iteration. The architecture of the AHC structure is shown in Figure 10.2.2 [8]. The AHC uses the instantaneous reinforcement signal to learn to map the state and input of the system to the expected total discounted reward. A reinforcement learning (RL) component is then developed to select an action a that maximizes the output of the AHC. It is similar to the policy iteration algorithm. First, the policy π is fixed and the AHC learns to find the expected state value. Then the value function V_π is fixed and RL starts to learn the optimal policy under the environment. Only alternating implementation can be guaranteed to converge to the optimal policy under appropriate conditions.

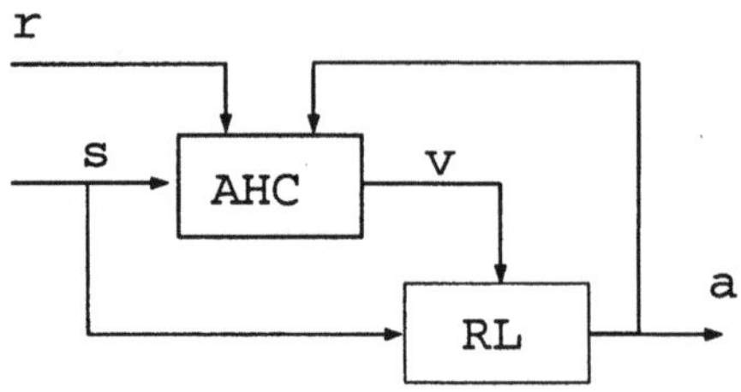

Figure 10.2.2: Architecture for the AHC [8]

TD(λ) is a commonly used method for implementing AHC. It uses the temporal difference (TD) of an immediate reward between current and the next state to estimate the total discounted rewards. A simple form of TD(λ) developed by [7] is known as TD(0). The update rule can be described as

$$V(s) := V(s) + \alpha(r + \gamma V(s') - V(s)). \tag{10.2.7}$$

For every visited state s, the state value $V(s)$ is updated toward the direction of

$$r + \gamma V(s') - V(s).$$

TD(0) looks only one step ahead when adjusting value estimation. $V(s)$ is adjusted by the r, the instantaneous reward received and $V(s')$, the estimated value of the next state. The major difference between TD and the valuate iteration is that TD samples from the real system rather than the model. If the learning rate α is adjusted properly, and the policy is fixed, the TD(0) is guaranteed to converge to the optimal value function.

TD(λ) is an extension of the TD(0) method. Tesauro [9–11] used this algorithm to train his neural network algorithm TD-Gammon to play backgammon and reach a strong master level. This application demonstrates the powerful learning ability of this TD(λ) method. Many times a simple linear neural network is used to approximate the value function $V(s)$, which may take the following form

$$V(s) := w(t) * x(t). \tag{10.2.8}$$

where $w(t)$ and $x(t)$ represent the weights and the system states, respectively. The TD(λ) weight update form can be described as

$$\Delta w(t) := \alpha(v(t+1) - V(t)) \sum_{k=1}^{t} \lambda^{t-k} \Delta_w V(k). \tag{10.2.9}$$

There are two limiting cases. If $\lambda = 1$, then it is roughly equivalent to the supervised learning rule, pairing the input state vector with an absolute final outcome. If $\lambda = 0$, then the value function is only updated based the most recent $\Delta_w V(k)$.

10.2.3 Q-learning

The algorithm stores Q values for every state-action pair, denoted as $Q(s,a)$, which stands for the expected discounted reinforcement of taking action a in state s. The update rule can be specified as

$$Q(s,a) = Q(s,a) + \alpha\left(r + \gamma \max_{a'} Q(s',a') - q(s,a)\right). \tag{10.2.10}$$

The optimal state value $V^*(s)$ is then the value of $Q^*(s,a)$ with the optimal action taken:

$$V(s) := \max_{a} Q(s,a). \tag{10.2.11}$$

Initially all Q values are set to arbitrary numbers. After the first action a is performed, the value is updated according to the above rule. Eventually, $Q(s,a)$ will converge to its optimal function $Q^*(s,a)$.

The Q-learning can learn from any sequence of experiences in which every action is tried in every state infinitely often. Its computational complexity per update is constant. If each action is executed in each state an infinite number of times on an infinite run and γ is decayed appropriately, the Q values will converge with probability 1 to Q^* [12].

10.3 Direct Neural Dynamic Programming (NDP)

The direct NDP [13] presented here is a model-free on-line learning control scheme.

The objective of an NDP controller is to optimize a desired performance measure by learning to create appropriate control actions through interaction with the environment. The learning process is advanced without requiring an explicit system model (such as the form $x(t+1) = f(x(t), u(t))$). Instead, the system information is implicitly "absorbed" by both the action and critic networks. The general schematic diagram of NDP is described in Figure 10.3.1.

Our direct NDP design includes two approximators, the action and the critic, as building blocks. Both the action and critic networks are trained toward optimizing a global objective, namely the Bellman Equation for the critic network and an ultimate performance objective for the action network. During the learning process, the action network is constrained by the critic network to generate controls that optimize the future reward-to-go instead of only temporarily optimal solutions. In contrast to usual neural network applications, there is no readily available training sets of input-output pairs used for approximating $R(t)$ in terms of a least squares fit. Instead, both the control action u and the critic output J are updated according to an error function that changes from one time step to the next. The binary reinforcement signal $r(t)$ is provided from the external environment and may be as simple as either a "0" or a "-1" corresponding to "success" or "failure," respectively.

In our on-line learning control implementation, the controller is "naive" when it just starts to control, namely the action network and the critic network are both randomly initialized in their weights/parameters. Once a system state is observed, an

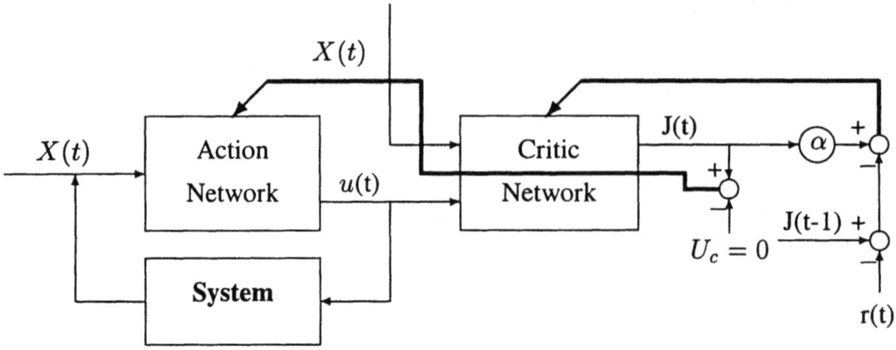

Figure 10.3.1: Schematic diagram for implementations of NDP

action will be subsequently produced based on the parameters in the action network. A "better" control value under the specific system state will lead to a more balanced equation of the principle of optimality. This set of system operations will be reinforced through memory or association between states and control output in the action network. Otherwise, the control value will be adjusted through tuning the weights in the action network in order to make the equation of the principle of optimality more balanced.

10.3.1 The critic network

The critic network is used to provide the discounted total reward-to-go as an approximation of $R(t)$. The approximation is given by

$$R(t) = t(t+1) + \alpha(t+2) + \cdots . \tag{10.3.1}$$

where $R(t)$ is the future accumulative reward-to-go value at time t and α is a discount factor for the infinite-horizon problem ($0 < \alpha < 1$).

We define the prediction error for the critic element as

$$e_c(t) = \alpha J(t) - [J(t-1) - r(t)] \tag{10.3.2}$$

and the objective function to be minimized in the critic network is

$$E_c(t) = \frac{1}{2} e_c^2(t). \tag{10.3.3}$$

The weight update rule for the critic network is a gradient-based adaptation given by

$$w_c(t+1) = w_c(t) + \Delta w_c(t), \tag{10.3.4}$$

$$\Delta w_c(t) = l_c(t) \left[-\frac{\partial E_c(t)}{\partial w_c(t)} \right], \tag{10.3.5}$$

$$-\frac{\partial E_c(t)}{\partial w_c(t)} = -\frac{\partial E_c(t)}{\partial J(t)} \frac{\partial (t)}{\partial w_c(t)}, \tag{10.3.6}$$

where $l_c(t) > 0$ is the learning rate of the critic network at time t, which usually decreases with time to a small value, and W_c is the weight vector in the critic network.

10.3.2 The action network

The principle in adapting the action network is to indirectly backpropagate the error between the desired ultimate objective, denoted by U_c, and the approximate function J from the critic network. Since we have defined "0" as the reinforcement signal for "success," Uc is set to "0" in our design paradigm and in our following case studies. In the action network, the state measurements are used as inputs to create a control as the output of the network. In turn, the action network can be implemented by either a linear or a nonlinear network, depending on the complexity of the problem. The weight updating in the action network can be formulated as follows. Let

$$e_a(t) = J(t) - U_c(t). \tag{10.3.7}$$

The weights in the action network are updated to minimize the following performance error measure:

$$E_a(t) = \frac{1}{2}e_a^2(t). \tag{10.3.8}$$

The update algorithm is then similar to the one in the critic network. By a gradient descent rule

$$w_a(t+1) = w_a(t) + \Delta w_a(t), \tag{10.3.9}$$

$$\Delta w_a(t) = l_a(t)\left[-\frac{\partial E_a(t)}{\partial w_a(t)}\right], \tag{10.3.10}$$

$$\frac{\partial E_a(t)}{\partial w_a(t)} = \frac{\partial E_a(t)}{\partial J(t)}\frac{\partial J(t)}{\partial u(t)}\frac{\partial u(t)}{\partial v(t)}\frac{\partial v(t)}{\partial w_a(t)}, \tag{10.3.11}$$

where $l_a(t) > 0$ is the learning rate of the action network at time t, which usually decreases with time to a small value, and W_a is the weight vector in the action network.

10.3.3 Direct NDP learning algorithms

Our on-line learning configuration introduced above involves two major components in the learning system, namely the action network and the critic network. In the following, we devise learning algorithms and elaborate on how learning takes place in each of the two modules. In our NDP design, both the action network and the critic network are multilayer feedforward networks. The neural network structure for the nonlinear multilayer critic network is shown in Figure 10.3.2.

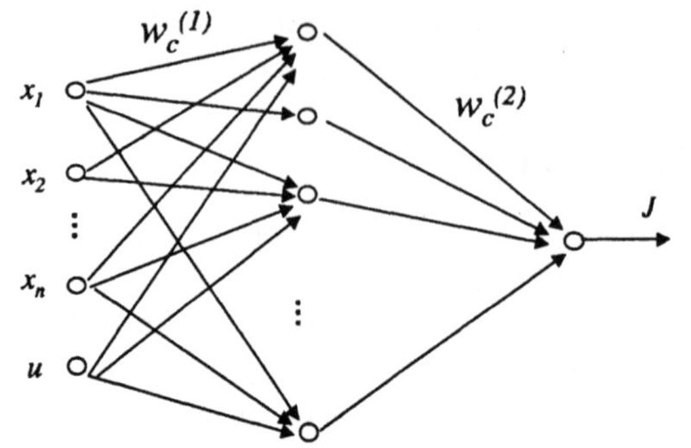

Figure 10.3.2: Schematic diagram for the implementation of a nonlinear critic network using a feedforward network with one hidden layer

In the critic network, the output $J(t)$ will be of the form

$$J(t) = \sum_{i=1}^{N_h} w_{c_i}^{(2)}(t)p_i(t), \tag{10.3.12}$$

$$p_i(t) = \frac{1-\exp(-q_i(t))}{1+\exp(-q_i(t))}, \; i = 1, \cdots, N_h, \tag{10.3.13}$$

$$q_i(t) = \sum_{j=1}^{n+1} w_{c_{ij}}^{(1)}(t)x_j(t), \; i = 1, \cdots, N_h, \tag{10.3.14}$$

where q_i is the ith hidden node input of the critic network; p_i is the corresponding output of the hidden node; N_h is the total number of hidden nodes in the critic network; and $n+1$ is the total number of inputs into the critic network including the analog action signal $u(t)$ from the action network.

By applying the chain rule, the adaptation of the critic network is summarized below.

(1) $\Delta w_c^{(2)}$ (hidden to output layer):

$$\Delta w_{c_i}^{(2)}(t) = l_c(t)\left[-\frac{\partial E_c(t)}{\partial w_{c_i}^{(2)}(t)}\right], \tag{10.3.15}$$

$$\frac{\partial E_c(t)}{\partial w_{c_i}^{(2)}(t)} = \frac{\partial E_c(t)}{\partial J(t)}\frac{\partial J(t)}{\partial w_{c_i}^{(2)}(t)} = \alpha e_c(t)p_i(t). \tag{10.3.16}$$

(2) $\Delta w_c^{(1)}$ (input to hidden layer):

$$\Delta w_{c_{ij}}^{(1)}(t) = l_c(t)\left[-\frac{\partial E_c(t)}{\partial w_{c_{ij}}^{(1)}(t)}\right], \tag{10.3.17}$$

$$\frac{\partial E_c(t)}{\partial w_{c_{ij}}^{(1)}(t)} = \frac{\partial E_c(t)}{\partial J(t)}\frac{\partial J(t)}{\partial p_i(t)}\frac{\partial p_i(t)}{\partial q_i(t)}\frac{\partial q_i(t)}{\partial w_{c_{ij}}^{(1)}(t)} \tag{10.3.18}$$

$$= \alpha e_c(t) w_{c_i}^{(2)}(t)\left[\frac{1}{2}\left(1-p_i^2(t)\right)\right] x_j(t). \tag{10.3.19}$$

Now let us investigate the adaptation in the action network, which is implemented by a feedforward network similar to the one in Figure 10.3.2 except that the inputs are the n measured states and the output is the action $u(t)$. The associated equations are

$$u(t) = \frac{1-\exp(-v(t))}{1+\exp(-v(t))}, \tag{10.3.20}$$

$$v(t) = \sum_{i=1}^{N_h} w_{a_i}^{(2)}(t) g_i(t), \tag{10.3.21}$$

$$g_i(t) = \frac{1-\exp(-h_i(t))}{1+\exp(-h_i(t))},\ i=1,\cdots,N_h, \tag{10.3.22}$$

$$h_i(t) = \sum_{j=1}^{n} w_{a_{ij}}^{(1)}(t) x_j(t),\ i=1,\cdots,N_h, \tag{10.3.23}$$

where v is the input to the action node, and g_i and h_i are the output and the input of the hidden nodes of the action network, respectively. Since the action network inputs the state measurements only, there is no $(n+1)$th term in (10.3.23) as in the critic network (see (10.3.14) for comparison). The update rule for the nonlinear multilayer action network also contains two sets of equations.

(1) $\Delta w_a^{(2)}$ (hidden to output layer):

$$\Delta w_{a_i}^{(2)}(t) = l_a(t)\left[-\frac{\partial E_a(t)}{\partial w_{a_i}^{(2)}(t)}\right], \tag{10.3.24}$$

$$\frac{\partial E_a(t)}{\partial w_{a_i}^{(2)}(t)} = \frac{\partial E_a(t)}{\partial J(t)}\frac{\partial J(t)}{\partial u(t)}\frac{\partial u(t)}{\partial v(t)}\frac{\partial v(t)}{\partial w_{a_i}^{(2)}(t)} \tag{10.3.25}$$

$$= e_a(t)\left[\frac{1}{2}\left(1-u^2(t)\right)\right] g_i(t)\sum_{i=1}^{N_h}\left[w_{c_i}^{(2)}(t)\frac{1}{2}\left(1-p_i^2(t)\right) w_{c_{i,n+1}}^{(1)}(t)\right]. \tag{10.3.26}$$

In the above equations, $\partial J(t)/\partial u(t)$ is obtained by changing variables and by the chain rule. $w_{c_{i,n+1}}^{(1)}$ is the weight associated with the input element from the action network.

(2) $\Delta w_a^{(1)}$ (input to hidden layer):

$$\Delta w_{a_{ij}}^{(1)}(t) = l_a(t)\left[-\frac{\partial E_a(t)}{\partial w_{a_{ij}}^{(1)}(t)}\right], \tag{10.3.27}$$

$$\frac{\partial E_a(t)}{\partial w_{a_{ij}}^{(1)}(t)} = \frac{\partial E_a(t)}{\partial J(t)}\frac{\partial J(t)}{\partial u(t)}\frac{\partial u(t)}{\partial v(t)}\frac{\partial v(t)}{\partial g_i(t)}\frac{\partial g_i(t)}{\partial h_i(t)}\frac{\partial h_i(t)}{\partial w_{a_{ij}}^{(1)}(t)} \tag{10.3.28}$$

$$= e_a(t)\left[\frac{1}{2}\left(1-u^2(t)\right)\right] w_{a_i}^{(2)}(t)\left[\frac{1}{2}\left(1-g_i^2(t)\right)\right] x_j(t) \times$$
$$\sum_{i=1}^{N_h}\left[w_{c_i}^{(2)}(t)\frac{1}{2}\left(1-p_i^2(t)\right) w_{c_{i,n+1}}^{(1)}(t)\right]. \tag{10.3.29}$$

Normalization is performed in both networks to confine the values of the weights into some appropriate range by

$$w_c(t+1) = \frac{w_c(t)+\Delta w_c(t)}{\|w_c(t)+\Delta w_c(t)\|_1}, \tag{10.3.30}$$

$$w_a(t+1) = \frac{w_a(t)+\Delta w_a(t)}{\|w_a(t)+\Delta w_a(t)\|_1}. \tag{10.3.31}$$

Several experiments were conducted to evaluate the effectiveness of our learning control designs. The parameters used in the simulations are defined as follows:

- $l_c(0)$: initial learning rate of the critic network
- $l_a(0)$: initial learning rate of the action network
- $l_c(t)$: learning rate of the critic network at time t which is decreased by 0.05 every 5 time steps until it reaches 0.005 and it stays at $l_c(f) = 0.005$ thereafter
- $l_a(t)$: learning rate of the action network at time t which is decreased by 0.05 every 5 time steps until it reaches 0.005 and it stays at $l_a(f) = 0.005$ thereafter
- N_c: internal cycle of the critic network
- N_a: internal cycle of the action network
- T_c: internal training error threshold for the critic network
- T_a: internal training error threshold for the action network
- N_h: number of hidden nodes

Note that the weights in the action and the critic networks were trained using their internal cycles, N_a and N_c, respectively. That is, within each time step the weights of the two networks were updated for at most N_a and N_c times, respectively, or stopped once the internal training error threshold T_a and T_c have been met.

10.4 A Control Problem: Helicopter Command Tracking

A helicopter is a sophisticated system with multiple inputs used to control a significant number of states. There exists a large amount of cross coupling between control inputs and states. Further, the system is highly nonlinear and changes significantly as a function of operating condition. For these reasons the helicopter serves as an excellent and challenging platform for testing approximate dynamic programming systems [14].

The helicopter's states are controlled by a main rotor and a tail rotor. There are three main rotor actuators whose positions, z_A, z_B, and z_C, control the position and orientation of a swash plate, which in turn controls the main rotor's blade angles as a function of rotational azimuth. There is a single tail rotor actuator position (z_D), which controls the tail rotor's blade angles. The aircraft states are numerous. For flight control purposes they are limited to the aircraft translational (u, v, w) and rotational (p, q, r) velocities and the aircraft orientation (θ, ϕ, ψ) for a total of nine states. The helicopter's longitudinal (u), lateral (v), and vertical velocities (w) are in ft/s. The helicopter's roll rate (p), pitch rate (q), and yaw rate (r) are in degrees/s. The helicopter's Euler angles, pitch (θ), roll (ϕ), and yaw (ψ) are in degrees. The states can be written in vector form as $\mathbf{x} = [u, v, w, p, q, r, \theta, \phi, \psi]$. The controls can be written in vector form as $\mathbf{u} = [z_A, z_B, z_C, z_D]$.

For simulation purposes we use a detailed helicopter model run at 50 Hz for evaluating our NDP controller's performance. The model, named FLYRT, is a sophisticated nonlinear flight simulation model of the Apache helicopter developed by Boeing over the past two decades.

10.4.1 Direct NDP mechanism for helicopter control

We now turn our attention toward applying the direct NDP framework to the tracking control of an Apache helicopter. The objective is to learn to create appropriate control actions solely by observing the helicopter states, evaluating the controller performance, and adjusting the neural networks accordingly.

Figure 10.4.1 outlines the direct NDP control structure applied to helicopter tracking control. It consists of a structured cascade of neural networks forming the action network, the critic network, and the trim network. The action network provides the controls required to drive the helicopter to the desired system state. The critic network approximates the cost function if an explicit cost function does not exist. The trim network provides nominal trim control positions as a function of the system desired operating condition.

In this chapter we introduce the concept of a structured cascade of artificial neural networks (ANN) as the action network. The explicit structure embedded in this ANN, lacking in earlier direct NDP designs [13], allows the NDP controller to more easily learn and take advantage of the physical relationships and dependencies of the system. To perform command tracking, a vector $\mathbf{x}_d$ as a function of time is specified first. For helicopters, it is well established that four of the states in the state vector $\mathbf{x}$ are explicitly controllable. In this experiment, we want to control the velocities u, v, and w and the aircraft's yaw, ψ. The rotational velocities and remaining Euler angles, pitch, and roll will be determined by NDP to achieve the specified tracking goal. We denote this new desired tracking vector $\mathbf{x}_d^0 = [u, v, w, \psi]$, a subset of the original desired state vector $\mathbf{x}_d$.

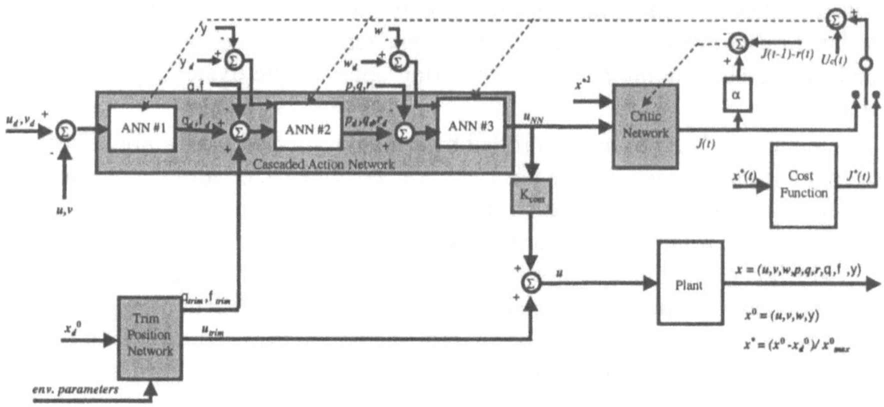

Figure 10.4.1: Direct NDP-based helicopter controller integrating three cascaded artificial neural networks in one action network, a trim network, and a critic network. Also allows for an explicit cost function to be used.

Such structure is similar to classic controllers for helicopters, providing for inner loop body rate control, attitude control, and outer loop velocity control. In this way we take advantage of the explicit relationships between body angular rates, attitudes, and translational velocities. The potential advantage of the structured ANN over classic design methodologies is that it permits full cross-axes control coupling that many single-input single-output (SISO) PID controller designs do not. However, the structured ANN does introduce a level of human knowledge/expertise to its implementation that is not transparent to nonexperts.

It is possible but rather cumbersome to show that a classic proportional controller can be equated to one instance (one set of weights) of our structured ANN if the network nonlinearities are removed (or linearized about the network operating point). Such a relationship between the two designs can be used to provide a good first guess of the action network weights should one want to apply this "expert" knowledge to the learning system. However, all our results shown in this chapter were obtained while the learning system was trained from scratch without using any expert knowledge.

Refering to Figure 10.4.1, the inputs to the first ANN are the longitudinal and lateral velocity errors, $u_{err} = u_d - u$ and $v_{err} = v_d - v$, respectively. The first ANN outputs are the resulting desired pitch and roll of the helicopter, θ_d and ϕ_d. The inputs to the second ANN are the errors in the aircraft attitudes, $\theta_{err} = \theta_d + \theta_{trim} - \theta$, $\phi_{err} = \phi_d + \phi_{trim} - \phi$ and $\psi_{err} = \psi_d - \psi$. The second ANN outputs are the desired roll, pitch, and yaw rates (p_d, q_d, r_d) of the helicopter as a function of the attitude errors. These are then summed with the actual angular rates to obtain the angular rate errors, $p_{err} = p_d - p$, $q_{err} = q_d - q$, and $r_{err} = r_d - r$. The angular rate errors and the vertical velocity error $w_{err} = w_d - w$ then form the inputs to the third ANN. The third ANN then computes the controls $\mathbf{u}_{NN} = [u_{NN,1}, u_{NN,2}, u_{NN,3}, u_{NN,4}]$

as a function of the angular rate and vertical velocity errors. The resulting $\mathbf{u}_{NN}$, which is normalized because of the ANN structure, is then scaled by the controller scaling gain K_{cont} and summed with the nominal trim control from the trim network. That is, $\mathbf{u} = K_{cont} * \mathbf{u}_{NN} + \mathbf{u}_{trim}$.

As described in previous sections, the objective of the NDP controller is to create a series of control actions to optimize a discounted total reward-to-go, which is rewritten below,

$$R(t) = r(t+1) + \alpha r(t+2) + \alpha^2 r(t+3) + \cdots . \tag{10.4.1}$$

Typically NDP has been applied to systems where explicit feedback, or instantaneous cost evaluation, is not available at each time step. In such cases the reinforcement signal, $r(t)$, takes a simple binary form with $r(t) = 0$ when the final event is successful (an objective is met), or $r(t) = -1$ if the final event is a failure (the objective is not met). For many real-world tracking problems, however, explicit feedback is continually available and so we can define a more informative quadratic reinforcement signal

$$r(t) = -\sum_{i=1}^{n} \left(\frac{(x_i^0 - x_{d,i}^0)}{x_{\max,i}^0} \right)^2 \tag{10.4.2}$$

where in our application $n = 4$, x_i^0 is the ith state variable of $x^0 = [u, v, w, \psi]$, and $x_d^0 = [u_d, v_d, w_d, \psi_d]$ is the desired state and $x_{\max,i}^0$ a normalization term.

The critic and action networks are then trained per Section 10.3. The equations governing the training of the action network are significantly more complex than prior work since backpropagation must be performed through ANN's 2 and 3 for the training of ANN 1. In essence, it is equivalent to training a six-layer network if we assume that each ANN has two layers of weights.

The objective is to train the controller to perform the specified maneuver regardless of the operating conditions or the vehicle initial conditions. The states of interest are the aircraft translational (u, v, w) and rotational (p, q, r) velocities and the aircraft Euler angles pitch (θ), roll (ϕ), and yaw (ψ). Failure criteria are used to bound allowed error for each state. The allowed errors, shown in Table 10.4.1, are initially large and decrease as a function of time to an acceptable minimum.

These failure criteria were chosen judiciously but no claims are made to their optimality. The results show that these criteria create a control system that can control the helicopter both in nominal conditions and when subjected to disturbances. Heuristic failure criteria is one of the advantages of NDP if one does not have an accurate account of the performance measure. This is also one characteristic of the NDP design that differs from other neural control designs. The critic network plays the role of working out a more precise account of the performance measure for credit/blame assignment derived from the heuristic criteria. If the networks have converged, an explicitly desired state has been achieved which is reflected in the U_c term in the NDP structure.

The trim network is a neural network, or lookup table, that is trained (programmed) to schedule the aircraft's nominal actuator position and aircraft orientation as a func-

Table 10.4.1: Failure criterion for helicopter stabilization

Aircraft State	Initial Allowed Error	Final Allowed Error	Error Rate
u, v, w	20 ft/s	4 ft/s	-0.8 (ft/s)/s
p, q, r	30°/s	6°/s	-1.2^o/s/s
θ, ϕ, ψ	30°	6°	-1.2^o/s

tion of operating conditions. It is a critical element in nonlinear flight control system design.

10.4.2 Direct NDP-based tracking control results

This section presents results showing the performance of the direct NDP controlling the Apache helicopter for a variety of maneuvers. This is a complex MIMO nonlinear control system design problem. It provides a realistic test bed for how well an ADP algorithm can generalize.

Characteristic to prior NDP research, the performance of the direct NDP is summarized statistically in tables. Fourteen maneuvers are considered: seven accelerations from hover to 50 ft/s at various accelerations and seven decelerations from 100 ft/s to 50 ft/s at various decelerations. Each maneuver is tested in three wind conditions: case (A) no wind, case (B) 10 ft/s step gust for 5 seconds, and case (C) turbulence simulated using a Dryden model with a spatial turbulence intensity of $\sigma = 5$ ft/s and a turbulence scale length of $L_W = 1750$ ft. In addition to the tabular statistics provided, both statistical and typical time history plots of the aircraft states are provided for two cases, a hover to 50 ft/s at 5 ft/s^2 maneuver with turbulence and a 100 ft/s to 50 ft/s at -4 ft/s^2 maneuver with a step gust.

The statistical success of the direct NDP to learn to control the helicopter is evaluated for the five flight conditions. For each flight condition 100 runs were performed to evaluate the direct NDP performance, where for each run initial weights in each network were set randomly. Each run consists of up to 5000 attempts (trials) to learn to successfully control the system. An attempt is deemed successful if the helicopter stays within the failure criteria bounds described in Table 10.4.1 for the entire flight duration (50 seconds). After each failed trial, the system is restarted with the same initial state as the previous trial, but with the network weights at the previous trial's termination. If the controller successfully controls the helicopter within 5000 trials the run is considered successful, if not, the run is considered a failure.

Table 10.4.2 statistically summarizes the learning ability of the direct NDP controller to perform a hover to 50 ft/s maneuver at a number of different accelerations. Results for 100 ft/s to 50 ft/s decelerations at a number of deceleration rates are provided in Table 10.4.3. In both cases results are provided for the three wind conditions cited above. The success percentage reflects the percentage of runs for which the direct NDP system successfully learns to control the helicopter. The average number of trials is what it takes the direct NDP system to learn to control the helicopter. Standard deviations from successful runs for various maneuvers are also used to demonstrate the (in)consistency of the learning control performance.

Table 10.4.2: Learning statistics for hover to 50 ft/s maneuver at various accelerations for three wind conditions

Cond.	Acceleration (ft/s^2)	2	3	4	5	6	7	8
Case A	Success percentage	94%	62%	67%	65%	66%	66%	74%
	Average # trials	1600	2019	2115	1950	1983	2028	1870
	Learning deviation	214	339	324	307	293	306	252
Case B	Success percentage	96%	66%	76%	70%	50%	57%	53%
	Average # trials	1367	1720	1770	1874	2173	1970	2419
	Learning deviation	191	280	255	275	381	337	400
Case C	Success percentage	95%	98%	97%	85%	60%	56%	58%
	Average # trials	642	824	1126	1843	1842	2145	2379
	Learning deviation	115	128	165	263	313	333	403

Table 10.4.3: Learning statistics for 100 ft/s to 50 ft/s maneuver at various decelerations for three wind conditions

Cond.	Acceleration (ft/s^2)	2	3	4	5	6	7	8
Case A	Success percentage	98%	90%	85%	80%	84%	76%	73%
	Average # trials	759	1700	1610	2114	1516	1800	2045
	Learning deviation	105	227	226	291	219	248	292
Case B	Success percentage	99%	85%	74%	71%	77%	76%	78%
	Average # trials	1260	1460	1650	1979	2030	1950	1737
	Learning deviation	181	215	229	298	295	264	239
Case C	Success percentage	100%	98%	93%	93%	97%	89%	91%
	Average # trials	778	1258	1373	1489	1236	1350	1677
	Learning deviation	105	180	183	200	162	186	220

Figure 10.4.2 shows both the statistical average state error and error deviation over all successful runs and a typical plot of the controller performance for a hover to 50 ft/s maneuver at an aggressive 5 ft/s^2 acceleration in the presence of turbulence. Figure 10.4.3 shows both the statistical average state error and error deviation and a typical plot of the controller performance for a 100 ft/s to 50 ft/s maneuver at 4 ft/s^2 deceleration in the presence of a step gust. Helicopter and control dynamics are similar for the other maneuvers once the learning controller becomes stabilized in learning.

The neural network parameters used during training are provided in Table 10.4.4. The learning rates, β, for the action network and critic network are scheduled to decrease linearly with time (typically over a few seconds). In every time frame the weight equations are updated until either the error has sufficiently converged ($E < E_{tol}$) or the internal update cycles (N_{cyc}) of the weights have been reached. N_h is the number of hidden nodes in the neural networks.

It is worth mentioning that a comprehensive analysis on the convergence performance of an entire NDP system in general does not exist, neither does an analytical framework on the relationship between the performance of the NDP learning controller versus the learning parameters. It has been argued [13] that updating individual networks alone, action or critic for example, may be viewed as a stochastic approximation problem and therefore, conditions similar to the Robbins-Monro algorithm may be used as guidelines in scheduling the learning parameters. Quantitatively, we have observed that the direct NDP learning parameters do impact the learning ability of the learning controller. For example, the learning rate for action networks can be tuned to perform different maneuvers with different system outcomes. This is illustrated by Table 10.4.5 which shows the direct NDP system performance for learning rates (β_a) of 0.2 and 0.02 for both more aggressive and less aggressive accelerations. Lower learning rates improve the success for more aggressive maneuvers but decrease the learning ability (increase the number of trials) required to learn for less aggressive maneuvers.

Table 10.4.4: Neural network parameter values for training

Parameter	α	$\beta_a(t_0)$	$\beta_a(t_f)$	$\beta_c(t_0)$	$\beta_c(t_f)$	$N_{cyc,a}$
Value	0.95	0.02	0.02	0.1	0.01	200
Parameter	$N_{cyc,c}$	$E_{tol,a}$	$E_{tol,c}$	N_h	K_{cont}	
Value	100	0.005	0.1	6	2.5	

Table 10.4.5: Learning statistics for hover to 50 ft/s maneuvers as a function of learning rate

β_a	Acceleration (ft/s^2)	2	6	8
0.02	Success percentage	94%	66%	74%
	Average # trials	1600	1983	1870
	Learning variance	214	293	252
0.2	Success percentage	100%	80%	20%
	Average # trials	248	1708	2809
	Learning variance	71	421	439

It is interesting to note is that despite what one may expect, overall the direct NDP controller more reliably and more quickly learns to control the helicopter in the presence of turbulence. This is clearly evident in Tables 10.4.2 and 10.4.3. The learning

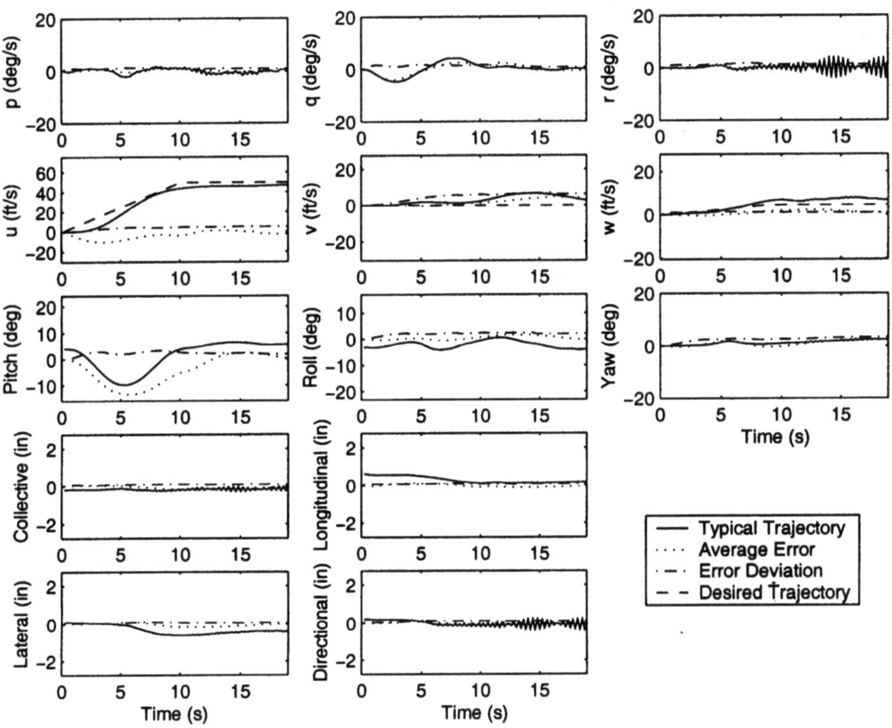

Figure 10.4.2: Statistical and typical state and control trajectories of the helicopter for a hover to 50 ft/s maneuver at 5 ft/s^2 acceleration in turbulence. Turbulence is simulated using a Dryden model with a spatial turbulence intensity of $\sigma = 5$ ft/s and a turbulence scale length of $L_W = 1750$ ft.

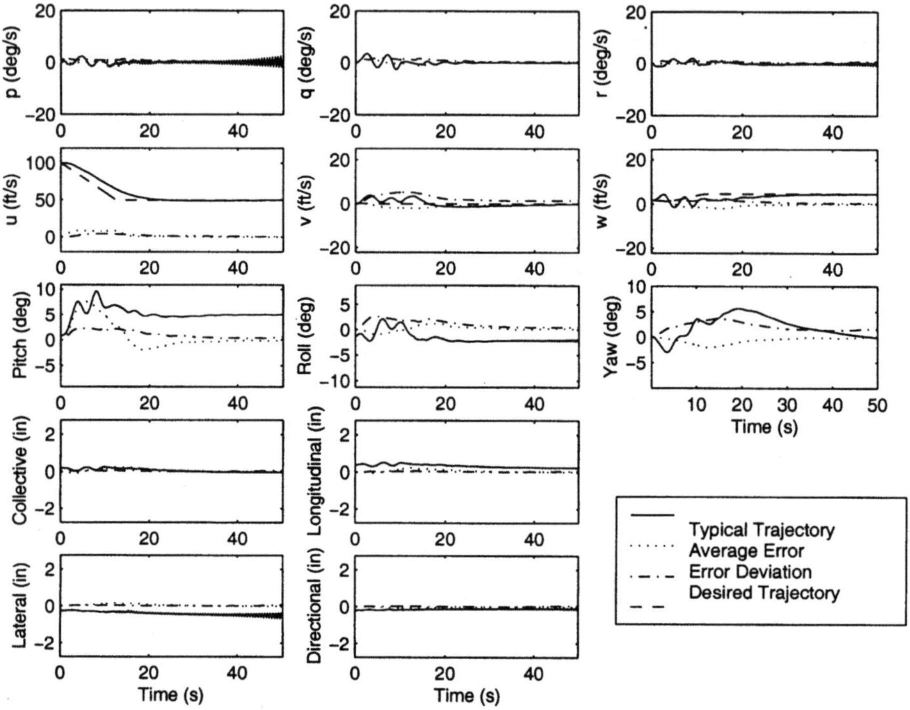

Figure 10.4.3: Statistical and typical state and control trajectories of the helicopter for a 100 ft/s to 50 ft/s maneuver at -4 ft/s^2 acceleration in a step gust. The step gust is 10 ft/s in magnitude and has a 5-second duration.

performance improvement can be attributed to the sustained larger excitation, due to turbulence, to both the neural network inputs and the cost function evaluation when there is turbulence. As a result, the network weights change more in (10.3.11) and thus the learning system explores more of the solution space and is less likely to become trapped in local minima that do not provide an adequate control solution. This suggests that in applications where turbulence or other excitation is not natural, it may be prudent to create an artificial equivalent to improve learning performance.

Also note that, as with previous NDP designs, a large number of trials must occur to successfully learn to perform the maneuver. Further, the more aggressive the maneuver (the higher the acceleration), the more trials required. This is not surprising for a learning system that is learning from experience without any a priori system knowledge. The ramification is that this training is done off-line (i.e., not in a real helicopter), where failures can be afforded, until the controller is successfully trained. Once trained, the neural network weights are frozen and the controller structure shown in Figure 10.4.1 can be implemented in a helicopter. Limited authority on-line training can then be performed to improve system performance.

10.5 Conclusions

In this chapter we consider a class of learning decision and control problems in terms of optimizing a performance measure over time with the following constraints. First, a model of the environment or the system that interacts with the learner is not available a priori. The environment/system can be stochastic, nonlinear, and subject to change. Second, learning takes place "on-the-fly" while interacting with the environment. Third, even though measurements from the environment are available from one decision and control step to the next, a final outcome of the learning process from a generated sequence of decisions and controls comes as a delayed signal in an indicative "win" or "loose" format. Our direct NDP was motivated by reinforcement learning as described in [8] using a simple and model-free framework. Our approximation-based action network approach was inspired by a series of designs developed in heuristic dynamic programming (HDP) [15] and variations of HDP in recent years [16–18]. Our NDP design has shown its promise as a scalable paradigm within the ADP family of algorithms. Our results are among the first to address the generalization issue, which has been discussed extensively in the reinforcement learning community.

Bibliography

[1] L. P. Kaelbling, M. L. Littman, and A. W. Moore, "Reinforcement learning: A survey," *J. Artificial Intelligence Research*, vol. 4, pp. 237–285, 1996.

[2] R. Bellman, *Dynamic Programming*, Princeton University Press, Princeton, NJ, 1957.

[3] D. P. Bertsekas, *Dynamic Programming: Deterministic and Stochastic Models*, Prentice-Hall, Englewood Cliffs, NJ, 1987.

[4] D. P. Bertsekas and D. A. Castanon, "Adaptive aggregation for infinite horizon dynamic programming," *IEEE Trans. Automatic Control*, vol. 34, no. 6, pp. 589–598, 1989.

[5] M. L. Littman, T. L. Dean, and L. P. Kaelbling, "On the complexity of solving Markov decision problems," *Proc. Eleventh Annual Conference on Uncertainty in Artificial Intelligence*, pp. 394–402, Montreal, Quebec, Canada, 1995.

[6] M. L. Puterman, *Markov Decision Processes–Discrete Stochastic Dynamic Programming*, John Wiley, New York, 1994.

[7] R. S. Sutton, "Learning to predict by the methods of temporal difference," *Machine Learning*, vol. 3, pp. 9–44, 1988.

[8] A. G. Barto, R. S. Sutton, and C. W. Anderson, "Neuron like adaptive elements that can solve difficult learning control problems," *IEEE Trans. Systems, Man, and Cybernetics*, vol. 13, pp. 834–847, 1983.

[9] G. Tesauro, "Practical issues in temporal difference learning," *Machine Learning*, vol. 8, pp. 257–277, 1992.

[10] G. Tesauro, "TD-Gammon, a self-teaching backgammon program, achieves master-level play," *Neural Computation*, vol. 6, no. 2, pp. 215–219, 1994.

[11] G. Tesauro, "Temporal difference learning and TD-Gammon," *Commun. ACM*, vol. 38, no. 3, pp. 58–67, 1995.

[12] C. J. Watkins, *Learning from Delayed Rewards*, Ph.D. Thesis, King's College, Cambridge, UK, 1989.

[13] J. Si and Y. Wang, "Online learning control by association and reinforcement," *IEEE Trans. Neural Networks*, vol. 12, no. 2, pp. 349–360, 2000.

[14] R. Enns and J. Si, "Helicopter trimming and tracking control using direct neural dynamic programming," submitted to *IEEE Trans. Neural Networks*.

[15] P. Werbos, "Advanced forecasting methods for global crisis warning and models of intelligence," *General System Yearbook*, vol. 22, pp. 25–38, 1977.

[16] P. Werbos, "A menu of design for reinforcement learning over time," in *Neural Networks for Control*, W. T. Miller III, R. S. Sutton, and P. J. Werbos, Eds., Chapter 3, MIT Press, Cambridge, MA, 1990.

[17] P. Werbos, "Neuro-control and supervised learning: An overview and valuation," in *Handbook of Intelligent Control*, D. White and D. Sofge, Eds., Chapter 3, Van Nostrand, New York, 1992.

[18] P. Werbos, "Approximate dynamic programming for real-time control and neural modeling," in *Handbook of Intelligent Control*, D. White and D. Sofge, Eds., Chapter 13, Van Nostrand. New York, 1992.

Chapter 11

Online Approximation-Based Aircraft State Estimation*

Jay Farrell, Manu Sharma, and Marios Polycarpou

Abstract: This chapter derives algorithms to estimate the aircraft state vector and to approximate the nonlinear forces and moments acting on the aircraft. The forces and moments are approximated as functions of (a portion of) the state vector over the aircraft operating envelope. The force and moment approximators are developed using the "nondimensionalized coefficient" representation that is standard in the aircraft literature. A Lyapunov-like function is used to prove the convergence of the estimator state and to discuss conditions sufficient for the convergence of the approximated force and moment functions.

11.1 Introduction

The past few decades have witnessed the development of a number of nonlinear control methodologies for application to advanced flight vehicles. These methods potentially offer both performance enhancement as well as reduced development times by dealing with the complete dynamics of the vehicle rather than point designs. Feedback linearization, in its various forms, is perhaps the most commonly employed nonlinear control method in this arena [3, 7, 11–13, 17]. In addition to feedback linearization, a number of other nonlinear methods have also been investigated for flight control. Khan and Lu [2] present a nonlinear model predictive control approach that relies on a Taylor series approximation to the system's dif-

*This research supported by Barron Associates, Inc. through Air Force PRDA 01-03-VAK, University of California–Riverside, and University of Cincinnati.

ferential equations. Optimal control techniques are applied to control load-factor in [10]. Prelinearization theory and singular perturbation theory are applied for the derivation of inner and outer loop controllers in [12]. The main drawback to the nonlinear control approaches mentioned above is that, as model-based control methods, they require accurate knowledge of the plant dynamics. This is of significance in flight control since aerodynamic parameters always contain some degree of uncertainty. Furthermore, certain flight regimes (e.g., high angle-of-attack conditions) are inherently fraught with highly nonlinear aerodynamic effects that are difficult to predict accurately preflight. Although some of the above approaches are robust to small modeling errors, they are not intended to accommodate significant unanticipated errors that can occur in the event of failure or battle damage. In such an event, the aerodynamics can change rapidly and deviate significantly from the model used for control design. Uninhabited air vehicles (UAVs) are particularly susceptible to such events since there is no pilot onboard.

The main motivation for this work is to learn during flight, as a function of flight condition, the aerodynamic parameters of the vehicle. An anticipated benefit from this approach is that an initial controller could be developed using a lower fidelity model than required by current methods, thereby offering a cost savings. Flight using this preliminary controller and the methods herein would provide increasingly accurate vehicle models which could be incorporated directly into higher fidelity nonlinear vehicle controllers.

This chapter presents an algorithm to approximate the nonlinear aircraft model during flight. This on-line approximation problem is formulated as a learning augmented estimation problem [4, 5]. The standard nonlinear aircraft model is discussed in Section 11.2. That model includes three unknown force and three unknown moment functions. We manipulate the aircraft model into a matrix vector form that will be convenient for algorithm development. Section 11.3 discusses the unknown forces and moments and their representation as "nondimensional coefficient"' functions over an operating envelope denoted by $\mathcal{D}$. Although the three forces and moments could be approximated directly herein, we choose to approximate the "nondimensional coefficient"' functions. Although there are typically many more than six "nondimensional coefficient"' functions, they are standard in the aircraft literature and, more importantly, they are functions of fewer variables. The approximation-based state estimators and estimator error dynamics are presented in Section 11.4. The parameter update laws for the on-line approximators are presented in Section 11.5. The Lyapunov-like analysis is presented in Section 11.6 and is summarized in the form of a theorem. Implementation-related issues are discussed in Section 11.7, and final conclusions are drawn in Section 11.8. The approach that we present is applicable to both piloted and UAVs. In the case of UAVs, the approximated functions may be used in prediction error-based adaptive control systems to maintain tracking for a vehicle after it has suffered faults or battle damage; however, we do not consider issues related to vehicle control in this chapter. See, e.g., [8, 9, 14, 15, 18]. The approach herein develops on-line approximations to the aircraft force and moment functions. Such off-line approximations have a long history in the aerodynamics community. Trankle et al. review off-line system identification methods in [21]. An overview, with several examples, of off-line spline-based system identification based on data partitioning is presented in [6]. Finally, the article [20] performs off-line estimation of additive functional corrections to the aircraft model. The corrections are multidimensional cubic splines (see [20], p. 1294), which is interesting relative to the approach herein where such model corrections are approximated on-line.

11.2 Aircraft Dynamics and Model Structure

Consider the following representation of the dynamics of a fixed-wing aircraft [1, 19] (see Appendix for the aircraft notation):

$$\dot{\chi} = \frac{1}{mV\cos\gamma}[D\sin\beta\cos\mu + Y\cos\beta\cos\mu + L\sin\mu + T(\sin\alpha\sin\mu - \cos\alpha\sin\beta\cos\mu)] \tag{11.2.1a}$$

$$\dot{\gamma} = \frac{1}{mV}[-D\sin\beta\sin\mu - Y\cos\beta\sin\mu + L\cos\mu + T(\sin\alpha\cos\mu + \cos\alpha\sin\beta\sin\mu)] - \frac{g}{V}\cos\gamma \tag{11.2.1b}$$

$$\dot{V} = \frac{1}{m}(T\cos\alpha\cos\beta - D\cos\beta + Y\sin\beta) - g\sin\gamma \tag{11.2.1c}$$

$$\dot{\mu} = -\frac{g\tan\beta\cos\gamma\cos\mu}{V} + \frac{P_s}{\cos\beta} + \frac{1}{mV}[D\sin\beta\tan\gamma\cos\mu + Y\cos\beta\tan\gamma\cos\mu + L(\tan\beta + \tan\gamma\sin\mu) + T(\sin\alpha\tan\gamma\sin\mu + \sin\alpha\tan\beta - \cos\alpha\sin\beta\tan\gamma\cos\mu)] \tag{11.2.1d}$$

$$\dot{\alpha} = \frac{-1}{mV\cos\beta}[L + T\sin\alpha] + \frac{g\cos\gamma\cos\mu}{V\cos\beta} + Q - P_s\tan\beta \tag{11.2.1e}$$

$$\dot{\beta} = \frac{1}{mV}[D\sin\beta + Y\cos\beta - T\cos\alpha\sin\beta] + \frac{g\cos\gamma\sin\mu}{V} - R_s \tag{11.2.1f}$$

$$\dot{P} = (c_1 R + c_2 P)Q + c_3\bar{L} + c_4\bar{N} \tag{11.2.1g}$$

$$\dot{Q} = c_5 PR - c_6(P^2 - R^2) + c_7\bar{M} \tag{11.2.1h}$$

$$\dot{R} = (c_8 P - c_2 R)Q + c_4\bar{L} + c_9\bar{N} \tag{11.2.1i}$$

where the c_i coefficients for $i = 1, \cdots, 9$ are defined on p. 80 in [19] and

$$\begin{bmatrix} P_s \\ R_s \end{bmatrix} = \begin{bmatrix} \cos\alpha & \sin\alpha \\ -\sin\alpha & \cos\alpha \end{bmatrix} \begin{bmatrix} P \\ R \end{bmatrix}. \tag{11.2.2}$$

Tables 11.9.1–11.9.3 in Appendix define the constants, variables, and functions used in the above equations. Let x denote the state vector $[\chi, \gamma, V, \mu, \alpha, \beta, P, Q, R]$. In this chapter, the aerodynamic forces (D, Y, L) and aerodynamic moments $(\bar{L}, \bar{M}, \bar{N})$ of the aircraft equations will be assumed to be unknown.

To simplify the notation of the following sections, it will be convenient to define the model of (11.2.1a–11.2.1i) in a matrix-vector form.

The flight path angles will be represented as $x_1 = [\chi, \gamma]^T$ with dynamics

$$\dot{x}_1 = \begin{bmatrix} \dot{\chi} \\ \dot{\gamma} \end{bmatrix} = A_1(x)\, f_1(x) + B_1(x)\, T - \begin{bmatrix} 0 \\ \cos\gamma/V \end{bmatrix} g \quad (11.2.3)$$

where

$$A_1(x) = \begin{bmatrix} \frac{1}{mV\cos\gamma} & 0 \\ 0 & \frac{1}{mV} \end{bmatrix} \begin{bmatrix} \sin\beta\cos\mu & \cos\beta\cos\mu & \sin\mu \\ -\sin\beta\sin\mu & -\cos\beta\sin\mu & \cos\mu \end{bmatrix}, \quad (11.2.4)$$

$$B_1(x) = \begin{bmatrix} \frac{1}{mV\cos\gamma} & 0 \\ 0 & \frac{1}{mV} \end{bmatrix} \begin{bmatrix} (\sin\alpha\sin\mu - \cos\alpha\sin\beta\cos\mu) \\ (\sin\alpha\cos\mu + \cos\alpha\sin\beta\sin\mu) \end{bmatrix}, \quad (11.2.5)$$

and $f_1(x) = [D(x), Y(x), L(x)]^T$.

The velocity will be represented as $x_2 = [V]$ with dynamics

$$\dot{x}_2 = \dot{V} = A_2(x)\, f_1(x) + B_2(x)\, T - g\sin\gamma. \quad (11.2.6)$$

where

$$A_2(x) = \frac{1}{m}\begin{bmatrix} -\cos\beta & \sin\beta & 0 \end{bmatrix} \quad (11.2.7)$$

and

$$B_2(x) = \frac{1}{m}\cos\alpha\cos\beta. \quad (11.2.8)$$

The wind axis angles will be represented as $x_3 = [\mu, \alpha, \beta]$ with dynamics

$$\dot{x}_3 = \begin{bmatrix} \dot{\mu} \\ \dot{\alpha} \\ \dot{\beta} \end{bmatrix} = F_3(x) + A_3(x)\, f_1(x) + B_3(x)\, T \quad (11.2.9)$$

where

$$A_3(x) = \frac{1}{mV}\begin{bmatrix} \sin\beta\tan\gamma\cos\mu & \cos\beta\tan\gamma\cos\mu & (\tan\beta + \tan\gamma\sin\mu) \\ 0 & 0 & -1/\cos\beta \\ \sin\beta & \cos\beta & 0 \end{bmatrix},$$

$$B_3(x) = \frac{1}{mV}\begin{bmatrix} (\sin\alpha\tan\gamma\sin\mu + \sin\alpha\tan\beta - \cos\alpha\sin\beta\tan\gamma\cos\mu) \\ -\sin\alpha/\cos\beta \\ -\cos\alpha\sin\beta \end{bmatrix},$$

and

$$F_3(x) = \begin{bmatrix} -\tan\beta\cos\mu \\ \cos\mu/\cos\beta \\ \sin\mu \end{bmatrix} \frac{g\cos\gamma}{V} + \begin{bmatrix} P_s/\cos\beta \\ Q - P_s\tan\beta \\ -R_s \end{bmatrix}. \quad (11.2.10)$$

Finally, the body axis angular rates will be represented as $x_4 = [P, Q, R]$ with dynamics

$$\dot{x}_4 = \begin{bmatrix} \dot{P} \\ \dot{Q} \\ \dot{R} \end{bmatrix} = F_4(x) + A_4 f_4(x, \delta) \tag{11.2.11}$$

where

$$F_4(x) = \begin{bmatrix} (c_1 R + c_2 P) Q \\ c_5 PR - c_6 (P^2 - R^2) \\ (c_8 P - c_2 R) Q \end{bmatrix}, \tag{11.2.12}$$

$$A_4 = \begin{bmatrix} c_3 & 0 & c_4 \\ 0 & c_7 & 0 \\ c_4 & 0 & c_9 \end{bmatrix}, \tag{11.2.13}$$

where δ is the vector of surface deflections and

$$f_4(x, \delta) = [\bar{L}(x, \delta), \bar{M}(x, \delta), \bar{N}(x, \delta)]^T.$$

11.3 Approximator Definition

The aircraft dynamics involve three forces denoted by $f_1(x)$ and three moments denoted by $f_4(x)$. In the aircraft literature (see, e.g., [19]), these functions are "built-up" using "nondimensional coefficients," for example:

$$D = \bar{q}S\left(C_{D_o} + \sum_{i=1}^{6} C_{D_{\delta_i}} \delta_i\right) \tag{11.3.1a}$$

$$Y = \bar{q}S\left(C_{Y_o} + C_{Y_\beta}\beta + C_{Y_P}\frac{bP}{2V} + \sum_{i=1}^{6} C_{Y_{\delta_i}} \delta_i\right) \tag{11.3.1b}$$

$$L = \bar{q}S\left(C_{L_o} + C_{L_\alpha}\alpha + \sum_{i=1}^{6} C_{L_{\delta_i}} \delta_i\right) \tag{11.3.1c}$$

$$\bar{L} = \bar{q}Sb\left(C_{\bar{L}_o} + C_{\bar{L}_\beta}\beta + C_{\bar{L}_P}\frac{bP}{2V} + C_{\bar{L}_R}\frac{bR}{2V} + \sum_{i=1}^{m} C_{\bar{L}_{\delta_i}} \delta_i\right) \tag{11.3.1d}$$

$$\bar{M} = \bar{q}S\bar{c}\left(C_{\bar{M}_o} + C_{\bar{M}_Q}\frac{\bar{c}Q}{2V} + \sum_{i=1}^{m} C_{\bar{M}_{\delta_i}} \delta_i\right) \tag{11.3.1e}$$

$$\bar{N} = \bar{q}Sb\left(C_{\bar{N}_o} + C_{\bar{N}_\beta}\beta + C_{\bar{N}_P}\frac{bP}{2V} + C_{\bar{N}_R}\frac{bR}{2V} + \sum_{i=1}^{m} C_{\bar{N}_{\delta_i}} \delta_i\right). \tag{11.3.1f}$$

The forces, moments, and nondimensional coefficients are functions of the operating condition; however, the arguments have been dropped for compactness of notation.

Although the forces and moments are dependent on many state variables, the nondimensional coefficients will typically only depend on a few of the state variables. For example, $\bar{L}$ is a function of $\bar{q}$, P, V, R, β, and δ, but the coefficient $C_{\bar{L}_R}$ may be accurately approximated as being only a function of $x_r = [M, \alpha]^T$, where M is the Mach number. Let $x_r \in \Re^d$ and $x \in \Re^n$, with $d \ll n$.

The above representation contains 52 nondimensional coefficients. We wish to simplify the representation for the purpose of deriving the on-line function approximation algorithms. For the algorithm derivation, we will use the expression

$$D = \bar{q}S\left(C_{D_o} + C_{D_y}y\right) \quad (11.3.2a)$$
$$Y = \bar{q}S\left(C_{Y_o} + C_{Y_y}y\right) \quad (11.3.2b)$$
$$L = \bar{q}S\left(C_{L_o} + C_{L_y}y\right) \quad (11.3.2c)$$
$$\bar{L} = \bar{q}Sb\left(C_{\bar{L}_o} + C_{\bar{L}_y}y\right) \quad (11.3.2d)$$
$$\bar{M} = \bar{q}S\bar{c}\left(C_{\bar{M}_o} + C_{\bar{M}_y}y\right) \quad (11.3.2e)$$
$$\bar{N} = \bar{q}Sb\left(C_{\bar{N}_o} + C_{\bar{N}_y}y\right) \quad (11.3.2f)$$

where y should be considered as a generic scalar variable that will be replaced later (see Section 11.7) with a particular variable that is of interest. This notation is analogous to the standard aircraft representation approach where, for example,

$$C_{L_y} = \frac{\partial C_L}{\partial y}.$$

We will proceed to develop an algorithm for approximating each of the 12 coefficient functions in (11.3.2a–11.3.2f). Once this algorithm is available, it is straightforward to extend that algorithm to approximate any of the specific 52 coefficient functions of the aircraft model.

Given the model structure of (11.2.1a–11.2.1i) with the unknown forces and moments represented as in (11.3.2a–11.3.2f), the model accuracy can be improved by estimating the nondimensional coefficient functions over the operating region $\mathcal{D}$. Therefore, let

$$\hat{D} = \bar{q}S\left(\hat{C}_{D_o} + \hat{C}_{D_y}y\right) \quad (11.3.3a)$$
$$\hat{Y} = \bar{q}S\left(\hat{C}_{Y_o} + \hat{C}_{Y_y}y\right) \quad (11.3.3b)$$
$$\hat{L} = \bar{q}S\left(\hat{C}_{L_o} + \hat{C}_{L_y}y\right) \quad (11.3.3c)$$
$$\hat{\bar{L}} = \bar{q}Sb\left(\hat{C}_{\bar{L}_o} + \hat{C}_{\bar{L}_y}y\right) \quad (11.3.3d)$$
$$\hat{\bar{M}} = \bar{q}S\bar{c}\left(\hat{C}_{\bar{M}_o} + \hat{C}_{\bar{M}_y}y\right) \quad (11.3.3e)$$
$$\hat{\bar{N}} = \bar{q}Sb\left(\hat{C}_{\bar{N}_o} + \hat{C}_{\bar{N}_y}y\right) \quad (11.3.3f)$$

where $\hat{C}_{D_o}$, $\hat{C}_{D_y}$, $\hat{C}_{Y_o}$, $\hat{C}_{Y_y}$, $\hat{C}_{L_o}$, $\hat{C}_{L_y}$, $\hat{C}_{\bar{L}_o}$, $\hat{C}_{\bar{L}_y}$, $\hat{C}_{\bar{M}_o}$, $\hat{C}_{\bar{M}_y}$, $\hat{C}_{\bar{N}_o}$, and $\hat{C}_{\bar{N}_y}$ are the functions to be approximated. Each of these functions will be represented as a row vector of parameters times a column vector of basis elements. For example,

$$\hat{C}_{D_y}(x_r) = \hat{\theta}_{D_y}^T \phi_{D_y}(x_r) \tag{11.3.4}$$

where $\phi_{D_y}(x_r)$ is a prespecified vector function and $\hat{\theta}_{D_y}$ will be estimated online to improve the accuracy of the function. The other nondimensional coefficients are represented similarly. Implicit in this approach is the assumption that $C_{D_y} = \theta_{D_y}^T \phi_{D_y}(x)$, where $\theta_{D_y}^T$ is known to exist but has an unknown value. We use the symbol $\tilde{\theta}_{D_y} = \hat{\theta}_{D_y} - \theta_{D_y}$ to denote the parameter estimation error.

The vector function $\phi_{D_y}(x_r)$ is referred to as the regressor vector for $C_{D_y}(x_r)$. Many candidate regressors are available: polynomials, splines, radial basis functions, wavelets, neural networks, etc. For the theoretical algorithm development presented herein the choice of the regressor functions is immaterial; however, implementation performance can critically depend on this choice. Also, each nondimensional coefficient could have a unique regressor vector. For simplicity of notation, we will assume that a single regressor vector is used for all the nondimensional coefficients, i.e.,

$$\phi(x_r) = \phi_{D_0}(x_r) = \cdots = \phi_{\bar{N}_y}(x_r).$$

Therefore,

$$\hat{D} = \bar{q}S[1,\ y]\begin{bmatrix}\hat{\theta}_{D_o}^T \\ \hat{\theta}_{D_y}^T\end{bmatrix}\phi(x_r) \tag{11.3.5a}$$

$$\hat{Y} = \bar{q}S[1,\ y]\begin{bmatrix}\hat{\theta}_{Y_o}^T \\ \hat{\theta}_{Y_y}^T\end{bmatrix}\phi(x_r) \tag{11.3.5b}$$

$$\hat{L} = \bar{q}S[1,\ y]\begin{bmatrix}\hat{\theta}_{L_o}^T \\ \hat{\theta}_{L_y}^T\end{bmatrix}\phi(x_r) \tag{11.3.5c}$$

$$\hat{\bar{L}} = \bar{q}Sb[1,\ y]\begin{bmatrix}\hat{\theta}_{\bar{L}_o}^T \\ \hat{\theta}_{\bar{L}_y}^T\end{bmatrix}\phi(x_r) \tag{11.3.5d}$$

$$\hat{\bar{M}} = \bar{q}S\bar{c}[1,\ y]\begin{bmatrix}\hat{\theta}_{\bar{M}_o}^T \\ \hat{\theta}_{\bar{M}_y}^T\end{bmatrix}\phi(x_r) \tag{11.3.5e}$$

$$\hat{\bar{N}} = \bar{q}Sb[1,\ y]\begin{bmatrix}\hat{\theta}_{\bar{N}_o}^T \\ \hat{\theta}_{\bar{N}_y}^T\end{bmatrix}\phi(x_r). \tag{11.3.5f}$$

After the algorithm has been derived, it will be straightforward to extend it to accommodate different regressors for the different functions. Since the designer selects the vector $\phi(x_r)$, we will assume that it has been selected so that $||\phi(x_r)||$ is bounded for any $x_r \in \mathcal{D}$.

11.4 On-Line Approximation-Based Estimation

Each of the following subsections presents a learning-based estimator for a portion of the aircraft state. The prediction errors are defined as

$$\left.\begin{array}{lll} e_V = V - \hat{V}, & e_\chi = \chi - \hat{\chi}, & e_\gamma = \gamma - \hat{\gamma}, \\ e_\mu = \mu - \hat{\mu}, & e_\alpha = \alpha - \hat{\alpha}, & e_\beta = \beta - \hat{\beta}, \\ e_P = P - \hat{P}, & e_Q = Q - \hat{Q}, & e_R = R - \hat{R}. \end{array}\right\} \quad (11.4.1)$$

These subsections also determine and simplify various error quantities that will be required in the Lyapunov analysis. Let

$$\tilde{D} = \hat{D} - D, \qquad \tilde{Y} = \hat{Y} - Y, \qquad \tilde{L} = \hat{L} - L, \quad (11.4.2)$$

$$\tilde{\bar{L}} = \hat{\bar{L}} - \bar{L}, \qquad \tilde{\bar{M}} = \hat{\bar{M}} - \bar{M}, \qquad \tilde{\bar{N}} = \hat{\bar{N}} - \bar{N}. \quad (11.4.3)$$

Assuming that the representations of (11.3.5a–11.3.5f) are capable of representing the true forces and moments, then

$$\tilde{D} = \bar{q}S[1,\ y]\begin{bmatrix} \tilde{\theta}^T_{D_o} \\ \tilde{\theta}^T_{D_y} \end{bmatrix}\phi(x_r), \quad \tilde{Y} = \bar{q}S[1,\ y]\begin{bmatrix} \tilde{\theta}^T_{Y_o} \\ \tilde{\theta}^T_{Y_y} \end{bmatrix}\phi(x_r), \quad (11.4.4)$$

$$\tilde{L} = \bar{q}S[1,\ y]\begin{bmatrix} \tilde{\theta}^T_{L_o} \\ \tilde{\theta}^T_{L_y} \end{bmatrix}\phi(x_r), \quad \tilde{\bar{L}} = \bar{q}Sb[1,\ y]\begin{bmatrix} \tilde{\theta}^T_{\bar{L}_o} \\ \tilde{\theta}^T_{\bar{L}_y} \end{bmatrix}\phi(x_r), \quad (11.4.5)$$

$$\tilde{\bar{M}} = \bar{q}S\bar{c}[1,\ y]\begin{bmatrix} \tilde{\theta}^T_{\bar{M}_o} \\ \tilde{\theta}^T_{\bar{M}_y} \end{bmatrix}\phi(x_r), \quad \tilde{\bar{N}} = \bar{q}Sb[1,\ y]\begin{bmatrix} \tilde{\theta}^T_{\bar{N}_o} \\ \tilde{\theta}^T_{\bar{N}_y} \end{bmatrix}\phi(x_r), \quad (11.4.6)$$

where $\tilde{\theta}_* = \hat{\theta}_* - \theta_*$ with $*$ representing each of the subscripts of the nondimensional coefficients.

Note that each of the following subsections develops an error equation of the form

$$\dot{e}_i + L_i e_i = -A_i(x)\tilde{f}$$

where L_i is a positive-definite diagonal design matrix, $A_i(x)$ is known, e_i is an available signal, and $\tilde{f}$ is a vector function that is to be estimated.

11.4.1 Flight path angle estimator

The implementation equations for the flight path angle estimator are

$$\dot{\hat{x}}_1 = \begin{bmatrix} \dot{\hat{\chi}} \\ \dot{\hat{\gamma}} \end{bmatrix} = \begin{bmatrix} L_\chi e_\chi \\ L_\gamma e_\gamma - \frac{g\cos\gamma}{V} \end{bmatrix} + A_1(x)\hat{f}_1(x) + B_1(x)T. \quad (11.4.7)$$

For analysis, the flight path angle estimator error dynamics are

$$\dot{e}_1 = \begin{bmatrix} \dot{e}_\chi \\ \dot{e}_\gamma \end{bmatrix} = -L_1\, e_1 - A_1(x)\begin{bmatrix} \tilde{D} \\ \tilde{Y} \\ \tilde{L} \end{bmatrix} \quad (11.4.8)$$

where $e_1 = [e_\chi, e_\gamma]^T = x_1 - \hat{x}_1$, $L_1 = \text{diag}(L_\chi, L_\gamma)$ and we have assumed that the thrust command T is achievable by the vehicle.

11.4.2 Velocity estimator

The implementation equations for the velocity estimator are

$$\dot{\hat{x}}_2 = \dot{\hat{V}} \quad = \quad L_V e_V + B_2(x)T + A_2(x)\hat{f}_1(x) - g\sin\gamma. \tag{11.4.9}$$

Assuming that the filtered thrust command T is achievable by the vehicle, for analysis, the velocity estimator error dynamics are

$$\dot{e}_2 = \dot{e}_V \quad = \quad -L_2 e_2 - A_2(x) \begin{bmatrix} \tilde{D} \\ \tilde{Y} \\ \tilde{L} \end{bmatrix} \tag{11.4.10}$$

where $e_2 = e_V = x_2 - \hat{x}_2$, and $L_2 = L_V$.

11.4.3 Wind axis angle estimator

The implementation equations for the wind axis angle estimator are

$$\dot{\hat{x}}_3 = \begin{bmatrix} \dot{\hat{\mu}} \\ \dot{\hat{\alpha}} \\ \dot{\hat{\beta}} \end{bmatrix} = \begin{bmatrix} L_\mu e_\mu \\ L_\alpha e_\alpha \\ L_\beta e_\beta \end{bmatrix} + F_3(x) + A_3(x)\hat{f}_1(x) + B_3(x)T. \tag{11.4.11}$$

For analysis, the wind axis angle estimator error dynamics are

$$\dot{e}_3 = \begin{bmatrix} \dot{e}_\mu \\ \dot{e}_\alpha \\ \dot{e}_\beta \end{bmatrix} \quad = \quad -L_3 e_3 - A_3(x) \begin{bmatrix} \tilde{D} \\ \tilde{Y} \\ \tilde{L} \end{bmatrix}$$

where $e_3 = [e_\mu, e_\alpha, e_\beta]^T = x_3 - \hat{x}_3$, and $L_3 = \text{diag}(L_\mu, L_\alpha, L_\beta)$.

11.4.4 Body angular rate estimator

The implementation equations for the body angular rate estimator are

$$\dot{\hat{x}}_4 = \begin{bmatrix} \dot{\hat{P}} \\ \dot{\hat{Q}} \\ \dot{\hat{R}} \end{bmatrix} \quad = \quad \begin{bmatrix} L_P e_P \\ L_Q e_Q \\ L_R e_R \end{bmatrix} + F_4(x) + A_4(x)\hat{f}_4(x, \delta). \tag{11.4.12}$$

For analysis, the body angular rate estimator error dynamics are

$$\dot{e}_4 = \begin{bmatrix} \dot{e}_P \\ \dot{e}_Q \\ \dot{e}_R \end{bmatrix} \quad = \quad -L_4 e_4 - A_4(x)\tilde{f}_4(x, \delta) \tag{11.4.13}$$

where $e_4 = [e_P, e_Q, e_R]^T = x_4 - \hat{x}_4$, and $L_4 = \text{diag}(L_P, L_Q, L_R)$.

11.5 Prediction Error-Based Learning

The prediction error-based parameter update laws are

$$\dot{\hat{\theta}}_{D_o} = \Gamma_{D_o} \bar{q}S \; \phi \, \hat{e}_D \qquad \dot{\hat{\theta}}_{D_y} = \Gamma_{D_y} \bar{q}S \; \phi \, \hat{e}_D \, y \tag{11.5.1a}$$

$$\dot{\hat{\theta}}_{Y_o} = \Gamma_{Y_o} \bar{q}S \; \phi \, \hat{e}_Y \qquad \dot{\hat{\theta}}_{Y_y} = \Gamma_{Y_y} \bar{q}S \; \phi \, \hat{e}_Y \, y \tag{11.5.1b}$$

$$\dot{\hat{\theta}}_{L_o} = \Gamma_{L_o} \bar{q}S \; \phi \, \hat{e}_L \qquad \dot{\hat{\theta}}_{L_y} = \Gamma_{L_y} \bar{q}S \; \phi \, \hat{e}_L \, y \tag{11.5.1c}$$

$$\dot{\hat{\theta}}_{\bar{L}_o} = \Gamma_{\bar{L}_o} \bar{q}Sb \, \phi \, \hat{e}_{\bar{L}} \qquad \dot{\hat{\theta}}_{\bar{L}_y} = \Gamma_{\bar{L}_y} \bar{q}Sb \, \phi \, \hat{e}_{\bar{L}} \, y \tag{11.5.1d}$$

$$\dot{\hat{\theta}}_{\bar{M}_o} = \Gamma_{\bar{M}_o} \bar{q}S\bar{c} \, \phi \, \hat{e}_{\bar{M}} \qquad \dot{\hat{\theta}}_{\bar{M}_y} = \Gamma_{\bar{M}_y} \bar{q}S\bar{c} \, \phi \, \hat{e}_{\bar{M}} \, y \tag{11.5.1e}$$

$$\dot{\hat{\theta}}_{\bar{N}_o} = \Gamma_{\bar{N}_o} \bar{q}Sb \, \phi \, \hat{e}_{\bar{N}} \qquad \dot{\hat{\theta}}_{\bar{N}_y} = \Gamma_{\bar{N}_y} \bar{q}Sb \, \phi \, \hat{e}_{\bar{N}} \, y \tag{11.5.1f}$$

where, if a_{ij} is the j-th column of A_i (i.e., $A_i = [a_{i1}, a_{i2}, a_{i3}]$), then

$$\hat{e}_D = a_{11}^T e_1 + a_{21}^T e_2 + a_{31}^T e_3 \tag{11.5.2a}$$

$$\hat{e}_Y = a_{12}^T e_1 + a_{22}^T e_2 + a_{32}^T e_3 \tag{11.5.2b}$$

$$\hat{e}_L = a_{13}^T e_1 + a_{23}^T e_2 + a_{33}^T e_3 \tag{11.5.2c}$$

$$\hat{e}_{\bar{L}} = a_{41}^T e_4 \tag{11.5.2d}$$

$$\hat{e}_{\bar{M}} = a_{42}^T e_4 \tag{11.5.2e}$$

$$\hat{e}_{\bar{N}} = a_{43}^T e_4. \tag{11.5.2f}$$

This choice of the parameter update laws will be motivated in the following section, to cause the time derivative of a certain Lyapunov function to be negative semidefinite.

11.6 Lyapunov Analysis: Prediction Error

To analyze the stability of this learning-based estimator, consider the Lyapunov function

$$\begin{aligned} \mathcal{V} = \frac{1}{2}\Bigg(& \sum_{i=1}^{4} e_i^T e_i + \tilde{\theta}_{D_o}^T \Gamma_{D_o}^{-1} \tilde{\theta}_{D_o} + \tilde{\theta}_{Y_o}^T \Gamma_{Y_o}^{-1} \tilde{\theta}_{Y_o} + \tilde{\theta}_{L_o}^T \Gamma_{L_o}^{-1} \tilde{\theta}_{L_o} \\ & + \tilde{\theta}_{\bar{L}_o}^T \Gamma_{\bar{L}_o}^{-1} \tilde{\theta}_{\bar{L}_o} + \tilde{\theta}_{\bar{M}_o}^T \Gamma_{\bar{M}_o}^{-1} \tilde{\theta}_{\bar{M}_o} + \tilde{\theta}_{\bar{N}_o}^T \Gamma_{\bar{N}_o}^{-1} \tilde{\theta}_{\bar{N}_o} \\ & + \tilde{\theta}_{D_y}^T \Gamma_{D_y}^{-1} \tilde{\theta}_{D_y} + \tilde{\theta}_{Y_y}^T \Gamma_{Y_y}^{-1} \tilde{\theta}_{Y_y} + \tilde{\theta}_{L_y}^T \Gamma_{L_y}^{-1} \tilde{\theta}_{L_y} \\ & + \tilde{\theta}_{\bar{L}_y}^T \Gamma_{\bar{L}_y}^{-1} \tilde{\theta}_{\bar{L}_y} + \tilde{\theta}_{\bar{M}_y}^T \Gamma_{\bar{M}_y}^{-1} \tilde{\theta}_{\bar{M}_y} + \tilde{\theta}_{\bar{N}_y}^T \Gamma_{\bar{N}_y}^{-1} \tilde{\theta}_{\bar{N}_y} \Bigg). \end{aligned} \tag{11.6.1}$$

Then,

$$\begin{aligned}
\frac{d\mathcal{V}}{dt} &= -\sum_{i=1}^{4} e_i^T L_i e_i \\
&-\sum_{i=1}^{3} e_i^T \left(a_{i1}(\bar{q}S)[1,\ y] \begin{bmatrix} \tilde{\theta}_{D_o}^T \\ \tilde{\theta}_{D_y}^T \end{bmatrix} \phi + a_{i2}(\bar{q}S)[1,\ y] \begin{bmatrix} \tilde{\theta}_{Y_o}^T \\ \tilde{\theta}_{Y_y}^T \end{bmatrix} \phi \right. \\
&\left. + a_{i3}(\bar{q}S)[1,\ y] \begin{bmatrix} \tilde{\theta}_{L_o}^T \\ \tilde{\theta}_{L_y}^T \end{bmatrix} \phi \right) - e_4^T \left(a_{41}[1,\ y] \begin{bmatrix} \tilde{\theta}_{\bar{L}_o}^T \\ \tilde{\theta}_{\bar{L}_y}^T \end{bmatrix} b \right. \\
&\left. + a_{42}[1,\ y] \begin{bmatrix} \tilde{\theta}_{\bar{M}_o}^T \tilde{\theta}_{\bar{M}_y}^T \end{bmatrix} \bar{c} + a_{43}[1,\ y] \begin{bmatrix} \tilde{\theta}_{\bar{N}_o}^T \\ \tilde{\theta}_{\bar{N}_y}^T \end{bmatrix} b \right) (\bar{q}S)\phi \qquad (11.6.2) \\
&+\tilde{\theta}_{D_o}^T \Gamma_{D_o}^{-1} \dot{\tilde{\theta}}_{D_o} + \tilde{\theta}_{Y_o}^T \Gamma_{Y_o}^{-1} \dot{\tilde{\theta}}_{Y_o} + \tilde{\theta}_{L_o}^T \Gamma_{L_o}^{-1} \dot{\tilde{\theta}}_{L_o} \\
&+\tilde{\theta}_{\bar{L}_o}^T \Gamma_{\bar{L}_o}^{-1} \dot{\tilde{\theta}}_{\bar{L}_o} + \tilde{\theta}_{\bar{M}_o}^T \Gamma_{\bar{M}_o}^{-1} \dot{\tilde{\theta}}_{\bar{M}_o} + \tilde{\theta}_{\bar{N}_o}^T \Gamma_{\bar{N}_o}^{-1} \dot{\tilde{\theta}}_{\bar{N}_o} \\
&+\tilde{\theta}_{D_y}^T \Gamma_{D_y}^{-1} \dot{\tilde{\theta}}_{D_y} + \tilde{\theta}_{Y_y}^T \Gamma_{Y_y}^{-1} \dot{\tilde{\theta}}_{Y_y} + \tilde{\theta}_{L_y}^T \Gamma_{L_y}^{-1} \dot{\tilde{\theta}}_{L_y} \\
&+\tilde{\theta}_{\bar{L}_y}^T \Gamma_{\bar{L}_y}^{-1} \dot{\tilde{\theta}}_{\bar{L}_y} + \tilde{\theta}_{\bar{M}_y}^T \Gamma_{\bar{M}_y}^{-1} \dot{\tilde{\theta}}_{\bar{M}_y} + \tilde{\theta}_{\bar{N}_y}^T \Gamma_{\bar{N}_y}^{-1} \dot{\tilde{\theta}}_{\bar{N}_y} \qquad (11.6.3)
\end{aligned}$$

which for the parameter update laws of the previous section yields

$$\frac{d\mathcal{V}}{dt} = -\sum_{i=1}^{4} e_i^T L_i e_i. \qquad (11.6.4)$$

Therefore, $\dot{\mathcal{V}}$ is negative semidefinite. From this point, the proof of the following theorem is completed by application of Barbalatt's Lemma [16].

Theorem 11.6.1. *For the aircraft dynamics described in (11.2.1), where the force and moment functions discussed in Section 11.3 are not accurately known, then for the state estimators described in (11.4.7), (11.4.9), (11.4.11), (11.4.12) with the parameter update laws of (11.5.1a)–(11.5.1f):*

(1) *All 12 of the parameter error vectors (i.e.,* $\tilde{\theta}_{D_o}, \cdots, \tilde{\theta}_{\bar{N}_y}$*) are bounded.*

(2) *The four prediction error vectors (i.e.,* $e_i,\ i = 1, \cdots, 4$*) each have bounded norms.*

(3) *The four prediction error vectors (i.e.,* $e_i,\ i = 1, \cdots, 4$*) are each elements of* $\mathcal{L}_2$ *(i.e., square integrable).*

11.7 Implementation Issues

The previous sections presented design and analysis results for a somewhat simplified situation. The two main simplifications are (1) only two nondimensional coefficients are considered for each unknown function and (2) the unknown functions are assumed to be perfectly representable by the (scaled) sum of nondimensionalized coefficients. These issues and the issues of regressor selection and convergence of the approximated functions are discussed in this section.

11.7.1 Example coefficient adaptation laws

To simplify the presentation of the previous sections, the previous sections presented algorithms for the approximation of only 12 "nondimensional coefficients' (e.g., $C_{D_o}(x_r), \cdots, C_{\bar{N}_y}(x_r)$) whereas a typical aircraft model may include on the order of 52 such nondimensionalized coefficients. Parameter estimation laws for a typical aircraft model can be easily derived from the cases that are given. The theoretical results of Theorem 11.6.1 can still be attained. Parameter adaptation laws for a few example coefficients are presented below:

$$\begin{aligned}
\dot{\theta}_{\bar{L}_{\delta_i}} &= \Gamma_{\bar{L}_{\delta_i}} \bar{q}Sb\, \phi\, \hat{e}_{\bar{L}}\delta_i \\
\dot{\theta}_{\bar{M}_{\delta_i}} &= \Gamma_{\bar{M}_{\delta_i}} \bar{q}S\bar{c}\, \phi\, \hat{e}_{\bar{M}}\delta_i \\
\dot{\theta}_{\bar{N}_{\delta_i}} &= \Gamma_{\bar{N}_{\delta_i}} \bar{q}Sb\, \phi\, \hat{e}_{\bar{N}}\delta_i \\
\dot{\theta}_{L_\alpha} &= \Gamma_{L_\alpha} \bar{q}S\ \phi\, \hat{e}_L\alpha.
\end{aligned}$$

Proceeding similarly, additional nondimensional coefficients can be added to build-up a model (e.g., see (11.3.1a)–(11.3.1f)) sufficient for accurate representation of the vehicle dynamics. Inclusion of too many nondimensional coefficients will complicate the process of guaranteeing accurate approximation of each coefficient functio (see below). Inclusion of too few nondimensional coefficients may result in inherent model errors that are too large for the desired level of performance.

11.7.2 Inherent model error

In (11.4.4)–(11.4.6), it was implicitly assumed that each force and moment approximation error could be represented exactly as parameter error, e.g.,

$$\tilde{D} = \bar{q}S[1,\ \delta_1, \cdots, \delta_m] \begin{bmatrix} \tilde{\theta}^T_{D_o} \\ \tilde{\theta}^T_{D_{\delta_1}} \\ \vdots \\ \tilde{\theta}^T_{D_{\delta_m}} \end{bmatrix} \phi(x_r).$$

No matter how many "nondimensional' coefficients are added to the force and moment representations, there will always exist residual (or inherent) model errors that cannot be represented as parameter errors within the specified model structure. Such inherent model errors can cause the estimated parameters to drift within a linear space defined in part by the regressor vector $\phi(x_r)$.

In the presence of such inherent model errors, convergence of the prediction errors to a (small) bounded neighborhood of the origin can be proven by inserting deadzones into the adaptive laws as

$$\hat{e}_D = a_{11}^T e_{1\epsilon} + a_{21}^T e_{2\epsilon} + a_{31}^T e_{3\epsilon} \quad (11.7.1)$$

$$\hat{e}_Y = a_{12}^T e_{1\epsilon} + a_{22}^T e_{2\epsilon} + a_{32}^T e_{3\epsilon} \quad (11.7.2)$$

$$\hat{e}_L = a_{13}^T e_{1\epsilon} + a_{23}^T e_{2\epsilon} + a_{33}^T e_{3\epsilon} \quad (11.7.3)$$

$$\hat{e}_{\bar{L}} = a_{41}^T e_{4\epsilon} \quad (11.7.4)$$

$$\hat{e}_{\bar{M}} = a_{42}^T e_{4\epsilon} \quad (11.7.5)$$

$$\hat{e}_{\bar{N}} = a_{43}^T e_{4\epsilon} \quad (11.7.6)$$

where

$$[e_{i\epsilon}]_j = \begin{cases} [e_{i\epsilon}]_j - \epsilon_{ij}, & \text{if } [e_{i\epsilon}]_j > \epsilon_{ij} \\ 0, & \text{if } -\epsilon_{ij} \leq [e_{i\epsilon}]_j \leq \epsilon_{ij} \\ [e_{i\epsilon}]_j + \epsilon_{ij}, & \text{if } [e_{i\epsilon}]_j < -\epsilon_{ij} \end{cases} \quad (11.7.7)$$

for $i = 1, \cdots, 4$ and $[x]_j$ denotes the j-th component of vector x. The parameter ϵ_{ij} must be an upper bound on the magnitude of the j-th row of A_i times the inherent approximation error.

11.7.3 Regressor selection

Selection of the approximator regressor vectors is a critical implementation issue that is often not given sufficient attention. If, for example, $\phi(x_r)$ is simply selected to be a constant, i.e., $\phi(x_r) = 1$, then the resulting approach would estimate the local linearized nondimensional coefficients at the present operating point. As the operating point changed, the estimated parameters would also have to change to keep them accurate for the local model. Such a choice of regressor would not allow the approximator to estimate the nondimensional coefficients as functions of the operating point. Therefore, it should be clear that the selection of $\phi(x_r)$ limits the function approximation accuracy that can be achieved.

If $x_r \in \Re^d$ and $\phi(\cdot) : \Re^d \mapsto \Re^N$, then N and the characteristics of each element of ϕ determine the approximation accuracy that can be achieved. The regressor vector should have the property that increasing N decreases the inherent approximation error. Many approximators have such properties: polynomials, splines, radial basis functions, and neural networks.

An additional property that greatly simplifies the implementation and analysis is that each element of the regressor vector should have local support. If the support of $[\phi(x_r)]_j$ is denoted by

$$Supp_j = \left\{ x_r \in \mathcal{D} \middle| [\phi(x_r)]_j \neq 0 \right\}$$

and $\lambda(\mathcal{S})$ denotes the measure of set $\mathcal{S}$, then local support requires

$$\lambda(Supp_j) \ll \lambda(\mathcal{D})$$

with $Supp_j$ connected. Local support allows the approximated function to converge rapidly in a neighborhood of the present operating point without affecting the approximation in regions of the operating envelope that are distant from the current operating point.

Finally, it is important to remember that the regressor vector can be distinct for different nondimensional coefficients. For example, if for a certain aircraft $\phi(\alpha, M)$ is sufficient for the majority of coefficients, but the surface with deflection δ_1 is placed physically in front of the surface with deflection δ_3, then the effectiveness of surface 3 may be a function of the deflection of surface 1. Therefore, for example,

$$\hat{C}_{\bar{L}_{\delta_3}}(\alpha, M, \delta_1) = \hat{\theta}^T_{\bar{L}_{\delta_3}} \phi_{\bar{L}_{\delta_3}}(\alpha, M, \delta_1) \tag{11.7.8}$$

may be an appropriate structure for approximating the roll moment dependence on δ_3.

11.7.4 Function convergence

Although Theorem 11.6.1 states that the prediction errors converge to zero (or a small neighborhood of the origin), the parameter errors are only shown to be bounded. Therefore, since $\phi(x_r)$ is assumed bounded, the error in the approximated functions is bounded; however, nothing has been stated about the convergence of the approximated functions. It can be shown that the approximated functions will converge if certain excitation conditions are satisfied [9]. These conditions are much easier to satisfy when the regressor vectors are selected to have local support.

11.8 Conclusions

This chapter has discussed design, analysis, and implementation issues for an aircraft state estimator incorporating on-line approximation of the aerodynamic coefficients as functions over an operating envelope $\mathcal{D}$. Several previous results in the literature have incorporated on-line adaptation of linearized aerodynamic coefficients, which has been shown to be a special case of the results derived herein.

Since the optimal linearized aerodynamic coefficients change with operating point, such linear adaptive methods are incapable of storing model information as a function of operating point. On-line approximation of the aerodynamic coefficients as a function over the operating envelope is a form of spatial memory; each time the aircraft enters a new region of the operating envelope, the approximation uses the last parameters that it had previously estimated while in that region. At the initialization of any flight, prior knowledge about the approximated functions (possibly from a prior flight) can be used, either by initializing the approximator parameter vectors or by additive known functions; however, such prior information is not required.

11.9 Appendix: Aircraft Notation

Table 11.9.1: Constant definitions

Symbol	Meaning
m	Mass
g	Vertical gravity component
c_i	Rotational inertia parameters defined on p. 80 in [19]
L_*	Estimator gain for variable *
Γ_*	Parameter adaptation matrix for coefficient *
b	Wing span
$\bar{c}$	Wing mean geometric chord
S	Surface area

Table 11.9.2: Variable definitions

Variable	Definition
χ	ground track angle
γ	climb angle
V	speed
μ	bank angle
α	angle of attack
β	side slip
P	body axis roll rate
Q	body axis pitch rate
R	body axis yaw rate
M	Mach number
P_s	stability axis roll rate
R_s	stability axis yaw rate
δ_i	Deflection of the i-th control surface
T	Thrust
$\overline{q}$	Dynamic pressure

Table 11.9.3 Function fefinitions

Symbol	Definition
D	Body axis drag force. This function is unknown
Y	Body axis side force. This function is unknown
L	Body axis lift force. This function is unknown
$\bar{L}$	Stability axis roll moment. This function is unknown
$\bar{M}$	Stability axis pitch moment. This function is unknown
$\bar{N}$	Stability axis yaw moment. This function is unknown
$\hat{D}$	Approximated body axis drag force
$\hat{Y}$	Approximated body axis side force
$\hat{L}$	Approximated body axis lift force
$\hat{\bar{L}}$	Approximated stability axis roll moment
$\hat{\bar{M}}$	Approximated stability axis pitch moment
$\hat{\bar{N}}$	Approximated stability axis yaw moment
f_*	Portion of the dynamic equation for variable * that contains at least one of the six unknown functions
F_*	Portion of the dynamic equation for variable * that contains known functions
$\hat{f}_*$	Approximation to f_*

Bibliography

[1] *Recommended Practice for Atmospheric and Space Flight Vehicle Coordinate Systems*, AIAA/ANSI R-004-1992.

[2] M. A. Khan and P. Lu, "New technique for nonlinear control of aircraft," *J. Guidance, Control, and Dynamics*, vol. 17 no. 5, pp. 1055–1060, 1994.

[3] M. Azam and S. N. Singh, "Invertibility and trajectory control for nonlinear maneuvers of aircraft," *J. Guidance, Control, and Dynamics*, vol. 17, no. 1, pp. 192–200, Jan.-Feb. 1994.

[4] W. L. Baker and J. A. Farrell, "Connectionist learning systems for control," *Proc. SPIE Conference on Optical Engineering*, Boston, MA, Nov. 1990.

[5] W. L. Baker and J. A. Farrell, "An introduction to connectionist learning control systems," in *Handbook of Intelligent Control: Neural Fuzzy, and Adaptive Approaches*, D. A. White and D. A. Sofge, Eds., Chapter 2, Van Nostrand Reinhold, New York, 1992.

[6] J. G. Batterson and V. Klein, "Partitioning of flight data for aerodynamic modeling of aircraft at high angles of attack," *J. Aircraft*, vo. 26, no. 4, pp. 334–339, 1989.

[7] D. J. Bugajski, D. F. Enns, and M. R. Elgersma, "A dynamic inversion based control law with application to high angle of attack research vehicle," *AIAA-90-3407-CP*, pp. 826–839, 1992.

[8] J. Y. Choi and J. A. Farrell, "Nonlinear adaptive control using networks of piecewise linear approximators," *IEEE Trans. Neural Networks*, vol. 11, no. 2, pp. 390–401, 2000.

[9] J. Farrell, "Stability and approximator convergence in nonparametric nonlinear adaptive control," *IEEE Trans. Neural Networks*, vol. 9, no. 5, pp. 1008–1020, Sept. 1998.

[10] W. L. Gerrard, D. F. Enns, and A. Snell, "Nonlinear longitudinal control of a supermaneuverable aircraft," *Proc. 1989 American Control Conference*, vol. 1, pp. 142–147, 1989.

[11] S. H. Lane and R. F. Stengel, "Flight control design using nonlinear inverse dynamics," *Automatica*, vol. 31, no. 4, pp. 781–806, 1988.

[12] P. K. A. Menon, M. E. Badget, R. A. Walker, and E. L. Duke, "Nonlinear flight test trajectory controllers for aircraft," *J. Guidance, Control, and Dynamics*, vol. 10, no. 1, pp. 67–72, 1987.

[13] G. Meyer, R. Su, and L. R. Hunt, "Application of nonlinear transformations to automatic flight control," *Automatica*, vol. 20, no. 1, pp. 103–107, 1984.

[14] M. M. Polycarpou, "Stable adaptive neural control scheme for nonlinear systems," *IEEE Trans. Automatic Control*, vol. 41, no. 3, pp. 447–451, 1996.

[15] M. Polycarpou and M. Mears, "Stable adaptive tracking of uncertain systems using nonlinearly parameterized on-line approximators," *Int. J. Control*, vol. 70, no. 3, pp. 363–384, May 1998.

[16] J.-J. Slotine and W. Li, *Applied Nonlinear Control*, Prentice-Hall, Englewood Cliffs, NJ, 1991.

[17] S. A. Snell, D. F. Enns, and W. L. Garrard, Jr., "Nonlinear inversion flight control for a supermaneuverable aircraft," *J. Guidance, Control, and Dynamics*, vol. 14, no. 4, pp. 976–984, 1992.

[18] M. Sharma, A. J. Calise, and J. E. Corban, "Application of an adaptive autopilot design to a family of guided munitions," *Proc. AIAA Guidance, Navigation, and Control Conference*, AIAA-2000-3969, 2000.

[19] B. L. Stevens and F. L. Lewis, *Aircraft Control and Simulation*, John Wiley, New York, 1992.

[20] T. L. Trankle and S. D. Bachner, "Identification of a nonlinear aerodynamic model of the F-14 aircraft," *J. Guidance, Control, and Dynamics*, vol. 18, no. 6, pp. 1292–1297, 1995.

[21] T. L. Trankle, J. H. Vincent, and S. N. Franklin, "System identification of nonlinear aerodynamic models," in *Advances in the Techniques and Technology of the Application of Nonlinear Filters and Kalman Filters*, AGARDograph No. 256, Mar. 1982.

Chapter 12

Evolutionary Multiobjective Optimization: Qualitative Analysis and Design Implementation

Gary G. Yen

Abstract: In this chapter, the author proposes a novel evolutionary approach to multiobjective optimization problems–dynamic multiobjective evolutionary algorithm (DMOEA). In DMOEA, a cell-based rank and density estimation strategy is proposed to efficiently compute dominance and diversity information when the population size varies dynamically. In addition, a population growing strategy and a population declining strategy are designed to determine if an individual will survive or be eliminated based on some qualitative indicators. Meanwhile, an objective space compression strategy is devised to continuously refine the quality of the resulting Pareto front. By examining the selected performance metrics on a recently designed benchmark function, DMOEA is found to be competitive with, or even superior to, five state-of-the-art MOEAs in terms of keeping the diversity of the individuals along the trade-off surface, tending to extend the Pareto front to new areas and finding a well-approximated Pareto optimal front. Moreover, DMOEA is evaluated by using different parameter settings on the chosen test function to verify its robustness of converging to an optimal population size, if it exists. From simulation results, DMOEA has shown the potential of autonomously determining the optimal population size, which is found insensitive to the initial population size chosen.

12.1 Introduction

In the past decade, several multiobjective evolutionary algorithms (MOEAs) have been proposed and applied in multiobjective optimization problems (MOPs) [1]. These algorithms share the common purpose: searching for a uniformly distributed, near-optimal, and well-extended Pareto front for a given MOP. However, this ultimate goal is far from being accomplished by the existing MOEAs as documented in the literature, e.g., [1]. In one respect, most of the MOPs are very complicated and require the computational resources to be homogeneously distributed in a high-dimensional search space. On the other hand, those better-fit individuals generally have strong tendencies to restrict searching efforts within local areas because of the "genetic drift" phenomenon [2], which adversely results in the loss of diversity. Additionally, most of the existing MOEAs adopt a fixed population size to initiate the evolutionary process. As pointed out in [3], evolutionary algorithms may suffer from premature convergence if the population size is too small, whereas an overestimated population size will result in a heavy burden on computation and a long waiting time for fitness improvement.

In the case of single objective (SO) optimization, several methods of determining an optimal population size from different perspectives have been proposed [3–5]. Since the purpose of solving a SO problem is to search for a single optimal solution at the final generation, the distribution characteristic of the final population is not a primary issue of concern. However, in order to solve MOPs, an MOEA needs to approximate a set of nondominated solutions that produces a uniformly distributed Pareto front. In general, the size of the final Pareto set yielded by most MOEAs is bounded by the size of the initial population chosen ad hoc. As indicated in [6], the exact trade-off surface of an MOP is often unknown a priori. It is difficult to estimate an optimal number of individuals necessary for effective exploration of the solution space as well as a good representation of the trade-off surface. This difficulty implies that an "estimated" size of the initial population is not appropriate in a real-world application. Therefore, a dynamic population size autonomously adjusted by the on-line characteristics of population trade-off and density distribution information seems to be more efficient and effective than a constant population size in terms of avoiding premature convergence and unnecessary computational complexity.

As highlighted in [6], the issue of dynamic population in MOEAs has not been well researched yet. Although in some elitism-based MOEAs, the main population and elitist archive are separated and updated by exchanging elitists between them, the size of the main population or the sum of the main population and the archive remains fixed [7, 8]. Therefore, either an estimated size of the initial population is needed in some of these algorithms, or a maximum size of the archive is predetermined [9]. Tan, Lee, and Khor proposed an incrementing multiobjective evolutionary algorithm (IMOEA) [6], which devises a fuzzy boundary local perturbation technique and a dynamic local fine-tuning method in order to achieve broader neighborhood explorations and eliminate gaps and discontinuities along the Pareto front. However, this algorithm adopts a heuristic method to estimate the desired population size, $dps(n)$, at generation n according to the approximated trade-off hyperareas of current generation, but not to the dominance and density information of the entire objective space. Therefore, the computation workload may be wrongly determined if the approximation of the $dps(n)$ value is inaccurate, which may force IMOEA to adjust the grid density to reach the incorrect "optimal" population size. Moreover, IMOEA is relatively complicated and not compared with those most recently designed MOEAs (e.g., PAES, SPEA II, NSGA-II, and RDGA). Its robustness of converging to an optimal population size needs to be further examined by various initial populations.

In this chapter, the author proposes a novel dynamic multiobjective evolutionary algorithm (DMOEA). In DMOEA, a cell-based rank and density calculation strategy is designed. A given MOP (in problem domain) will be converted into a bi-objective optimization problem (in algorithm domain) in terms of individual's rank and density values [10]. Meanwhile, a population growing strategy and a population declining strategy are devised based on the converted fitness. Three types of qualitative indicators–age, health, and crowdedness–are associated with each individual in order to determine the likelihood of eliminating an individual. In addition, an objective space compression strategy is introduced, and the resulting Pareto front is continuously refined based on different steady states. Due to space limitation, only one recently designed test function is used to examine the efficiency and effectiveness of the proposed DMOEA. For a fair comparison, five representative MOEAs (PAES [9], SPEA II [7], NSGA-II [8], RDGA [10], and IMOEA [6]) are also tested by the chosen benchmark problem. By examining four performance metrics and the resulting Pareto fronts, DMOEA is found to be competitive with, or even superior to, the five selected MOEAs in terms of keeping the diversity of the individuals along the trade-off surface, tending to extend the Pareto front to new areas and finding a well-approximated Pareto optimal front. Moreover, from simulation results, DMOEA shows the potential to autonomously determine the optimal population size, which is found insensitive to the initial population size chosen.

This chapter is organized as follows. Section 12.2 reviews existing literature on MOEAs. Section 12.3 describes the proposed cell-based rank and density calculation strategy and introduces a population growing strategy and a population declining strategy. An objective space compression strategy is designed and elaborated. In Section 12.4, four performance measures are adopted and DMOEA with five state-of-the-art MOEAs are used to produce experimental results for a chosen test function. The performance metrics of the resulting Pareto fronts are examined to compare the performance of DMOEA and the chosen algorithms. In Section 12.5, DMOEA with different parameter settings is exploited by the chosen test function to show its robustness of converging to an optimal population size independent of the initial population. Finally, Section 12.6 provides some pertinent concluding remarks.

12.2 Evolutionary Multiobjective Optimization Algorithms

Generally, the approximation of the Pareto-optimal set involves two competing objectives: the distance to the true Pareto front is to be minimized, while the diversity of the generated solutions is to be maximized [8]. To address the first objective, a Pareto-based fitness assignment method is usually designed in order to guide the search toward the true Pareto front [1]. For the second objective, some successful MOEAs provide density estimation methods to preserve the population diversity [8]. In addition, several other techniques have been adopted, such as elitism scheme [7–9], crowded comparison [7, 11], and archive truncation [8]. These methods and techniques can be found in four state-of-the-art MOEAs–PAES, NSGA-II, SPEA II, and RDGA, which are briefly reviewed in the following. Furthermore, as the first MOEA that is designed with a dynamic population size, IMOEA [6] is also included and analyzed.

12.2.1 Pareto archive evolutionary strategy (PAES)

As a local search algorithm that simulates a random mutation hill-climbing strategy, PAES may represent the simplest possible yet effective algorithm capable of generating diverse solutions in the Pareto optimal set [9]. In PAES, a pure mutation operation is adopted to fulfill the local search scheme. A reference archive of previously found nondominated solutions is updated at each generation in order to identify the dominance ranking of all the resulting solutions. Although (1+1)-PAES is originated as the simplest version, PAES can also generate λ mutants by mutating one of the μ current solutions, which is called $(\mu + \lambda)$-PAES [9]. Since PAES does not perform a population-based search, only tournament selection can be applied to determine the survivors of the next generation. It is worthy to mention that although the archive size has to be predetermined, PAES implements a population incrementing scheme by continuously adding new nondominated individuals to the archive.

12.2.2 Nondominated sorting genetic algorithm II (NSGA-II)

NSGA-II [7] was advanced from its origin, NSGA [12]. In NSGA-II, a nondominated sorting approach is used for each individual to create a Pareto rank, and a crowding distance assignment method is applied to implement density estimation. In fitness assignment between two individuals, NSGA-II prefers the individual with a lower rank value, or the individual located in a region with fewer number of points if both of the points belong to the same front. Therefore, by combining a fast nondominated sorting approach, an elitism scheme and a parameterless sharing method with the original NSGA, NSGA-II claimed to produce a better spread of solutions in some testing problems [7]. However, a heuristically chosen initial population size is needed in NSGA-II.

12.2.3 Strength Pareto evolutionary algorithm II (SPEA II)

Similar to NSGA-II, SPEA II [8] is an enhanced version of SPEA [13]. In SPEA II, instead of calculating standard Pareto rank, each individual in both main population and elitist archive is assigned a strength value, which incorporates both dominance and density information. On the basis of the strength value, the final rank value is determined by the summation of the strengths of the points that dominate the current individual. Meanwhile, a kth nearest neighbor density estimation method is applied to obtain the density value of each individual. The final fitness value for each individual is the sum of rank and density values. In addition, a truncation method is used in elitist archive in order to maintain a constant number of individuals contained in the archive. In the experimental results, SPEA II shows better performances than SPEA over all of the test functions considered therein. Nevertheless, SPEA II also used an "estimated" population size and a fixed size of elitist archive.

12.2.4 Rank-density-based genetic algorithm (RDGA)

RDGA [10] proposes a new ranking scheme–automatic accumulated ranking strategy (AARS) and a cell-based adaptive density estimation method. By its design, the original problem domain is simplified by converting high-dimensional multiple objectives into two objectives (in algorithm domain) to minimize the individual rank value and maintain the population density value. Afterwards, a fitness assignment scheme borrowed from vector evaluated genetic algorithm (VEGA [14]) is applied to minimize population rank and density values independently. In addition, a "forbidden region" concept is introduced in RDGA to prevent the survival of any offspring with a low density value but a high rank value. By examining the selected performance indicators, RDGA is found to be competitive with SPEA II and NSGA-II. Although the elitism scheme is applied in RDGA, it does not have a built-in mechanism to deploy a dynamic population size either.

12.2.5 Incrementing multiobjective evolutionary algorithm

As the first MOEA equipped with a dynamic population strategy, IMOEA [6] devised a method of fuzzy boundary local perturbation with interactive local fine-tuning for a broader neighborhood exploration to increase the population size with competent offspring. In IMOEA, considering an m-dimension objective space, the desired population size $dps(n)$, with the desired population size per unit volume, ppv (a user-specified parameter), and an approximated number of tradeoff hyperareas, $A_{to}(n)$, discovered by the population at generation n, is defined as

$$lowbps \le dps(n) = ppv \times A_{to}(n) \le upbps, \tag{12.2.1}$$

where $lowbps$ and $upbps$ are the lower and upper bounds for the desired population size $dps(n)$, respectively. In addition, IMOEA applied the method derived in [15] to estimate the approximated number of hyperareas by

$$A_{to}(n) \approx \frac{\pi^{(m-1)/2}}{(\frac{m-1}{2})!} \times \left(\frac{d(n)}{2}\right)^{m-1} \tag{12.2.2}$$

where m is the number of objectives and $d(n)$ is the diameter of the hypersphere at generation n. Therefore, based on the difference between the resulting population size and estimated desired population size $dps(n)$, IMOEA adaptively filled in or filtered out individuals according to their rank and density status. In the simulation results, NSGA and SPEA are compared with IMOEA on two test functions, and IMOEA shows better performance than the other two in terms of selected indicators. Unfortunately, none of the more advanced MOEAs (e.g., PAES, SPEA II, NSGA-II, and RDGA) were used to compare with IMOEA, and the robustness of IMOEA on different initial population size is not carefully examined.

12.3 Dynamic Multiobjective Evolutionary Algorithm (DMOEA)

As discussed in Section 12.2, an ideal Pareto front resulting from an MOEA possesses two characteristics: close to the true Pareto front and uniformly distributed.

A well-extended Pareto front can be assured if a dynamic population is used to regulate the population size and preserve the diversity throughout the evolutionary process. Unfortunately, these two characteristics are competing since the diversity preservation process may degrade the quality of the resulting Pareto front [8]. In one respect, in the spirit of evolutionary algorithm, MOEA exploits the "genetic drift" effect to converge the solution to each of the optimal points. On the other hand, the "genetic drift" phenomenon must be avoided to sketch a uniformly sampled trade-off surface for the final Pareto front. This dilemma is very difficult to solve by MOEAs with a fixed population size, since they have to homogeneously distribute the predetermined computation resource to all the possible directions in the objective space. If the predetermined population size is too small, there will not be enough schemas for an MOEA to exploit. The "genetic drift" phenomenon may result in a nonuniformly distributed Pareto front. Meanwhile, choosing an excessive initial population size will not be desirable since it may require a large computation resource and result in an extremely long running time. Moreover, a fixed population size will have great difficulty obtaining a Pareto front with a desired resolution because the shape and size of the true Pareto front are unknown a priori for most of the MOPs. To cope with this contradiction, a DMOEA is proposed.

Similar to other advanced MOEAs [7, 8, 10], DMOEA also converts the original MOP into a bi-objective optimization problem in algorithm domain [10], minimizing the individual rank value and maintaining the individual density value. However, as adding or removing an individual will affect the rank and density values of other individuals, the rank and density values of each individual need to be recalculated after the population has been updated. This recalculation will exert even more computation resource as the population size grows and declines dynamically. Therefore, to solve this problem, we design a cell-based rank and density calculation scheme.

12.3.1 Cell-based rank and density calculation scheme

In DMOEA, the original m-dimensional objective space is divided into $K_1 \times K_2 \times \cdots \times K_m$ cells (i.e., grids), thus the cell width in the ith objective dimension, d_i, can be computed as

$$d_i = \frac{\max_{\mathbf{x}\in X} f_i(\mathbf{x}) - \min_{\mathbf{x}\in X} f_i(\mathbf{x})}{K_i}, \quad i = 1, \cdots, m, \tag{12.3.1}$$

where d_i is the width of the cell in the ith dimension, f_i refers to the ith objective, K_i denotes the number of cells designated for the ith dimension (i.e., in Figure 12.3.1,

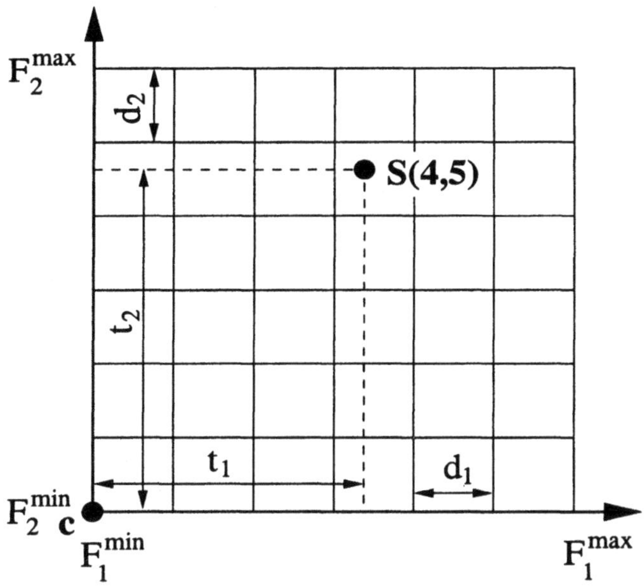

Figure 12.3.1: Illustration of cell-based rank and density estimation scheme

$K_1 = 6$ and $K_2 = 6$), **x** is taken from the whole decision space X, and we denote

$$F_i^{\min} = \min_{\mathbf{x}\in X} f_i(\mathbf{x}), \quad F_i^{\max} = \max_{\mathbf{x}\in X} f_i(\mathbf{x}), \quad i = 1, \cdots, m. \tag{12.3.2}$$

The grid scales K_i, $i = 1, \cdots, m$, are chosen heuristically. It is worthy to note that PAES [9] and PESA (Pareto-envelope-based selection algorithm) [16] also uses an adaptive grid-based density estimation method and a quadtree-encoding technique is designed to quickly find the location of an individual. However, DMOEA does not apply quadtree-encoding in searching for the cell in which an individual is located. As shown in Figure 12.3.1, point **c** is denoted as the origin point of current objective space. In other word, **c** is the cross point of all the lower boundaries of an m-dimensional objective space. The position of **c** is denoted as $(F_1^{\min}, F_2^{\min}, \cdots, F_m^{\min})$. For a newly generated individual **S**, whose position is $(s_1, s_2, \cdots, s_m)$ in the objective space, the distance between point **S** and point **c** will be measured in each dimension in the objective space as $[t_1, t_2, \cdots, t_m]$, where

$$t_i = s_i - F_i^{\min}, \quad i = 1, \cdots, m. \tag{12.3.3}$$

Therefore, the "home address" of individual s in the ith dimension is calculated as

$$h_i = \text{mod}(t_i, d_i) + 1, \quad i = 1, \cdots, m, \tag{12.3.4}$$

where function mod(x, y) represents the modulus (integer part) after division x/y. Therefore, by this setting, finding the grid location (home address) of a single solution requires only m "division" operations, and there is no comparison needed in

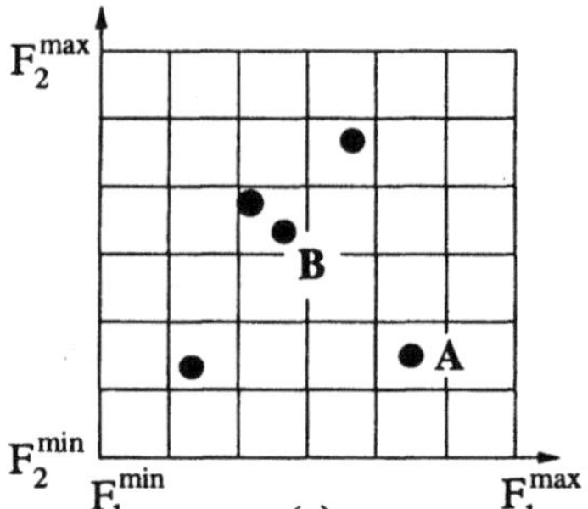

(a)

1	2	4	5	6	6
1	2	4	**4**	6	6
1	2	**2**	4	5	5
1	2	2	2	3	3
1	**1**	2	2	**2**	3
1	1	1	1	1	1

(b)

0	0	0	0	0	0
0	0	0	**1**	0	0
0	0	**2**	0	0	0
0	0	0	0	0	0
0	**1**	0	0	**1**	0
0	0	0	0	0	0

(c)

Figure 12.3.2: (a) Initial population and the location of each individual, (b) rank value matrix of initial population, and (c) density value matrix of initial population

this technique, as opposed to $l \times m$ comparisons needed by quadtree encoding in PAES [9] (l represents the level of subdivisions of PAES). In terms of computational complexity, "comparison" is considered to be expensive compared to algebraic operations (i.e., addition/subtraction and multiplication/division).

After the objective space has been determined and divided, as shown in Figure 12.3.2(a), the center position of each cell will be obtained first, and two matrices are set up to store the rank and density values of each cell, which initially have all 1 and 0, respectively. For the purpose of visualization, m is chosen to be 2 in Figure 12.3.2. Second, each individual of the initial population will search for its nearest cell center, identify this cell as its "home address," and consider the other individuals who share the same "home address" as its "family members." Thus as shown in Figure 12.3.2, for each of these "homes," the number of "family members" who dwell in it will be counted and saved as the density value of the "home." In addition, the rank values of the cells that are dominated by any of these "homes" will be increased by the density values of those "homes." Third, when an offspring is generated and accepted (individual **C** in Figure 12.3.3(a)), its "home address" can be easily located by following the second step, the density value of its "home" will increase by one, and the rank values of the cells dominated by its "home" are increased by one. Meanwhile, if an old individual (individual **B** in Figure 12.3.2(a)) is removed, its "home" will be notified, the density value of its "home" will be decreased by one, and the rank values of the cells dominated by its "home" are decreased by one, correspondingly. Therefore, at each generation, an individual can access its "home address" and then obtain the corresponding rank and density values. The "home address" is merely a "pointer" to inform an individual where to find its rank and density values. For instance, as shown in Figure 12.3.3, the "home address," rank, and density values of individual **A** are (5,2), 2, and 1, respectively. Therefore, if the estimated objective space is properly chosen so that a newly generated or a removed individual does not change the boundaries of the range of the current objective space, the size of each cell will not change, which means an individual's "home address" will never change if this individual is not removed. By this design, the original objective of searching

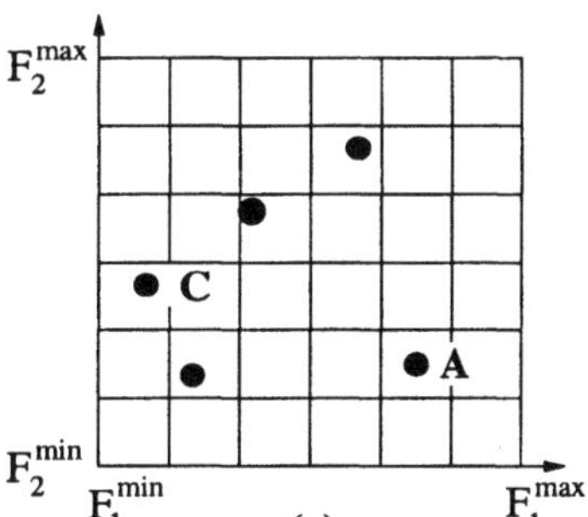

(a)

2	3	4	5	6	6
2	3	4	**4**	6	6
2	3	**3**	4	5	5
1	3	3	3	4	4
1	**1**	2	2	**2**	3
1	1	1	1	1	1

(b)

0	0	0	0	0	0
0	0	0	**1**	0	0
0	0	**1**	0	0	0
1	0	0	0	0	0
0	**1**	0	0	**1**	0
0	0	0	0	0	0

(c)

Figure 12.3.3: (a) New population and the location of each individual, (b) rank value matrix of new population, and (c) density value matrix of new population

for a well-extended, near-optimal, and uniformly distributed Pareto front has been converted to locate as many optimal "home addresses" as possible, each of which contains *ppv* number of these individuals.

Although the genetic operations (i.e., crossover and mutation) are still performed by genotype chromosomes, the fitness evaluation of the quality of the resulting offspring is based on its location in the rank and density matrices. By this method, the procedure of updating rank and density matrices is irrelevant to the procedure of a fitness evaluation on an individual. In one respect, as each crossover or mutation operation can only produce at most two new individuals, the computational load on updating the rank and density matrices will be trivial for each generation. On the other hand, when two individuals are compared, they just need to provide their "home addresses," and the current rank or density values of their "home addresses" can be evaluated to determine which individual is better-fitted. Therefore, no matter how large the population size is, the computation effort for both matrices' updating and fitness evaluation will not be greatly affected. The proposed cell-based rank and density estimation design has provided an efficient way in applying a dynamic population size in evolutionary process.

12.3.2 Cell rank and health indicator

Once the rank and density values of each cell have been obtained by using the method described in Section 12.3.1, two qualitative indicators that are associated with rank and density values are designed to determine whether a cell is "comfortable" enough for an individual to reside. These are health and crowdedness indicators. The health indicator is associated with cell rank and is discussed in this subsection, while the crowdedness indicator is derived from cell density and will be elaborated in the next subsection.

In DMOEA, we use the rank value of a cell to derive a health indicator in order to measure the dominance relationship of the concerned cell comparing to the others. Assume at generation n, a cell c has a rank value $\text{rank}(c, n)$: the health value of cell

c at generation n is given as

$$H(c,n) = \frac{1}{\text{rank}(c,n)}. \tag{12.3.5}$$

From (12.3.5), a cell with the rank value 1, which is regarded as the healthiest, will have an H value equal to 1, and a cell with a higher rank value will have a lower H value that is closer to 0. Therefore, an H value implies the Pareto rank information of a cell. The relationship between a cell's rank value and H value is nonlinear. In one aspect, H values drop very fast if rank values are greater than 1, which results in a significant difference between dominated and nondominated cells in terms of the health condition. On the other hand, H values also saturate very fast, which means that all the dominated cells are assigned very low H values (near zero) if their rank values are very high. This characteristic will be used by the individual elimination scheme discussed later.

12.3.3 Cell density and crowdedness indicator

Referring to [6], consider an m-dimension objective space and the desired population size $dps(n)$. The desired population size per unit volume *ppv* and the number of explored trade-off hyperareas $A_{to}(n)$ are discovered at generation n defined as in (12.2.1). Therefore, with a given population size per unit volume, *ppv*, the optimal population size can be obtained if an MOEA can correctly discover all the trade-off hyperareas for an MOP. In DMOEA, instead of estimating the trade-off hyperareas $A_{to}(n)$ at each generation [6], we concentrate on searching for a uniformly distributed and well-extended final set of trade-off hyperareas and ensure that each of these areas contains *ppv* number of nondominated individuals. Therefore, by using DMOEA, the optimal population size and final Pareto front will be found simultaneously at the final generation. As discussed in Section 12.3.1, the density value of a cell is defined as the number of the individuals located in it. The finer the resolution of the cell, the better performance DMOEA can provide. A crowdedness value is associated with each cell to show current density information of the concerned cell. Assume at generation n, the density value of cell c is $density(c,n)$, the crowdedness value of cell c is defined as

$$D(c,n) = \begin{cases} \frac{density(c,n)}{ppv}, & \text{if} \quad density(c,n) > ppv \\ 1, & \text{otherwise.} \end{cases} \tag{12.3.6}$$

Therefore, by using the crowdedness indicator, we can obtain information about how congested each cell is, comparing to the desired *ppv* value.

12.3.4 Population growing strategy

In general, if an MOEA has a fixed population size, a "replacement" scheme is always applied. In this scheme, in order to keep the population size unchanged, a newborn offspring will replace one of its parents if its fitness value is higher than that of the parent. However, this scheme brings up the problem that some of the replaced parents may still be very valuable and have not been well exploited before replacement. Although some MOEAs adopt an elitist archive in addition to the main population to store some of the nondominated individuals generated during the evolutionary process, the problem caused by the replacement scheme is still not completely resolved. Therefore, DMOEA applies two independent strategies–population growing strategy and population declining strategy. The first strategy focuses on pure population increment to ensure that each individual survives enough generations so that it can contribute its valuable schemas. Meanwhile a population declining strategy is also enforced to prevent the population size from growing excessively. The second strategy will be discussed in the next subsection.

Because exploring the cells with minimum rank values and maintaining these cells densities to a desired value are two converted objectives of DMOEA, crossover and mutation operations need to be devised to fit both of the purposes. The fitness assignment scheme of DMOEA is borrowed from RDGA [10]; the main population is equally divided into two subpopulations that are responsible for minimizing the rank value and maintaining the density value, respectively. For crossover, a reproduction pool with a fixed number of selected parents is set up. A cellular genetic algorithm (GA) [17] is then modified and applied to explore the new search area by "diffusion": each selected parent performs crossover with a randomly selected individual located in the immediate neighboring cell that dominates the concerned cell. Note that the original cellular GA is applied in the decision space whereas DMOEA uses diffusion scheme in the objective space. This idea is inspired by particle swarm optimization (PSO) [18], where an individual shares its information with the leading individuals in order to locate its moving direction. If a resulting offspring is located in a cell with a better fitness (a lower rank value or a lower density value) than its selected parents, it will be kept to the next generation; otherwise, it will not survive. The mutation operation is analogous. As a result, this strategy guarantees that a newborn individual will have a better fitness value than at least one of its parents, which helps DMOEA cover all the unexplored cells in the objective space.

However, as DMOEA encourages an offspring to land in a sparse area whose unexplored cells have higher chances to be visited, it is expected that some offspring will tend to move toward an opposite direction to the true Pareto front where the cells close to the Pareto front are crowded. Obviously, these movements are harmful to the population converging to the Pareto front. To prevent "harmful" offspring from surviving and affecting the evolutionary direction and speed, a forbidden region concept is proposed in the offspring generating scheme for the density subpopulation, thereby preventing the "backward drifting" effect. The forbidden region includes all the cells dominated by the selected parent. Therefore, an offspring located in the forbidden region will have higher value of home addresses in each dimension and will not survive in the next generation, thus the selected parent will not be replaced.

12.3.5 Population Declining Strategy

As discussed in Section 12.3.4, a population declining strategy is necessary to prevent the population size from growing unboundedly. In DMOEA, whether an individual will be removed or not depends on its health and crowdedness indicators as defined in Sections 12.3.2 and 12.3.3. Moreover, to ensure that each generated individual has enough lifespan to contribute its valuable schemas, an age indicator is also designed in DMOEA. For an individual in the initial population, its age value is assigned to be 1, and its age will increase by 1 if the individual survives to the next generation. Similarly, the age of a newborn offspring is 1 and grows generation by generation as long as it lives. Assume at generation n, an individual y has an age value, $age(y, n)$, its age indicator $A(y, n)$ given by

$$A(y,n) = \begin{cases} \frac{age(y,n)-A_{th}}{n}, & \text{if } \; age(y,n) > A_{th} \\ 0, & \text{otherwise} \end{cases} \tag{12.3.7}$$

where A_{th} is a prespecified age threshold. (12.3.7) implies that an individual will not be eliminated if its age is less than A_{th}. To ensure that an eliminated individual has a low fitness value, DMOEA periodically removes three types of individuals with different likelihood.

A. Likelihood of removing the most unhealthy individuals

At generation n, find a set Y_r that contains all the individuals with the highest rank value $r_{\max}$. Therefore, if $r_{\max}$ is larger than 1, the likelihood of individual $y_i \in Y_r$ being eliminated is given by

$$l_1^i = (1 - H(c_i, n))^2 \times A(y_i, n), \tag{12.3.8}$$

where $H(c_i, n) = \frac{1}{r_{\max}}$ denotes the health indicator value of the cell c_i that contains individual y_i at generation n.

B. Likelihood of removing the unhealthy individuals from the most crowded cells

At generation n, find a set Y_d that contains all the individuals with the highest density value, and then find a set $Y_{dr} \subseteq Y_d$ that includes all the individuals with the highest cell rank value $r_{d\max}$. In addition, denote the pure Pareto rank of individual $y_i \in Y_{dr}$ to be r_{di}. Therefore, if $r_{d\max}$ is greater than 1, the likelihood of individual y_i being eliminated is given by

$$l_2^i = (1 - H(c_i, n))^2 \times \left(1 - \frac{1}{r_{di}}\right)^2 \times (D(c_i, n) - 1) \times A(y_i, n), \tag{12.3.9}$$

where $H(c_i, n) = \frac{1}{r_{d\max}}$ and $D(c_i, n)$ represent the health and crowdedness values of the cell c_i that contains individual y_i at generation n. It is noted that $R_{dr} = \{r_{di}\}$ represents the local rank value of the individuals of set Y_{dr} and is calculated by the pure Pareto ranking scheme proposed by Goldberg [19]. Although all the individuals located in the same cell share the same cell rank value, they may still have local dominance relationships. Therefore, by measuring local rank values among all the individuals in one cell, DMOEA can determine the likelihood of eliminating an individual more precisely.

C. Likelihood of removing nondominated individuals from the most crowded cells

At generation n, if $r_{\max}$ is equal to 1, find a set Y_{rc} that contains all the individuals with the highest density value, and their local pure Pareto rank values of individual $y_i \in Y_{rc}$ to be r_{di}. Therefore, the likelihood of individual y_i being eliminated is given by

$$l_3^i = (D(c_i, n) - 1) \times \left(1 - \frac{1}{r_{di}}\right)^2 \times A(y_i, n), \qquad (12.3.10)$$

where $D(c_i, n)$ represents the crowdedness value of the cell c_i that contains individual y_i at generation n.

To determine whether an individual y_i will be eliminated, three random numbers between [0, 1] are generated to compare with the concerned likelihood, l_1^i, l_2^i, and l_3^i, according to the situation of the given individual. If the likelihood is larger than the corresponding random number, the selected individual will be removed from the population. Otherwise, the selected individual will survive to the next generation. Therefore, from (12.3.8)–(12.3.10), we can draw some observations as follows.

(1) Because the age indicator $A(y_i, n)$ influences all of three likelihoods, l_1, l_2, and l_3 will be 0 if the age of the concerned individual is smaller than the age threshold A_{th}. This implies that if an individual is not old enough, it will not be eliminated from the population no matter how high its rank and density values are.

(2) At each generation, DMOEA will remove those most unhealthy individuals according to likelihood l_1, based on their rank values and ages. Assume the age indicator of an individual y is

$$A(y, n) \approx 1,$$

without considering the effects of other indicators. When an unhealthy individual in the set Y_r has a very high $r_{\max}$ value, it will have a very high likelihood (l_1) being eliminated, since it is too far away from the current Pareto front. Moreover, as $r_{\max}$ drops and gets closer to 1, l_1 will decrease very fast, and the concerned individual will not be removed easily because it is very likely to be evolved into an elitist in the future.

(3) Because all the individuals in the same cell share the fixed computation resource (or "living resource"), the individuals located in a crowded cell have to compete much harder for the limited resource than those located in a sparse cell. Therefore, another elimination scheme based on the crowdedness indicator is designed in the DMOEA in order to remove some unhealthy individuals that stay in the most crowded areas. From (12.3.9), at each generation, if an individual belongs to the set Y_{dr}, it will have the likelihood of l_2 being eliminated based on its age, health, and local rank value and density condition. From this scheme, the population tends to be homogeneously distributed by eliminating the redundant individuals.

(4) After every individual has converged into a Pareto point, another elimination scheme is implemented based on l_3 values. Therefore, the resulting trade-off hyperareas $A_{to}(n)$ are counted, and the final population is truncated to ensure each cell contains only *ppv* number of individuals; thus the optimal population size can be calculated by (12.2.1).

12.3.6 Objective space compression strategy

Although the cell-based rank and density calculation scheme discussed in Section 12.3.1 can significantly improve the efficiency of DMOEA during its evolutionary process, it cannot guarantee the accuracy of the resulting Pareto front since an individual's rank value is represented by the rank value of its "home address," not by its own dominance status. Because the size of the true Pareto front is generally unknown, the boundaries of the objective are usually selected to be very large, which may be far away from the true Pareto front, to ensure that the entire true Pareto front is covered by the estimated objective space. In this case, if the predetermined cell scales $K_1, \cdots, K_m$ are not chosen to be correspondingly large enough, the size of a cell will be too spacious comparing to the true Pareto front, which may result in an inaccurate Pareto optimal set. This phenomenon can be illustrated in Figure 12.3.4(a), where the rank value for both cells **A** and **B** is 1 since they contain the true Pareto front. In this case, all the resulting individuals located in cells **A** and **B** are non-dominated solutions according to the proposed cell-based rank calculation scheme. However, if we examine these individuals by using the pure Pareto ranking method, we will find that most of these individuals are dominated points. To address this problem, we can either increase the cell scales $K_1, \cdots, K_m$ to a very large number or adaptively compress the objective space. Nevertheless, the first method will increase the computation time because it leaves too many redundant empty cells when the resulting Pareto front approaches the true Pareto front. Therefore, an objective space compression strategy is designed to adjust the size of the objective space and make it suitable to search for the true Pareto front with high precision. Assume at generation n, the upper and lower boundaries of the ith dimension of the objective space and current population are $F_i^{\max}$, $F_i^{\min}$, $P_i^{\max}$ and $P_i^{\min}$, respectively. Then three criteria are evaluated:

(1) maximum cell rank value of all the individuals is 1;

(2) either $(F_i^{\max} - P_i^{\max}) > 0.1(F_i^{\max} - F_i^{\min})$ or

$$(P_i^{\min} - F_i^{\min}) > 0.1(F_i^{\max} - F_i^{\min}); \text{ and} \tag{12.3.11}$$

(3) minimum age value of all the individuals is greater than the predefined age threshold A_{th}.

The ratio 0.1 in (12.3.11) is chosen heuristically, which implies the objective space will be compressed if there is at least 10% space in any dimension. Therefore, if all of the above three criteria are satisfied, the objective space compression process will be activated. As a result, a refined upper boundary of the objective space is defined as

$$F_i^{\max} = (P_i^{\max} + F_i^{\max})/2, \tag{12.3.12}$$

which means the distance between the updated upper boundary of the objective space and the upper boundary of the current population has decreased to half of its original value. Similarly,

$$F_i^{\min} = (P_i^{\min} + F_i^{\min})/2, \tag{12.3.13}$$

which means the distance between the updated lower boundary of the objective space and the lower boundary of the current population has decreased to half of its original value. The rationale for introducing the first criterion is to ensure the approximated area of the true Pareto front has been discovered before the objective space is compressed, which can avoid incorrect truncation of potential nondominated cells. Moreover, after a compression strategy is performed, the cell rank and density value will not remain the same as before, and the "home address" of each individual may change correspondingly. For these reasons, the objective space is not compressed at each generation. It only occurs when the three criteria listed above are satisfied. By continuously compressing the objective space as necessary, the resulting nondominated set can be tuned, and a more extended and homogeneously distributed Pareto front can be obtained.

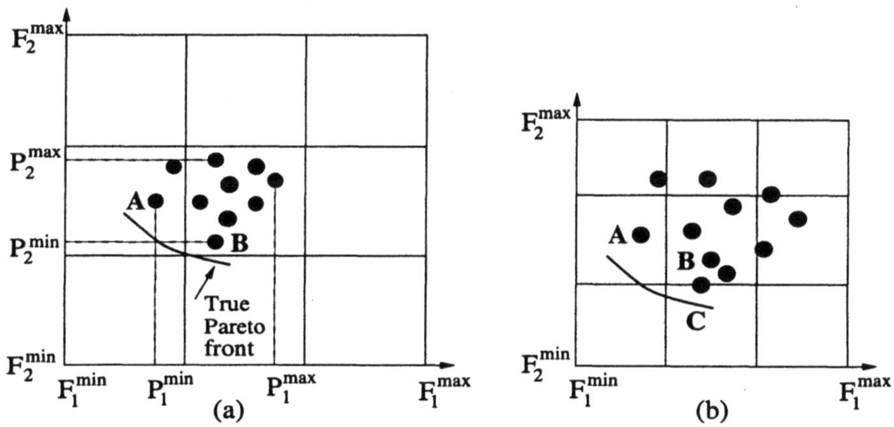

Figure 12.3.4: (a) The objective space and resulting Pareto front before compression and (b) the compressed objective space and newly explored nondominated cell

12.3.7 Stopping criteria and final refinement

Based on all the techniques introduced in Sections 12.3.1–12.3.6, we can determine whether DMOEA should stop by examining the following criteria.

(a) The rank value of all cells is 1.

(b) The objective space cannot be compressed anymore.

(c) Each resulting nondominated cell contains *ppv* individuals.

When all three criteria are met, the resulting nondominated set can hardly be refined by DMOEA any further. At this stage, DMOEA will keep running, and the cell-based rank calculation scheme will be replaced by the pure Pareto ranking scheme [19], whereas the cell-based density calculation scheme remains unchanged. The reason of this step is because another criterion ("the Pareto ranks of all resulting individuals are equal to 1") should be satisfied as well to guarantee that there is no dominance relationship among the resulting Pareto solutions at the final generation.

12.4 Comparative Study

In order to validate the proposed DMOEA and compare its performance to other advanced MOEAs, a recently designed benchmark problem is tested by six MOEAs: PAES, SPEA II, NSGA-II, RDGA, IMOEA, and the proposed DMOEA, in the comparative study. Each of the algorithms runs for 50 times to obtain the statistical results. For each test function, DMOEA will run with the initial population size equal to 2 and achieve an approximated desired population size dps. Afterwards, for each of 50 runs, an initial population with dps individuals is randomly generated and used by each of three population-based MOEAs (i.e., NSGA-II, SPEA II, and RDGA), while only one initial individual is generated for PAES according to its design procedure [9]. For IMOEA, its initial population size is also set to be 2 for a fair comparison. We use three indicators derived from the final generations of 50 runs to benchmark the comparison results via statistical Box plots. They are average individual rank value, average individual density value, and average individual distance. As discussed in Sections 12.2.1–12.2.5, for an individual, different ranking schemes will produce different rank values, which will be used in their respective fitness evaluations and selections. However, for a fair comparison in terms of ranking indicator among different MOEAs, we use Goldberg's pure Pareto ranking method [19] to recalculate the rank value for each individual resulting from each applied MOEA. Meanwhile, the average individual density value is calculated as the mean value of all the individual density values. The procedure of obtaining the average density value from the final results by all six MOEAs is summarized below. First, the final setting of density grid by DMOEA is obtained. Second, we run all the other algorithms and get the final result. Third, by filling the final results from different MOEAs into the specific density grid derived from the first step, we obtain the density values. Therefore, this calculation ensures that no preference to DMOEA will be made since it only implies that a certain grid division method will be applied to the results from all the chosen algorithms. Furthermore, because the rank is a relative value, it must be stated that we cannot guarantee that the final population will be a true Pareto set even if all its individuals have rank values equal to 1. For this reason, we use the "final average individual distance" as the third indicator to show how far the nondominated points on the resulting final Pareto front PF_{final} are away from the true Pareto front PF_{true}, where PF_{true} is known a priori for the given test function in this chapter. This indicator was originally introduced by Van Veldhuizen and Lamont [20], where a result of $G = 0$ indicates the convergence

$$PF_{final} = PF_{true};$$

any other value indicates PF_{final} deviates from PF_{true}.

Moreover, in order to compare the dominance relationship between the two final populations resulting from two different MOEAs, the coverage of two sets (C value) [13] is measured to show how the final population of one algorithm dominates the final population of another algorithm. Function C maps the ordered pair (X_i, X_j) to the interval [0, 1], where X_i and X_j denote the final populations resulting from the algorithms i and j, respectively. The value $C(X_i, X_j) = 1$ means that all points

Table 12.4.1: General parameter settings for both test functions

Common parameter	Chromosome length	Offspring generated per generation	Crossover rate	# individuals perform mutation	# binary bit flipped
DMOEA	15 × dec_num	10	0.7	1	1
IMOEA	15 × dec_num	10	0.7	1	1
NSGA-II	15 × dec_num	10	0.7	1	1
PAES	15 × dec_num	10	0.7	10	1
RDGA	15 × dec_num	10	0.7	1	1
SPEA II	15 × dec_num	10	0.7	1	1

* dec_num represents the number of decision variables

in X_j are dominated by, or equal to, points in X_i. The opposite, $C(X_i, X_j) = 0$, represents the situation when none of the points in X_j is covered by the set X_i.

Therefore, four indicators represent quantitative measures that describe the quality of the final result of selected MOEAs. The average individual rank value shows the dominated relationship between different individuals; the average individual density value illustrates how well the population diversity is preserved; the average individual distance measures distance between PF_{final} and PF_{true}; and the C value compares the domination relationship of a pair of MOEAs. All the values of the four indicators evaluated at the final generation are illustrated by Box plots to provide the statistical comparison results. Function F_1 has a twelve-dimensional decision and a three-dimensional objective space [21], and its true Pareto front is a surface (the first quadrant of a unit sphere) instead of a curve. Table 12.4.1 lists the parameters that are set to be unchanged for the test function. Crossover rate and maximum generation are chosen heuristically in Table 12.4.1. Moreover, it is worthy to note that although (1+1) PAES may produce better performances in some test functions according to [9], we select (1+10) PAES (which has 10 mutants rather than one mutant) to run the simulation because DMOEA and the other population-based MOEAs have 10 offspring generated at each generation. Therefore, by using (1+10) PAES, we can compare the performances of different MOEAs with the same evaluations (10 in this study) at each generation. For a fair statistical comparison, 50 runs were performed for each MOEA.

$$\text{Minimize } f_1(x), f_2(x), rmand\ f_3(x)$$

where

$$f_1(x) = (1 + g(x)) \cos \frac{\pi x_1}{2} \cos \frac{\pi x_2}{2}, \tag{12.4.1}$$

$$f_2(x) = (1 + g(x)) \cos \frac{\pi x_1}{2} \sin \frac{\pi x_2}{2}, \tag{12.4.2}$$

Table 12.4.2: Specific parameter settings for test function F_1

	Internal population size	Population size (archive)	L	ppv	K_1, K_2, K_3	A_{th}	t_{dom}
DMOEA	2	-	-	3	(20,20,20)	10	-
IMOEA	2	-	-	3	(20,20,20)	-	2
NSGA-II	1800	0*	-	-	-	-	2
PAES	1	1800	4	-	-	-	-
RDGA	1200	600	-	-	(20,20,20)	-	-
SPEA II	1200	600	-	-	-	-	-

* Instead of using archive, NSGA-II combines the populations from two consecutive generations

$$f_3(x) = (1 + g(x)) \sin \frac{\pi x_1}{2}, \tag{12.4.3}$$

$g(x) = \sum_{i=1}^{12} (x_i - 0.5)^2$, subject to $0 \le x_i \le 1, i = 1, \cdots, 12$.

According to [21], although NSGA-II can locate most of the population at its final generation on the true Pareto front, the resulting nondominated individuals are not homogeneously distributed, which implies that this test function presents a great challenge for MOEAs in searching for a good representation of the true Pareto front when it is a surface instead of a curve. Table 12.4.2 lists the specific parameter settings for function F_1. The age threshold is chosen to be 10 in DMOEA. At the final generation, DMOEA results in about 1800 individuals as the approximated optimal population size. Based on this estimation, the initial population size for NSGA-II, RDGA, and SPEA II is chosen to be 1800.

We have done simulation studies for six algorithms: DMOEA, IMOEA, NSGA-II, PAES, RDGA, and SPEA II. Simulation results show that DMOEA produces a more accurate and homogeneously distributed Pareto front compared to the other advanced MOEAs. Indeed, if the initial population size is correctly chosen, the MOEAs with the fixed population size (i.e., NSGA-II, RDGA, and SPEA II) also yield to a competent Pareto front in terms of rank, density, and distance values. In addition, because the true Pareto front is a surface instead of a curve, it is difficult for the resulting nondominated sets from any two MOEAs to cover each other. As the result, the C values are relatively low. In particular, because the Pareto points resulting from DMOEA have the lowest average individual distance values and a converged average individual density value, they are very competitive. This fact has made the resulting Pareto front of all the other five MOEAs difficult to cover, and $C(X_2, X_1), C(X_3, X_1), C(X_4, X_1), C(X_5, X_1)$, and $C(X_6, X_1)$ values are all near zero.

12.5 Robustness Study

From the description in Section 12.3, the performance of DMOEA may be affected by several parameters such as the initial population size P_0, age threshold A_{th}, population size per unit volume *ppv*, and grid scales $K_1, \cdots, K_m$. Among these parameters, the initial population size and the age threshold are the most important while the other two types of parameters are mostly determined by users based on their preferences and requirements in the resolution of the resulting Pareto front. In general, a user may not clearly understand the design mechanism of DMOEA and just randomly select an initial population size and the age threshold. Therefore, the relationship between these two parameters and the performances of the final Pareto front needs to be characterized in order to study the robustness of DMOEA based on these two parameters. For this reason, DMOEA is run for six settings of P_0 and A_{th} on the test function described in Section 12.4. These settings are: $P_0 = 2, A_{th} = 10$; $P_0 = 2, A_{th} = 30$; $P_0 = 2, A_{th} = 100$; $P_0 = 30, A_{th} = 10$; $P_0 = 100, A_{th} = 10$; and $P_0 = 500, A_{th} = 10$.

We have done simulation studies for the given six settings over 50 runs on the chosen test function. Simulation shows the evolutionary trajectories of the population size, average individual density value, and average individual distance value resulting from DMOEA. The results indicate that no matter which setting we select on DMOEA, the population size, average individual density, and average individual distance all converge to a constant value at the final generation. This implies different combinations of initial population size and that the age threshold may not change the resulting optimal population size and qualities of the final Pareto front. However, convergence speed may vary according to different settings. In particular, when initial population size or age threshold values are chosen to be relatively high, the convergence speed will be slow due to the high population size generated in the middle of the evolutionary process. Nevertheless, based on the objective compression strategy, this high population size only exists before the first compression process occurs. Meanwhile, according to the cell-based rank and density fitness assignment scheme described in Section 12.3.1, the computational complexity will not increase remarkably when the population size increases, thus the computation time will not alter very much even the population size is extraordinary high. Table 12.5.1 shows the average computation time for test function F_1 with 10,000 generations from IMOEA, PAES, NSGA-II, RDGA, SPEA II, and DMOEA with six settings over 50 runs.

The "CPUTIME" command from MATLAB (version 6.1) is used to measure the time elapsed for each MOEA implemented in MATLAB. Each MOEA is running in a HP computer with dual 2 GHz processors and 1 GByte RAM. It is worthy to note the time shown in Table 12.5.1 provides only a relative measure among chosen MOEAs based on the complexity of the algorithms. From Table 12.5.1, we can observe that among all chosen MOEAs, DMOEA demands the shortest running time. The improvement is significant compared to the other state-of-the-art MOEAs. In addition, different settings will not change the computation efforts of DMOEA and makes the final result of DMOEA robust in terms of both efficiency and effectiveness.

Table 12.5.1: Comparison results of computation time of F_1 from selected MOEAs and DMOEA with different settings

	IMOEA	PAES	NSGA-II	RDGA	SPA II
Time (min)	106	133	251	684	407

	DMOEA (2,10)	DMOEA (2,30)	DMOEA (2,100)	DMOEA (30,10)	DMOEA (100,10)	DMOEA (500,10)
	25	25	25	26	26	26

12.6 Conclusions

In this chapter, a new DMOEA is proposed. DMOEA can be characterized as (a) simplifying the computational complexity by using a cell-based rank and density fitness estimation scheme, (b) effectively refining individual's rank and density values by objective compression strategy, (c) adaptively increasing or decreasing population size based on rank and density status of current population, and (d) converging to an optimal (desired) population size by exploring all the nondominated hyperareas with a fixed number of Pareto points. From the comparative study, DMOEA has shown its potential in producing statistically competitive or even superior results with the five state-of-the-art MOEAs (IMOEA, PAES, NSGA-II, RDGA, and SPEA II) on one recently designed multiobjective optimization benchmark problem, which is used to exploit various complications in finding near-optimal, well-extended, and uniformly distributed true Pareto fronts.

Furthermore, to validate the robustness of DMOEA on different parameter settings, DMOEA is examined by the chosen test function with six pairs of parameter settings involving initial population sizes and age thresholds. From the simulation results, DMOEA is shown to be insensitive to the selection of both parameters in terms of running time and discovering a stable optimal population size with a constant cell density and a converged distance value. In addition, since the computational complexity of DMOEA is much lower than the other state-of-the-art MOEAs, DMOEA can be a potential candidate in solving some time-critical or on-line MOPs. However, as the test function used in this chapter is still far from covering all the challenging characteristics of MOPs, a more profound study in applying DMOEA to real-world MOPs is absolutely necessary in future work.

Bibliography

[1] C. M. Fonseca and P. J. Fleming, "An overview of evolutionary algorithms in multiobjective optimization," *Evolutionary Computation*, vol. 3, pp. 1–16, 1995.

[2] S. W. Mahfoud, "Genetic drift in sharing methods," *Proc. 1st IEEE Congress on Evolutionary Computation*, pp. 67–72, 1994.

[3] J. Arabas, Z. Michalewicz, and J. Mulawka, "GAVaPS–A genetic algorithm with varying population size," *Proc. 1st IEEE Congress on Evolutionary Computation*, pp. 73–74, 1994.

[4] N. Zhuang, M. Benten, and P. Cheung, "Improved variable ordering of BDDS with novel genetic algorithm," *Proc. IEEE International Symposium Circuits and Systems*, pp. 414–417, Monterey, CA, 1996.

[5] J. Grefenstette, "Optimization of control parameters for genetic algorithms," *IEEE Trans. Systems, Man and Cybernetics*, vol. 16, pp. 122–128, 1986.

[6] K. Tan, T. Lee, and E. Khor, "Evolutionary algorithms with dynamic population size and local exploration for multiobjective optimization," *IEEE Trans. Evolutionary Computation*, vol. 5, pp. 565–588, 2001.

[7] K. Deb, S. Agrawal, A. Pratap, and T. Meyarivan, "A fast elitist non-dominated sorting genetic algorithm for multi-objective optimization: NSGA-II," *IEEE Trans. Evolutionary Computation*, vol. 6, pp. 182–197, 2002.

[8] E. Zitzler, M. Laumanns, and L. Thiele, *SPEA2: Improving the Strength Pareto Evolutionary Algorithm*, Technical Report TIK-Report 103, Swiss Federal Institute of Technology, 2001.

[9] J. D. Knowles and D. W. Corne, "Approximating the non-dominated front using the Pareto archived evolutionary strategy," *Evolutionary Computation*, vol. 8, pp.149–172, 2000.

[10] H. Lu and G. G. Yen, "Rank-density based multiobjective genetic algorithm," *Proc. 9th IEEE Congress on Evolutionary Computation*, pp. 944–949, 2002.

[11] C. M. Fonseca and P. J. Fleming, "Multiobjective optimization and multiple constraint handling with evolutionary algorithms–Part I: A unified formulation," *IEEE Trans. Systems, Man and Cybernetics*, vol. 28, pp. 26–37, 1998.

[12] N. Srinivas and K. Deb, "Multi-objective function optimization using non-dominated sorting genetic algorithms," *Evolutionary Computation*, vol. 2, pp. 221–248, 1994.

[13] E. Zitzler and L. Thiele, "Multiobjective evolutionary algorithms: A comparative case study and the strength Pareto approach," *IEEE Trans. Evolutionary Computation*, vol. 3, pp. 257–271, 1999.

[14] J. D. Schaffer, "Multiple objective optimization with vector evaluated genetic algorithms," *Proc. 1st International Conference on Genetic Algorithms*, pp. 93–100, 1985.

[15] K. Tan, T. Lee, and E. Khor, "Evolutionary algorithms with goal and priority information for multi-objective optimization," *Proc. 6th IEEE Congress on Evolutionary Computation*, pp. 106–113, 1999.

[16] D. W. Corne and J. D. Knowles, "The Pareto-envelope based selection algorithm for multiobjective optimization," *Proc. 6th International Conference on Parallel Problem Solving from Nature*, pp. 839–848, 2000.

[17] T. Krink and R. K. Ursem, "Parameter control using agent based patchwork model," *Proc. 7th IEEE Congress on Evolutionary Computation*, pp. 77–83, 2000.

[18] X. Hu and R. C. Eberhart, "Multiobjective optimization using dynamic neighborhood particle swarm optimization," *Proc. 9th IEEE Congress on Evolutionary Computation*, pp. 1677–1681, 2002.

[19] D. E. Goldberg, *Genetic Algorithms in Search, Optimization and Machine Learning*, Addison-Wesley, Reading, MA, 1989.

[20] D. A. Van Veldhuizen and G. B. Lamont, "On measuring multiobjective evolutionary algorithm performance," *Proc. 7th IEEE Congress on Evolutionary Computation*, pp. 204–211, 2000.

[21] K. Deb, L. Thiele, M. Laumanns, and E. Zitzler, "Scalable multi-objective optimization test problems," *Proc. 9th IEEE Congress on Evolutionary Computation*, pp. 825–890, 2002.

Chapter 13

Set-Membership Adaptive Filtering

Yih-Fang Huang

Abstract: Based on a bounded-error assumption, set-membership adaptive filtering (SMAF) offers a viable alternative to traditional adaptive filtering techniques that are aimed to minimize an ensemble (or a time) average of the errors. This chapter presents an overview of the principles of SMAF, its features, and some applications. Highlighting the novel features of SMAF includes data-dependent selective update of the parameter estimates. Simulation experiences have shown that the SMAF algorithms normally use less than 20% of the data for updating parameter estimates while achieving performance comparable to traditional algorithms.

13.1 Introduction

Adaptive filtering is a field that has been studied extensively over the last few decades. There is a huge body of literature on theory, algorithms, and applications of adaptive filtering, see, e.g., [1–4] and the references therein. The discussions of this chapter will focus on recursive algorithms with a linear-in-parameter filter model.

Generally speaking, at the core of adaptive filtering is parameter estimation which provides updates of the filter coefficients (like the transfer function coefficients in the case of linear filters) to track the variation of the underlying process and/or the environment. The parameter estimation algorithms of most adaptive filtering schemes have been derived with an objective function that is usually defined in terms of the errors or the likelihood function. Among others, the most noted recursive algorithms with linear discrete-time model are that of recursive least-squares (RLS) algorithm (see, e.g., [5]) and that of least-mean-squares (LMS) (see, e.g., [2, 6]).

The RLS algorithm aims to minimize the time-average of the squared error. On the other hand, the LMS algorithm seeks to minimize the ensemble average of the error (namely, the mean-squared error) and it is implemented with a stochastic gradient search method. Both of those algorithms have been studied extensively and there are many extensions available to suit different applications.

Generally speaking, the RLS algorithm enjoys faster convergence yet requires more computational efforts. On the other hand, implementation of the LMS algorithm is rather simple, yet it does not converge fast enough in many applications [3]. Another issue is that of tracking, and the performance of those two algorithms in tracking depends primarily on the application situation.

Set-membership adaptive filtering (SMAF) [7, 8] is different from the aforementioned in that the estimation algorithm is not derived with the conventional notion of objective function, and it employs a bounded error (noise) assumption. The basic idea underlying SMAF is that the difference between the filter output and the desired output must be bounded (in magnitude). In terms of an equation error model, this amounts to a bounded noise assumption. From another point of view, this amounts to a performance specification that filter designers would like to define [8]. In essence, any parameter vector that results in compliance with the prespecified bound can be taken as a legitimate estimate of the filter coefficient. The goal of SMAF is thus to find a set of parameter vectors, termed "membership set," that is consistent with the filter model, the bounded error specification, and measurements. This is in contrast to conventional approaches (e.g., RLS and LMS) that render only one estimate at a time. Thus SMAF can potentially offer more robust filtering performance, which can be very useful in addressing robust control problems [9–11].

One of the distinguishing features of SMAF is the selective update of parameter estimates. This feature can potentially lead to significant complexity reduction, both in hardware and in computation. Furthermore, it offers a good deal of convenience in addressing performance-complexity trade-offs. More interestingly, this feature can also be properly explored for parallel adaptive architecture for multiple access systems. Some results on these topics have been given in [7].

After two decades of continued research, many SMAF algorithms have been proposed and studied. The discussion of this chapter will focus on two groups of SMAF algorithms: the OBE algorithms and SM-NLMS algorithms. This chapter is organized as follows. The next section presents a problem formulation for SMAF, followed by the OBE algorithms and discussions. Section 13.4 presents the SM-NLMS algorithm which resembles the conventional NLMS (normalized LMS) algorithm, except for a time-variant step size. Conclusions are given in Section 13.5.

13.2 Problem Formulation

With a linear discrete-time filter structure, we can formulate the adaptive filter as follows

$$y(k) = \theta^T \phi(k) \tag{13.2.1}$$

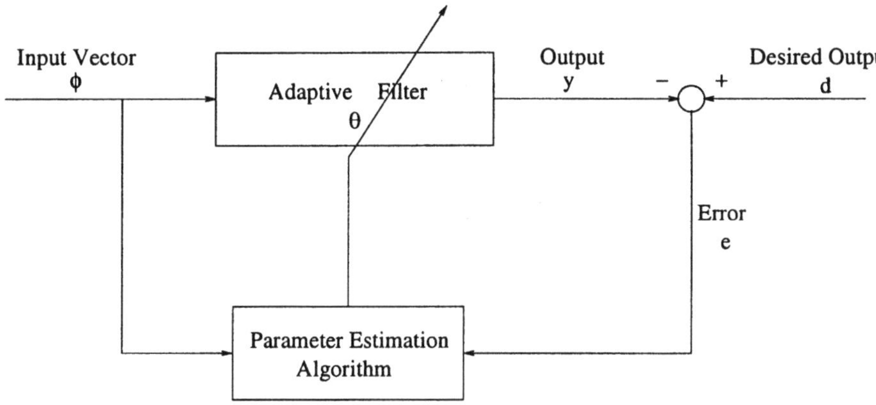

Figure 13.2.1: A basic adaptive filter structure

where $\theta = [\theta_1 \theta_2 \cdots \theta_N]^T$ is the filter parameter vector to be determined (or estimated) and $\phi(k)$ is the (measurable) input vector to the filter. Depending on what the input vector $\phi(k)$ consists of, (13.2.1) represents either a finite impulse response (FIR) or infinite impulse response (IIR) filter structure. If an IIR structure is used, $\phi(k)$ consists of previous outputs, i.e., $y(k-1), y(k-2), \cdots$. In any case, the filter coefficient θ is time variant, adapting to the changing environment.

As shown in Figure 13.2.1, which depicts the basic building blocks of a typical adaptive filter, the desired (or a reference) output $d(k)$ is used to adjust the filter parameter θ. In most applications, $d(k)$ is assumed to be known. In system identification (see, e.g., [12–14]), $d(k)$ is usually formulated by

$$d(k) = \theta_o^T \phi(k) + v(k) \tag{13.2.2}$$

where $d(k)$ is the measurable system output, θ_o is the presumed true parameter vector, and $v(k)$ is a noise process. The vector $\phi(k)$ in both (13.2.1) and (13.2.2) can be the same. In this case, the purpose of the adaptive filter is to identify the system parameter θ_o. If the underlying system is modeled as an autoregressive (AR) process with exogenous inputs (ARX), or an autoregressive moving average (ARMA) process, then $\phi(k)$ is simply a regressor vector; see, e.g., [15]. In this framework, it is common to assume that the system order N is known.

In practice, there are many applications for which the system order is not known and there exists no true system parameter to speak of. Adaptive equalization and adaptive echo cancellation in telecommunication are two notable examples (see, e.g., [16, 17]). In those cases, it is more practical and realistic to consider $d(k)$ as simply the desired output or the training sequence, as opposed to the output of any filter with a known order. Our presentation for SMAF will focus on adaptive filtering for those applications where the filter designer's primary objective is to make $y(k)$ as good a replica of $d(k)$ as possible, regardless of how $d(k)$ has come about.

In SMAF, the objective is to design an adaptive filter so that the error between the filter output and the desired output meets a bounded-error (say, bounded in magnitude) specification. In particular, SMAF requires that

$$|e(k)| = |d(k) - y(k)| \leq \gamma \quad \text{for all } k \tag{13.2.3}$$

where γ is some prespecified constant. Combining (13.2.1) and (13.2.3) yields a set at time instant k

$$\mathcal{H}_k = \{\theta \in R^N : |d(k) - \theta^T \phi(k)|^2 \leq \gamma^2\}. \tag{13.2.4}$$

The set $\mathcal{H}_k$ is sometimes referred to as the "constraint set." Given a set of data pairs $(\phi(k), d(k)), k = 1, 2, \cdots$, if the desired output sequence $d(k)$ has been generated by a time-invariant model structure, then the parameter vector to be estimated in (13.2.1) must lie in the intersection of the constraint sets for all time instants. In particular, one can define the exact membership set Ω_k as

$$\Omega_k \triangleq \cap_{i=1}^{k} \mathcal{H}_i. \tag{13.2.5}$$

Hence any point in Ω_k is a legitimate estimate for θ. Note that Ω_k is a sequence of monotone nonincreasing sets. Specifically, $\Omega_k \supseteq \Omega_{k+1} \supseteq \Omega_{k+2} \cdots$, for all k.

From a geometric point of view, (13.2.4) represents the region between two parallel hyperplanes in the parameter space Θ. Therefore, Ω_k would be a polygon in Θ. One of the important goals in SMAF is to obtain an effective analytical description of this polygon. In practice, however, it is often more convenient to find some analytically tractable outer bounding sets, e.g., ellipsoids [18], for Ω_k.

Intuitively, if the data pairs, $(\phi(k), d(k)), k = 1, 2, \cdots$, are rich enough, Ω_k will likely be small when k is large. Thus, it is likely that for some k_o, $\Omega_{k_o+1} = \Omega_{k_o}$, because $\Omega_{k_o} \subset \mathcal{H}_{k_o+1}$, as illustrated in Figure 13.2.2, and the filter coefficient θ does not have to be updated. This results in the so-called data-dependent selective updates for the filter parameter. In essence, the filter adapts only when necessary. The selective update feature of SMAF offers many benefits and opens up new avenues for exploration (see, e.g., [19]). It is a significant departure from conventional recursive adaptive filtering algorithms like RLS and LMS in which the estimates of the filter coefficients are updated at each data point, even when there are no benefits to those updates.

The choice of γ in (13.2.3) can be critical in the implementation of an SMAF algorithm. For example, if γ is set too low, it can result in an empty Ω_k for some k. Fortunately, studies have shown that performance of most SMAF algorithms, especially those that use the outer bounding sets to track the membership sets, is relatively insensitive to the exact values of γ [18, 20]. Moreover, there are algorithms that can estimate or tune the value of γ; see, e.g., [21, 22].

13.3 Optimal Bounding Ellipsoid (OBE) Algorithms

Before presenting any particular SMAF algorithm, we shall make some remarks on the development and evolution of SMAF. The fundamental idea of SMAF basically

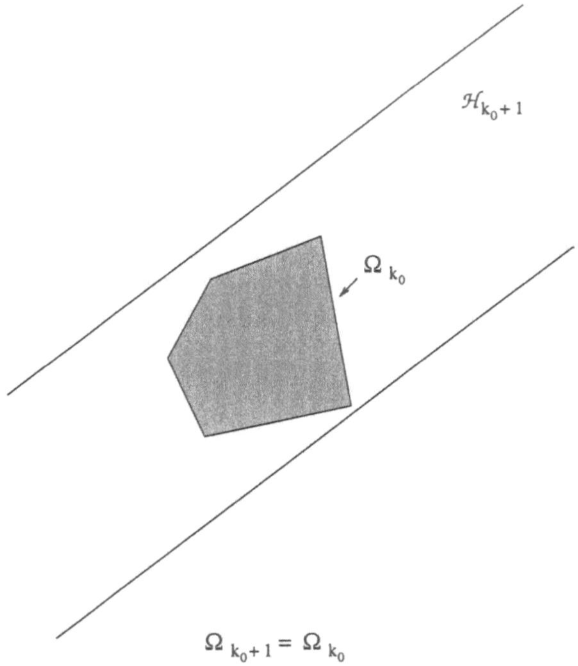

Figure 13.2.2: A membership set contained in the constraint set

evolved from set-membership identification (SMI), which was motivated by the work by Schweppe [23].

Schweppe was perhaps the first to use the term "membership set" and he employed a bounded-noise assumption to address a state estimation problem. Schweppe's idea was to assume that the process noise and the output measurement noise (in the state-space model) are both constrained by either an energy-type or an instantaneous magnitude-type bound. He then derived ellipsoids to bound the membership set, the smallest closed set of states consistent with the measurement, bounded-noise assumption, and the state-space model. The equations that govern the resulting state estimator bear a good deal of similarity to those of the Kalman filter. An important point to add is that, in Schweppe's algorithm, the choice of bounding ellipsoid is not unique, nor is it optimal in any sense or even the smallest in size.

Bertsekas and Rhodes [24] and Schlaepfer and Schweppe [25] then followed with more detailed treatments on the subject. Fogel [26] may have been the first one to consider (energy-)bounded noise in addressing the problem of system identification. This was then followed by Fogel and Huang [18], who derived OBE that enclose the membership sets recursively, based on a presumed known instantaneous magnitude noise bound. One of the main contributions of Fogel and Huang's paper was the optimization of the bounding ellipsoids, which ensures efficiency of estimation and leads to data-dependent selective updates of parameter estimates. In fact, Fogel and Huang are the first ones to point out the possibility of selective updates and their benefits. Their work was followed by a multitude of publications by other researchers on the topics of SMI and its applications; see [27] for a survey.

The initial motivation for Fogel and Huang to derive the OBE algorithms was to get around the difficulty of analytically tracking the exact membership set. The simplicity of formulating ellipsoids with quadratic equations and the fact that the constraint set $\mathcal{H}_k$ in (13.2.4) is a degenerate ellipsoid make it appealing to consider ellipsoidal bounding sets.

The basic principles of the OBE algorithms are outlined as follows. Let E_0 be an initial ellipsoid large enough to include all possible values for θ that render an adaptive filter satisfying the constraint (13.2.3). In particular, let

$$E_0 = \{\theta \in R^N : [\theta - \theta(0)]^T P^{-1}(0)[\theta - \theta(0)] \leq \sigma^2(0)\} \tag{13.3.1}$$

where $\theta(0)$ is the center of the ellipsoid, $P(0)$ is a positive-definite matrix that, along with $\sigma^2(0)$, defines the size of the ellipsoid. Without loss of generality, one can assume that $\theta(0) = \underline{0}$, i.e., the ellipsoid is centered at the origin. To ensure that this initial ellipsoid is large enough, one can define $P(0) = \frac{1}{\epsilon} I_{N\times N}$ and $\sigma^2(0) = 1$, where ϵ is a very small number (say, 10^{-5}) and $I_{N\times N}$ is an identity matrix.

At the time instant $k = 1$, the constraint set $\mathcal{H}_1$, resulting from the data pair $(\phi(1), d(1))$, will intersect E_0 if the bound γ in (13.2.3) has been chosen properly. The objective of an OBE algorithm is to find an ellipsoid E_1 (optimized in some sense) that encloses $\mathcal{H}_1 \cap E_0$, which contains the exact membership set Ω_1. This search for the OBE is repeated for subsequent data, i.e., $(\phi(k), d(k)), k = 2, 3, \cdots$.

In general, at time instant $k - 1$, the OBE E_{k-1} is described by

$$E_{k-1} = \{\theta \in R^N : [\theta - \theta(k-1)]^T P^{-1}(k-1)[\theta - \theta(k-1)] \leq \sigma^2(k-1)\}. \tag{13.3.2}$$

Kahn [28] showed that an ellipsoid E_k which tightly contains $E_{k-1} \cap \mathcal{H}_k$ is given by a linear combination of (13.2.4) and (13.3.2). In particular,

$$\begin{aligned} E_k = \{\theta \in R^N\colon\ & [\theta - \theta(k-1)]^T P^{-1}(k-1)[\theta - \theta(k-1)] \\ & + \lambda_k[y(k) - \theta^T \phi(k)]^2 \leq 1 + \lambda_k \gamma^2\} \end{aligned} \tag{13.3.3}$$

for some $\lambda_k \in [0, \infty)$, with $\sigma^2(k-1) \equiv 1$ for all k.

However, the tight bounding ellipsoid given in (13.3.3) is not unique and nor is it optimum in terms of the size of ellipsoid. Furthermore, the monotone nonincreasing property, i.e., $E_k \supseteq E_{k+1} \supseteq E_{k+2} \cdots$ no longer holds, since there are infinitely many such tight bounding ellipsoids. Thus Fogel and Huang [18] proposed to optimize the ellipsoid E_k with respect to λ_k, resulting in OBE. Clearly, different optimization criteria will result in different OBE algorithms. Two optimization criteria were suggested in [18]: one optimizes the ellipsoid in terms of its volume while the other one optimizes in terms of the trace of the positive-definite matrix $P(k)$. Note that the length of a semi-axis of the ellipsoid is directly proportional to the square root of an eigenvalue of $P(k)$. Thus minimizing the $Tr\{P(k)\}$ is like minimizing the dominant dimensions of the ellipsoid.

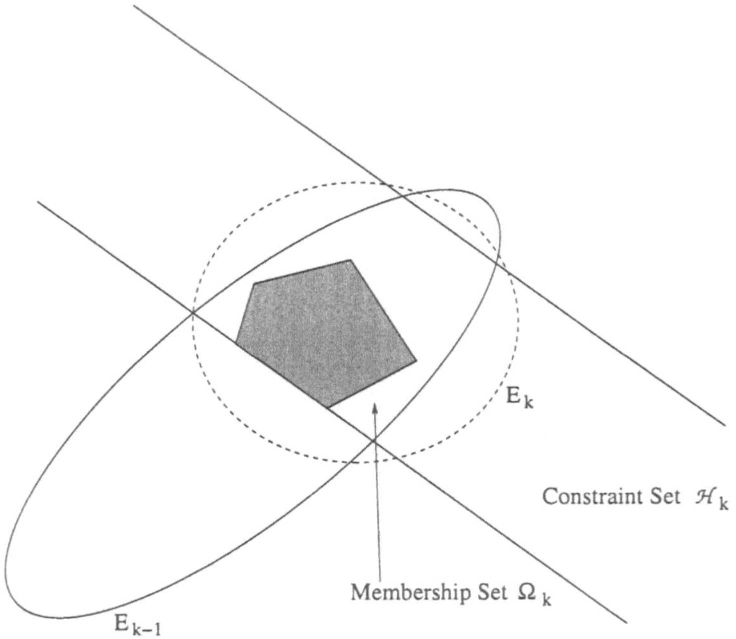

Figure 13.3.1: Construction of an OBE

It is clear from (13.3.2) that if $\lambda_k = 0$ for some k, then the bounding ellipsoid remains unchanged and there is no update of the parameter estimates. The volume minimizing OBE is often termed the F-H/OBE algorithm. Figure 13.3.1 depicts the construction of the OBE in a two-dimensional space.

Dasgupta and Huang [20] proposed a somewhat different formulation by taking a convex combination of (13.2.4) and (13.3.2) and constraining λ_k to be in $[0,1)$. In addition, they did not require $\sigma^2(k) \equiv 1$, for all k, in their formulation for the ellipsoid. Specifically,

$$E_k = \{\theta \in R^N : (1-\lambda_k)[\theta - \theta(k-1)]^T P^{-1}(k-1)[\theta - \theta(k-1)] \quad (13.3.4)$$
$$+\lambda_k [y(k) - \theta^T \phi(k)]^2 \leq (1-\lambda_k)\sigma^2(k-1) + \lambda_k \gamma^2\}.$$

By some algebraic manipulations on (13.3.4), an expression for E_k can be obtained as

$$E_k = \{\theta \in R^N : [\theta - \theta(k)]^T P^{-1}(k)[\theta - \theta(k)] \leq \sigma^2(k)\} \quad (13.3.5)$$

where

$$P^{-1}(k) = (1-\lambda_k)P^{-1}(k-1) + \lambda_k \phi(k)\phi^T(k) \quad (13.3.6)$$
$$\sigma^2(k) = \sigma^2(k-1) + \lambda_k \gamma^2 - \frac{\lambda_k(1-\lambda_k)[y(k) - \phi^T(k)\theta(k-1)]^2}{1 - \lambda_k + \lambda_k \phi^T(k)P(k-1)\phi(k)} \quad (13.3.7)$$
$$\theta(k) = \theta(k-1) + \lambda_t P(k)\phi(k)[y(k) - \phi^T(k)\theta(k-1)]. \quad (13.3.8)$$

Using the matrix inversion lemma [5] on $P^{-1}(k)$ in (13.3.6) yields

$$P(k) = \frac{1}{1-\lambda_k}\left[P(k-1) - \frac{\lambda_k P(k-1)\phi(k)\phi^T(k)P(k-1)}{1-\lambda_k+\lambda_k\phi^T(k)P(k-1)\phi(k)}\right]. \qquad (13.3.9)$$

The above equations (13.3.6)–(13.3.9) characterize the update of the bounding ellipsoids. The optimization criterion proposed in [20] minimizes $\sigma^2(k)$ with respect to λ_k, resulting in the so-called D-H/OBE algorithm.

Even though the SMAF algorithms render a set of feasible estimates at each time instant, many practical problems require one parameter estimate at any given time instant. For OBE algorithms, the bounding ellipsoid's center $\theta(k)$ can be taken to be that point estimate whenever needed. It is seen that the formulations for $\theta(k)$ and $P(k)$, (13.3.8), and (13.3.9), are strikingly similar to those of RLS with a forgetting factor. The key difference is that, for OBE, this forgetting factor is optimized at every time instant and it is often zero, implying no update of the parameter estimates. In particular, denote $\delta(k) \triangleq d(k) - \theta^T(k-1)\phi(k)$, the optimal value for λ_k in D-H/OBE is zero whenever $\sigma^2(k-1) + \delta(k) \leq \gamma^2$ [20].

One may argue that, for RLS, a dead zone can be defined so that whenever the prediction error falls inside the zone, no update will be taken. However, that procedure will result in a suboptimal performance. On the other hand, for OBE the optimal results are obtained with the selective updates.

The selective update feature of the OBE algorithms is further illustrated in Figure 13.3.2. One can see that the adaptive filter structure consists of two modules, an information evaluator followed by an updating processor. If the information evaluation involves little computation, as is the case for most SMAF algorithms, and if the frequency of update is small, then there is a good deal of unused computational resource to be explored. U-SHAPE [19], an adaptive equalization architecture that explores the selective update feature of SMAF algorithms, was presented with some detailed queueing analysis.

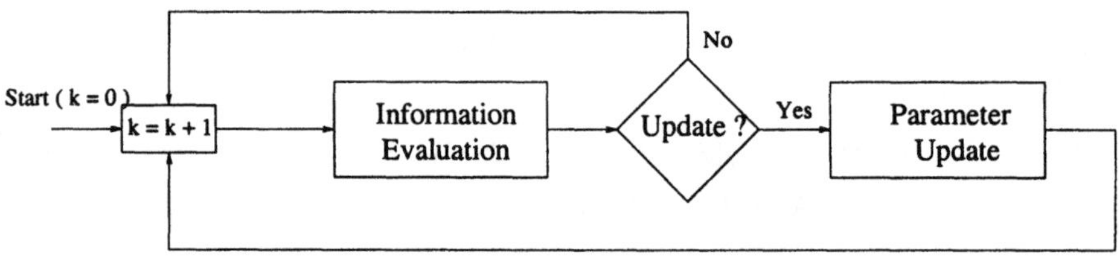

Figure 13.3.2: Selective update of SMAF

There is a good deal of analytical results relevant to the convergence of the D-H/OBE algorithm. Three aspects of convergence were presented in [20]:

- Cessation of updates, i.e., $\lim_{k\to\infty} \lambda_k = 0$;
- Convergence of size of ellipsoids, i.e., $\lim_{k\to\infty} \sigma^2(k) \in [0, \gamma^2]$; and
- Convergence of parameter estimates to a constant, i.e.,

$$\lim_{k\to\infty} \|\theta(k) - \theta(k-1)\| = 0.$$

Note that the SMAF algorithms, particularly the OBE algorithms, offer a convenient mechanism for performance-complexity trade-offs. In particular, if the bound γ is set higher, there may be fewer updates, thus less computational complexity, and the resulting performance in terms of filter output errors will be worse [7]. On the other hand, if γ is set lower, more frequent updates of parameter estimates will occur, thus more computation and better error performance. Another interesting property of the OBE algorithms is that they usually enjoy good tracking properties (see, e.g., [29, 30]).

Over the last two decades, there have been many variations of the OBE algorithm and SMI algorithms; see, e.g., [31–33]. Among others, the survey papers [34, 35] provide useful insights into the field up to 1990. Some conference proceedings, special issues of archival journals, and edited books (see, e.g., [27, 36, 37]) also offer a broader coverage of the subject field. Combettes [38] presented a thorough mathematical treatment on the foundation of the set-membership estimation.

13.4 Set-Membership NLMS (SM-NLMS) Algorithm

In the previous section, OBE algorithms have been presented as SMAF algorithms. Those OBE algorithms are seen as similar in formulation to RLS algorithms with a forgetting factor. As such, the computational complexity of OBE algorithms is in the same order of RLSs, even though the OBE algorithms do not update every time. A simpler alternative to OBE algorithms is the SM-NLMS algorithm [7, 8], which is like a set-membership counterpart of the conventional NLMS algorithm.

The SM-NLMS algorithm can be derived with the notion of optimal bounding spheroids. It can also be derived with a point-wise approach that takes projections of the previous estimate on the constraint set at each time and seeks to minimize the Euclidean distance between the existing parameter estimate and the new one. This is like building on the minimum distance principle.

The set-bounding approach is to compute outer approximations of the membership set by minimal volume spheroids. A spheroid S_k in the N-dimensional parameter space is described by

$$S_k = \{\theta \in R^N : ||\theta - \theta(k)|| \leq \sigma^2(k)\}, \tag{13.4.1}$$

where $\theta(k)$ is the center of the spheroid and $\sigma(k)$ is its radius. Since any measure of the size of a spheroid (e.g., volume and surface area) is directly proportional to its radius, one can unambiguously minimize the size of the bounding spheroid by minimizing the radius. Note that this is not really the case when ellipsoids are used, as in the previous section, instead of spheroids.

The formulations for the resulting SM-NLMS are given by the following recursions [7, 8]:

$$\theta(k) = \theta(k-1) + \lambda_k \frac{\delta(k)\phi(k)}{\phi^T(k)\phi(k)} \tag{13.4.2}$$

$$\sigma^2(k) = \sigma^2(k-1) - \lambda_k^2 \frac{|\delta(k)|^2}{\phi^T(k)\phi(k)} \tag{13.4.3}$$

where the prediction error $\delta(k)$ and the gain λ_k are given by

$$\begin{aligned}\delta_k &= d(k) - \theta^T(k-1)\phi(k) \\ \lambda_k &= \begin{cases} 1 - \frac{\gamma}{|\delta(k)|} & \text{if } |\delta(k)| > \gamma \\ 0, & \text{otherwise.} \end{cases}\end{aligned} \tag{13.4.4}$$

It may be observed that the SM-NLMS recursion, although derived using a completely different philosophy, is identical to the recursion in the conventional NLMS algorithm, except for the assignment to the gain term λ_k. The gain λ_k is a constant in conventional NLMS, whereas it is adaptive in SM-NLMS, reflecting the information provided by the current data set. The variable step size is optimal in the sense of minimizing the bounding spheroids or of minimum distance. Furthermore, the SM-NLMS algorithm guarantees a nonincreasing parameter error, as noted in [7, 8].

Implementation of the SM-NLMS algorithm can now be summarized as follows. Choose $\theta(0) = \underline{0}$, and $\sigma^2(0) \gg 0$ such that the feasible set $\Theta \subset S_0$, the initial spheroid. The algorithm at time k is given by

$$\begin{aligned}&\delta(k) = d(k) - \theta^T(k-1)\phi(k) \\ &\text{If } |\delta(k)| > \gamma, \\ &\qquad \lambda_k = 1 - \tfrac{\gamma}{|\delta(k)|} \\ &\qquad \theta(k) = \theta(k-1) + \lambda_k \tfrac{\delta(k)\phi(k)}{\|\phi(k)\|^2} \\ &\qquad \sigma^2(k) = \sigma^2(k-1) - \lambda_k^2 \tfrac{|\delta(k)|^2}{\|\phi(k)\|^2} \\ &\text{else, (no update!)} \\ &\qquad \theta(k) = \theta(k-1) \\ &\qquad \sigma^2(k) = \sigma^2(k-1).\end{aligned}$$

The SM-NLMS has been employed for many interesting applications in communications systems [7, 39].

13.5 Conclusions

This chapter has presented the principles of SMAF. Some SMAF algorithms and their features are presented. It is noted that SMAF features data-dependent selective updates of parameter estimates and offers performance comparable to what can be achieved with conventional algorithms like RLS and LMS. It is noted that SMAF, with a convenient complexity-performance trade-off feature, can be applied to many practical problems, including robust control, adaptive equalization, and multiuser detection. It can also adapt to changing environments more effectively than RLS and LMS and offer performance comparable to what RLS and LMS can provide.

Bibliography

[1] G. C. Goodwin and K. S. Sin, *Adaptive Filtering, Prediction and Control*, Prentice-Hall, Englewood Cliffs, NJ, 1982.

[2] B. Widrow and S. D. Stearns, *Adaptive Signal Processing*, Prentice-Hall, Englewood Cliffs, NJ, 1985.

[3] S. Haykin, *Adaptive Filter Theory*, 4th ed., Prentice-Hall, Englewood Cliffs, NJ, 2001.

[4] P. S. R. Diniz, *Adaptive Filtering–Algorithms and Practical Implementation*, Kluwer, Boston, MA, 1997.

[5] L. Ljung and T. Söderström, *Theory and Practice of Recursive Identification*, The MIT Press, Cambridge, MA, 1983.

[6] M. L. Honig and D. G. Messerschmitt, *Adaptive Filters–Structures, Algorithms, and Applications*, Kluwer, Boston, MA, 1984.

[7] S. Gollamudi, *Nonlinear Mean-Square Multiuser Detection and Set-Membership Estimation for Digital Communications*, Ph.D. Dissertation, Department of Electrical Engineering, University of Notre Dame, Notre Dame, IN, 1999.

[8] S. Gollamudi, S. Nagaraj, S. Kapoor, and Y. F. Huang, "Set-membership filtering and a set-membership normalized LMS algorithm with an adaptive step size," *IEEE Signal Processing Letters*, vol. 5, no. 5, pp. 111–114, May 1998.

[9] B. Wahlberg and L. Ljung, "Hard frequency-domain model error bounds from least-squares like identification techniques," *IEEE Trans. Automatic Control*, vol. AC-37, pp. 900–912, July 1992.

[10] G. Goodwin, M. Gevers, and B. Ninness, "Quantifying the error in estimated transfer functions with application to model order selection," *IEEE Trans. Automatic Control*, vol. AC-37, pp. 913–928, July 1992.

[11] R. L. Kosut, M. K. Lau, and S. P. Boyd, "Set-membership identification with parametric and nonparametric uncertainty," *IEEE Trans. Automatic Control*, vol. AC-37, pp. 929–941, July 1992.

[12] K. J. Åström and P. Eykhoff, "System identification–A survey," *Automatica*, vol. 7, pp. 123–167, 1971.

[13] I. D. Landau, *System Identification and Control Design*, Prentice-Hall, Englewood Cliffs, NJ, 1990.

[14] L. Ljung, *System Identification–Theory for the User*, 2nd ed., Prentice-Hall, Upper Saddle River, NJ, 1999.

[15] D. Graupe, *Time Series Analysis, Identification, and Adaptive Filtering*, 2nd ed., R. E. Krieger Publishing Company, Malabar, FL, 1989.

[16] J. G. Proakis, *Digital Communications*, 2nd ed., McGraw-Hill, New York, 1989.

[17] S. Haykin, *Communication Systems*, 4th ed., John Wiley, New York, 2001.

[18] E. Fogel and Y. F. Huang, "On the value of information in system identification–Bounded noise case," *Automatica*, vol. 18, pp. 229–238, Mar. 1982.

[19] S. Gollamudi, S. Kapoor, S. Nagaraj, and Y. F. Huang, "Set-membership adaptive equalization and an updator-shared implementation for multiple channel communications systems," *IEEE Trans. Signal Processing*, vol. 46, no. 9, pp. 2372–2385, Sept. 1998.

[20] S. Dasgupta and Y. F. Huang, "Asymptotically convergent modified recursive least squares with data dependent updating and forgetting factor for systems with bounded noise," *IEEE Trans. Information Theory*, vol. 33, pp. 383–392, 1987.

[21] D. Maksarov and J. P. Norton, "Tuning of noise bounds in state bounding," *Proc. Symposium on Modeling, Analysis and Simulation* (IMACS Multiconference), pp. 837–842, July 1996.

[22] T. M. Lin, M. Nayeri, and J. R. Deller, Jr., "Consistently convergent OBE algorithm with automatic selection of error bounds," *Int. J. Adaptive Control and Signal Processing*, vol. 12, pp. 305–324, June 1998.

[23] F. C. Schweppe, "Recursive state estimation: Unknown but bounded errors and system inputs," *IEEE Trans. Automatic Control*, vol. AC-13, no. 1, pp. 22–28, Feb. 1968.

[24] D. P. Bertsekas and I. B. Rhodes, "Recursive state estimation for a set-membership description of uncertainty," *IEEE Trans. Automatic Control*, vol. AC-16, no. 2, pp. 117–128, Apr. 1971.

[25] F. M. Schlaepfer and F. C. Schweppe, "Continuous-time state estimation under disturbances bounded by convex sets," *IEEE Trans. Automatic Control*, vol. AC-17, no. 2, pp. 197–205, Apr. 1972.

[26] E. Fogel, "System identification via membership set constraints with energy constrained noise," *IEEE Trans. Automatic Control*, vol. AC-24, pp. 752, 1979.

[27] M. Milanese, J. P. Norton, and E. Walter, Eds., *Bounding Approaches to System Identification*, Plenum, London, 1996.

[28] W. Kahn, "Circumscribing an ellipsoid about an intersection of two ellipsoids," *Canadian Mathematical Bulletin*, vol, 11, no. 3, pp. 437–441, 1968.

[29] Y. F. Huang and J. R. Deller,"On the tracking capabilities of optimal bounding ellipsoid algorithms," *Proc. 30th Annual Allerton Conference on Commununication, Control, and Computing*, pp. 50–59, Monticello, IL, Sept. 1992.

[30] A. K. Rao and Y. F. Huang, "Tracking characteristics of an OBE parameter estimation algorithm," *IEEE Trans. Signal Processing*, vol. 41, no.3, pp. 1140–1148, Mar. 1993.

[31] J. R. Deller and S. F. Odeh, "SM-WRLS algorithms with an efficient test for innovation," *Proc. 9th IFAC/IFORS Symposium on Identification and System Parameter Estimation*, pp. 1044–1049, Budapest, July 1991.

[32] J. R. Deller, M. Nayeri, and S. F. Odeh, "Least square identification with error bounds for real-time signal processing and control," *Proc. IEEE*, vol. 81, pp. 813–849, 1993.

[33] S. Nagaraj, S. Gollamudi, S. Kapoor, and Y. F. Huang, "BEACON: An adaptive set-membership filtering technique with sparse updates," *IEEE Trans. Signal Processing*, vol. 47, no. 11, pp. 2928–2941, Nov. 1999.

[34] J. R. Deller, Jr., "Set-membership identification in digital signal processing," *IEEE ASSP Magazine*, vol. 6, pp. 4–22, Oct. 1989.

[35] E. Walter and H. Piet-Lahanier, "Estimation of parameter bounds from bounded-error data: A survey," *Mathematics and Computers in Simulation*, vol. 32, pp. 449–468, 1990.

[36] E. Walter, Ed., Special Issue on Parameter Identification Bound, *Mathematics and Computers in Simulation*, vol. 32, Dec. 1990.

[37] *Proc. 9th IFAC/IFORS Symposium on Identification and System Parameter Estimation*, Budapest, July 1991.

[38] P. M. Combettes, "The foundations of set-theoretic estimation," *Proc. IEEE*, vol. 81, pp. 182–208, 1993.

[39] S. Gollamudi and Y. F. Huang, "Iterative nonlinear MMSE multiuser detection," *Proc. IEEE International Conference on Acoustics, Speech and Signal Processing*, Phoenix, AZ, 1999.

PART III

POWER SYSTEMS AND CONTROL SYSTEMS

Chapter 14

Trajectory Sensitivity Theory in Nonlinear Dynamical Systems: Some Power System Applications*

M. A. Pai and Trong B. Nguyen

Abstract: Trajectory sensitivity analysis (TSA) has been applied in control system problems for a long time in such areas as optimization and adaptive control. Applications in power systems in conjunction with Lyapunov/transient energy functions first appeared in the 1980s. More recently, TSA has found applications on its own by defining a suitable metric on the trajectory sensitivities with respect to the parameters of interest. In this chapter we present theoretical as well as practical applications of TSA for dynamic security applications in power systems. We also discuss the technique to compute critical values of any parameter that induces stability in the system using trajectory sensitivities.

14.1 Introduction

Security in power systems became an issue after the Northeast blackout in 1965 [1]. Since then a lot of research has investigated both static and dynamic aspects of security. While a lot of success has been achieved on the static front [2], such is not the

*This research was supported by the National Science Foundation under Grant ECS 00-00474 and the Grainger Foundation.

case with dynamic security. Dynamic security assessment (DSA) in power systems comprises the following main tasks: contingency selection/screening, security evaluation, contingency ranking, and limit computation. Dynamic simulation has historically been the main tool and currently, in combination with heuristics and some form of learning, it remains the tool for DSA in the energy control centers. Intensive research since the 1960s in applying Lyapunov's direct method has resulted in useful algorithms in the form of transient energy function (TEF) technique [3–5] and single machine equivalent (SIME) technique [6]. Artificial neural network and artificial intelligence-based techniques have also been applied [7, 8]. Of these the TEF and SIME techniques are considered most promising by the research community. In the deregulated environment, the existing transmission system often operates at its limit due to inadequate capacity and multilateral transactions. In addition, power systems must be operated to satisfy the transient stability constraints for a set of contingencies. In these situations, DSA plays a crucial role. The repetitive nature of the dynamic simulation for DSA must be replaced by a procedure where complex models can be handled in a more direct manner. The aim of on-line DSA is to assess the stability of the system to a set of predefined contingencies. These contingencies are user specified or chosen automatically through some procedure such as a filtering process. For each contingency where the system is stable, it can also provide a security margin based on the technique used. For instance, if critical clearing time is computed, $t_{cr} - t_{cl}$ is the margin. On the other hand, if the TEF is used, then $V_{cr} - V_{cl}$ is the margin. The security margin can be used to provide the operators with guidelines to improve system security while at the same time maintaining economic operation. This is known as security-constrained optimization or preventive rescheduling (see Figure 14.1.1). The literature on preventive control is largely tied to the TEF method, namely, to enhance the stability margin as defined by the difference between critical energy V_{cr} and energy at clearing V_{cl}. However, the need for computation of the unstable equilibrium point and the absence of an analytical closed form of the energy function make it difficult to apply for larger and more complex models of machines.

In this chapter we review some recent results in applying trajectory sensitivity techniques [9] to DSA and preventive control. As will be shown, this new technique has several advantages over all the other techniques:

(a) No restriction on complexity of the model.
(b) Extension to systems with discrete events is possible.
(c) Information other than mere stability can be obtained.
(d) Limits to any parameter in the system affecting stability can be studied.
(e) Identification of weak links in the transmission network is possible.
(f) Preventive strategies can be incorporated easily.

However, the above advantages are obtained at the expense of increased computational cost. This question will also be addressed in this chapter.

The chapter outline is as follows. In Section 14.2 we will explain the derivation of the basic theory of trajectory sensitivity analysis (TSA) for the differential algebraic equation (DAE) form of the system model. The overall approach to DSA using the TSA technique will be explained. In Section 14.3 we will use the TSA technique

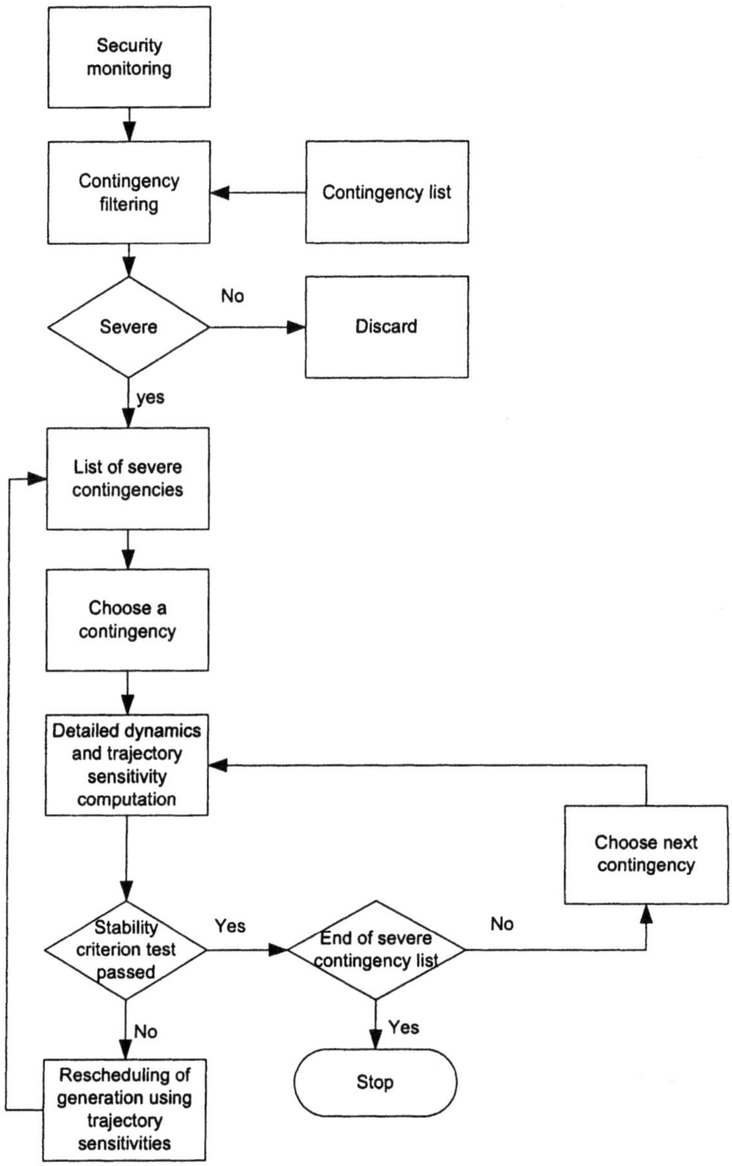

Figure 14.1.1: DSA scheme using trajectory sensitivities

to compute critical parameter values such as clearing time, mechanical input power, and line reactance. In Section 14.4 we will explain the dynamic security constrained dispatch problem and its application [10]. In Section 14.5 we will use the TSA technique to find weak links, vulnerable relays, and the electrical centers of the system for a given fault. This information is useful in proper islanding of the system in a self-healing way [11].

14.2 Trajectory Sensitivity Analysis (TSA)

Sensitivity theory in dynamic systems has a long and rich history well documented in the books by Frank [12] and Eslami [13]. It can be traced back to the work of Bode [14] in designing feedback amplifiers where feedback is used to cancel the effect of unwanted disturbances and parameter variations. The concept of sensitivity matrix using state-space analysis can be used [15]. While the bulk of the work is in the area of linear time-invariant (LTI) systems, the fundamental theory is applicable to nonlinear systems as well. Applications of sensitivity theory or more specifically TSA to nonlinear dynamic systems are few. The books of Tomovic [15], Tomovic and Vukobratovic [16], and Cruz [17] contain control-oriented applications.

There have been applications of sensitivity theory to power systems but mostly for linear systems such as eigenvalue sensitivity [18]. From a stability point of view the theory has been applied for computing sensitivity of the energy margin while using the TEF method [3, 4]. Applications of trajectory sensitivity to nonlinear models of power systems are somewhat recent [19].

The application in the linear system arises from the linearization of a nonlinear system around an equilibrium point. Stability of the equilibrium point is evaluated through eigenvalues. In trajectory sensitivity analysis, we linearize around a nominal trajectory and try to interpret the variations around that trajectory. It is a challenge to develop a metric for those variations and relate it to the stability of the nominal trajectory. In [15] the author hints at the intimate connection between trajectory sensitivity analysis and Lyapunov stability but it is not quantified. In this chapter we make an attempt to do so.

14.3 Trajectory Sensitivity Theory for DAE Model

The development in this section follows [20, 21], where the application to hybrid systems is discussed. A DAE system is a special case of the more general hybrid systems. A fairly accurate description of the power system model is represented by a set of differential algebraic equation of the form

$$\dot{x} = f(x, y, \lambda) \tag{14.3.1}$$

$$0 = \begin{cases} g^-(x, y, \lambda) & s(x, y, \lambda) < 0 \\ g^+(x, y, \lambda) & s(x, y, \lambda) > 0 \end{cases} \tag{14.3.2}$$

and a switching occurs when $s(x, y, \lambda) = 0$.

In the above model, x are the dynamic state variables such as machines' rotor angles, and velocities; y are the algebraic variables such as load bus voltage magnitudes and angles; and λ are the system parameters such as line reactances, generator mechanical powers, and fault clearing time. Note that the state variables x are continuous while the algebraic variables y can undergo step changes at the switching instants.

The initial conditions for (14.3.1) and (14.3.2) are given by $x(t_0) = x_0$ and $y(t_0) = y_0$,where y_0 satisfies $g(x_0, y_0, \lambda) = 0$. For compactness of notation, the following definitions are used

$$\underline{x} = \begin{bmatrix} x \\ \lambda \end{bmatrix}, \ \underline{f} = \begin{bmatrix} f \\ 0 \end{bmatrix}.$$

With these definitions, (14.3.1) and (14.3.2) can be written in a compact form as

$$\dot{\underline{x}} = \underline{f}(\underline{x}, y) \tag{14.3.3}$$

$$0 = \begin{cases} g^-(\underline{x}, y) & s(\underline{x}, y) < 0 \\ g^+(\underline{x}, y) & s(\underline{x}, y) > 0. \end{cases} \tag{14.3.4}$$

The initial conditions for (14.3.1) and (14.3.2) are

$$\underline{x}(t_0) = \underline{x}_0, \ \ y(t_0) = y_0. \tag{14.3.5}$$

We divide the time interval as consisting of nonswitching subintervals and switching instants for which the sensitivity model is now developed.

14.3.1 Trajectory sensitivity calculation for nonswitching periods

This section gives the analytical formulae for calculating sensitivities $\underline{x}_{\underline{x}_0}(t)$ and $y_{\underline{x}_0}(t)$ on the nonswitching time intervals as discussed in [21]. On these intervals, the DAE systems can be written in the form

$$\dot{\underline{x}} = \underline{f}(\underline{x}, y), \ \ 0 = g(\underline{x}, y). \tag{14.3.6}$$

Differentiating (14.3.6) with respect to the initial conditions $\underline{x}_0$ yields

$$\dot{\underline{x}}_{\underline{x}_0} = \underline{f}_{\underline{x}}(t)\underline{x}_{\underline{x}_0} + \underline{f}_y(t)y_{\underline{x}_0}, \ \ 0 = g_{\underline{x}}(t)\underline{x}_{\underline{x}_0} + g_y(t)y_{\underline{x}_0} \tag{14.3.7}$$

where $\underline{f}_{\underline{x}}$, $\underline{f}_y$, $g_{\underline{x}}$, and g_y are time-varying matrices calculated along the system trajectories.

Initial conditions for $\underline{x}_{\underline{x}_0}$ are obtained by differentiating (14.3.5) with respect to $\underline{x}_0$ as

$$\underline{x}_{\underline{x}_0}(t_0) = I \tag{14.3.8}$$

where I is the identity matrix.

Using (14.3.8) and assuming that $g_y(t_0)$ is nonsingular along the trajectories, the initial condition for $y_{\underline{x}_0}$ can be calculated from (14.3.7) as

$$y_{\underline{x}_0}(t_0) = -[g_y(t_0)]^{-1} g_{\underline{x}}(t_0). \tag{14.3.9}$$

Therefore, the trajectory sensitivities can be obtained by solving (14.3.6) simultaneously with (14.3.7) using any numerical method with (14.3.5), (14.3.8), and (14.3.9) as the initial conditions.

At switching instants, it is necessary to calculate the jump conditions that describe the behavior of the trajectory sensitivities at the discontinuities. Since we are considering time instants, which do not depend on the states, the sensitivities of the states will be continuous whereas those of the algebraic are not. When the trajectory sensitivities are known, the perturbed trajectories can be estimated by first-order approximation without redoing simulation as $\Delta\underline{x}(t) \approx \underline{x}_{\underline{x}_0}(t)\Delta\underline{x}_0$ and $\Delta y(t) \approx y_{\underline{x}_0}(t)\Delta\underline{x}_0$.

14.3.2 Computing critical values using energy function metric

In the literature, trajectory sensitivities have been used [4] to compute the energy margin sensitivity with respect to system parameters, such as interface line flow and system loading using TEF methods. In these cases, the critical energy ν_{cr}, which is the energy at the controlling u.e.p., depends on the parameters. Therefore, computation of $\frac{\partial \nu_{cr}}{\partial t_{cl}}$ is necessary while using the TEF method. This is a computationally difficult task. On the other hand, because the energy function $\nu(x)$ is used here only as a metric to monitor the system sensitivity for different t_{cl}, we can avoid the computation of ν_{cr} and use $\nu(x)$ directly.

The process of estimating critical values of parameters will be illustrated using the clearing time t_{cl}. However, the process is appropriate for any parameter that can induce instability, such as mechanical power P_M. We can use the sensitivity $\frac{\partial \nu}{\partial t_{cl}}$ to estimate t_{cr} directly. With the classical model for machines, the energy function $\nu(x)$ for a structure-preserving model is computed as follows.

The post fault power system can be represented by the DAE model in the center of angle reference frame [3] as

$$\dot{\theta}_{n_0+i} = \tilde{\omega}_{g_i}, \qquad i = 1, \cdots, m \tag{14.3.10}$$

$$M_i\dot{\tilde{\omega}}_{g_i} = P_{M_i} \sum_{j=1}^{n} B_{n_0+i,j} V_{n_0+i} V_j \sin\,(\theta_{n_0+i} - \theta_j) - \frac{M_i}{M_T} P_{COA}, \quad i = 1, \cdots, m \tag{14.3.11}$$

$$P_{d_i} + \sum_{j=1}^{n} B_{ij} V_i V_j \sin(\theta_i - \theta_j) = 0, \qquad i = 1, \cdots, n_0 \tag{14.3.12}$$

$$Q_{d_i}(V_i) - \sum_{j=1}^{n} B_{ij} V_i V_j \cos(\theta_i - \theta_j) = 0, \qquad i = 1, \cdots, n_0 \tag{14.3.13}$$

where m is the number of machines, n_0 is the number of buses in the system, and

$$P_{COA} = \sum_{i=1}^{m} \left(P_{M_i} - \sum_{j=1}^{n} B_{ij} V_i V_j \sin(\theta_i - \theta_j) \right).$$

We assume constant real power loads and voltage-dependent reactive power load as

$$Q_{di}(V_i) = Q_{di}^s \left(\frac{V_i}{V_i^s} \right)^{\alpha} \tag{14.3.14}$$

where Q_i^s and V_i^s are the nominal steady state reactive power load and voltage magnitude at the ith bus, and α is the reactive power load index.

The corresponding energy function is established as [3]

$$\begin{aligned} \nu(\tilde{\omega}_g, \theta, V) \;=\; & \frac{1}{2} \sum_{i=1}^{m} M_i \tilde{\omega}_{g_i}^2 - \sum_{i=1}^{m} P_{M_i} (\theta_{n_0+i} - \theta_{n_0+i}^s) + \sum_{i=1}^{n_0} P_{d_i} (\theta_i - \theta_i^s) \\ & - \frac{1}{2} \sum_{i=1}^{n_0} B_{ii} (V_i^2 - V_i^{s2}) + \sum_{i=1}^{n_0} \frac{Q_{d_i}^s}{\alpha V_i^{s\alpha}} (V_i^{\alpha} - V_i^{s\alpha}) \\ & - \sum_{i=1}^{n-1} \sum_{j=i+1}^{n} B_{ij} (V_i V_j \cos\theta_{ij} - V_i^s V_j^s \cos\theta_{ij}^s) \end{aligned} \tag{14.3.15}$$

where $\theta_{ij} = \theta_i - \theta_j$.

The sensitivity S of the energy function $\nu(x)$ with respect to clearing time ($\lambda = t_{cl}$) is obtained by taking partial derivatives of (14.3.15) with respect to t_{cl} as

$$\begin{aligned} S \;=\; & \frac{\partial \nu}{\partial t_{cl}} = \sum_{i=1}^{m} M_i \tilde{\omega}_{g_i} \frac{\partial \tilde{\omega}_{g_i}}{\partial t_{cl}} - \sum_{i=1}^{m} P_{M_i} \frac{\partial \theta_{n_0+i}}{\partial t_{cl}} + \sum_{i=1}^{n_0} P_{d_i} \frac{\partial \theta_i}{\partial t_{cl}} - \sum_{i=1}^{n_0} B_{ii} V_i \frac{\partial V_i}{\partial t_{cl}} \\ & + \sum_{i=1}^{n_0} Q_{d_i}^s \frac{V_i^{\alpha-1}}{V_i^{s\alpha}} \frac{\partial V_i}{\partial t_{cl}} - \sum_{i=1}^{n-1} \sum_{j=i+1}^{n} B_{ij} \Bigg(V_j \cos\theta_{ij} \frac{\partial V_i}{\partial t_{cl}} \\ & + V_i \cos\theta_{ij} \frac{\partial V_j}{\partial t_{cl}} - V_i V_j \sin\theta_{ij} \frac{\partial \theta_{ij}}{\partial t_{cl}} \Bigg). \end{aligned} \tag{14.3.16}$$

The partial derivatives of $\tilde{\omega}_{g_i}$, θ, and V with respect to t_{cl} are the sensitivities obtained by integrating the dynamic system and the sensitivity system as discussed earlier.

The sensitivity $S = \frac{\partial \nu}{\partial t_{cl}}$ is computed for two different values of t_{cl}, which are chosen to be less than t_{cr}. Since we are computing only first-order trajectory sensitivities, the two values of t_{cl} must be less than t_{cr} by at most 20%. This might appear to be a limitation of the method. However, extensive experience with the system generally will give us a good estimate of t_{cr}. Because the system under consideration is stable, the sensitivity S will display larger excursions for larger t_{cl} [9]. Since sensitivities generally increase rapidly with increases in t_{cl}, we plot the reciprocal of

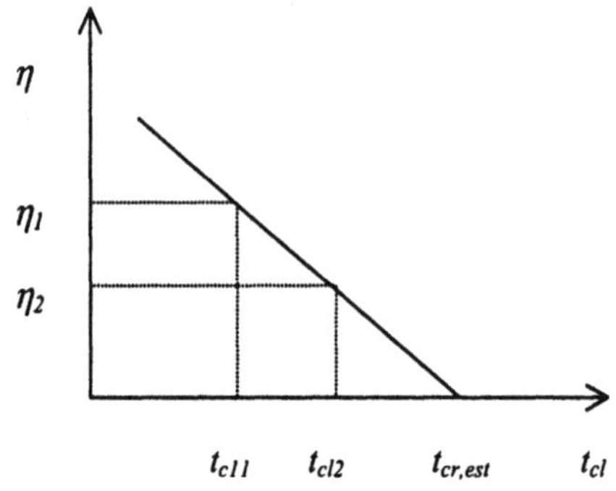

Figure 14.3.1: Estimate of t_{cr}

the maximum deviation of S over the post fault period as $\eta = \frac{1}{\max(S)-\min(S)}$. A straight line is then constructed through the two points (t_{cl1}, η_1) and (t_{cl2}, η_2). The estimated critical clearing time $t_{cr,est}$ is the intersection of the constructed straight line with the time-axis in the (t_{cl}, η)-plane as shown in Figure 14.3.1. As discussed later, this linearity is valid for a small region around t_{cr}.

14.3.3 Computation of critical clearing time [22]

In this section we outline an approach using trajectory sensitivity information directly instead of via the energy function to estimate the critical clearing time. To motivate this approach, let us consider a single machine infinite bus (SMIB) system described by

$$\begin{array}{l} M\ddot{\delta} + D\dot{\delta} = P_M, \ \ 0 < t \leq t_{cl} \\ M\ddot{\delta} + D\dot{\delta} = P_M - P_{em}\sin\delta, \ \ t > t_{cl} \end{array} \qquad \delta(0) = \delta_0, \ \ \dot{\delta}(0) = 0. \tag{14.3.17}$$

The corresponding sensitivity equations are

$$\begin{array}{l} M\ddot{u} + D\dot{u} = 0, \ \ 0 < t \leq t_{cl} \\ M\ddot{u} + D\dot{u} = (-P_{em}\cos\delta)u, \ \ t > t_{cl} \end{array} \qquad u(0) = 0, \ \ \dot{u}(0) = 0 \tag{14.3.18}$$

where $u = \frac{\partial \delta}{\partial t_{cl}}$.

If we plot the phase plane portrait of the system for two values of t_{cl}, one small and the other close to t_{cr}, and monitor the behavior of sensitivities in the $(u, \dot{u})$-plane, we observe that the sensitivity magnitudes increase much more rapidly as t_{cl} approaches t_{cr}. Also, the trajectories in the $(u, \dot{u})$-plane can cross each other since the system (14.3.18) is time varying, whereas that is not the case for the system (14.3.17), which is an autonomous system. Qualitatively, both trajectories in the (δ, ω)-plane and the $(u, \dot{u})$-plane give the same information about the stability of the system, but the sensitivities seem to be stronger indicators because of their rapid changes in magnitude as t_{cl} increases. Hence, we can associate sensitivity information with the stability level of the system for a particular clearing time. When the system is very close to instability, the sensitivity reflects this situation much more quickly. This qualitative relationship has been discussed for the general nonlinear dynamic systems by Tomovic [15]. One possible measure of proximity to instability may be through some norm of the sensitivity vector. The Euclidean norm is one such possibility. For the single machine system, if we plot the norm $\sqrt{u^2 + \dot{u}^2}$ as a function of time for different values of t_{cl}, one can get a quick idea about the system stability as shown in Figures 14.3.2 and 14.3.3. For a stable system, although the sensitivity norm tends to become a small value eventually, it transiently assumes a very high value when t_{cl} is close to t_{cr}.

Thus we associate with each value of t_{cl} the maximum value of the sensitivity norm. The procedure to calculate the estimated value of t_{cr} is the same as described in the previous section but using the sensitivity norm instead of the energy function sensitivity. Here, the sensitivity norm for an m-machine system is defined as

$$S_N = \sqrt{\sum_{i=1}^{m} \left(\left(\frac{\partial \delta_i}{\partial t_{cl}} - \frac{\partial \delta_j}{\partial t_{cl}} \right)^2 + \left(\frac{\partial \omega_i}{\partial t_{cl}} \right)^2 \right)}$$

where the jth machine is chosen as the reference machine.

The norm is calculated for two values of $t_{cl} < t_{cr}$. For each t_{cl}, the reciprocal η of the maximum of the norm is calculated. A line through these two values of η is then extrapolated to obtain the estimated value of t_{cr}. If other parameters of interest are chosen instead, the technique will give an estimate of critical values of those parameters.

Since this technique does not require computation of the energy function, it can be applied to power systems without any restriction on system modeling. This is a major advantage of this technique.

14.3.4 Numerical examples

We consider three systems and applications of both energy function-based and direct sensitivity-based metrics. These are the 3-machine, 9-bus [23, 24]; the 10-machine, 39-bus [3]; and the 50-machine, 145-bus [4] systems. Results are presented in Tables 14.3.1 and 14.3.2 and Figure 14.3.4.

For the 50-machine system, the estimated value of clearing time for a self-clearing fault at bus 58 using the sensitivity norm technique is computed. The corresponding values of η for different values of t_{cl} are shown in Figure 14.3.4. We note that if the two values of t_{cl} are chosen in the close range of t_{cr} = 0.315 s, the estimated value of t_{cr} will be quite accurate. On the other hand, picking arbitrary values of t_{cl} may give erroneous results. Since computing sensitivities is computationally extensive, choosing good values of t_{cl} requires judgment and experience.

14.3.5 Computation of critical loading of generator

Next, the sensitivity norm technique is used to estimate the critical value of generator loading, or equivalently, the mechanical input power P_M. Two simulations for

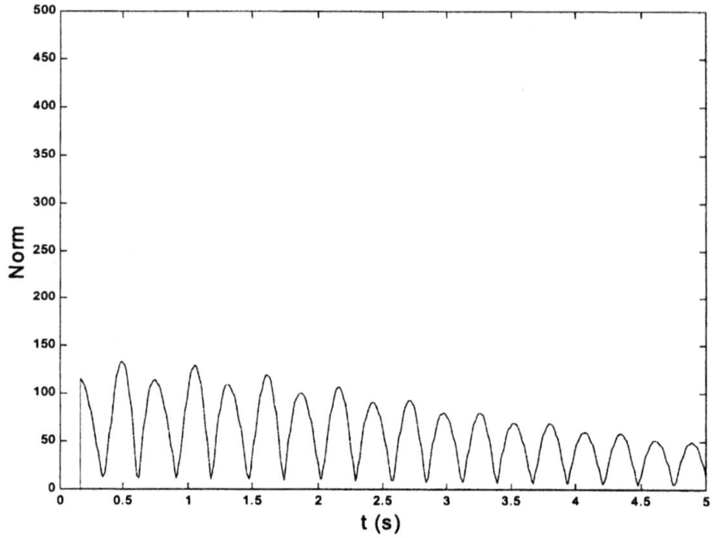

Figure 14.3.2: Sensitivity norm for small t_{cl} (≈50% of t_{cr})

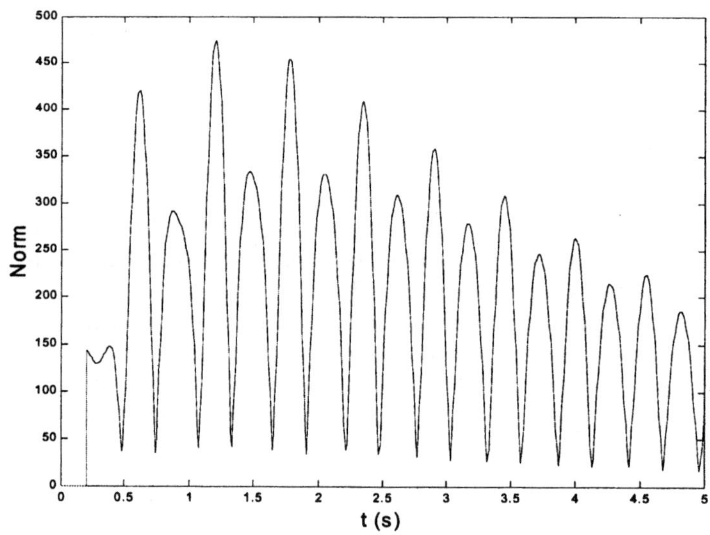

Figure 14.3.3: Sensitivity norm for t_{cl} close to t_{cr} (≈80% of t_{cr})

Table 14.3.1: Results for the 3-machine system

Faulted Bus	TEF Sensitivity	Sensitivity Norm	Actual
	$t_{cr,est}$ (s)	$t_{cr,est}$ (s)	t_{cr} (s)
5	0.354	0.352	0.352
8	0.333	0.333	0.334

Table 14.3.2: Results for the 10-machine system

Faulted bus	Line Tripped	Sensitivity Norm	Actual
		$t_{cr,est}$ (s)	t_{cr} (s)
4	4–5	0.210	0.212
15	15–16	0.204	0.206
17	17–18	0.169	0.168
21	21–22	0.122	0.125

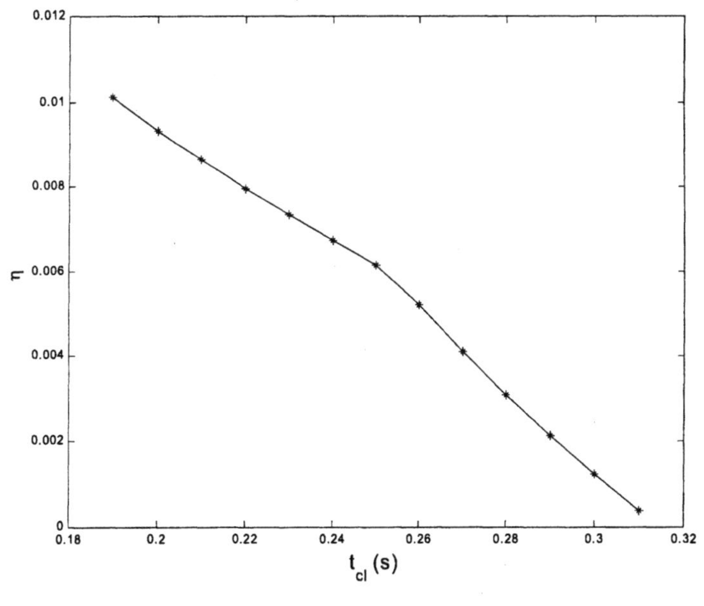

Figure 14.3.4: Estimate t_{cr} for fault at bus 58 using norm sensitivity

two values of P_M are carried out. The change from normal operating values in P_M is distributed uniformly among all loads in the system, so that the loading of the rest of the generators is unchanged. The sensitivity norm is calculated for the two specified values of P_M and then extrapolated to obtain the estimated value of the critical P_M for the chosen generator.

14.3.6 The 10-machine system

A fault is simulated in the system at bus 21 of the 10-machine system and cleared at $t_{cl} = 0.1$ s by tripping the line 21–22. The estimated results for a few generators are shown in Table 14.3.3.

Table 14.3.3: Estimated value of critical input power P_M vs. the actual value

Machine Number	Sensitivity Norm $P_{Mcr,est}$ (pu)	Actual $P_{M,cr}$ (pu)
3	10.7	10.4
5	6.3	6.4
8	12.4	12.2

14.3.7 The 50-machine system

A self-clearing fault is simulated at bus 58 and cleared at $t_{cl} = 0.15$ s. Applying the proposed technique, the results obtained for critical value of P_M is as shown in Table 14.3.4. To validate the results it was verified that with the critical value of P_M the system becomes unstable.

Table 14.3.4: Estimated value of critical input power P_M vs. the actual value

Machine Number	Sensitivity Norm $P_{M,est}$ (pu)	Actual $P_{M,act}$ (pu)
4	22.9	22.3
5	17.0	16.5
7	4.3	4.2
12	10.0	9.6

14.3.8 Computation of critical impedance of a transmission line

The norm sensitivity technique is used to estimate the critical value of a line reactance. The 3-machine system is used to illustrate the technique. A fault is simulated at bus 7 and cleared at $t_{cl} = 0.08$ s by tripping the line 5–7. Figure 14.3.5 shows the corresponding values of η for different values of reactance of the line 8–9. The critical value of the reactance of line 8–9 is 0.246 pu. It can be seen from Figure 14.3.5

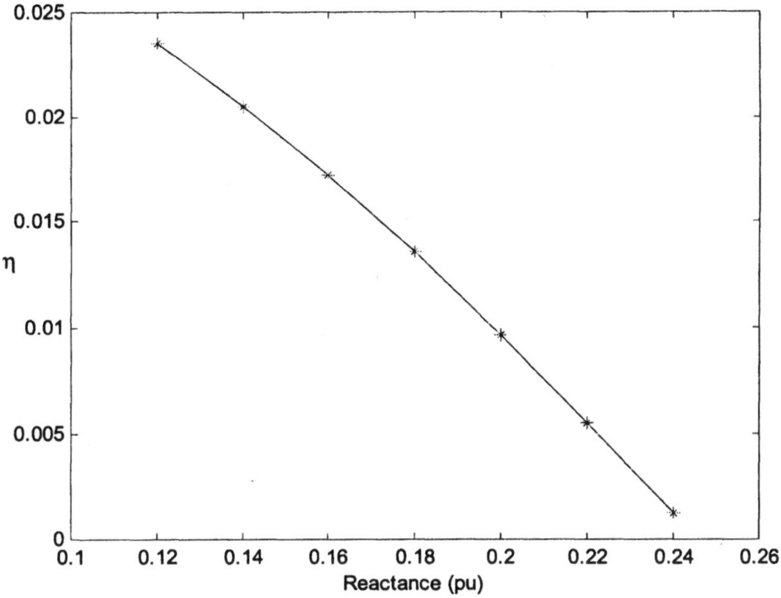

Figure 14.3.5: Estimate critical value of the reactance of the line 8–9

that the estimated value of the critical reactance is quite accurate if the two values of the line reactance are picked in the close range of the actual critical value.

Knowing the critical value of a line reactance is very important in controlling power flow path in the system by the variable impedance devices. Such devices belong to a type called flexible ac transmission systems (FACTS) that can be very useful in controlling the stability of power systems [25].

14.4 Stability-Constrained Optimal Power Flow Formulation [10]

While [26] formulates the problem in a different way, here we use the relative rotor angles to detect the system stability/instability of the power system. To check the stability of the system for a credible contingency, relative rotor angles are monitored at each time step during dynamic simulation. The sensitivities are also computed at the same time. Although sensitivity computation requires extensive computational effort, an efficient method to compute sensitivities is available by making effective use of the Jacobian common to both the system and sensitivity equations [27]. This will reduce the computational burden considerably. We propose that when the relative rotor angle $\delta_{ij} = \delta_i - \delta_j > \pi$ for a given contingency, the system is considered unstable. This is an extreme case as pointed out in [6], and one can choose an angle difference less than π depending on the system. Here i and j refer to the most and least advanced generators, respectively. The sensitivities of the rotor angles at this instant are used to compute the amount of power needed to be shifted from the most advanced generator (generator i) to the least advanced one (generator j) according to the formulae

$$\Delta P_{i,j} = \left. \frac{\delta_{ij} - \delta_{ij}^0}{\frac{\partial \delta_{ij}}{\partial P_i}} \right|_{\max \delta_{ij} = \pi} \tag{14.4.1}$$

$$P_i^{new} = P_i^0 - \Delta P_{i,j} \text{ and } P_i^{new} = P_j^0 + \Delta P_{i,j} \tag{14.4.2}$$

where P_i^0, P_j^0, and δ_{ij}^0 are the base loading of generators i and j, and the relative rotor angle of the two, respectively, at the solution of the optimal power flow (OPF) problem. $\frac{\partial \delta_{ij}}{\partial P_i}$ is the sensitivity of relative rotor angle with respect to the output of the ith generator; P_i in this case is the parameter λ in equations stated in Section 14.3.

After shifting the power from generator i to generator j according to (14.4.2), the system is secure for that contingency but it is not an optimal schedule. We can improve the optimality by introducing new power constraints on generation. The OPF problem with new constraints is then resolved to obtain the new operating point for the system. The detailed algorithm is discussed in [10].

14.5 Transmission Protection System Vulnerability to Angle Stability

Power system protection at the transmission system level is based on distance relaying. Distance relaying serves the dual purpose of apparatus protection and system protection. Significant power flow oscillations can occur on a transmission line or a network due to major disturbances such as faults, subsequent clearing, and load rejection. They are related to the swings in the rotor angles of synchronous generators. If the rotor angles settle down to a new stable equilibrium point, the disturbance is classified as stable. Otherwise, it is unstable. Hence, for stable disturbances, power swings die down with time.

In this section, we discuss the application of TSA to detect vulnerable relays in the system. For each contingency we compute a quantity called "branch impedance sensitivity," which will be used to identify weak links in the system. Its main advantage is that it can handle systems of any degree of complexity in terms of modeling and can be used as an on-line DSA tool.

Distance relays are used to detect swings and take appropriate action (tripping or blocking) depending on the nature of swing (stable or unstable). The reason is that a change in transmission line power flow translates into a corresponding change in the impedance seen by the relay. Relay operation after fault clearing depends on the power swing and proximity of the relay to an electrical center.

The apparent impedance seen by a relay on a transmission line connecting nodes i and j having flow $P_{ij} + jQ_{ij}$ is given by

$$Z_{app} = \left[\frac{P_{ij}}{P_{ij}^2 + Q_{ij}^2} + j \frac{Q_{ij}}{P_{ij}^2 + Q_{ij}^2} \right] |V_i|^2 . \tag{14.5.1}$$

As $|V_i|$ is only a scalar, it cannot differentiate between the quadrants in the R-X plane. Thus, the location of Z_{app} depends on the direction of P and Q flows. Clearly, swings are severe when P_{ij} and/or Q_{ij} are large and V_i small. Under such circumstance Z_{app} is small, and hence can cause a relay to trip.

Work based upon Lyapunov stability criterion has been reported in [28] to rank relays according to the severity of swings. In [29] relay margin is used as a measure of how close a relay is from issuing a trip command. Basically, it is the ratio of the time of longest consecutive stay of a swing in zone to its time dial setting. For relays that see swing characteristics outside of their zone settings, the relay space margin is used. It is defined as the smallest distance between the relay characteristic and the swing trajectory in R-X plane. To identify the most vulnerable relay, the magnitude of the ratio of swing impedance to line impedance is used as a performance parameter. The most vulnerable relay corresponds to one with minimum ratio, where the search space extends over all the relays and time instants of simulation. Reference [30] discusses the challenges of relaying in the restructured power system operation scenario. In such a scenario, there will be varying power flow patterns dictated by the market conditions. During the congestion period, relay margins and relay space margins will be reduced. This may pose challenges to system protection design.

The vulnerability of a relay to power swings is directly dependent on the severity of oscillations in power flow observed in the primary transmission line and adjacent transmission lines covered by backup zones. Hence, the problem of assessment of transmission protection system vulnerability to power swings translates into assessment of oscillations in power flow on a transmission system due to disturbance.

We will show that branch impedance trajectory sensitivity (BITS) (that is, trajectory sensitivity of rotor angles to branch impedance) can be used to locate the electrical center [9] in the transmission system and rank transmission system distance relays according to their vulnerability to tripping on swings.

14.5.1 Electrical center and weakest link in a network

For power systems that essentially behave as a two-area system under instability, the out-of-step relaying can be explained by considering the equivalent generators connected by a tie line. When the two generators fall out of step, i.e., the angular difference between the two generator voltage phasors is 180°, they create a voltage zero point on the connecting circuit. This is known as the electrical center. A distance relay perceives it as a solid 3-phase short-circuit and trips the line.

One way to locate the electrical center [31] in a power system is to create a fault with fault clearing time greater than the critical clearing time of the circuit breaker to make the resulting post fault system unstable. Through transient stability simulations, one identifies the groups of accelerating and decelerating machines in the

Table 14.5.1: Normalized BITS norm (nominal or 100% loading)

Line	Normalized Sensitivity	RSM	$\lvert Z_{min} \rvert$
29–26	1	0.0231	0.0658
26–27	0.1630	0.1759	0.1841
26–25	0.1511	0.1870	0.2578
26–28	0.0111	0.4842	0.5077

system. The subnetwork interconnecting such groups will contain the electrical center. For the relays contained in the subnetwork, by simulating the power swing on the R-X plane one can locate the relays for which the power swing intersects the transmission line impedance. This line contains the electrical center of the system. The electrical center will also be observed by the backup relays depending upon their zone 2 and 3 setting. Usually the relays near the electrical center are highly sensitive to power swings. A well-known property to characterize an electrical center is that the network adjoining the electrical center has low voltages. Such characterization is qualitative and can be used as a screening tool. Since natural splitting of the system due to operation of distance relays takes place at the electrical center, we refer to such a line (or a transformer) as the "weakest link" in the network.

To compress the information of all rotor angles with respect to a given line, for an m-machine system the following norm first introduced in [22] is used in the numerical examples section. The BITS norm is computed as follows:

$$S_N = \sqrt{\sum_{i=1}^{m}\left(\frac{\partial \alpha_i}{\partial x}\right)^2 + \sum_{i=1}^{m}\left(\frac{\partial \omega_i}{\partial x}\right)^2} \tag{14.5.2}$$

where $\alpha_i = \delta_i - \delta_j$ and the jth machine is chosen as the reference, and x is the transmission line reactance.

14.5.2 Numerical examples

A 10-machine, 39-bus system [3] is used for illustrative examples. The following two cases are considered.

Case 1: Fault at bus 28

The fault is cleared at t_{cl} = 0.06 s by tripping line 28–29 simultaneously from both ends. Table 14.5.1 captures the normalized indices for a subset of lines, which have high, medium, and low sensitivities. BITS sensitivity norm is computed for all the lines and normalized with respect to the maximum one. Therefore, after normalization, the line with max BITS norm has value 1. For all other lines, BITS norm is less than or equal to 1. We would characterize line 29–26 as the most vulnerable line because its rank is 1, and its absolute peak norm of 30 822 is also very high. The normalized indices for all other lines are much lower than that of line 29–26. Hence,

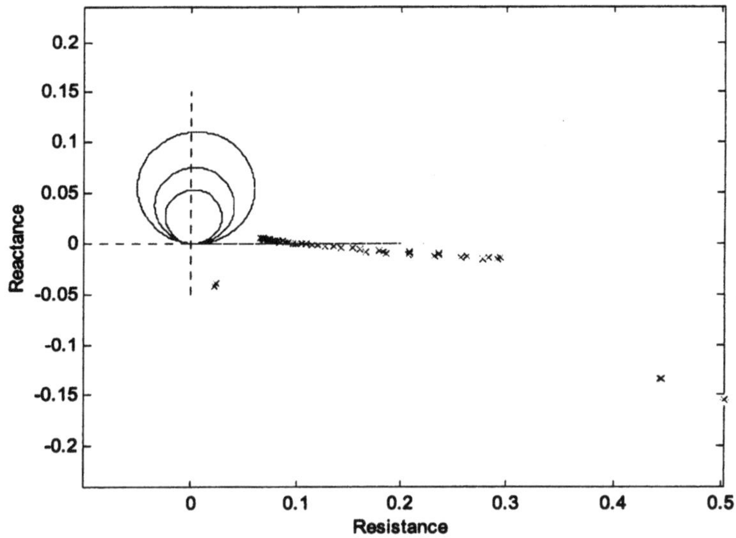

Figure 14.5.1: Swing curve for relay on line 29–26 (stable)

other lines are less vulnerable to the swing. These inferences have been confirmed by using the relay space margin (RSM) concept. Figures 14.5.1 and 14.5.2 show the swing curves for relay on lines 29–26 and 26–27, respectively.

As line 29–26 is the weakest link in the post fault system, it has a high possibility of developing the electrical center in case a fault leads to instability (because of larger t_{cl} or system loadings). To confirm the location of the electrical center, t_{cl} was increased from 0.06 s to 0.07 s to create system instability. The small increase of 0.01 s in fault clearing time induced instability. This indicates that the existing system is working close to its stability limits.

The swing trajectory for the relay on line 29–26 located near bus 29 for t_{cl} = 0.07 s is shown in Figure 14.5.3. As the trajectory cuts the transmission line characteristic in zone 1, the location of electrical center on line 29–26 is thus established. This is clearly an unstable case. The relative rotor angles in this case are show in Figure 14.5.4 where machine 9 is the unstable machine.

Table 14.5.2 summarizes actual BITS norms as well as the results for various loading conditions (80%, 90%, and 100% of the system load). It can be seen that as system loading increases, the trajectory sensitivity also increases. As the system approaches a dynamical stability limit, the trajectory sensitivity for the critical line jumps from 145.17 for the 90% loading case to 30 822 for the 100% loading case. Thus, a high peak value of maximum sensitivity can be used as an indicator of reduced stability margins. Also, note that the sensitivity norm changes rapidly when the system loading approaches the stability boundary. Therefore, it can be used as an indicator for system stability margin.

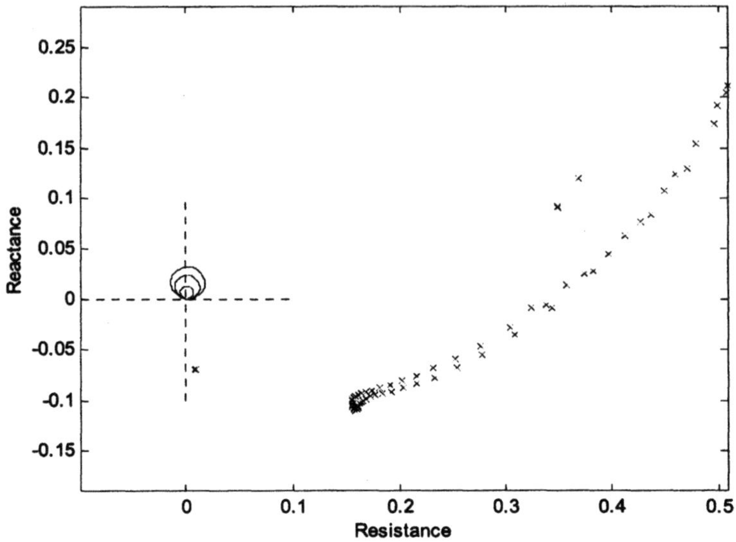

Figure 14.5.2: Swing curve for relay on line 26–27 (stable)

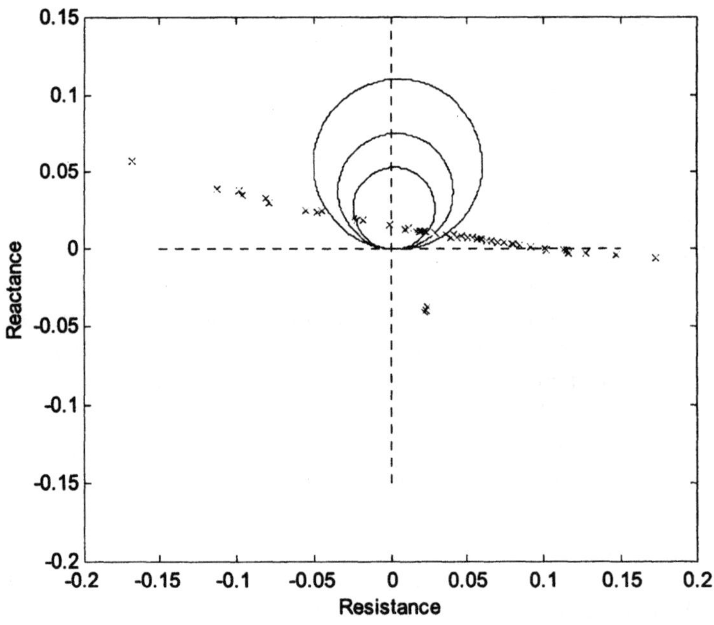

Figure 14.5.3: Electrical center location on line 29–26 (unstable)

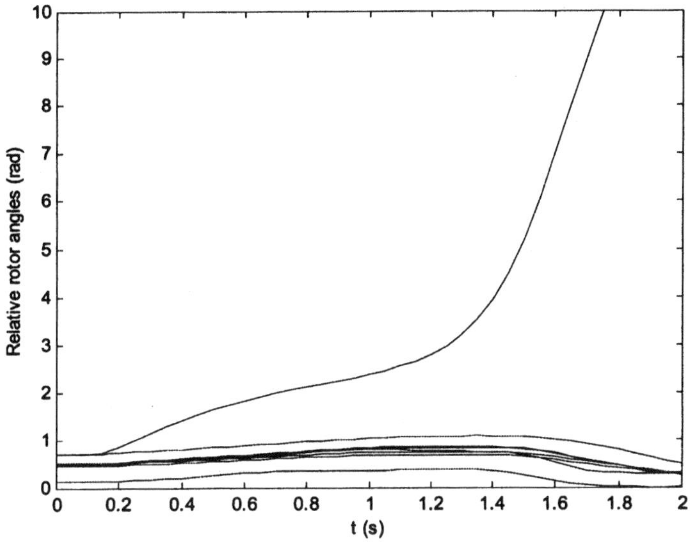

Figure 14.5.4: Relative rotor angles for the unstable case

Table 14.5.2: Absolute norm for different loading conditions

Line	80% load	90% load	100% load
29–26	90.2730	145.1799	30822
26–27	25.6782	31.7447	5022
26–25	9.4081	20.0233	4656
26–28	5.8828	6.8147	343.09

Table 14.5.3: Normalized BITS norm with 100 % load

Line	Normalized Sensitivity	RSM	$\lvert Z_{min} \rvert$
1–2	1	0.1600	0.1900
1–39	0.9680	0.1694	0.2036
2–25	0.6196	0.2337	0.2036
14–15	0.5409	0.3201	0.3310

Case 2: Fault at bus 4

The fault is cleared at $t_{cl} = 0.1$ s by tripping line 4–5 simultaneously from both ends. For this scenario, results similar to the case discussed earlier are summarized in Tables 14.5.3 and 14.6.1. Electrical center is located on line 1–2. The results replicate the pattern of behavior discussed in the previous case. From Table 14.6.1 it can be seen that peak sensitivity norm is quite low (60 in comparison to 30 822 of Table 14.5.2), and the sensitivity norm does not change much with various loading conditions. It indicates that for this fault, the loading conditions are such that the system is still far away from the stability boundary. In fact, the critical clearing time for this fault with the nominal loading condition is 0.29 s.

This section has developed the concept of maximum rotor angle BITS as a tool for assessing transmission protection system vulnerability to angle stability problems. It is used for applications in power systems such as

(a) Ranking transmission lines and relays as vulnerability to swings. The ranking can help to improve the relay coordination in the presence of power swings.

(b) Determining location of electrical center. Because transmission lines adjoining the electrical center have low voltages, other devices such as FACTS could be used to strengthen system stability.

(c) Indicating stability margin. When BITS is getting higher, it indicates that the operating point of the system is moving closer to the system stability boundary.

14.6 Conclusions

This chapter summarizes a recent approach taken for DSA of power systems using TSA. It is independent of model complexity and requires some a priori knowledge of the system criticality. In power systems, years of experience provide this. Future research will involve applying this technique to relieve congestion in deregulated systems. Future research should extend the application to exploring the connection between stability and sensitivity theory in hybrid dynamical systems, an emerging area of research in control [32]. In power systems, examples include state-dependent variation of taps in tap changing transformers or switching of FACTS devices for which the theory of trajectory sensitivity analysis is already available [21].

Table 14.6.1: Absolute BITS norm for various loading conditions

Line	90% load	100% load	110% load
1–2	53.7342	57.4645	60.6352
1–39	51.1508	55.6236	58.0612
2–25	29.1845	35.6055	53.0692
14–15	26.5545	31.0798	36.3755

Bibliography

[1] G. L. Wilson and P. Zarakas, "Anatomy of a blackout," *IEEE Spectrum*, vol. 15, no. 2, pp. 38–46, Feb. 1978.

[2] O. Alsac and B. Stott, "Optimal load flow with steady state security," *IEEE Trans. Power Apparatus and Systems*, vol. PAS-93, pp. 745–754, 1974.

[3] M. A. Pai, *Energy Function Analysis for Power System Stability*, Kluwer, Norwell, MA, 1989.

[4] A. A. Fouad and V. Vittal, *Power System Transient Stability Analysis Using the Transient Energy Function Method*, Prentice-Hall, Englewood Cliffs, NJ, 1991.

[5] H. D. Chiang, C. C. Chu, and G. Cauley, "Direct stability analysis of electric power system using energy functions: Theory, applications, and perspective," *Proc. IEEE*, vol. 83, no. 11, pp. 1497–1529, Nov. 1995.

[6] M. Pavella, D. Ernst, and D. Ruiz-Vega, *Transient Stability of Power Systems: A Unified Approach to Assessment and Control*, Kluwer, Norwell, MA, 2000.

[7] K. W. Chan, A. R. Edwards, R. W. Dunn, and A. R. Daniels, "On-line dynamic security contingency screening using artificial neural networks," *IEE Proc. Generation, Transmission, and Distribution*, vol. 147, no. 6, pp. 367–372, Nov. 2000.

[8] Y. Mansour, E. Vaahedi, and M. A. El-Sharkawi, "Dynamic security contingency screening and ranking using neural networks," *IEEE Trans. Neural Networks*, vol. 8, no. 4, pp. 942–950, July 1997.

[9] T. B. Nguyen, *Dynamic Security Assessment of Power Systems using Trajectory Sensitivity Approach*, Ph.D. Dissertation, University of Illinois, Urbana-Champaign, 2002.

[10] T. B. Nguyen and M. A. Pai, "Dynamic security-constrained rescheduling of power systems using trajectory sensitivities," *IEEE Trans. Power Systems*, to appear.

[11] H. You, V. Vittal, and Z. Yang, "Self-healing in power systems: An approach using islanding and rate of frequency decline based load shedding," *IEEE Trans. Power Systems*, to appear.

[12] P. M. Frank, *Introduction to System Sensitivity Theory*, Academic Press, New York, 1978.

[13] M. Eslami, *Theory of Sensitivity in Dynamic Systems*, Springer-Verlag, New York, 1994.

[14] H. W. Bode, *Network Analysis and Feedback Amplifier Design*, Van Nostrand, New York, 1945.

[15] R. Tomovic, *Sensitivity Analysis of Dynamic Systems*, McGraw-Hill, New York, 1963.

[16] R. Tomovic and M. Vukobratovic, *General Sensitivity Theory*, Elsevier, New York, 1972.

[17] J. B. Cruz, Jr., Ed., *System Sensitivity Analysis*, Dowden, Hutchingson and Ross, Stroudsburg, PA, 1973.

[18] G. C. Verghese, I. J. Perez-Arriaga, and F. C. Scheweppe, "Selective modal analysis with applications to electric power systems, Part I and II," *IEEE Trans. Power Apparatus and Systems*, PAS-101, pp. 3117–3134, Sept. 1982.

[19] M. J. Laufenberg and M. A. Pai, "A new approach to dynamic security assessment using trajectory sensitivities," *IEEE Trans. Power Systems*, vol. 13, no. 3, pp. 953–958, Aug. 1998.

[20] I. A. Hiskens and M. Akke, "Analysis of the Nordel power grid disturbance of January 1, 1997 using trajectory sensitivities," *IEEE Trans. Power Systems*, vol. 14, no. 3, pp. 987–994, Aug. 1999.

[21] I. A. Hiskens and M. A. Pai, "Trajectory sensitivity analysis of hybrid systems," *IEEE Trans. Circuits and Systems Part I: Fundamental Theory and Applications*, vol. 47, no. 2, pp. 204–220, Feb. 2000.

[22] T. B. Nguyen, M. A. Pai, and I. A. Hiskens, "Sensitivity approaches for direct computation of critical parameters in a power system," *Int. J. Electrical Power and Energy Systems*, vol. 24, no. 5, pp. 337–343, 2002.

[23] P. W. Sauer and M. A. Pai, *Power System Dynamics and Stability*, Prentice-Hall, Upper Saddle River, NJ, 1998.

[24] K. R. Padiyar, *Power System Dynamics Stability and Control*, BS Publications, Hyderabad, India, 2002.

[25] N. G. Hingorani and L. Gyugyi, *Understanding FACTS Concepts and Technology of Flexible AC Transmission Systems*, IEEE Press, New York, 2000.

[26] D. Gan, R. J. Thomas, and R. Zimmerman, "Stability-constrained optimal power flow," *IEEE Trans. Power Systems*, vol. 15, pp. 535–540, May 2000.

[27] D. Chaniotis, M. A. Pai, and I. A. Hiskens, "Sensitivity analysis of differential-algebraic systems using the GMRES method–Application to power systems," *Proc. IEEE International Symposium on Circuits and Systems*, pp.117–120, Sydney, Australia, May 2001.

[28] C. Singh and I. A. Hiskens, "Direct assessment of protection operation and non-viable transients," *IEEE Trans. Power Systems*, vol. 16, no. 3, pp. 427–434, Aug. 2001.

[29] F. Dobraca, M. A. Pai, and P. W. Sauer, "Relay margins as a tool for dynamical security analysis," *Int. J. Electrical Power and Energy Systems*, vol. 12, no. 4, pp. 226–234, Oct. 1990.

[30] J. S. Thorp and A. G. Phadke, "Protecting power systems in the post-restructuring era," *IEEE Computer Applications in Power*, vol. 12, no. 1, pp. 33–37, Jan. 1999.

[31] P. Kundur, *Power System Stability and Control*, McGraw-Hill, New York, 1994.

[32] A. N. Michel and B. Hu, "Toward a stability theory of general hybrid dynamical systems," *Automatica*, vol. 35, pp. 371–384, 1999.

Chapter 15

Emergency Control and Special Protection Systems in Large Electric Power Systems*

Vijay Vittal

Abstract: This chapter addresses the topic of designing a corrective control strategy after large disturbances. When a power system is subjected to large disturbances such as simultaneous loss of several generating units or major transmission lines, and the vulnerability analysis indicates that the system is approaching a catastrophic failure, control actions need to be taken to limit the extent of the disturbance. In our approach, the system is separated into smaller islands at a slightly reduced capacity. The basis for forming the islands is to minimize the generation-load imbalance in each island, thereby facilitating the restoration process. An analytical approach to forming the islands using a two-time-scale procedure is developed. Then by exploring a carefully designed load-shedding scheme based on the rate of frequency decline, we limit the extent of the disruption and are able to restore the system rapidly. We refer to this corrective control scheme as "controlled islanding" followed by load-shedding based on the rate of frequency decline. A new class of controls called special protection systems is now finding greater acceptance in power systems with their enhanced need to provide reliability in a competitive restructured utility environment. These systems are triggered only when large disturbances occur. They provide a degree of safety in preventing cascading outages. The special protection systems be presented and their potential for greater use will be examined.

*This research is sponsored by U. S. Department of Defense and Electric Power Research Institute through the Complex Interactive Networks/Systems Initiative, WO 8333-01.

15.1 Introduction

Power systems are being operated closer to the stability limit nowadays as deregulation introduces many more economic objectives for operation. As open access transactions increase, weak connections, unexpected events, hidden failures in protection system, human errors, and other reasons may cause the system to lose balance and even lead to catastrophic failures. As a result several innovative emergency control procedures and special protection systems are being introduced to maintain the reliability of the system. The control approaches fall under the category of corrective control and are initiated after the occurrence of the disturbance.

This chapter addresses the topic of designing a corrective control strategy after large disturbances. When a power system is subjected to large disturbances such as simultaneous loss of several generating units or major transmission lines, and the vulnerability analysis indicates that the system is approaching a catastrophic failure, control actions need to be taken to limit the extent of the disturbance.

In our approach, the system is separated into smaller islands at a slightly reduced capacity. The basis for forming the islands is to minimize the generation-load imbalance in each island, thereby facilitating the restoration process. An analytical approach to forming the islands using a two-time-scale procedure is developed. Then by exploring a carefully designed load-shedding scheme based on the rate of frequency decline, we limit the extent of the disruption and are able to restore the system rapidly. We refer to this corrective control scheme as "controlled islanding" followed by load-shedding based on the rate of frequency decline.

Subsumption architecture [1], which is used in the field of controlled robots, is adopted here to identify the hierarchies of the various controls, protection, and communication systems between various agents in the deregulated electric utility environment. The architecture is based on the premise that storing models of the world is dangerous in dynamic and unpredictable environments because representations may be incorrect or outdated. It defines layers of finite state machines (FSMs) that are augmented with timers. Sensors feed information into FSMs at all levels. The FSMs of the lowest level are control actuators. The FSMs of the higher levels may inhibit (attenuate the signal of one output wire) or suppress (attenuate the signal on all output wires) output values of the FSMs on the layers below them. In this way, a hierarchy of progressively refined behaviors may be established.

Agents designed using the subsumption architecture do not use symbol manipulation in a fixed manner to represent processing. They also have no global knowledge and are generally decentralized. The agents are nonprogrammable single-purpose devices because of their lack of symbolism and global knowledge. However, they have the advantage of rapid response for dealing with dynamic and unpredictable events.

A load-shedding scheme based on the subsumption model is designed with consideration of certain criteria. The proposed scheme is tested on a 179-bus 20-generator test system and shows very good performance.

15.2 Controlled Islanding

We employ a two-time-scale method to determine the groups of the generators with slow coherency. This method considers the structural characteristics of the power system to determine the interactions of the various generators and find the strong and weak couplings. The method is implemented by running the Dynamic Reduction Program 5.0 (DYNRED) software obtained from the EPRI software center. Through the selection of the two-time-scale option, the coherent groups of generators can be obtained on any power system. We also develop an automatic islanding program to fully support the application of the theory.

15.2.1 Slow coherency

In the controlled islanding self-healing approach, determinating the islands for a given operating condition is the critical step. A reasonable approach to islanding can result in significant benefit to the corrective control actions that follow the islanding procedure. In determining the islands, the inherent structural characteristics of the system should be considered. In addition, the choice of these islands should not be disturbance dependent. The two-time-scale method we employed is an application of the singular perturbation method in power systems [2–4]. The method assumes the state variables of an nth-order system are divided into r slow states y, and $(n-r)$ fast states z, in which the r slowest states represent r groups with the slow coherency. The user provides an estimate for the number of groups. However, the automatic islanding program takes into account the mismatch between generation and load and availability of the tie lines to form islands and appropriately combines groups when islands cannot be formed. Both the linearized and nonlinear power system models can be used to apply the two-time-scale method. In the linearized model, we start from the basic classical second-order electromechanical model of an n-machines power system [5]:

$$\dot{\delta}_i = \Omega(\omega_i - 1) \tag{15.2.1}$$

$$2H_i\dot{\omega}_i = -D_i(\omega_i - 1) + (P_{mi} - P_{ei}), \quad i = 1, 2, \cdots, n \tag{15.2.2}$$

where δ_i is the rotor angle of machine i in radians, ω_i is the speed of machine i in per unit (pu), P_{mi} is the mechanical input power of machine i in pu, P_{ei} is the electrical output power of machine i in pu, H_i is the inertia constant of machine i in seconds, D_i is the damping constant of machine i in pu, and Ω is the base frequency in radians per second.

If we neglect damping and line conductance and we linearize the system dynamic equation around an equilibrium point $(\delta^*, 1)^T$, we obtain

$$\ddot{X} = -(1/2)\Omega H^{-1}KX = AX \tag{15.2.3}$$

$$x_i = \Delta\delta_i, \quad i = 1, 2, \ldots, n \tag{15.2.4}$$

$$H = \text{diag}(H_1, H_2, \cdots, H_n) \tag{15.2.5}$$

$$K = (k_{ij}) = (V_i V_j B_{ij} \cos(\delta_i - \delta_j)|_{\delta^*}), \;\; j \neq i \tag{15.2.6}$$

where V_i is the voltage of bus i (per unit), and B_{ij} is the pre-fault network's susceptance between bus i and bus j.

The states x_i and x_j of a system of (15.2.3) are slow coherent if and only if they are coherent with respect to a set of r slowest modes σ_s of the system, or if $x_i(t) - x_j(t) = z_{ij}(t)$ where $z_{ij}(t)$ contains none of the r slow modes. Note in general $x_i(t)$, $x_j(t)$ will contain all the modes of the systems. Some modes will be more dominant than others. Slow coherency is manifested when the difference between the two states $z_{ij}(t)$ does not contain any of the slow modes. When the system has r slow modes, it is defined to be r decomposable. A 3-area 5-machine system is used below in (15.2.7)–(15.2.9) as an illustration [2]. x^1 contains the reference states or machines and x^2 contains all the other machines in the system. The matrix L_g is called a grouping matrix, which has only one 1 in each row with all the other elements being 0. It provides the grouping information. For each entry that contains a 1 in the matrix, the row number represents each machine state in x^2 and the column number corresponds to a reference state in x^1, which are grouped together. For example,

$$x^1 = (x_1, x_2, x_4)^T \tag{15.2.7}$$

$$x^2 = (x_3, x_5)^T \tag{15.2.8}$$

$$L_g = \begin{array}{c} \\ x_3 \\ x_5 \end{array} \begin{array}{c} \begin{array}{ccc} x_1 & x_2 & x_4 \end{array} \\ \begin{bmatrix} 1 & 0 & 0 \\ 0 & 1 & 0 \end{bmatrix} \end{array}. \tag{15.2.9}$$

This indicates there are three areas. These areas are composed of machines 1 and 3, machines 2 and 5, and machine 4 alone, in which machines 1, 2, and 4 are reference machines. Then

$$x^2(t) - L_g x^1(t) = z^2(t) \tag{15.2.10}$$

where $z^2(t)$ contains none of the r slowest modes. Then define the transformation

$$\begin{bmatrix} x^1 \\ z^2 \end{bmatrix} = \begin{bmatrix} I & 0 \\ -L_g & I \end{bmatrix} \begin{bmatrix} x^1 \\ x^2 \end{bmatrix}. \tag{15.2.11}$$

Substituting the transformation (15.2.11) into (15.2.3), we obtain

$$\begin{bmatrix} \ddot{x}^1 \\ \ddot{z}^2 \end{bmatrix} = \begin{bmatrix} B_1 & A_{12} \\ R(L_g) & B_2 \end{bmatrix} \begin{bmatrix} x^1 \\ z^2 \end{bmatrix} \tag{15.2.12}$$

where

$$B_1 = A_{11} + A_{12} L_g \tag{15.2.13}$$

$$B_2 = A_{22} - L_g A_{12} \tag{15.2.14}$$

$$R(L_g) = A_{22}L_g - L_g A_{11} - L_g A_{12} L_g + A_{21}. \tag{15.2.15}$$

$A_{11}, A_{12}, A_{21}, A_{22}$ are the submatrices of A conformal with x^1, x^2. When $R(L_g) = 0$ and $|\lambda_j(B_1)| < |\lambda_i(B_2)|$, such an L_g is called "dichotomy" and is denoted by L_d. Normally L_d is not a grouping matrix. In such a case, a near r decomposable system is applied. Compare different combinations of the selection of slow variable x^1 and get different solutions of L_d. Use L_d with min $||L_d - L_g||$ as the approximation and then let the largest number of each row be 1 and all the other elements be 0. Let

$$V = \begin{bmatrix} V_1 \\ V_2 \end{bmatrix} \tag{15.2.16}$$

where V contains the r columns of the eigensubspace of the matrix A. The $r \times r$ matrix V_1 is nonsingular, being a basis of the eigensubspace of the slow modes. Then

$$L_d = V_2 V_1^{-1}. \tag{15.2.17}$$

It can be proven that L_d is the unique dichotomy solution of the Riccati equation (15.2.15). If two coherent machines are in x^1, then V_1 will be singular. In the near r decomposable system, V_1 will be near singular. Thus we aim to find the r largest and most linearly independent rows of V. Gaussian elimination with complete pivoting is employed to find the r most independent vectors of V. Permutation is done in the elimination, and the first r steps provide V_1. A grouping algorithm is provided below:

(1) Choose the number of areas r.

(2) Compute a basis matrix V for a given ordering of the x variables containing slow modes.

(3) Apply Gaussian elimination with complete pivoting to V and obtain the set of reference machines. Each group will then have one and only one reference machine.

(4) Compute L_d for the set of reference machines chosen in step 3. Then determine the group that each generator belongs to from the matrix L_d by comparing the row of each generator with the row of the reference machine.

Given the reference machines, the above grouping algorithm provides a method to get the generator groupings with slow coherency. By selecting the r slowest modes, an objective can be achieved to have the weakest connections between the areas, which will be discussed in the next section. So it provides a complete procedure to determine the generators in each island.

15.2.2 Slow coherency and weak connection

Slow coherency solves the problem of identifying theoretically the weakest connection in a complex power system network. Previous work shows groups of generators with slow coherency may be determined using Gaussian elimination on the eigensubspace matrix after selection of r slowest modes σ_a. In [2], it has been proven through linear analysis that with selection of the r slowest modes, the aggregated system will have the weakest connection between groups of generators.

Furthermore, linear dynamic networks can take various two-time-scale forms. Power system linearized dynamics take the weak connection form [2],

$$\ddot{X} = M^{-1}KX = (M^{-1}K^I + M^{-1}K^E)X \tag{15.2.18}$$

where

$$M = \text{diag}(m_1, m_2, \cdots, m_n) \tag{15.2.19}$$

and K^I is a block matrix representing internal connections. K^E represents external connections. With the selection of the r slowest modes, $M^{-1}K^E$ has much smaller elements compared with the block matrix $M^{-1}K^I$. Suppose the system is partitioned into two areas, and the system equations can be written as

$$\varepsilon \begin{bmatrix} d^2X_1/dt^2 \\ d^2X_2/dt^2 \end{bmatrix} = \begin{bmatrix} d^2X_1/d\tau^2 \\ d^2X_2/d\tau^2 \end{bmatrix} = \begin{bmatrix} A_{11} + \varepsilon A_{11}^T & \varepsilon A_{12} \\ \varepsilon A_{21} & A_{22} + \varepsilon A_{22}^T \end{bmatrix} \begin{bmatrix} X_1 \\ X_2 \end{bmatrix}$$

$$= \begin{bmatrix} A_{11} & 0 \\ 0 & A_{22} \end{bmatrix} \begin{bmatrix} X_1 \\ X_2 \end{bmatrix} + \varepsilon \begin{bmatrix} A_{11}^T & A_{12} \\ A_{21} & A_{22}^T \end{bmatrix} \begin{bmatrix} X_1 \\ X_2 \end{bmatrix}. \tag{15.2.20}$$

Here, $\varepsilon \to 0$ is a small number. $\tau = (t - t_0)/\varepsilon$ is a new time variable that is much larger than the original time scale. X_1 and X_2 are n_1 and n_2 vectors. Let the rows of P_i and Q_i span the left null and row spaces of A_{ii}, respectively, and the columns of V_i and W_i span the right null and range spaces, $i = 1, 2$. Use the following transformation to obtain the explicit separated form

$$\begin{bmatrix} Y_1 \\ Y_2 \end{bmatrix} = \begin{bmatrix} P_1X_1 \\ P_2X_2 \end{bmatrix} \tag{15.2.21}$$

$$\begin{bmatrix} Z_1 \\ Z_2 \end{bmatrix} = \begin{bmatrix} Q_1X_1 \\ Q_2X_2 \end{bmatrix} \tag{15.2.22}$$

$$\begin{bmatrix} \ddot{Y}_1 \\ \ddot{Y}_2 \end{bmatrix} \begin{bmatrix} P_1A_{11}^TV_1 & P_1A_{12}V_2 \\ P_2A_{21}V_1 & P_2A_{22}^TV_2 \end{bmatrix} \begin{bmatrix} Y_1 \\ Y_2 \end{bmatrix} \tag{15.2.23}$$

$$\begin{bmatrix} \ddot{Z}_1 \\ \ddot{Z}_2 \end{bmatrix} = \begin{bmatrix} Q_1A_{11}W_1 + \varepsilon Q_1A_{11}^TW_1 & \varepsilon Q_1A_{12}W_2 \\ \varepsilon Q_2A_{21}W_1 & Q_2A_{22}W_2 + \varepsilon Q_2A_{22}^TW_2 \end{bmatrix} \begin{bmatrix} Z_1 \\ Z_2 \end{bmatrix}. \tag{15.2.24}$$

Equations (15.2.23) and (15.2.24) are the aggregated slow model and fast model. This system contains two subsystems and can be generalized for more than two subsystems. It shows some important features of the dynamics with the weak connection

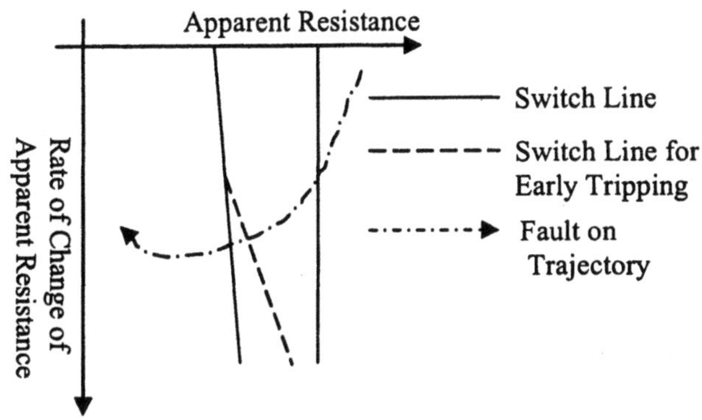

Figure 15.2.1: R-Rdot out of step relay switch lines

form. At first, each subsystem has weak connections with the other subsystem. Each subsystem also has its own fast and slow state variables. Second, fast variables in each subsystem have weak connections with fast variables in other subsystems so that the connection can be omitted. Third, slow variables of each subsystem have strong connections to the other subsystems, which should be treated together. In all, for each subsystem, there are fast local models with weak connections to the other local fast models and slow models forming a slow "core," which describes the aggregate dynamics, with strong connection to the other subsystems.

The weak connection form best states the reason for islanding based on slow coherency. That is, when the disturbance happens, it is required to separate in the transient time scale the fast dynamics, which could propagate the disturbance very quickly, through islanding on the weak connections. While in the transient time scale, the slow dynamics will mostly remain constant or change slowly on the tie lines between the areas. In other words, once fast dynamics are detected on the tie lines, it means fast dynamics are being propagated through these weak connections. As a summary, the slow coherency-based grouping method has the following explicit advantages:

(1) Slow coherency is independent of initial condition and disturbance.

(2) The two-time-scale weak connection form inherently describes the oscillation feature of large-scale power systems: the fast oscillation within a group of machines and the slow oscillation between the groups via weak tie lines.

(3) The slow coherency method also preserves the features of the coherency-based grouping. It is independent of the size of the disturbance and the generator model detail.

The technique is implemented by running DYNRED. Through the selection of the two-time-scale option, the coherent groups of generators can be obtained on any power system. The islands are formed by slow coherency considering the modes of the oscillation of the linearized state space model of the system. This is reasonable since the islands contain the generators with similar swing characteristics and oscillation frequencies unrelated to the size and location of the disturbance. It is also reasonable to assume that the same group of generators will show similar behavior (slow coherency) following the disturbance, since the automatic islanding occurs quite early after the disturbance (0.2 seconds after the initiation of the disturbance). We have also verified by nonlinear simulation that the grouping identified by the slow coherency approach is maintained in the nonlinear simulation beyond this time instant. It has also been observed that, in nonlinear simulations following disturbances, the fundamental modes of oscillation that describe the inherent structural characteristics of the systems can be clearly identified. Hence, the fundamental modes of the oscillation determined by the linear analysis provide a convenient approach to detect the structural characteristics of the system.

15.2.3 Islanding

Having determined the coherent groups of the generators, we still have two questions to answer: Where and when to form the islands? We develop an automatic islanding program to fully support the application of theory. The program is dedicated to searching for the optimum cut sets after we have the grouping information. The optimum cut set is obtained considering the least generation-load imbalance. The approach begins with the characterization of the network structure or connectivity using the adjacent link table data structure [6]. Then through a series of reduction processes, the program forms a small network and performs an exhaustive search on it to get all the possible cut sets.

With the information of the coherent groups of generators and the exact locations of where to form the islands, the R-Rdot out of step relay developed at Bonneville Power Administration (BPA) is deployed to form the islands [7, 8]. The relay only requires local measurements and makes tripping decisions using settings based on various offline contingency simulations. It shows much better performance than the conventional out of step relay, which is actually the impedance relay. Besides the impedance, the new relay uses the information of the rate of change of the impedance or resistance and gets better results in practice. Different switching lines make sure different corrective control actions are taken based on the level of the seriousness of the disturbance. The switching lines are shown in Figure 15.2.1 [7, 8]. When a fault trajectory enters into the range defined by the switching lines, the tripping action will take place.

15.3 Load-Shedding Scheme

Controlled islanding divides the power system into islands. Some of these islands are load rich and others may be generation rich. Generally, in a load-rich island, the situation is more severe. The system frequency will drop because of the generation shortage. If the frequency falls below a certain set point, e.g., 57.5 Hz, the generation protection system will begin operation and trip the generator, further reducing the generation in the island and making the system frequency decline even worse. In the worst case, the entire island will blackout. In a load deficient island, either intentional or forced generator tripping will reduce the gap between the generation and the load. As a result, we put more effort to save the load-rich island and develop a new two-layer load-shedding scheme to perform the task [9].

15.3.1 Load-shedding scheme under subsumption model

In the literature, there exist two kinds of load-shedding schemes: load shedding based on frequency decline and load-shedding based on rate of frequency decline [10, 11]. The first approach [10] has mostly conservative settings because of the lack of information regarding the magnitude of the disturbance. Although this approach is effective in preventing inadvertent load-shedding in response to small disturbances with relatively longer time delay and lower frequency threshold, it is not able to distinguish between the normal oscillations of the power system and the large disturbances on the power system. Thus, the approach is prone to shedding less load. This is not beneficial to the quick recovery of the island and may lead to further cascading events. The second approach [11] avoids these shortcomings by utilizing the frequency decline rate as a measure of the load shortage. Thus it has a faster response time compared to the other scheme.

The idea of the load-shedding based on the rate of change of frequency can be traced back to as early as the 1960s [12, 13]. Issues of hardware implementations in the form of relays were discussed and resolved in the 1970s and 1980s. In [13], the leakage occurring in the fast Fourier transform is advantageously used to detect the fluctuations in the fundamental frequency of a power system so that it can optimally estimate the mean frequency and its average rate of decline and determine the appropriate amount of load to be shed. The idea was then adopted in an isolated power system [14, 15]. In the United Kingdom, the principle of the rate of change of frequency used for load-shedding is referred to as ROCOF. In [16], an adaptive load-shedding scheme is developed that uses information including the system demand, spinning reserve, system kinetic energy, amount of lower-priority load available for shedding elsewhere, and locally measured rate of change of frequency.

We develop a load-shedding scheme based on the rate of frequency decline, which can identify the magnitude of the disturbance. At the same time, we incorporate the conventional load-shedding scheme into our subsumption model to form a new two-level load-shedding scheme as shown in Figure 15.3.1.

Normally the relay will operate the conventional load-shedding scheme. The conventional load-shedding scheme has longer time delays and lower frequency thresholds, which can be used to prevent inadvertent load-shedding in response to small disturbances. If the system disturbance is large and exceeds the signal threshold, the second layer will be activated and send an inhibition signal to the first layer and the load-shedding scheme based on the rate of frequency decline will take effect. This layer will shed more load quickly at the early steps to prevent the cascading events in the island. This can greatly enhance the system's ability to withstand large disturbances. In all, the new two-level load-shedding scheme has the following explicit features:

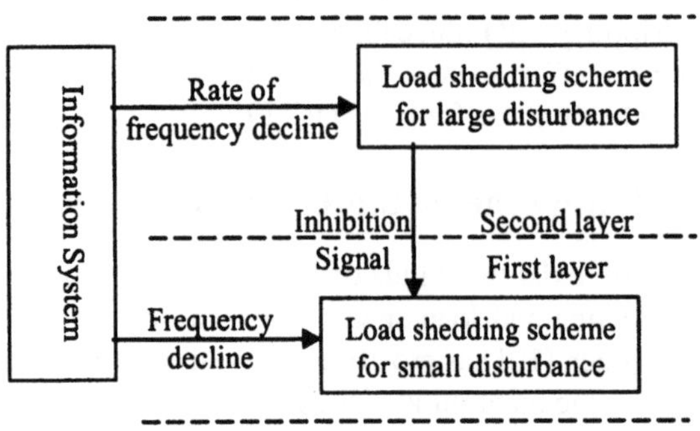

Figure 15.3.1: The subsumption model for the load-shedding scheme

(1) Suitable for large and small disturbances.
(2) Suitable for self-healing when combined with islanding in power system recovery.

15.3.2 Determination of the magnitude of the disturbance

A variable that measures the magnitude of the disturbance should be determined in order to make the subsumption approach feasible. From an intuitive analysis [17], the rate of frequency decline at the beginning of the disturbance can accurately reflect the magnitude of the disturbance. From chapter 3 of [18], we have

$$\frac{df_i}{dt} = -\frac{60 \times P_{sik}}{2H_i}\left(P_{L\Delta}(0^+)\Big/\sum_{i=1}^{n} P_{sik}\right), \quad i = 1, 2, \cdots, n. \tag{15.3.1}$$

Define

$$\bar{f} = \sum_{i=1}^{n}(H_i f_i)\Big/\sum_{i=1}^{n} H_i. \tag{15.3.2}$$

In (15.3.1), we add all the equations to obtain

$$\frac{d\bar{f}}{dt} = \sum_{i=1}^{n}\left(H_i \cdot \frac{df_i}{dt}\right)\Big/\sum_{i=1}^{n} H_i = -60 \times P_{L\Delta}\Big/\sum_{i=1}^{n} 2H_i \tag{15.3.3}$$

where f_i is the frequency of generator i in Hz, $\frac{d\bar{f}}{dt}$ is the average rate of frequency decline in Hz/second, P_{sik} is the synchronizing power coefficient between generator i and the disturbance node k in pu (see chapter 3 of [18]), $P_{L\Delta}$ is the magnitude of the disturbance in pu, H_i is the inertia of generator i in pu, and ω_i is the rotor speed of each generator i in pu. We define

$$m_i = \frac{df_i}{dt} \tag{15.3.4}$$

$$m_0 = \frac{d\bar{f}}{dt}. \tag{15.3.5}$$

Substituting (15.3.5) into (15.3.2), we have

$$m_0 = -60 \times P_{L\Delta} \Big/ \sum_{i=1}^{n} 2H_i. \tag{15.3.6}$$

The equation can be alternatively written as

$$P_{L\Delta} = -m_0 \times \sum_{i=1}^{n} 2H_i/60. \tag{15.3.7}$$

Since H_i is constant, the magnitude of the disturbance can be directly related to the average rate of system frequency decline. Hence, m_0 can be an indicator of the severity of the disturbance. The rate of frequency decline at the beginning of the disturbance can be used as the input signal of the second layer. Once the threshold of $P_{L\Delta}$ to activate the second layer is decided, the corresponding m_0 can be calculated. When the disturbance occurs, we measure m_i at each bus and compare it with m_0. If m_i is greater than m_0, the second layer is activated; otherwise the conventional load-shedding scheme is used.

By using m_i at each bus to decide the amount of load that should be shed locally, the system oscillations after the disturbance can be reduced. We know that at the beginning of the disturbance, the impact of disturbance is shared immediately by the generators according to their synchronizing power coefficients with respect to the bus at which the disturbance occurs [18]. Thus the machines electrically close to the point of impact will pick up the greater share of the load regardless of their size. On the other hand, standards [19] and guides [20] give a fairly strict regulation on tolerable frequency deviations. The area between 59.5 Hz and 60.5 Hz is the area of unrestricted time operating frequency limits. The areas above 60.5 Hz and below 59.5 Hz are areas of restricted time operating frequency limits. In Subsection 15.3.3 we also note that the system frequency is not allowed to drop below 57 Hz. Hence, on detection of the system frequency dropping below 59.5 Hz, the load-shedding schemes should trigger the corrective control ensuring that the system frequency will not drop below 57 Hz. Although the disturbance is ultimately shared according to the inertia of each machine, sometimes the frequency of some generators near the disturbance can drop below 57 Hz before reaching the final state. Using the value of frequency at each bus, the buses whose frequencies drop quickly are likely to have more load-shed locally; this can reduce the frequency deviation and system oscillations.

15.3.3 Determination of the threshold $P_{L\Delta}$

Considering the governor protection system limitation and regional operation criteria, we define $P_{L\Delta}$ as the minimum load deficit that can drive the system average frequency below 57 Hz. This frequency threshold is chosen because it is widely recognized that the system is not allowed to operate below 57 Hz. There are three main reasons why the system cannot operate below 57 Hz.

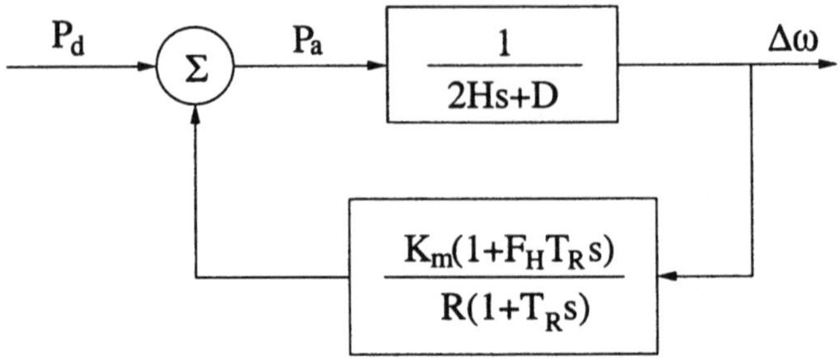

Figure 15.3.2: The reduced model of reheat unit for frequency disturbance

(1) *Coordination with the governor-turbine system*

Underfrequency operating limitations imposed by the manufacturers of turbine-generator units are primarily concerned with the avoidance of resonant frequencies and turbine blade fatigue. Since fatigue effects are cumulative, the limitation is defined in terms of total accumulated times of operation within specified frequency ranges. Turbine manufacturers provide limitations of various turbines to frequency variation. Based on this data it is very reasonable to choose 57 Hz as system operation limit [17].

(2) *Coordination with the plant auxiliary system*

Nuclear units having a pressurized water reactor steam supply use special underfrequency protection for their primary system reactor coolant pumps. For these units, this protection will trip the coolant pumps and shut down the reactor at the fixed time of 0.25s and a pickup setting of 57.0 Hz [21].

(3) *Coordination with existing operation criteria*

According to the North East Power Coordinating Council (NPCC) Standard, the generation rejection should be deployed immediately if system frequency drops below 57 Hz [22].

To find $P_{L\Delta}$, we use a reduced model for a reheat unit for frequency disturbance as shown in Figure 15.3.2 [18].

In Figure 15.3.2, K_m is mechanical power gain factor. We use a typical value of 0.95. H is inertia constant in seconds, typically 4.0 seconds. F_H is high-pressure power fraction, typically 0.3. D is damping factor, typically 1.0. T_R is reheat time constant in seconds, typically 8.0 seconds. R is fraction of the reheat turbine, typically 0.559. P_d is disturbance power in pu.

We use typical system data to compute the minimum load deficit that can drive the system to the minimum frequency of 57 Hz (representing the worst-case scenario).

From Figure 15.3.2,

$$\Delta\omega = \left(\frac{R\Omega_n^2}{DR + K_m}\right)\left(\frac{(1 + T_R s)P_d}{s^2 + 2\Omega_n^2\lambda s + \Omega_n^2}\right) \qquad (15.3.8)$$

where

$$\Omega_n^2 = \frac{DR + K_m}{2HRT_R} \tag{15.3.9}$$

$$\lambda = \left(\frac{2HR + (DR + K_m F_H)T_R}{2(DR + K_m)}\right)\Omega_n.$$

If P_d is a step function, then

$$P_d(t) = P_{step}u(t). \tag{15.3.10}$$

Using this reduced model and normalizing, we learn that the lowest system average frequency for this disturbance is 57 Hz when $P_d = P_{L\Delta} = 0.3P_{sys}$. So we choose $0.3P_{sys}$ as the threshold value of $P_{L\Delta}$ for the new load-shedding scheme. This value of $P_{L\Delta}$ is used in (15.3.6) to determine the limiting threshold for m_0.

15.3.4 Frequency threshold, step size, and time delay

The frequency threshold should be chosen carefully. First, it should not be too close to normal frequency in order to avoid tripping on severe but nonemergency frequency swings. On the other hand, it is more effective to shed load earlier.

The step size is an important variable in load-shedding. Conventionally, the amount of load-shed at each step is increased while the system frequency decreases. This choice is reasonable for those schemes that use the frequency as the criterion to shed load because before the system deteriorates, it is unreasonable to shed too much load if the disturbance is unknown. It has also been observed that for large disturbances, such schemes may be insufficient to arrest system frequency decline [23]. Our second layer of load-shedding scheme, as stated before, will only take action when the disturbances are large enough to cause the system frequency to drop below 57 Hz. So instead of increasing the step size while the system frequency is decreasing, we set the first step to be the largest step size. This helps the system recovery in the case of large disturbance. Since the first layer of the new load-shedding scheme will mainly deal with small disturbances, the conventional philosophy is adopted for this layer, or the load is shed only based on frequency decline. For the steps of load shedding, the following three facts have been observed [23]:

(1) Frequency steps must be far enough apart to avoid overlap of shedding due to (intentional or inherent) time delay.

(2) The number of steps does not have very great impact on the effect of load shedding.

(3) Generally, the threshold of the last step of load-shedding is chosen no less than 58.3 Hz.

Time delay is very important for load-shedding schemes to avoid overlapping and unexpected action for small frequency oscillations. Generally, for the conventional load-shedding scheme, the delay time for the first step is usually very long to avoid unexpected actions due to small frequency oscillations. For the following steps, the more the frequency declines, the quicker the action. For the new scheme, to prevent sharp frequency declines following a large disturbance, we set the delay time for the first step of the second layer as 0 cycles.

Finally, the two layers of load-shedding scheme are developed as shown in Table 15.3.1. When the disturbance occurs, we measure m_i or the rate of frequency decline at each bus and compare it with m_0 calculated from $P_{L\Delta}$. If m_i is greater than m_0, the new load-shedding scheme shown in the second row of Table 15.3.1 is deployed. Twenty percent of the total load is shed with 0 cycle delay in the first step. The character C in the table means cycle. Otherwise, the conventional load-shedding scheme is used, which is shown in the second row.

Table 15.3.1: Step size and delay time of the two layers as percentage of the total load

	59.5 Hz	59.3 Hz	58.8 Hz	58.6 Hz	58.3 Hz
$m_i > m_0$	20% (0C)		5% (6C)	4% (12C)	4% (18C)
$m_i < m_0$		10% (28C)	15% (18C)		

15.4 Simulation Result

The load-shedding scheme is tested on a 179-bus, 29-generator test system. The system has a total generation of 61410 MW and 12325 Mvar. It has a total load of 60785 MW and 15351 Mvar. The simulation is made using a detailed generator model with governors, exciter, and power system stabilizers. In the first case, the system is islanded by experience. In the second case, the DYNRED program in the PSAPAC package was chosen to form four groups of generators based on slow coherency. With the help of the automatic islanding program, we determine the cut sets of the island with consideration of the least generation-load imbalance and topology requirements. In the third case, a different fault is chosen from the previous two cases. Islands are formed based on the slow coherency approach.

In the first and the second case, to test the system response to a severe contingency, three 500 kV transmission lines in the system are tripped simultaneously shown as in Figure 15.4.1. This corresponds to a catastrophic transmission failure where an incident takes out all the three transmission lines simultaneously. The arrow shows the location where the three transmission lines are disconnected. These lines are connected between buses:

(1) Bus 83 – Bus 168

(2) Bus 83 – Bus 170

(3) Bus 83 – Bus 172.

Simulations conducted on the system indicate that this disturbance will result in the system being unstable. In these simulations, no conventional protection settings

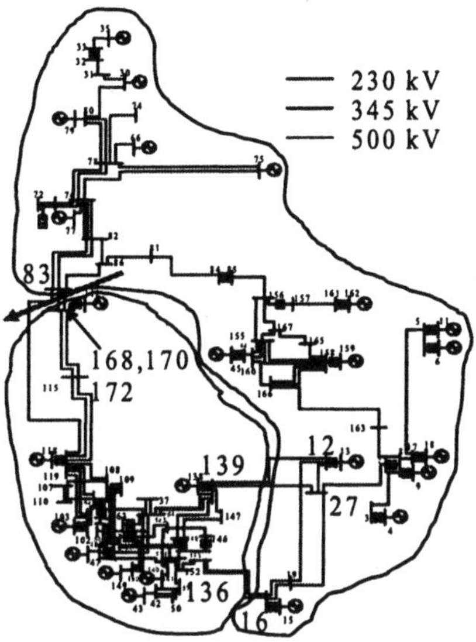

Figure 15.4.1: Case 1–Two islands for 179-bus system based on experience

were considered. To save the system from an impending blackout, we split the system into two islands 0.2 seconds after the contingency. In the first case, the islands are formed by experience. The following lines are tripped:

(1) Bus 139 – Bus 12

(2) Bus 139 – Bus 27

(3) Bus 136 – Bus 16 (1 and 2).

After islanding, the system is divided into two islands as shown in Figure 15.4.1. The two islands can be characterized as the northeast island, which is generation rich, and the southwest island, which is load rich. In the southwest island, some of the buses have m_i smaller than m_0. So the conventional load-shedding scheme is deployed at these buses. For the other buses at which m_i is larger than m_0, the load-shedding scheme based on the rate of frequency decline is deployed.

In the simulation, underfrequency load-shedding with various schemes is performed in the southwest island to maintain acceptable frequency. Simulations are conducted using EPRI's Extended Transient-Midterm Stability Program (ETMSP). Figure 15.4.2 shows the frequency responses of a typical generator in the southwest island in four situations.

Curve 1 and Curve 2 show that, following the disturbance, the system will lose stability without any self-healing strategy or only with islanding. Curve 3 and Curve 4 give a comparison between the two load-shedding schemes. To maintain stability of the system, less load needs to be shed with the new load-shedding scheme than the old scheme. At the same time, the system experiences smaller frequency excursions under the new scheme than the old scheme.

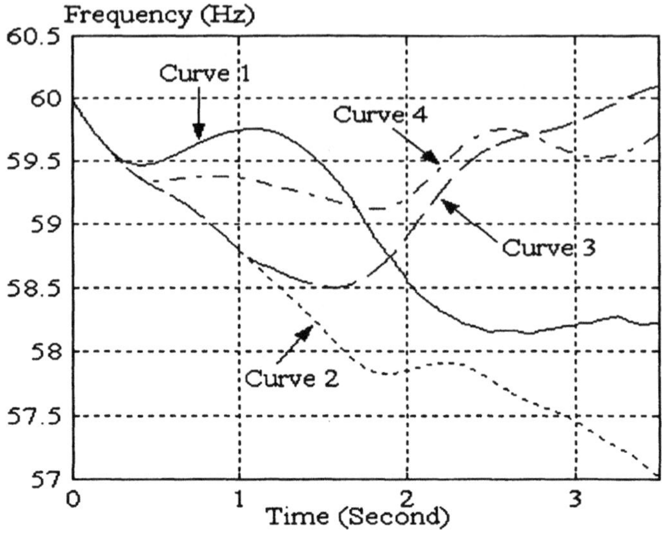

Figure 15.4.2: Frequency response of generator 118 after the contingency of the 179-bus system. Curve 1: Without self-healing. Curve 2: Islanding with no load-shedding. Curve 3: Islanding followed by load-shedding based on frequency difference. Curve 4: Islanding followed by load-shedding based on the rate of frequency decline.

In determining the islands using slow coherency we specified the initial estimate of the groups to be 4. Based on the grouping and the fault location in the second case, in order to create the islands for such a large disturbance, the automatic islanding program selects the following lines to be tripped:

(1) Bus 133 – Bus 108

(2) Bus 134 – Bus 104

(3) Bus 29 – Bus 14.

As mentioned before, the lines are determined by the automatic islanding program. Nonlinear fault-on simulations show that the generators within the same group show coherency with each other. Simulations indicate that BPA's R-Rdot out of step relays are able to quickly identify the out of step situation on those tie lines and initiate local and remote trippings. After islanding, the system is divided into three areas. These areas are shown in Figure 15.4.3. The arrow shows the location where the three transmission lines are disconnected. The three islands in Figure 15.4.3 can be characterized as the northeast island, the central island and the south island. Figure 15.4.4 shows the frequency responses of the same generator as in Figure 15.4.2 in the central island in four situations.

The south island and the central island are load-rich areas. The other island is the

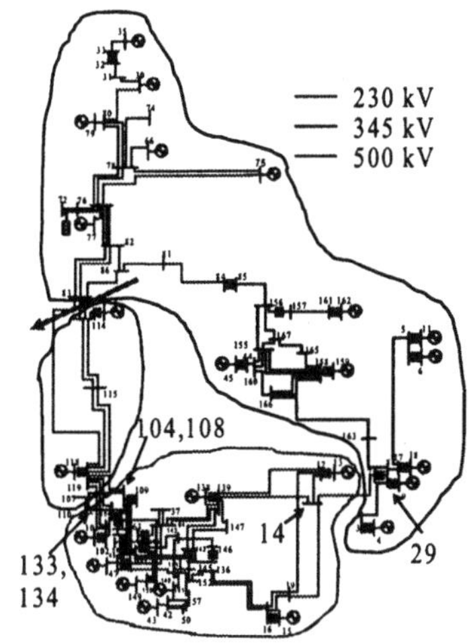

Figure 15.4.3: Case 2–Three islands for 179-bus system based on slow coherency

generation-rich area. In the south island, all the buses have m_i less than m_0 and the conventional load-shedding scheme is deployed. Simulations indicate that no load needs to be shed in the south island according to our load-shedding scheme. The frequency recovers through the coordination of the generators' governors and exciters. In the central island, the new load-shedding scheme is deployed. Similar conclusions can be obtained for this case as in the previous one from the four frequency response curves.

In the third case, four lines are tripped simultaneously:

(1) Bus 12 – Bus 139

(2) Bus 27 – Bus 139

(3) Bus 16 – Bus 136 (1 and 2).

To save the system from an impending blackout, we split the system into two islands 0.2 seconds after the contingency. The islands are determined by the slow coherency. In order to create the island, the following lines are tripped:

(1) Bus 133 – Bus 108

(2) Bus 134 – Bus 104.

The two islands are shown in Figure 15.4.5. Figure 15.4.6 shows the frequency responses of one typical generator in the south island in four situations. Table 15.5.1 shows the amount of load that needs to be shed for system recovery under the two load-shedding schemes in the three cases considered.

In the cell that shows the amount of load-shed, the first percentage is the ratio of the load-shed compared to the total system load. The second percentage is the ratio of the load-shed compared to the island load. It is observed that the load-shedding scheme based on the rate of frequency decline sheds much less load than the conventional load-shedding scheme in all three cases.

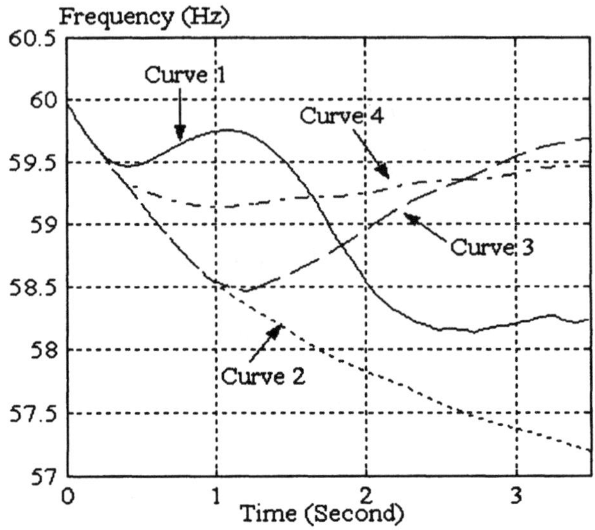

Figure 15.4.4: Frequency response of generator 118 after the contingency of the 179-bus system. Curve 1: Without self-healing. Curve 2: Islanding with no load-shedding. Curve 3: Islanding followed by load-shedding based on frequency difference. Curve 4: Islanding followed by load-shedding based on the rate of frequency decline.

In this chapter, a self-healing scheme for large disturbances with concentration on a new load-shedding scheme is described in detail. The scheme is tested on the 179-bus sample system and shows very good performance. The new two-level load-shedding scheme raises the stability performance of the system by shedding less load compared to the conventional load-shedding scheme. Since the tripping action does not require many calculations and the islanding information can be obtained offline, the speed in the real-time implementation should mostly depend on the speed of communication devices and switching actions. In order to facilitate restoration, the islands are formed by minimizing the generation-load imbalance.

15.5 Other Forms of System Protection Schemes

The description provided above details the application of a system protection scheme or remedial action scheme (RAS) for frequency instability after a large disturbance. Other forms of RASs are becoming quite prevalent and are integral parts of corrective control measures. Practices around the world have been detailed recently in [24].

A large variety of RASs are now in service. They can be classified as follows:

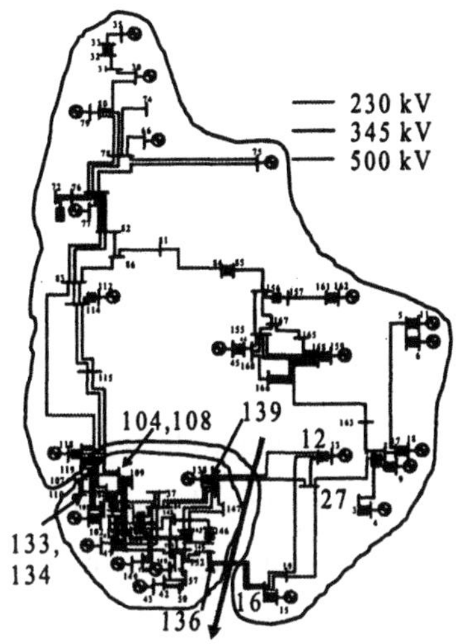

Figure 15.4.5: Case 3–Two islands for 179-bus system based on slow coherency

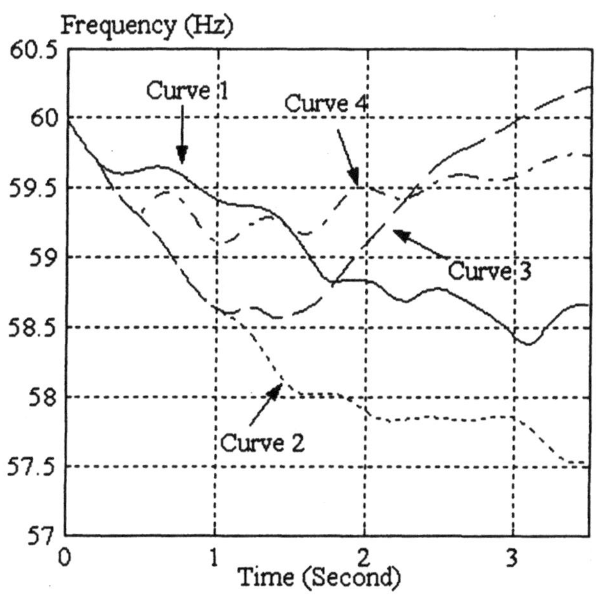

Figure 15.4.6: Frequency response of generator 43 after the contingency of the 179-bus system. Curve 1: Without self-healing. Curve 2: Islanding with no load-shedding. Curve 3: Islanding followed by load-shedding based on frequency difference. Curve 4: Islanding followed by the new load-shedding.

- Generation rejection
- Turbine fast valving/generator runback
- Gas turbine/pump storage start-up
- Actions on the automatic generation control such as setpoint changes
- Underfrequency load-shedding
- Undervoltage load-shedding
- Remote load-shedding
- HVDC power modulation
- Automatic shunt switching of reactive elements
- Controlled opening of interconnection/area islanding
- Tap changers blocking and setpoint adjustment
- Quick increase of generator voltage setpoint.

These RASs specifically address the following types of instability:

- Transient angle instability
- Small signal angle instability
- Frequency instability
- Voltage instability.

Table 15.5.1: Comparison of the two load-shedding schemes in three cases

Cases	Generation Load Imbalance (MW)	Load Shed with Conventional Scheme (MW)	Load Shed with New Scheme (MW)
No. 1	Generation 16,265 Load 22,679	6937 (11.4% 30.6%)	5698 (9.4% 25.1%)
No. 2	Central Island: Generation 5118 Load 7006 South Island: Generation 15,477 Load 17,373	1810/0 (3.0%/0% 25.8%/0%)	1450/0 (2.4%/0% 20.7%/0%)
No. 3	Generation 11,148 Load 15,674	5127 (8.4% 32.7%)	3672 (6.0% 23.4%)

The RAS described above can be initiated by a local detection system or by a wide area detection system. Detection is considered local when all information required by the decision-making process is available at the same location where the action is performed. Local RASs are generally considered to be more dependable because they do not have to rely on telecommunication facilities for their operation and are also considered the most secure because their actions are generally limited and localized.

In a wide area detection system, the action is initiated by information acquired at several locations geographically dispersed in the system. This type of RAS is generally used to mitigate complex and large phenomena that may jeopardize the integrity of the entire power system. As a result, these schemes are designed with a higher level of complexity than local RASs and are closely intertwined with telecommunication facilities. In these schemes dependability is the primary concern because the impact of unintended operation can be catastrophic.

Acknowledgments

The author would like to acknowledge the help of his graduate students Haibo You and Xiaoming Wang in conducting this research.

Bibliography

[1] R. Brooks, *Subsumption Architecture*, Available at web site http://www.mit.edu/people/brooks, 2001.

[2] J. H. Chow, *Time-Scale Modeling of Dynamic Networks with Applications to Power Systems*, Lectures Notes in Control and Information Sciences 46, Springer-Verlag, Berlin, 1982.

[3] M. Amin, C. C. Liu, G. T. Heydt, A. Phadke, and V. Vittal, "Development of analytical and computational methods for the strategic power infrastructure defense (SPID) system," *Annual Report*, Jan. 2001.

[4] J. R. Winkelman, J. H. Chow, B. C. Bowler, B. Avramovic, and P. V. Kokotovic, "An analysis of interarea dynamics of multi-machine systems," *IEEE Trans. Power Apparatus and Systems*, vol. PAS-100, no. 2, Feb. 1981.

[5] R. Podmore, "Identification of coherent generators for dynamic equivalents," *IEEE Trans. Power Apparatus and Systems*, vol. PAS-97, pp. 1344–1354, 1978.

[6] M. S. Tsai, "Development of islanding early warning mechanism for power systems," *Proc. IEEE Power Engineering Society Summer Meeting*, vol. 1, pp. 22–26, 2000.

[7] C. W. Taylor, J. M. Haner, L. A. Hill, W. A. Mittelstadt, and R. L. Cresap, "A new out-of-step relay with rate of change of apparent resistance augmentation," *IEEE Trans. Power Apparatus and Systems*, vol. PAS-102, no. 3, Mar. 1983.

[8] J. M. Haner, T. D. Laughlin, and C. W. Taylor, "Experience with the R-Rdot out-of-step relay," *IEEE Trans. Power Systems*, vol. PWRS-1, no. 2, Apr. 1986.

[9] Z. Yang, *A New Automatic Under-Frequency Load Shedding Scheme*, M.S. Thesis, Department of Electrical and Computer Engineering, Iowa State University, 2001.

[10] D. W. Smaha, C. R. Rowland, and J. W. Pope, "Coordination of load conservation with turbine-generator underfrequency protection," *IEEE Trans. Power System Apparatus and System*, vol. PAS-99, no. 3, pp. 1137–1145, May/June 1980.

[11] G. S. Grewal, J. W. Konowalec, and M. Hakim, "Optimization of a load shedding scheme," *IEEE Industry Application Magazine*, pp. 25–30, July/Aug. 1998.

[12] C. J. Drukin, Jr., E. R. Eberle, and P. Zarakas, "An underfrequency relay with frequency decay compensation," *IEEE Trans. Power Apparatus and Systems*, vol. PAS-88, no. 6, pp. 812–819, June 1969,

[13] A. A. Girgis and F. M. Ham, "A new FFT-based digital frequency relay for load shedding," *IEEE Trans. Power Apparatus and Systems*, vol. PAS-101, no. 2, pp. 433–439, Feb. 1982.

[14] M. M. Elkateb and M. F. Dias, "New technique for adaptive-frequency load shedding suitable for industry with private generation," *Generation, Transmission and Distribution, IEE Proc. C*, vol. 140, no. 5, pp. 411–420, Sept. 1993.

[15] M. M. Elkateb and M. F. Dias, "New proposed adaptive frequency load shedding scheme for cogeneration plants," *Proc. Fifth International Conference on Developments in Power System Protection*, 1993.

[16] J. G. Thompson and B. Fox, "Adaptive load shedding for isolated power systems," *Generation, Transmission and Distribution, IEE Proc.*, vol. 141, no. 5, Sept. 1994.

[17] P. M. Anderson and M. Mirheydar, "An adaptive method for setting underfrequency load shedding relays," *IEEE Trans. Power Systems*, vol. 7, No. 2, pp. 720–729, May 1992.

[18] P. M. Anderson and A. A. Fouad, *Power System Control and Stability*, IEEE, New York, 1994,

[19] *American National Standard Requirements for Synchronous Machines*, ANSI C50.10-1977, New York, 1977.

[20] *IEEE Guide for Abnormal Frequency Protection of Power Generating Plants*, IEEE Standard C37.106-1988, New York, 1988.

[21] P. Kundur, *Power System Stability and Control,* McGraw-Hill, New York, 1993.

[22] *NPCC Emergency Operation Criteria*, Available at http://www.npcc.org.

[23] S. Lindahl, G. Runvik, and G. Stranne, "Operational experience of load shedding and new requirements on frequency relay," *Developments in Power System Protection*, Conference Publication No. 434, IEE, pp. 262–265, 1997.

[24] *System Protection Schemes in Power Networks*, CIGRE Task Force Report 38.02.19, June 2001.

Chapter 16

Power System Stability: New Opportunities for Control*

Anjan Bose

Abstract: The electric power generation-transmission-distribution grid in developed countries constitutes a large system that exhibits a range of dynamic phenomena. Stability of this system needs to be maintained even when subjected to large low-probability disturbances so that the electricity can be supplied to consumers with high reliability. Various control methods and controllers have been developed over time for this purpose. New technologies in the area of communications and power electronics, however, have raised the possibility of developing much faster and more wide-area stability control that can allow safe operation of the grid closer to its limits. This chapter presents a conceptual picture of these new stability control possibilities.

16.1 Introduction

The power system networks in North America and Europe are the largest man-made interconnected systems in the world. The Eastern Interconnection in North America, which stretches from the East Coast almost to the Rocky Mountains, is the largest in

*The work described in this chapter was partially coordinated by the Consortium for Electric Reliability Technology Solutions, and funded by the Assistant Secretary of Energy Efficiency and Renewable Energy, Office of Distributed Energy and Electricity Reliability, Transmission Reliability Program of the U.S. Department of Energy under under Interagency Agreement No. DE-AI-99EE35075 with the National Science Foundation. It was also partially funded by the Power System Engineering Research Center, which is a National Science Foundation IUCRC Center supported by about 40 power industry organizations.

terms of geographic area covered, total installed generation capacity, and total miles of transmission lines. Moreover, all the rotating generators in one network rotate synchronously, producing alternating current at the same frequency, that is, all the generators operate together in dynamic equilibrium. Any unbalance in the energy distribution of the system caused by disturbances tends to perturb the system. Large disturbances, usually caused by short circuits of high-voltage equipment, can make the power system become unstable.

Large power systems exhibit a large range of dynamical characteristics, very slow to very fast, and various controllers have been developed over time to control various phenomena. Many of the controls are on-off switches (circuit breakers) that can isolate short-circuited or malfunctioning equipment or shed load or generation. Others are discrete controllers such as tap-changers in transformers or switching of capacitor/reactor banks. Still others use continuous control such as voltage controllers and power system stabilizers in rotating generators or the newer power electronic controls in flexible AC transmission systems (FACTS) devices. FACTS refers to modern electronic devices like high-voltage DC transmission or static VAR controllers (SVC) that can control power flows or voltage.

However, all the controls (especially the fast ones) are local controls; that is, the input and control variables are in the same locale (substation). Most dynamic phenomena in the power system, on the other hand, are regional or sometimes system-wide. Thus designers of power system control have been constrained to handle system-wide stability problems with local controllers. The only system-wide control in the power system is the balancing of the slowly changing system electrical load by adjusting generation levels. This slow dynamical phenomenon allows a slow communication system to reach all the generators in the system in time for the adjustments to be effective.

The only other way to implement nonlocal control has been to dedicate a communication channel between the input variable in one substation to the control variable in another, an expensive proposition that has limited its use.

The tremendous breakthroughs in computer communications of the last decade, both in cost and bandwidth, have opened opportunities yet to be fully utilized in the control of power systems. The availability of many new control devices such as FACTS devices and accurate time-synchronizing signals through the global positioning system is also a factor in this new equation. It is certainly possible now to design fast system-wide controls. However, much research and development are needed to bring such designs to fruition.

In this chapter, we first survey the state of the art in stability control of power systems. Then we outline the new technologies that can be brought to bear on this problem. Finally we lay out a possible development path for system-wide controls in which simple extensions of existing controls can start helping power system operations right away. We will extend to concepts that require significant time and effort to control more complex phenomena. The goal, as always, is to provide more efficient operation, that is, to be able to transmit more power over existing transmission lines with more flexibility.

For more detailed overviews of the existing controls please refer to [1–4].

16.2 Power System Stability

A power system is a complex conglomeration of equipment all connected together electrically. A simple description of the power system and its model is given first so that power system stability and related control can be discussed conceptually without getting bogged down in the details. Of course, readers must be cautioned that often the feasibility of proposed controls depends on these details and implementation of such controls in a power system is a complex undertaking.

16.2.1 Power system model

A power system consists of a transmission network, the nodes of which can be connected to generators or distribution feeders. The transmission network is made up of three phase transmission lines that carry alternating current at 60 Hz (50 Hz in some countries). It is a meshed network, the mesh having developed over time and geography to provide adequate capacity to transmit the electric power from the generators to the distribution feeders. This transmission network can be modeled as a standard mesh circuit

$$I = Y \cdot V \tag{16.2.1}$$

with the frequent assumption that the three phases are balanced and thus only a single phase needs to be represented. A large transmission network can easily have thousands of nodes thus determining the size of (16.2.1). Also, the elements of (16.2.1) are complex quantities representing an AC circuit, which can also be written as an equation of real quantities of twice the size.

The current injections I are either from generators inputting power into the transmission network or from distribution feeders taking power out of the network to the electrical loads. As the main consideration is the movement of electric power from the generators to the loads, the current injections I are of less interest than the power injections S. Thus, instead of (16.2.1) the transmission network is often represented as

$$S = g(V) \tag{16.2.2}$$

where g represents the nonlinear functions that relate power injections to the node voltages. Note that the steady state representation of the power system is a very large set of nonlinear algebraic equations, the nonlinearities being multiplications of complex quantities.

Most of the generators and some of the loads are machines that rotate at 60 Hz at steady state. (To be more precise, the electrical field in these machines rotates at 60 Hz. The actual mechanical rotational speed depends on the design of the machine itself, with synchronous machines having speeds that are a multiple of 60 Hz and induction machines slightly off from such multiples of 60 Hz.) The system, however, is never at steady state because the loads are always varying as customers constantly change their electricity consumption. Thus the power system is continuously subjected to random perturbations. If these changes are relatively small compared to the inertia of the total rotational mass of all the machines, the machine rotational speed does not deviate much from synchronous speed. As long as the small imbalances between generation and load are continually corrected, the system will stay close to synchronous speed (that is, be in synchronism).

If the disturbance is large–loss of a large generator or load, short circuit on a high-voltage transmission line or substation–it is quite possible that some of the machines will deviate significantly from 60 Hz. In some cases machines deviating from synchronous speed may become unstable (that is, not recover and lose synchronism). Obviously this is not desirable, but large disturbances do happen, albeit infrequently, and the power system has to be designed and operated so that such credible disturbances do not actually disrupt power supply to customers. The usual reliability target for North America is to limit large disruptions of power supply to once in ten years.

The dynamics of a rotating machine can be represented in the usual manner as

$$M\omega' + D\delta' = P_a = P_m - P_e \qquad (16.2.3)$$

where δ is the deviation of the shaft rotational angle from synchronous, $\omega = \delta'$ is the deviation from synchronous speed, M is the rotational inertia, and D is the damping. Thus the movement of the shaft away from steady state 60 Hz is dependent on the accelerating power P_a, which is the difference between the mechanical and electrical powers P_m and P_e to the shaft. Equation (16.2.3) can be written as two first-order differential equations and so the dynamics of each rotating machine can be represented by two such differential equations.

This second-order model is somewhat approximate because the electrical power P_e is the real power injection at the electrical node where this rotating machine is connected and is related nonlinearly to the voltage at that node. But the terminal voltage of a generator is dependent on the speed of the machine and is not an independent variable. Moreover, to keep this voltage within certain limits it is controlled by the exciter, which can be modeled by differential equations. Sometimes, especially in those power systems that are prone to sustained oscillations, generators are fitted with a power system stabilizer (PSS), which is another feedback loop from the shaft speed to the exciter. On the mechanical side, P_m is controlled by the governor, which also has describable dynamics. Thus the dynamics of a rotating machine can be represented minimally by two differential equations and more accurately by up to a dozen differential equations. In practice, the details of the representation depend on the importance (size, proximity, etc.) of the machine and the sophistication of the controls in the machine. So the power system dynamics can then be represented as a set of nonlinear differential equations

$$X' = f(X) \qquad (16.2.4)$$

where X is the set of variables such as shaft angle, shaft speed, terminal voltage, and internal variables of the machine, exciter, and governor. This machine terminal voltage is the same as the node voltage, where the generator is connected, as used in (16.2.2). The electrical power P_e used in (16.2.4) is also the real component of the complex power S used in (16.2.2). Thus (16.2.2) representing the static network and (16.2.4) representing the dynamic machines are connected through the power injection and the voltage at the generator nodes.

To summarize, the power system model needed for the purposes of this chapter can be represented by the nonlinear differential and algebraic equations

$$X' = f(X, V, S) \tag{16.2.5}$$

$$S = g(V) \tag{16.2.6}$$

where the number of differential equations can vary from two to twelve per rotating machine and the number of algebraic equations are approximately twice the number of nodes. Thus a large interconnected power system will typically be represented by several hundred differential and several thousand algebraic equations.

The dynamic behavior of this system, as can be expected from the nature of the equations, is complex and varies quite a bit from system to system. The North American Eastern Interconnection and the West European Interconnection are characterized by short transmission lines connecting large load centers. These tend to be more stable than the North American Western Interconnection, which is characterized by long transmission lines. In general, large disturbances tend to make the system unstable and lose synchronism in a second or two; this is known as transient instability. The less stable systems, however, can exhibit undamped oscillations that can go unstable after many seconds. Such systems have natural oscillatory modes that are exhibited under certain operating conditions. Finally, some systems are now susceptible to voltage collapse, which can occur when the system is operating close to its voltage limits when the disturbance occurs.

16.2.2 Power system control

Given the complexity of the power system and its dynamic phenomena, various controls have been developed over time to control various phenomena. These developments have followed the availability of enabling hardware technologies (e.g., electronics, communications, microprocessors) as well as the evolution of control methodologies. In this section, a brief survey is presented of the various controls available today. The survey is neither comprehensive nor complete but is meant to provide a general feel for the technologies being used today and the phenomena that are being controlled.

A. Power system protection

It has always been necessary to protect electrical equipment from burning up due to a short circuit. From the humble fuse to today's microprocessor-based relay, protection gear and methodology have progressed to the point where protection can be looked upon as a fast method of control. The many types of protection technology are obviously outside the scope of this section and only a few applications that affect power system stability are mentioned.

When a fault (short circuit) occurs, the faulted equipment has to be isolated. A short circuit is characterized by very low voltages and very high currents, which can be detected and the faulted equipment identified. If the fault is on a shunt element, like a generator or a distribution feeder, the relay will isolate it by opening the connecting circuit breakers. If the fault is on a series element, like a transmission line or transformer, the breakers on both sides have to be opened to isolate it. The main characteristic of the protection system is that it operates quickly, often in tens of milliseconds, so as to protect the equipment from damage.

In addition, protective relays can be used to do such switching of circuit breakers under other circumstances. For example, if the frequency deviates much from 60 Hz or the voltage from nominal, this may be an indication that the system may be going into an unstable state. Generation or load can be shed (isolated) by the protective relays to correct the situation.

The main characteristics of protective systems are that they are fast and usually triggered by local variables. Circuit breakers that are switched are close to the detection points of the anomalous variables. This obviously makes sense when the purpose is to isolate faulted equipment. However, communication technology today makes it possible to open circuit breakers far from the detected anomalies which raises the possibility of remote or wide-area control.

B. Voltage control

As mentioned before, one way to control node voltages is to vary the excitation of the rotating generators. A feedback control loop changes the excitation current in the generator to maintain a particular node voltage. This control is very fast.

Other ways to control node voltage are to change the tap setting of a transformer connected to the node to switch shunt capacitors or reactors at the nodes. These changes can be made manually by the operator or automatically by implementing a feedback control that senses the node voltage and activates the control. Unlike the generator excitation control, transformer taps and shunt reactances can only be changed in discrete quantities. Often these types of control scheme have time delays built into them to avoid excessive control actions.

More recently power electronic control devices have been introduced in the shunt reactance voltage control schemes. This makes the control much more continuous and often is done at a much faster time frame than the usual shunt switching. These SVCs are becoming more common.

Voltage control is always a local control. However, controlling the voltage at one node affects the neighboring nodes. In Europe, controllers that coordinate the voltage control over an area are being tried and such area controls may be introduced in the North American systems.

C. Transmission power flow control

Most power systems have free flowing transmission lines. Although power injections and node voltages are controlled quite closely, the power flow on each transmission line is usually not controlled. However, such control is feasible.

A phase-shifting transformer can control the power flow across it by changing the phase using taps. This has been used especially on the Eastern Interconnection in North America. The control is local, discrete and slow. A power electronic version is now under experimentation.

The major advantage of the AC transmission grid is its free flowing lines with relatively less control, so the wholesale control of every transmission line is not desirable nor contemplated. However, controls on some lines have always been necessary and some new advantages may be realized in the more deregulated power system when monitoring transactions between buyers and sellers have to be better controlled.

Flow over DC transmission lines is always controlled and the control is very fast. The number of DC lines is low.

D. Frequency control

Frequency is controlled by balancing the load with generation. The governors on every generator sense any change in the rotational speed and adjust the mechanical input power. This governor control is the primary control for maintaining frequency. A secondary control to set the governor setpoints is used to ensure that the steady state always returns to 60 Hz. The governor control is local at the generator and fast. The secondary control is done over the whole system by the central controller and is slow. This control is also known as automatic generation control (AGC) or load frequency control.

In North America and Western Europe, the frequency is controlled quite tightly whereas in many other places, even in developed economies like the United Kingdom and Scandinavia, frequency is allowed to vary over a wider range. As the deviation of frequency from 60 Hz is a symptom of the imbalance between generation and load, the frequency control performance requirement depends on how well one wants to control the power supply commitments made between seller and buyer.

E. Control center

As mentioned in the above sections most of the controls are local. The only area-wide control is the secondary frequency control or AGC. This is implemented as a feedback control loop in which the generator outputs and tie-line flows are measured and brought back to the control center, and the governor control setpoints are calculated and sent out to the generators from the control center. The data rate (both input and output) is between 2 and 4 seconds.

The control center performs many other functions, although AGC is the only automatic feedback control function. The main function is real-time data acquisition from all over the grid so that the operator can monitor its operation. Another is the manual operation of controls such as opening or closing circuit breakers and changing transformer taps. These functions are jointly known as the supervisory control and data acquisition (SCADA) and the control center is often referred to as SCADA.

These SCADA-AGC functions at central control centers evolved in the earlier part of the last century but in the 1960s their implementation was accomplished with digital computers. Remote terminal units (RTU) were positioned in every substation and generating station to gather local data, and this data was then transmitted from the RTUs to the control center over communication lines, usually microwave channels but sometimes telephone lines. This scheme is shown in Figure 16.2.1. The data normally includes the switching statuses (on/off) of all the circuit breakers as well

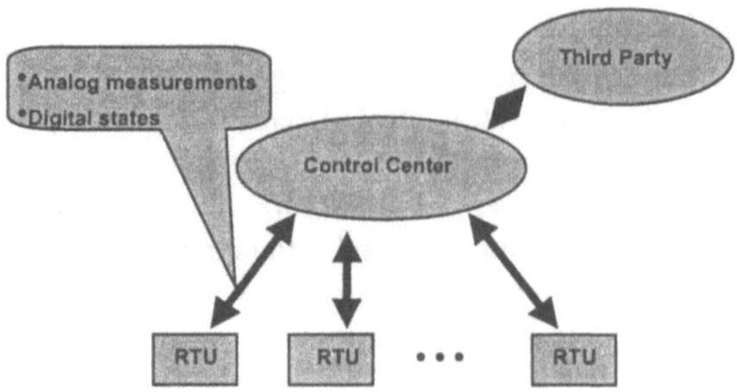

Figure 16.2.1: The control center has direct communication channels to the RTUs at each substation and generating station

as the current values of voltages and complex power (i.e., watts and vars). Although these control centers are quite separate from other computer systems, they do accumulate a large set of historical data that can be utilized for various engineering study and analysis. Thus it is quite common to have a network connection to third-party (usually engineering) computers.

As the computational power of the control centers grew, more functions have been added. The main one has been the state estimator, which calculates the real-time steady state model of the network. This real-time model can then be used for two kinds of calculations. One, known as security analysis, can study the effects of disturbances (contingencies) and can alert the operator if the postcontingency conditions violate limits. The other, usually using a family of analysis known as optimal power flow, can suggest better operational conditions. All these analytical tools provide better operational guidance than the old SCADA systems, and they are now known as energy management systems (EMS).

Another recent trend has been the increasing use of microprocessors and faster communication within the substations to gather more real-time data. This data gathered at the few milliseconds rate is stored at the substations but is too voluminous to be broadcast. Instead certain sequences of this data (say, after an emergency or disturbance) are then imported, increasingly, over some sort of network and used for study purposes. This is shown in Figure 16.2.2. Data is now being measured and gathered at the substations at a much faster rate than can be communicated to the control center, which are only capable of polling RTU data every few seconds. The excess data can be recorded at the substations and for now is gathered only after the fact for study.

Power system control can then be summarized as follows:

- Most automatic controls are local.
- At the generator there is the governor control of generator output, the exciter control of generator terminal voltage, and sometimes, PSS control. These are continuous fast feedback control.
- Node voltages can also be controlled by transformer taps and shunt reactances.

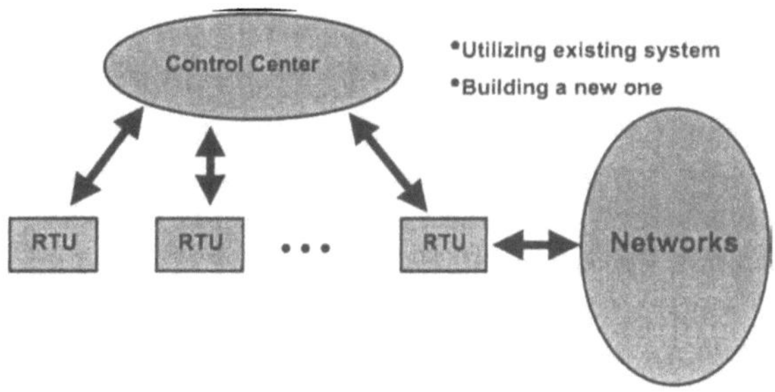

Figure 16.2.2: Substations today are measuring more real-time data than can be sent over the communication channel capacities to the control center. However, stored data can be moved slowly over other computer networks.

These are slow discrete controls but new continuous fast SVCs are becoming available for use.

- Where DC transmission is used, fast continuous control of line flow is available and new tools to do so on AC lines are becoming available. Slow controls using phase-shifting transformers are still being used in a few places.
- Protective relays that isolate faulted equipment operate locally but are very fast. With communication from other parts of the network, they have great potential for fast control.
- The secondary frequency control of generator governor setpoints is the only area-wide control used today. This slow control implemented through the central control center is discrete at the rate of a few seconds.
- Much more data at very fast rates is being gathered at the substations but the communication and control system to utilize this data for faster controls is lacking.

16.2.3 Stability limits

Power systems are designed and operated to survive large disturbances such as storms, lightning strikes, and equipment failures. This usually means that even though some power system equipment may be separated or isolated as a result of automatic protection and control actions, power supply to customers will not be disrupted, or at least very localized.

Operational limits for the transmission network can then be set using the following logic.

- The maximum power each transmission line can transmit is limited by its current carrying capacity, known as its thermal limit.
- The maximum loading of the transmission network is determined by any one of the lines hitting its thermal limit. The loading of each line in the power system is determined by (16.2.6), and so various combinations of generator and load injections at the nodes (that is, various operating conditions) can produce

this limiting condition. The SCADA continually checks for such thermal limit violations as the operating condition changes over time.

- However, operating the system at such a limiting condition is not prudent because the loss of a transmission line or other equipment would probably overload some other lines. Thus the operating limit is not when the first line hits its thermal limit, but when the loss of any one piece of equipment makes a line hit its thermal limit. This is known as the $N - 1$ criterion for operation because any one of the N components of the power system can be lost without overloading any part of the system.
- If a disturbance or short circuit occurs, the first line of protection should isolate the faulted equipment. This is the rationale behind the $N - 1$ criterion. It is possible, however, that the same disturbance may cause instability (determined by equations (16.2.5) and (16.2.6)) in which case more than just the faulted equipment will be lost. In such a case the maximum loading of the system will have to be lowered so that instability does not occur. This loading limit is thus set by the stability criterion rather than the thermal loading. Those power systems that are stability limited are of interest in this chapter.
- If better controls can increase this stability limit, then the system can be loaded at a higher level. This provides better utilization of the transmission network. For example, the north-south power transfer along the West Coast of the United States is limited by stability, and any increase in this limit has a direct economic benefit by enabling more transactions between generators and customers.

16.3 Challenges and Opportunities

There are significant economic incentives to increase the transmission limits of existing systems. In fact, the major constraints of the deregulated power markets are the transmission system limits. Generation companies sell power to distribution companies (or directly to large customers) through bilateral agreements or auction markets. These transactions must flow over the transmission system. If the transmission capacity was higher than all possible power flows these transactions produce, the market would be ideal. This, however, is not the case because the transmission system was built when power companies were vertically integrated and sized for the expected power flows resulting from the planned operation of the generators. The transmission system was not designed to accommodate all buy-sell agreements between generators and consumers.

Thus all power transactions must be checked before hand to ensure that the flows are within capacity limits. As there may be hundreds of simultaneous transactions between generators and consumers, and because the effects of these transactions on the flows are not linear (16.2.2), all simultaneous transactions must be studied together to check whether transmission limits are violated. When they are, they cause the "congestion problem." If congestion is expected, all the transactions cannot be allowed. Different power systems have predetermined procedures about how and which transactions will have to be cut back. The procedures have to be fair to all parties and may require compensation to some consumer. An independent referee called a "security" coordinator has to make these decisions.

These security coordinators are known by various names: independent system operator, regional transmission operator, transmission system operator, and so forth. They discharge their responsibilities in two steps. First, they have to approve all transactions. Their main tool to check violations of transmission limits is the "power flow" program that solves (16.2.2). Their main tool to recommend corrective action in case of anticipated congestion is an optimal power flow (OPF) program that optimizes the correction procedure (e.g., minimum compensation) using (16.2.2) as constraints. Then, in real time, the independent system operator has to watch for unexpected emergencies that cause limit violations, and if it happens, take emergency corrective action. In such emergencies, ensuring the survivability of the power system is more important than minimizing cost.

So the transmission limits are the constraints that also limit the power markets. For systems that are thermally limited, the only way to raise limits is to build more transmission. For those systems that are stability limited, better controls could increase the stability limit. Thus our interest in this chapter is on better control of stability.

16.3.1 New technologies

Essentially, there are three relevant classes of technologies:

- Faster and cheaper computers
- Cheap broadband communications
- Better power electronic controls (also known as FACTS specifically developed to control the AC power system).

Some of these technologies are already in use in power systems as mentioned in Section 16.2. What we are proposing here is the development of new controls using a combination of these technologies. These controls will be significantly different in concept than the existing ones, and they will be fast and system-wide to dramatically increase stability limits.

A. Computers

Computers (or microprocessors) are embedded in everything: meters, protective relays, data concentrators, communication switches, etc. They are programmable; that is, the functions of the gadget in which they are embedded can be changed by software. Thus controls that utilize these components can be adapted through changed settings (simple) or changed logic (more difficult), providing flexibility in the design of this software.

Workstation computers are also much faster and cheaper than in the past. Very large calculations can be done very quickly. Such analysis can then be part of the control, bringing even more intelligence into the control loop. For example, if a control is devised to shed load to avoid instability, an optimal power flow could determine which loads are to be switched off.

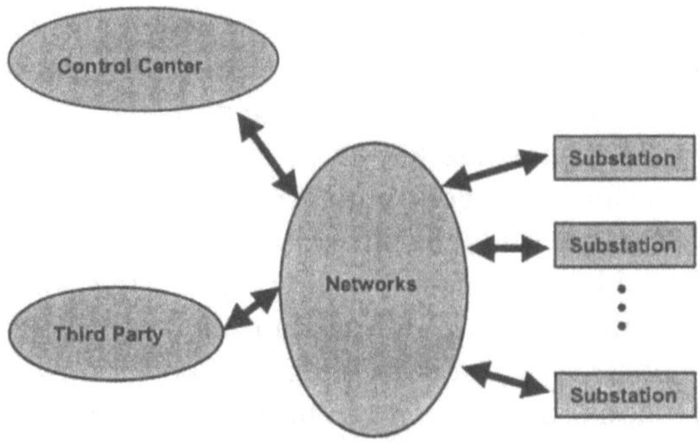

Figure 16.3.1: High bandwidth dedicated communication channels between substations and control center will allow faster area-wide controls

B. Communications

Electric power companies have always had their own communication systems. As mentioned in Section 16.2, these have mainly been microwave channels that connect every substation and generating station. The use of optical fiber is increasing at a tremendous rate. Optical fiber was first used within newer substations and generating stations, but the older ones are being rapidly retrofitted. The substations can now gather more real-time data at faster rates so that fast-appearing emergency conditions (such as lightning strikes) can be dealt with quickly. The captured data has to be stored locally for later transmission over communication networks (Figure 16.2.2).

Power companies also ran optical fiber along transmission towers to become communications providers capable of satisfying the ever-increasing demand for bandwidth. Although this venture has not panned out because of the glut of unused bandwidth, a broadband network is now easily available to the power companies. If this network bandwidth is broad enough, then all the data being collected at the substations can be transmitted in real time to other locations like the control center. In fact, a network can be envisioned (Figure 16.3.1) where the real-time data would be available to different computers depending on their function. The control center could be decentralized so that functions could be put in different places depending on need. With a network like this, the stark differentiation between centralized and local control would diminish and controllers could use the most appropriate data needed for control.

A communication network that can meet the varied needs of the power system would be much more complex than the simple star network used today to poll substation RTUs. Moreover, the control functions would not be concentrated at a central computer in the control center, but would be distributed over numerous computers in substations, generating stations, and engineering offices. Such distributed computer communication is being developed today for various applications and is shown conceptually in Figure 16.3.2. It shows that some of the functions (measurements or calculations) would be publishers of data while other functions using this data (applications, controls) would be subscribers. The network would be controlled by other computers called "quality of service managers" (QoS). Such middleware is being developed for other applications, although it has yet to be developed for the architecture appropriate for the power grid. It should be mentioned that, given the concern for the security of such critical infrastructures as the power grid, such computer communication systems must be secure from external intrusions and must be built into the QoS.

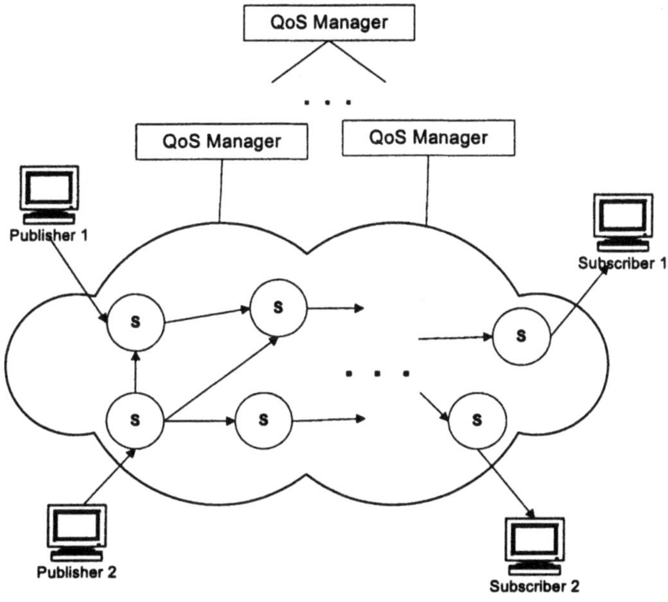

Figure 16.3.2: A distributed communication architecture that can be used to meet power system control requirements of the future

C. FACTS

FACTS devices available today were discussed in Section 16.2. Although they are different in detail by model and manufacturer, they fall into three classes:

- DC transmission controls
- SVC
- power flow controller.

In addition, special controllers can be built for specific purposes using the same principles. One major advantage of these controllers is their speed: control actions take place in milliseconds, which is in the same time frame as protection actions.

16.3.2 New controls

The proposed control concepts described here are all wide-area controls. Although local controls continue to be improved by newer technologies, the conceptual functionality of these local controls remain the same. The wide-area controls presented here would often take care of the local controllers, but would mainly improve the overall stability of the power system. The concepts are presented in the order of increasing complexity, implying that the ones presented first would be easiest to implement.

A. Frequency control

As noted before, frequency is controlled by balancing load with generation. The primary governor control at the generators is local while the secondary AGC that adjusts the governor setpoints is area-wide. The primary control is continuous whereas the secondary control is discrete, usually using 2- to 4-second sampling.

Given that all generators in a region are no longer owned by the same organization, this area-wide AGC will become more decentralized. The Federal Energy Regulatory Commission (FERC) ancillary service regulations do allow third-party AGC, but a new communication-computation-control scheme needs to be developed. As this control is quite slow (2- to 4-second sampling), feasibility of control is not a problem. The more complex communication scheme required is also not a problem; although a meshed communication network is required rather than the present star network, the bandwidth requirement remains modest. However, such a network introduces other modes of failure such as signal delays and the control has to be robust enough to handle them.

B. Regional voltage control

Voltage control in North America has always been local, although Europe is trying some regional control schemes. FERC recognizes voltage-VAR control as an ancillary service. Control schemes for such regional control need to be developed, but the schemes must ensure that such service can be quantified and paid for as an ancillary service. This type of control, like frequency control, is relatively slow, and so the feasibility of the control and communication is not an issue. The main hurdle is the selection of input and output variables that can handle all the varied operating conditions that the power system endures. Thus this challenge is a classical one of developing a practical robust controller.

C. Small signal stability control

Small signal instability occurs when a system perturbation, even a small one, excites a natural oscillatory mode of the power system. These oscillations are slow, usually under 1 Hz. The main method used today to guard against small signal instability is the off-line tuning of PSS. These PSS are local controllers on the generators. Thus local controllers are used to mitigate system oscillation modes, a procedure recognized to have significant disadvantages. New controllers that can use system-wide inputs (not necessarily more inputs per controller but input signals from further away) must be developed. Such remote signal inputs will obviously require communication channels that could be dedicated or that could use a more flexible communication mesh network.

Another control concept is to adaptively change the PSS setpoints according to the power system operating conditions. This would be analogous to the AGC by introducing a secondary control scheme that would periodically adjust the setpoints of the local PSS controllers as the system changes. The challenge here is that the determination of PSS setpoints requires large analytical calculations, which are today done off-line but will have to be done on-line in this case. The speed of calculation is not a major concern as changing the setpoints can be done quite infrequently, probably in minute increments.

D. Voltage stability control

Voltage instability occurs when a change in the power system causes an operating condition that is deficient in reactive power support. Guarding against such instability requires the anticipation of such contigencies that cause voltage instability and the taking of preventive action. New preventive control schemes that also include special protection schemes are needed to isolate those areas with VAR deficiencies.

This is not a traditional stability control that responds to a disturbance. This is an action plan to ensure that the system operating condition does not stray into an area where a perturbation can cause voltage instability. The control of the transient condition after a disturbance occurs is handled in the next section.

E. Transient stability control

The development of such a control scheme is by far the most difficult because a disturbance that can cause instability can only be controlled if a significant amount of computation (analysis) and communication can be accomplished very rapidly. This concept is approached in three increasingly difficult levels:

- the first uses off-line studies to manually adjust protective schemes that would operate only if the disturbance occurs,
- the second automatically adjusts these protective schemes with on-line calculations,
- the third and final directly operates the control actions after the disturbance occurs.

1. 'Soft-wired' remedial action schemes

An advance in this direction would be to generalize remedial action schemes (also known as special protection schemes) to control transient stability. These remedial action schemes are developed today from the results of voluminous off-line studies and are implemented with a "hard-wired" communication system. Thus the system values and statuses monitored and the breakers controlled cannot be modified. What is proposed here is the development of a generalized communication system that can enable the implementation of new remedial action schemes by software modification. Although a comprehensive communication scheme will be required in this type of control, the computation requirements will be modest as the control schemes are largely defined off-line.

2. On-line setting of remedial action schemes

A step forward would be to develop methods to control transient stability but with less dependence on off-line studies and more use of on-line computation. The main idea is to use more real-time data to determine what control is needed. It is proposed to develope soft-computing techniques using pattern recognition, neural networks, expert systems, etc., to process real-time data to decide the best control action. Of course, much off-line training of the software may still be required, but the expectation is that the control action would be much more efficient than that decided purely off-line.

3. Real-time control of transient stability

The objective here is to develop a global control for transient stability (with no off-line assists). For this to be feasible, the computation needed to determine the disturbance scenario and the necessary controls for stabilization has to be in the same timeframe as today's protection schemes (milliseconds). Whether this is indeed possible with today's technology is not known. However, the goal here would be to determine what kind of communication-computation structure will be needed to make this goal feasible.

16.4 Conclusions

A roadmap for the development of new controls for power system stability is presented. The challenges are significant but the economic payoff rewards are high. The ability to push the limits of the transmission system depends on how well stable operation can be maintained against disturbances. The improvements in computers, communications, and controllers are already being used in power systems in many ways. By combining them to develop area-wide controls for power systems, stability can be controlled better to increase transmission limits.

The roadmap outlined here ranks the conceptual developments by the difficulty of their implementation. The requirement of more complex computation or communication makes implementation more difficult, but several of the steps outlined above can be taken right now using existing technology without significant new development.

Bibliography

[1] P. M. Anderson and B. K. LeReverend, "Industry experience with special protection schemes: IEEE/CIGRE Committee Report," *IEEE Trans. Power Systems*, vol. 11, pp. 1166-1179, Aug. 1996.

[2] M. Begovic, D. Novosel, and M. Milisavljevic, "Trends in power system protection and control," *Proc. 1999 Hawaii International Conference on System Sciences*, Maui, HI, Jan. 1999.

[3] C. W. Taylor, *Power System Stability Controls*, The Electric Power Engineering Handbook, CRC and IEEE Press, New York, 2000.

[4] K. N. Stanton, J. C. Giri, and A. Bose, *Energy Management*, The Electric Power Engineering Handbook, CRC and IEEE Press, New York, 2000.

Chapter 17

Data Fusion Modeling for Groundwater Systems using Generalized Kalman Filtering

David W. Porter

Abstract: Engineering projects involving groundwater systems are faced with uncertainties because the earth is heterogeneous and data sets are fragmented. Current methods providing support for management decisions are limited by the data types, models, computations, or simplifications. Data fusion modeling (DFM) removes many of the limitations and provides predictive modeling to help close the management control loop. DFM is a spatial and temporal state estimation and system identification methodology that uses three sources of information: measured data, physical laws, and statistical models for uncertainty in spatial heterogeneities and for temporal variation in driving terms. Kalman filtering methods are generalized by introducing information filtering methods due to Bierman coupled with (1) a Markov random field representation for spatial variation and (2) the representer method for transient dynamics from physical oceanography. DFM provides benefits for waste management, water supply, and geotechnical applications.

17.1 Introduction

Engineering projects involving groundwater systems are faced with uncertainties because the earth is heterogeneous and typical data sets are fragmented and disparate. In theory, predictions provided by computer simulations using calibrated models constrained by geological boundaries provide answers to support management decisions, and geostatistical methods quantify safety margins. In practice, current methods are limited by the data types and models that can be included, computational demands, or simplifying assumptions.

Data fusion modeling (DFM) removes many of the current limitations and is capable of providing data integration and model calibration with quantified uncertainty for a variety of hydrological, geological, and geophysical data types and models. The benefits of DFM for waste management, water supply, and geotechnical applications are savings in time and cost through the ability to produce visual models that fill in missing data and predictive numerical models. Predictive models aid management optimization and help close the management control loop.

DFM has the ability to update field scale models in real time using PC or workstation systems. DFM is a spatial and temporal state estimation and system identification methodology that uses three sources of information: measured data, physical laws, and statistical models for uncertainty in spatial heterogeneities and for temporal variation in driving terms.

What is new in DFM is the solution of the causality problem in the data assimilation Kalman filter methods to achieve computational practicality. The Kalman filter is generalized by introducing information filter methods due to Bierman coupled with a Markov random field (MRF) representation for spatial variation and with the representer method from physical oceanography. A Bayesian penalty function is implemented with Gauss-Newton methods. This leads to a computational problem similar to numerical simulation of the partial differential equations (PDEs) of groundwater. In fact, extensions of PDE solver ideas to break down computations over space form the computational heart of DFM.

State estimates and uncertainties can be computed for heterogeneous hydraulic conductivity fields in multiple geological layers from the usually sparse hydraulic conductivity data and the often more plentiful head data. Further, a system identification theory has been derived based on statistical likelihood principles. A maximum likelihood theory is provided to estimate statistical parameters such as Markov model parameters that determine the geostatistical variogram.

Field scale application of DFM at the Department of Energy (DOE) Savannah River Site is presented and compared with manual calibration. A better fit to the data is obtained with less time spent on calibration. DFM calibration runs converge in less than one hour on a Pentium Pro PC for a three-dimensional model with more than 15,000 nodes. Run time is approximately linear with the number of nodes. Conditional simulation is used to quantify the statistical variability in model predictions such as contaminant breakthrough curves.

DFM has been applied at other sites for models with more than 100,000 nodes. Specific sites where DFM has been applied include DOE at Fernald, Hanford, and Pantex; Department of Defense (DOD) at the Massachusetts Military Reservation, Beale Air Force Base, and the Letterkenney Army Depot; and at a former manufactured gas plant in Marshalltown, Iowa.

17.2 Background

The DFM methodology builds on a foundation of work from hydrogeology, meteorology, image processing, and the data fusion communities in military, computer vision, and control application areas. DFM fits into the framework for hydrogeological decision analysis presented in Freeze et al. [17]. A stochastic methodology is described for waste management, water supply, and geotechnical applications. It is stated that uncertainty in heterogeneous hydraulic conductivity must be viewed as an autocorrelated spatial stochastic process in order to account for uncertainty in subsurface investigation. It is explained in [17] that the current Bayesian theory has limitations in the application of inverse modeling to take advantage of important data worth such as hydraulic head data used in conjunction with hydraulic conductivity data.

The methods that are most limited by data and model types are traditional inverse modeling methods such as those in Cooley [11] and Hill [22] and classical kriging methods such as those in Journal and Huijbregts [25]. Inverse modeling is performed using nonlinear regression in Hill [22]. In theory, a least squares fit to head data can be achieved with the incorporation of point data on conductivity as prior information. However, in order to avoid overparameterization, the estimation must be performed on a larger scale of variability than appropriate for the conductivity data as explained in Hill [22]. Kitanidis and Vomvoris [27] explain that the overparameterization problem can be solved using a random field representation that can account for spatial variability at various scales. The lack of a random field representation in traditional inverse modeling limits the data types that can be combined.

Classical kriging methods use nested variability structures to account for variability at different scales. However, the classical methods do not use groundwater numerical models, so there is no way to combine conductivity data with head data based on the physical relationship between them. The lack of a groundwater numerical model in classical kriging limits the data types that can be combined.

Numerous methods are able to work with a variety of data and model types, such as those in Van Geer, Te Stroet, and Yangxiao [38, 41], Zou and Parr [43], Eppstein and Dougherty [16], Daniel and Willsky [14], McLaughlin and Townley [30], Carrera and Glorioso [5], Carrera and Neuman [6–8], Medina and Carrera [29], Kitanidis and Vomvoris [27], Hoeksema and Kitinidis [23], Dagan [13], Gelhar [18], and Sun and Yeh [37]. Several unified theories exist for state estimation and system identification to support modeling conceptual iteration. The remaining issue is that computational demands require simplifying assumptions for application.

In other fields, Kalman filtering solved apparently similar computational problems, but this has not happened for groundwater systems. Part of the contribution of the results presented here is to recognize that the difficulty is the assumption of causality used by Kalman in his classical paper (Kalman [26]). The Kalman filter uses the idea of causality, that there is a past causing a future, in order to break down computations over time to achieve computational efficiency. But the spatial physics and statistics of groundwater are noncausal, so the Kalman filter cannot break down the computations over space the way it does over time for groundwater.

Results presented here provide the first solution to the causality problem that generalizes the Kalman filter to produce a spatial state estimation and system identification methodology that is computationally practical. The Kalman filter is generalized by introducing information filter methods due to Bierman (Bierman [3], Bierman and Porter [4], and Porter [31]) coupled with an MRF representation (Whittle [40], Clark and Yuille [10], and Chellappa and Jain [9]) for spatial variation and with the representer method (Bennett [1]) from physical oceanography for transient dynamics.

DFM includes a system identification theory based on statistical likelihood methods. A maximum likelihood theory is provided to estimate statistical parameters such as Markov model parameters that determine the geostatistical variogram by generalizing ideas in Levy and Porter [28].

DFM draws on the related application area of meteorological estimation as reviewed and described in Ghil [20] and Courtier et al. [12]. Bayesian estimation is accomplished using data assimilation methods that rely on Kalman filtering, adjoint, and variational techniques. Many of the same methods apply to hydrogeology, but practical applications are limited by computational demands. DFM also draws on the mathematically related field of image restoration and reconstruction for an understanding of the roles of causality and Markovian statistical models (Jain [24] and Sezan and Tekalp [35]). In addition, DFM draws on well-established methods from the military, computer vision, and control communities (Hall [21], Bierman [3], and Clark and Yuille [10]). Applications from these communities provide fundamental concepts and illustrate the value of mathematically combining data from multiple sources.

17.3 Mathematical Motivation

DFM builds on information filter methods that solve the same estimation problem that the Kalman filter solves but work in terms of statistical information rather than covariance (Bierman [3]). Time domain techniques are explained here in preparation for extension to spatial estimation. The physical and statistical models for the state x and the measurements z over time k used by the information filter are as follows:

$$x_{k+1} = \Phi_k x_k + u_k + w_k, \quad z_k = H_k x_k + v_k \tag{17.3.1}$$

where x_k is the physical and statistical state vector for ($k = 0, 1, 2, \cdots$), Φ_k is the state transition matrix, u_k is the known deterministic input, w_k is the white process noise with mean zero normalized to identity covariance, H_k is the measurement sensitivity matrix, and v_k is the white measurement noise with mean zero normalized to identity covariance.

To be compatible with computer implementation, the equations are rearranged as

$$-\Phi_k x_k + x_{k+1} = u_k + w_k, \quad H_k x_k = z_k - v_k. \tag{17.3.2}$$

The physical and statistical model corresponding to the first equation in (17.3.2) is interpreted as a data equation using the ideas of Bierman [3] and Duncan and Horn

[15]. The state is estimated by solving a regression problem where the deterministic input u is taken as a measurement, the process noise w is taken as measurement noise, and the model dynamics on the left-hand side of the equation are taken as measurement sensitivity. Remarkably, the regression estimate and estimate error covariance are the same as the Kalman filter estimate and estimate error covariance for the problem represented by (17.3.1). But the regression problem is simpler conceptually and well-established numerical methods exist for its efficient solution.

The regression estimate is formed by doing a least squares fit of the state to the measurements in (17.3.2). This is done by solving the generally overdetermined system that is the same as (17.3.2) except that there is no measurement noise on the right-hand side. The system will be overdetermined if sufficient measurements and prior information are available. Solution of the overdetermined system is achieved in the computer by manipulation of the information array given by

$$\begin{bmatrix} H_0 & & & & z_0 \\ -\Phi_0 & I & & & u_0 \\ & H_1 & & & z_1 \\ & -\Phi_1 & I & & u_1 \\ & & H_2 & & z_2 \\ & & -\Phi_2 & I & u_2 \\ & & & \ddots & \vdots \end{bmatrix} \tag{17.3.3}$$

where I is the identity matrix.

A family of solution methods exist to take advantage of different problem structures for computational efficiency (Bierman [3], Porter [31]). Banded sparse solution methods such as Householder reflections and Givens rotations break down the computations over time to achieve the Kalman filter solution. If the actual measurements are removed, the information filter solves the physical system. Consequently, the information filter can be viewed as a generalization of the numerical solver for the physical system. When we generalize to spatial estimation, we take advantage of this to use extensions of PDE solvers to achieve computationally efficient estimation.

The system of equations represented by (17.3.1) takes the same general form as the models used for the hydrogeological, meteorological, and image processing applications of Van Geer, Te Stroet, and Yangxiao [38, 41], Zou and Parr [43], Eppstein and Dougherty [16], Ghil [20], Courtier et al. [12], and Sezan and Tekalp [35]. Markov processes that are correlated in time and are inputs to the physical model or enter the measurements directly can be adjoined to the state vector where appropriate modifications are made to the state transition matrix, process noise, and measurement sensitivity matrix (Bierman [3]). Noncausal spatial correlations can be incorporated in a covariance matrix for process or measurement noise that is used to statistically normalize (17.3.1) to have noise terms with identity covariance. However, the statistical normalization of the spatial correlations can be computationally prohibitive for field scale parameterization. In later sections, it will be shown that DFM generalizes previous results to achieve computational practicality by adjoining MRF states to the state vector and keeping the process and measurement noises white with identity covariance. This generalizes the Kalman filter for noncausal spatial processes.

17.4 Approaches

State estimation and prediction is based on combining information from physical models, statistical models, and measurement models. Physical models for flow and transport provide relationships between dependent states (such as heads and contaminant concentrations) and parameter states described below. Physical models can include model errors or be treated as hard constraints.

Statistical models include models for spatial variation, temporal variation, and constants. Spatial variation in parameter states such as hydraulic conductivity or kinetic mass transfer parameters can be modeled using geostatistical MRF models. Temporal variation in parameter states such as recharging and contaminant source loading can be modeled using geostatistical Markov random processes. (See Snodgrass and Kitanidis [36] for the use of geostatistical models for contaminant source identification.)

Bayesian prior knowledge of constant parameter states such as trend, zonation, and individual parameters can be modeled using prior estimates and prior uncertainty. An alternative is to treat constant parameters as structural parameters with no prior knowledge except that they are fixed but unknown. Parameter states can include quantities such as hydraulic conductivity, storativity, recharge, kinetic mass transfer parameters, dispersivities, porosity, source loading, boundary conditions, initial conditions, and measurement error parameters.

Calibration targets and measurements can include water level and contaminant concentration time histories and data that can be interpreted as hydraulic conductivity measurements and particle pathline measurements. Measurement error models are additive and reflect measurement dynamics that can include bias and trend parameters and measurement noise.

The DFM method uses Bayesian state estimation methods accomplished through a combination of information filter and Markov process methods from Porter et al. [32] with the representer method from physical oceanography from Bennett [1]. The computationally intensive part of the approach involves simulations forward and backward in time where the number of simulations depends on modeling conditions.

A Bayesian penalty function is minimized using the Gauss-Newton method. For the Gauss-Newton method, options have been implemented for back tracking (Press et al. [33]) and the generalization of the Levinberg-Marquardt approach known as the model trust region method (Press et al. [33] and Vandergraft [39]). Inequality constraints can also be included to keep the state estimates physically reasonable. To make the ideas concrete, the simplest Gauss-Newton option is presented by replacing the nonlinear problem with a sequence of locally linearized problems.

17.4.1 Information filtering method

The DFM approach from Porter et al. [32] is described here. The following describes the MRF representation, models and data, Bayesian penalty function, and state estimation. The spatial autoregressive representation of an MRF for the spatial stochastic process f on a regular two-dimensional grid is given by (Whittle [40],

Clark and Yuille [10], and Chellappa and Jain [9])

$$f_{i,j} = \sum_{S} a_{k,l}\, f_{i+k,j+l} + w_{i,j} \tag{17.4.1}$$

where $a_{k,l}$ are Markov interpolation coefficients, $w_{i,j}$ is Markov process noise that is spatially white with variance q, and S is a local spatial region of support. The MRF extends to three dimensions, and the correlation structure can be anisotropic. The MRF provides a general and simple way to model spatial variability over any scale with only a local equation. This amazing property is behind MRF use in statistical physics and computer vision (Geman and Geman [19]). The Markov model has been extended to a nonuniform slightly deformed grid to be compatible with grids over which groundwater numerical models are often defined.

We use nested structures for spatial variability at different scales similarly to Kitanidis and Vomvoris [27] and Carrera and Glorioso [5] using ideas from Journal and Huijbregts [25]. At the observational scale, measurement noise and microstructure appear as white noise. At intermediate scales larger than observational but smaller than the aquifer, spatial variability is modeled as a stationary process using the MRF. At the aquifer scale, nonstationary effects are taken into account with a deterministic trend having unknown coefficients that are estimated as additional states in DFM. To keep the presentation simple, the trend terms are not included in the following description of the information filter method.

A baseline groundwater system is considered that has steady state flow with transient advective-dispersive transport for which point measurements of hydraulic head, hydraulic conductivity, and contaminant concentration are available, for which the log-conductivity is viewed as an MRF, and for which the finite difference representation is used. The baseline model avoids extraneous detail and can be generalized to include many other effects. Taking the Gauss-Newton viewpoint and linearizing the equation about the current estimate, the following is obtained:

$$\begin{array}{c} A_1\, h + A_2\, f = q_1 + w_1 \\ \text{(Steady Flow Model)} \\[1em] M_1\, h = z_1 - v_1 \\ \text{(Head Measurements)} \\[1em] M_2\, f = z_2 - v_2 \\ \text{(Conductivity Measurements)} \\[1em] A_3\, f = q_2 + w_2 \\ \text{(MRF)} \\[1em] A_{4,t+1}\, c_{t+1} + A_{5,t}\, c_t + A_{6,t}\, h + A_{7,t}\, f = q_{3,t} + w_{3,t} \\ \text{(Transient Transport Model)} \\[1em] M_{3,t}\, c_t = z_{3,t} - v_{3,t} \\ \text{(Contaminant Data)} \end{array} \tag{17.4.2}$$

where h is hydraulic head relative to the current estimate; f is log conductivity relative to the current estimate; A_1, A_2, and A_4 through A_7 are banded sparse matrices for flow and transport models; A_3 is the banded sparse matrix for the Markov model; q_1 and q_3 are sink/source and local linearization terms; q_2 is the Markov linearization term; w_1, w_2, and w_3 are process noises with zero means and normalized to identity covariances; M_1, M_2, and M_3 are measurement sensitivity matrices; z_1, z_2, and z_3 are measured data and local linearization terms; v_1, v_2, and v_3 are measurement noises with zero means and identity covariances; and c_t is contaminant concentration at time t relative to the current estimate.

Boundary and initial conditions are assumed known, although they can be estimated also. The measurement sensitivity matrices for the data perform an interpolation from the grid where the states are defined to the points where the measurements are made.

DFM estimation is accomplished by finding h, f, and c that minimize the Bayesian penalty function, J, defined as follows:

$$J = \underbrace{\|q_1 - A_1 h - A_2 f\|^2}_{\substack{\text{Penalizes flow} \\ \text{model errors}}} + \underbrace{\|z_1 - M_1 H\|^2}_{\substack{\text{Penalizes} \\ \text{head fit errors}}} + \underbrace{\|z_2 - M_2 f\|^2}_{\substack{\text{Penalizes} \\ \text{conductivity} \\ \text{fit errors}}} + \underbrace{\|q_2 - A_3 f\|^2}_{\substack{\text{Penalizes} \\ \text{excessive} \\ \text{conductivity} \\ \text{variations}}}$$

$$+ \sum_t \Big(\underbrace{\|q_{3,t} - A_{4,t+1} c_{t+1} - A_{5,t} c_t - A_{6,t} h - A_{7,t} f\|^2}_{\text{Penalizes transport model errors}} + \underbrace{\|z_{3,t} - M_{3,t} c_t\|^2}_{\substack{\text{Penalizes} \\ \text{contaminant} \\ \text{fit errors}}} \Big) \tag{17.4.3}$$

where $\|\cdot\|$ denotes the standard Euclidean norm. The penalty function can be viewed as a natural and intuitive formalization of the criteria under which manual trial-and-error model calibration is performed. There is no additional weighting of the terms in the penalty function since the measurement and process noises have already been normalized to identity covariance in (17.4.2). Notice that minimization of the penalty function fits the data subject to flow and transport model constraints like traditional inverse modeling but also subject to spatial variability constraints like kriging.

The wide sense random assumption is made that the first and second moments of the random terms are well defined, but no distributional assumptions are required. The wide sense assumption is adequate to compute the estimate error covariance within local linear perturbation as shown later. If the Gaussian assumption is made, then minimizing the Bayesian penalty function is equivalent to the nonlinear maximum a posteriori (MAP) estimator. In other words, DFM computes the state estimates that have the highest probability of being true given the measured data.

The minimization of the penalty function given by (17.4.3) is a least squares problem. By inspection, the solution of (17.4.3) is the equivalent to treating the model equations and the MRF equation in (17.4.2) as if they were equations for measured data q in combination with the actual measured data z. This equivalence is the data equation idea expressed in (17.3.2) that leads to the information array of (17.3.3) and a version of Kalman filtering over time. Consequently, the physical and statistical model terms of (17.4.2) can be reinterpreted as data equations that are equivalent to measured data. Now the square root information filter (SRIF) and conjugate gradient information filter (CGIF) solutions are presented.

In order to focus on the spatial aspects of estimation and on the combined study of point measurements of head and conductivity, attention is limited for the SRIF to steady flow with head and conductivity measurements and with an MRF model for log-conductivity. Using the kind of techniques for assembly that PDE solvers use, assemble the relevant parts of (17.4.2) in the banded sparse form:

$$\frac{\partial g_m}{\partial h,f}\begin{bmatrix} h_{m-b} \\ f_{m-b} \\ \vdots \\ h_m \\ f_m \\ \vdots \\ h_{m+b} \\ f_{m+b} \end{bmatrix} = z_m - v_m,$$

$$L_m \begin{bmatrix} f_{m-b} \\ \vdots \\ f_m \\ \vdots \\ f_{m+b} \end{bmatrix} = q_m + w_m \qquad (17.4.4)$$

where h_m and f_m are head and log-conductivity states in region m relative to the current estimate; g_m denotes nonlinear physical model and measurement equations; z_m denotes measurements, sources, and sinks, and linearization terms that are all treated as measurements; L_m denotes Markov coefficients; q_m is the Markov linearization term; and v_m and w_m are noises with zero mean and identity covariance.

The states in a region m of space have been defined so that the data equations break down into blocks with a bandwidth of b. For example, if a two-dimensional grid is used with physical and statistical models at any grid cell depending only on adjacent cells and with point data, then region m can be selected to be grid column m and the block bandwidth b is one. The partial derivative is shown in (17.4.4) to emphasize the local linearization in the nonlinear Gauss-Newton method. The equation for the MRF is shown separately to emphasize that Markov states have been adjoined to the state vector for spatial estimation in much the same way they are adjoined to states of the Kalman filter for temporal estimation.

For computer processing, (17.4.4) is expressed in an information array in which x denotes nonzero block matrices as shown below:

$$\begin{bmatrix} & & & \vdots & & & & & & \vdots \\ \text{x} & \cdot & \cdot & \text{x} & \cdot & \cdot & \text{x} & & & z_{m-1} \\ \text{x} & \cdot & \cdot & \text{x} & \cdot & \cdot & \text{x} & & & q_{m-1} \\ & \text{x} & \cdot & \cdot & \text{x} & \cdot & \cdot & \text{x} & & z_m \\ & \text{x} & \cdot & \cdot & \text{x} & \cdot & \cdot & \text{x} & & q_m \\ & & \text{x} & \cdot & \cdot & \text{x} & \cdot & \cdot & \text{x} & z_{m+1} \\ & & \text{x} & \cdot & \cdot & \text{x} & \cdot & \cdot & \text{x} & q_{m+1} \\ & & & & & \vdots & & & & \vdots \end{bmatrix} \tag{17.4.5}$$

The least squares solution of (17.4.4) is accomplished by manipulation of the information array (17.4.5). State estimation is broken down over space by proceeding recursively down the diagonal of the information array of (17.4.5) using a sequence of triangularizing rotations in the same general manner as state estimation is broken down over time in the SRIF in Bierman [3]. Also, the estimate error covariance can be computed using the same kind of ideas as used in Bierman [3].

For the full flow and transport formulation of (17.4.2) the CGIF method for DFM has less computational burden. Lumping the prior information terms in (17.4.2) for the flow model, transport model, and MRF model in terms subscripted by p and lumping the measurement information terms in (17.4.3) for head, conductivity, and contaminant data into terms subscripted by m gives the normal equation for the least squares estimate as

$$(H_p^T H_p + H_m^T H_m)x = H_p^T z_p + H_m^T z_m \tag{17.4.6}$$

where x is composed of head, conductivity, and contaminant states; the H matrices are model dynamics and measurement sensitivity matrices; and the z variables are measurements, source/sink terms, and linearization terms. Since the normal equations are sparse, they can be efficiently solved using iterative conjugate gradient methods. In particular, the CGNR method described in Saad [34] can be used. An ideal preconditioner is the prior statistical information matrix $H_p^T H_p$. The CGNR computations are dominated by the preconditioning calculation that is dominated by the equivalent of a forward and backward flow and transport simulation for each iteration. Further, the matrix on the left side of the preconditioned normal equation is of the form of an identity matrix plus a matrix with rank no greater than the number of measured data. In theory, the number of CGNR iterations is less than the number of measurements, and, in practice, the number of iterations should be much less than the number of measurements. However, there is no analogous way to compute the estimation uncertainty for the CGNR method as there is for the SRIF method.

17.4.2 Representer method

This section presents previously unpublished results for the representer method. The representer method is the computationally most powerful DFM approach. It is computationally practical for field-scale transient groundwater systems and provides both state estimates and quantification of estimate errors. The method is presented with full treatment of constant trend and measurement bias terms. The following presents the models, penalty function, data equations, state estimates, and estimate error uncertainties.

The physical model for flow and transport can be provided in any standard discretized form suitable for simulation and including interpolation of parameters from parameter grids and timelines. The model equations are denoted by

$$f(x, \theta) = 0 \tag{17.4.7}$$

where the vector f denotes the model equations stacked over space and time, the vector x denotes dependent states such as heads and concentrations, and the vector θ denotes independent parameter states. If the physical model is known with sufficient confidence, then it can be treated as a hard constraint for state estimation. If model errors are anticipated, then physical model errors can be incorporated as a process noise v_f in

$$f(x, \theta) = v_f. \tag{17.4.8}$$

Bennett [1] advocates this formulation to incorporate errors in the dynamics or forcing terms. The simplest prior statistics are a zero mean process noise that is white in space and time with a diagonal covariance

$$E v_f v_f^T = Q_f. \tag{17.4.9}$$

As stated by Bierman [3], a wide class of temporal random processes can be described or approximated by the following exponentially correlated process f

$$\dot{f} = -\frac{1}{\tau} f + w \tag{17.4.10}$$

where τ is the correlation time and w is a zero mean white noise process with noise density q. The first-order model will be assumed here, but it generalizes to higher orders (see Bierman [3]). Equation (17.4.10) can be used to model the temporal variation about a trend for parameters such as recharge and contaminant source loading. Discretizing (17.4.10) on a timeline t_l for $l = 0, 1, 2, \cdots$, gives the discrete Markov random process

$$\begin{aligned} f_{l+1} &= e^{\frac{-(t_{l+1}-t_l)}{\tau}} f_l + w_l \approx \left(1 - \tfrac{(t_{l+1}-t_l)}{\tau}\right) f_l + w_l \\ q_l &= .5\tau \left(1 - e^{\frac{-2(t_{l+1}-t_l)}{\tau}}\right) q \approx q(t_{l+1} - t_l) \end{aligned} \tag{17.4.11}$$

where w_l is a zero mean white noise process with variance q_l, and the approximations are good if the correlation time is much longer than the time step as assumed here. The term involving f_l on the right-hand side of (17.4.11) is the prediction of f_{l+1} given f_l and w_l is the prediction error. As will be seen later, state estimation selects states to minimize these prediction errors in combination with other errors.

A wide class of two- and three-dimensional spatial random fields can be described or approximated by the following first-order process f

$$f - \nabla \cdot (C \nabla f) = w \tag{17.4.12}$$

where the matrix C is a function of the correlation distances and w is a zero mean white noise input with noise density q. The first-order model is assumed here, but it readily generalizes to higher orders (see Yucel and Shumway [42] for ideas on the use of higher-order models in hydrology and geology). A finite difference discretization of (17.4.12) over a three-dimensional curvilinear grid gives the discrete MRF

$$f_{i,j,k} = \sum_{S} a_{l,m,n} f_{i+l,j+m,k+n} + w_{i,j,k} \tag{17.4.13}$$

where the a coefficients are defined over a region of support S that is the adjacent nodes and the integrated noise $w_{i,j,k}$ is a zero mean white noise process with variance $q_{i,j,k}$. It is assumed that the correlation distance is greater than the grid spacing in each direction. The summation on the right-hand side of (17.4.13) is an interpolation for $f_{i,j,k}$ given f on the adjacent nodes and $w_{i,j,k}$ is the interpolation error. As will be seen later, state estimation selects states to minimize these interpolation errors in combination with other errors.

Prior knowledge about constant parameters θ_c is provided as a prior estimate $\hat{\theta}_c$ and prior estimate error covariance $P_{\tilde{\theta}_c}$ that can be expressed as

$$\begin{aligned} \theta_c &= \hat{\theta}_c + \tilde{\theta}_c \\ E\tilde{\theta}_c\tilde{\theta}_c^T &= P_{\tilde{\theta}_c} \end{aligned} \tag{17.4.14}$$

where $\tilde{\theta}_c$ is the constant parameter estimate error. The prior covariance will usually be diagonal, but off-diagonal terms pose no problems. As will be seen later, state estimation selects states to minimize constant parameter estimate errors in combination with other errors.

In order to develop the state and structural parameter estimation relationships, the temporal variation, spatial variation, and constant parameter state models are stacked into

$$\begin{aligned} L_1\theta_t + L_2\theta_c &= v_{s1} \\ L_3\theta_s + L_4\theta_c &= v_{s2} \\ \theta_c &= s_3 + v_{s3} \end{aligned} \tag{17.4.15}$$

where θ_t is the whole value temporal (variation plus trend) parameter state, θ_s is the whole value spatial (variation plus trend) parameter state, θ_c is the constant parameter state, s_3 is the prior constant parameter estimate, L_1 to L_4 are clear from context, v_{s1} is the prediction error with diagonal covariance Q_{s1}, v_{s2} is the interpolation error with diagonal covariance Q_{s2}, and v_{s3} is the constant parameter estimate error with covariance $P_{\tilde{\theta}_c}$. Equation (17.4.15) is expressed in compact form for later derivations as

$$L\theta = s + v_s \tag{17.4.16}$$

where v_s has covariance Q_s and the definitions of variables are clear from context.

Calibration targets and measurements can include water level and contaminant concentration time histories and data that can be interpreted as hydraulic conductivity measurements and pathline measurements. The measurement equations take the form

$$h(x, \theta) = z - v_z \tag{17.4.17}$$

where z is an m-dimensional data vector, h is a function of the dependent states x and the parameter states θ, and v_z is zero mean white measurement noise with diagonal covariance R. Bias and trend measurement errors can be included in h.

The penalty function P is defined to be the sum-of-squares of the physical model errors, prediction errors in temporal variation, interpolation errors in spatial variation, prior estimate errors in constant parameters, and data fit errors. All of the errors are normalized by the expected statistical variation in the error to give P as

$$P = v_f^T Q_f^{-1} v_f + v_s^T Q_s^{-1} v_s + v_z^T R^{-1} v_z. \tag{17.4.18}$$

State estimates are selected that minimize (17.4.18) subject to the equality constraints of (17.4.8), (17.4.16), and (17.4.17). Substituting the constraints into the penalty function, this constrained least squares problem becomes an unconstrained least squares problem with the penalty

$$\begin{aligned} P &= f(x,\theta)^T Q_f^{-1} f(x,\theta) + (s - L\theta)^T Q_s^{-1} (s - L\theta) \\ &\quad + (z - h(x,\theta))^T R^{-1} (z - h(x,\theta)). \end{aligned} \tag{17.4.19}$$

The above is consistent with the viewpoint presented by McLaughlin and Townley [30]. In fact, (17.4.19) is a discretized version of (49) in [30] with a physical model error term included from Bennett [1]. The term in (17.4.19) that accounts for spatial parameter variation involves sparse terms in the matrix L and is weighted by diagonal terms in the matrix Q_s. This is because a Markov model is used for spatial variation. The comparable term in (49) of [30] is weighted by the matrix C_a that will in general be full and cause excessive computational burden. This reflects the fact that the Markov model represents large-scale variation with only local equations to achieve computational efficiency.

Slightly rearranging the penalty function of (17.4.19) suggests how the data equations can be formed as follows

$$\begin{aligned} P &= (0 - f(x,\theta))^T Q_f^{-1} (0 - f(x,\theta)) + (s - L\theta)^T Q_s^{-1} (s - L\theta) \\ &\quad + (z - h(x,\theta))^T R^{-1} (z - h(x,\theta)). \end{aligned} \tag{17.4.20}$$

Consider the nonlinear regression problem with the following measurement equations

$$\begin{aligned} f(x,\theta) &= 0 + v_f \\ L\theta &= s + v_s \\ h(x,\theta) &= z - v_z \end{aligned} \tag{17.4.21}$$

where the data are 0, s, and z in the three equations; the measurement noises are v_f, v_s, and v_z with covariances Q_f, Q_s, and R; and the left-hand sides of the equations are the measurement models. The solution of the regression problem is obtained by minimizing a least squares cost function that is the same as the Bayesian penalty function in (17.4.20). Further, perturbation analysis shows that the estimate error covariance is the same for the Bayesian estimation and regression problems. Consequently, the data equations are given by (17.4.21) and the Bayesian estimation and regression problems are equivalent.

As stated before, the nonlinear regression problem is solved using the Gauss Newton method. The nonlinear problem of (17.4.21) is replaced by the iterative solution of a sequence of linearized problems with the data equations for iteration r given by

$$\begin{aligned} &A_1 \delta x + A_2 \delta\theta = \delta f + v_f \\ &L\delta\theta = \delta s + v_s \\ &M_1 \delta x + M_2 \delta\theta = \delta z - v_z \end{aligned} \tag{17.4.22}$$

where δx, $\delta\theta$ are perturbation states referenced to state estimates from the previous iteration $\hat{x}_{r-1}, \hat{\theta}_{r-1}$ as

$$\begin{aligned} \delta x &= x - \hat{x}_{r-1} \\ \delta\theta &= \theta - \hat{\theta}_{r-1} \end{aligned} \tag{17.4.23}$$

where the linearization terms in the first and third equation in (17.4.22) are

$$\begin{aligned} \delta f &= -f(\hat{x}_{r-1}, \hat{\theta}_{r-1}) \\ A_1 &= \left.\frac{\partial\, f(x,\theta)}{\partial\, x}\right|_{x,\theta=\hat{x}_{r-1},\hat{\theta}_{r-1}} \\ A_2 &= \left.\frac{\partial\, f(x,\theta)}{\partial\, \theta}\right|_{x,\theta=\hat{x}_{r-1},\hat{\theta}_{r-1}} \\ \delta z &= z - h(\hat{x}_{r-1}, \hat{\theta}_{r-1}) \\ M_1 &= \left.\frac{\partial\, h(x,\theta)}{\partial\, x}\right|_{x,\theta=\hat{x}_{r-1},\hat{\theta}_{r-1}} \\ M_2 &= \left.\frac{\partial\, h(x,\theta)}{\partial\, \theta}\right|_{x,\theta=\hat{x}_{r-1},\hat{\theta}_{r-1}} \end{aligned} \tag{17.4.24}$$

and where the second equation in (17.4.22) is referenced to the previous estimate defining

$$\delta s = s - L\hat{\theta}_{r-1}. \tag{17.4.25}$$

For convenience later, the measurement matrix $[M_1\ M_2]$ is denoted by M.

For iteration r, the regression problem with data equations (17.4.22) is solved for the Gauss-Newton step $\delta\hat{x}$, $\delta\hat{\theta}$ to give the new state estimates

$$\hat{x}_r = \hat{x}_{r-1} + \delta\hat{x},\ \ \hat{\theta}_r = \hat{\theta}_{r-1} + \delta\hat{\theta}. \tag{17.4.26}$$

The representer method solves the regression problem for the linearized data equations of (17.4.22) to produce the Gauss-Newton step. A penalty function is defined using Lagrange multipliers as follows:

$$P = \frac{1}{2}\left(\begin{bmatrix} v_f \\ v_s \end{bmatrix}^T \begin{bmatrix} Q_f^{-1} & 0 \\ 0 & Q_s^{-1} \end{bmatrix} \begin{bmatrix} v_f \\ v_s \end{bmatrix} + \left(\delta z - M\begin{bmatrix} \delta x \\ \delta\theta \end{bmatrix}\right)^T R^{-1} \times \right.$$
$$\left.\left(\delta z - M\begin{bmatrix} \delta x \\ \delta\theta \end{bmatrix}\right)\right) - \lambda^T\left(\begin{bmatrix} \delta f \\ \delta s \end{bmatrix} - \begin{bmatrix} A_1 & A_2 \\ 0 & L \end{bmatrix}\begin{bmatrix} \delta x \\ \delta\theta \end{bmatrix} + \begin{bmatrix} v_f \\ v_s \end{bmatrix}\right). \tag{17.4.27}$$

Following Bennett [1], the necessary conditions for the solution are obtained by differentiating P. Eliminating v_f and v_s, and rearranging terms, gives what Bennett [1] calls the Euler-Lagrange equations

$$\begin{gathered} \begin{bmatrix} A_1 & A_2 \\ 0 & L \end{bmatrix}\begin{bmatrix} \delta x \\ \delta\theta \end{bmatrix} = \begin{bmatrix} \delta f \\ \delta s \end{bmatrix} + \begin{bmatrix} Q_f & 0 \\ 0 & Q_s \end{bmatrix}\lambda \\ \begin{bmatrix} A_1 & A_2 \\ 0 & L \end{bmatrix}^T \lambda = M^T R^{-1}\left(\delta z - M\begin{bmatrix} \delta x \\ \delta\theta \end{bmatrix}\right). \end{gathered} \tag{17.4.28}$$

The first equation provides the linearized physical and statistical models that can be solved by simulation running forward in time. These models are driven by a term involving the Lagrange multiplier in place of the physical and statistical model errors. The second equation provides an adjoint equation for the Lagrange multiplier that can be solved using simulation running backward in time.

The representer solution of Bennett [1] unravels the forward and backward relationships in (17.4.28). The solution is represented as the sum of a data-independent term and a data-dependent term that is an m-dimensional reparameterization. By superposition from the first equation in (17.4.28), the data-independent term denoted by $\begin{bmatrix} \delta\hat{x}_0 \\ \delta\hat{\theta}_0 \end{bmatrix}$ is the response of the first equation to the input $\begin{bmatrix} \delta f \\ \delta s \end{bmatrix}$ given by

$$\begin{bmatrix} \delta\hat{x}_0 \\ \delta\hat{\theta}_0 \end{bmatrix} = \begin{bmatrix} A_1 & A_2 \\ 0 & L \end{bmatrix}^{-1}\begin{bmatrix} \delta f \\ \delta s \end{bmatrix}. \tag{17.4.29}$$

This can be solved by simulation running forward in time.

The data-dependent term is the response of the first equation in (17.4.28) to the Lagrange multiplier term. The data and other quantities on the right-hand side of the adjoint equation for the Lagrange multiplier drive the equation only through the m columns of M^T. By superposition, the data-dependent term is a linear combination of the responses to the columns. Consequently, the data-dependent term must be in the m-dimensional subspace spanned by the responses to the columns. If the responses to the columns are denoted by the columns of the matrix Y, then the representer solution can be parameterized in terms of the m-dimensional parameter b as

$$\begin{bmatrix} \delta x \\ \delta\theta \end{bmatrix} = \begin{bmatrix} \delta\hat{x}_0 \\ \delta\hat{\theta}_0 \end{bmatrix} + Yb \tag{17.4.30}$$

and the columns of the matrix Y are called the "representers."

By definition, Y is given by the solution to

$$\begin{aligned}\begin{bmatrix} A_1 & A_2 \\ 0 & L \end{bmatrix} Y &= \begin{bmatrix} Q_f & 0 \\ 0 & Q_s \end{bmatrix} \Lambda \\ \begin{bmatrix} A_1 & A_2 \\ 0 & L \end{bmatrix}^T \Lambda &= M^T.\end{aligned} \tag{17.4.31}$$

This can be solved for each of the m columns by performing a simulation running backward in time for the second equation and a simulation running forward in time for the first equation. Since the Markov representation for spatial variability involves only local equations, the terms involving L in (17.4.31) are sparse and can be efficiently solved with iterative conjugate gradient methods such as those in Saad [34].

The parameter b can be found by substituting the representer solution into the Euler-Lagrange equations and solving to get

$$\hat{b} = (MY + R)^{-1} \left(\delta z - M \begin{bmatrix} \delta \hat{x}_0 \\ \delta \hat{\theta}_0 \end{bmatrix} \right). \tag{17.4.32}$$

This gives the final state estimate solution as

$$\begin{bmatrix} \delta \hat{x} \\ \delta \hat{\theta} \end{bmatrix} = \begin{bmatrix} \delta \hat{x}_0 \\ \delta \hat{\theta}_0 \end{bmatrix} + Y (MY + R)^{-1} \left(\delta\, z - M \begin{bmatrix} \delta \hat{x}_0 \\ \delta \hat{\theta}_0 \end{bmatrix} \right). \tag{17.4.33}$$

Equation (17.4.33) can be implemented by first solving (17.4.29) for the data-independent term by performing a simulation running forward in time. Then (17.4.31) can be solved for the m columns of Y a column at a time. The second equation in (17.4.31) can be solved by backward simulation in time for each column. Notice that the simulation starts at the time of the measurement and proceeds backward so simulations for early measurements will not run long. The first equation in (17.4.31) is solved by forward simulation in time for each column. Each column of the *MY* term in (17.4.32) can be accumulated as a sparse multiply as the column of Y is computed and then the elements of the column of Y for previous times can be discarded. The $\hat{b}$ term can be obtained by using Cholesky methods (Bertsekas [2]) to successively factor the principal submatrices of the matrix being inverted in (17.4.32). $\hat{b}$ in (17.4.32) is then computed using Cholesky factors with forward and backward substitution to apply the inverse to the prior data residuals. In order to avoid storing the columns of Y in (17.4.33), the $Y\hat{b}$ part is produced by postmultiplying the M^T term in (17.4.31) by $\hat{b}$ and performing two simulations backward and forward in time.

The above computations are dominated by $2m + 3$ simulations. The most storage that is required at any point for the $2m$ simulations is one backward simulation from the time of the measurement back in time and one forward simulation for the current time.

Formally, it would appear that hard physical constraints could be imposed by setting Q_f and δf to zero in the representer solution and iterating to nonlinear convergence. However, rigorous treatment requires returning to the nonlinear penalty function. The physical model can be imposed as a hard constraint with a Lagrange multiplier term. The necessary conditions are obtained by differentiation. Then the Newton method can be used and second-order terms set to zero in the spirit of the Gauss-Newton method. When this is done, the formal solution is in fact correct except that the dependent states are computed as a function of the independent states using simulation. The simulation is required to get back on the constraint surface after a linear iteration.

Both direct and Monte Carlo methods are provided to compute state estimate error uncertainties. From the first two equations of (17.4.22), the model noise terms produce a prior uncertainty P_0 in the states of

$$\begin{aligned} P_0 &= E\begin{bmatrix} \delta\tilde{x}_0 \\ \delta\tilde{\theta}_0 \end{bmatrix}\begin{bmatrix} \delta\tilde{x}_0 \\ \delta\tilde{\theta}_0 \end{bmatrix}^T \\ &= \begin{bmatrix} A_1 & A_2 \\ 0 & L \end{bmatrix}^{-1}\begin{bmatrix} Q_f & 0 \\ 0 & Q_s \end{bmatrix}\begin{bmatrix} A_1 & A_2 \\ 0 & L \end{bmatrix}^{-T}. \end{aligned} \tag{17.4.34}$$

The matrix of representers Y is actually the covariance between the states and the measurements given by

$$Y = P_0 M^T. \tag{17.4.35}$$

By Bayesian estimation relationships, the estimate error covariance is

$$E\begin{bmatrix} \delta\tilde{x} \\ \delta\tilde{\theta} \end{bmatrix}\begin{bmatrix} \delta\tilde{x} \\ \delta\tilde{\theta} \end{bmatrix}^T = P_0 - Y(MY+R)^{-1}Y^T. \tag{17.4.36}$$

Selected diagonal and off-diagonal elements of the covariance can be computed. As the representer matrix Y is being computed, elements needed to compute selected covariances are saved. Since the Cholesky factors of $MY + R$ are already computed, the selected elements of the right-hand term in (17.4.36) can be efficiently computed. For independent states, the selected elements of P_0 are available from the prior temporal and spatial autocorrelation functions and from the prior constant estimate error covariance. For dependent states, (17.4.34) can be used to obtain selected elements of P_0 by simulation backward and forward in time. Consequently, dependent states are the most costly for covariance computation.

Not all quantities are amenable to direct uncertainty calculation such as clean up time. For such quantities, Monte Carlo methods are appropriate. From the data equations (17.4.22) and the representer solution of (17.4.29) and (17.4.33), the following provides Monte Carlo equations for the estimate errors:

$$\begin{aligned} \begin{bmatrix} \delta\tilde{x} \\ \delta\tilde{\theta} \end{bmatrix} &= \begin{bmatrix} \delta x - \delta\hat{x} \\ \delta\theta - \delta\hat{\theta} \end{bmatrix} = \begin{bmatrix} A_1 & A_2 \\ 0 & L \end{bmatrix}^{-1}\begin{bmatrix} v_f \\ v_s \end{bmatrix} \\ &\quad - Y(MY+R)^{-1}\left(v_z + M\begin{bmatrix} A_1 & A_2 \\ 0 & L \end{bmatrix}^{-1}\begin{bmatrix} v_f \\ v_s \end{bmatrix}\right). \end{aligned} \tag{17.4.37}$$

The noise terms can be produced with random noise generators. The terms involving v_f, v_s can be computed with one simulation forward in time. The Cholesky factors of $MY + R$ are available from the representer solution. The terms on the right-hand side of (17.4.37) can be computed using two simulations forward and backward in time. Consequently, one Monte Carlo run can be performed with three simulations. For hard physical constraints, the above needs to be modified. If the dependent states are retained, then v_f is set to zero, only $\delta\tilde{\theta}$ is computed from (17.4.37), and $\delta\tilde{x}$ is produced by simulation of the hard physical constraint.

17.4.3 System identification

DFM uses a combination of residual analysis and likelihood theory for system identification as described in Porter et al. [32]. Under the Gaussian assumption and within linear perturbation, DFM uses the expectation maximization (EM) method of Levy and Porter [28] to provide maximum likelihood estimates. The EM iteration is usually intuitive and simple. For example, it may be desired to estimate the noise variance q for a Markov process of (17.4.1). If the log-conductivity f were directly measured noise free at all grid locations, then (17.4.1) could be solved for the process noise w at all locations and q would be estimated as the sample variance of the interpolation errors as follows:

$$\hat{q} = \frac{1}{N} \sum_{i,j} w_{i,j}^2. \tag{17.4.38}$$

It might seem reasonable to use the estimate for q from the current EM iteration in a state estimation method to estimate f. Then the estimate for f could be used in (17.4.1) and (17.4.38) to update the estimate of q. This is almost the answer that EM provides, but it would be biased downward due to estimate error in w. EM removes the bias by adding in the estimate error variance in w computed by the state estimator to give the estimate for q:

$$\hat{q} = \frac{1}{N} \sum_{i,j} \left(\hat{w}_{i,j}^2 + E(w_{i,j} - \hat{w}_{i,j})^2 \right). \tag{17.4.39}$$

The above procedure is iterated to convergence.

17.5 Application at Savannah River Site

DFM was used at the DOE Savannah River Site to calibrate a groundwater flow and transport model for the Old Burial Ground. The Old Burial Ground is located in the central portion of the Savannah River Site and was a solid radioactive and hazardous waste burial location. The contaminant of interest is tritium because it is a geochemically conservative tracer that has been monitored for several years.

The DFM results were compared to an existing model calibrated via the typical trial-and-error method. By jointly processing available data, DFM produced a better

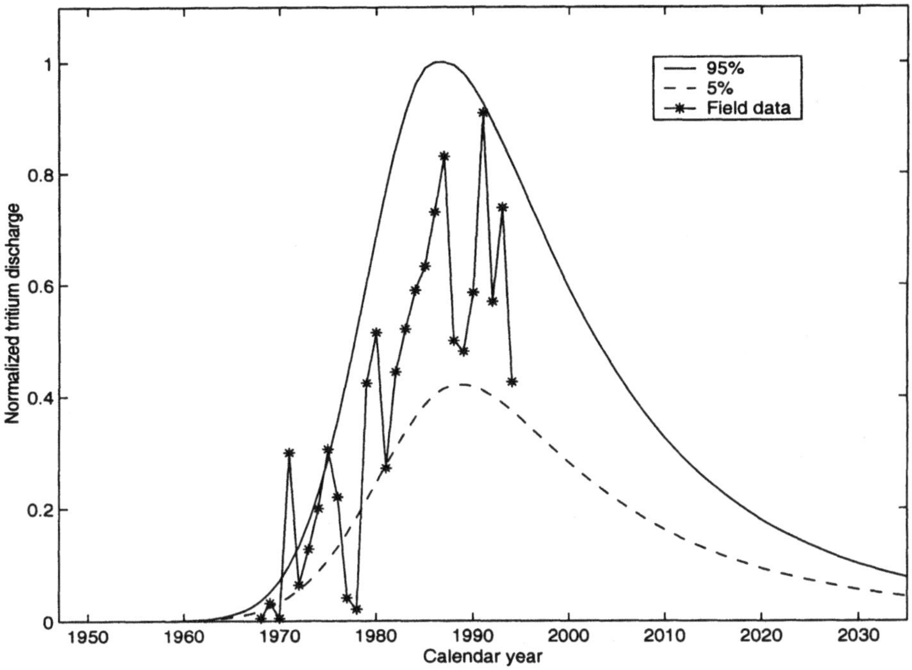

Figure 17.5.1: Tritium breakthrough curve with 5% and 95% confidence bounds

calibration in less time and also quantified uncertainties. Specific benefits are the following:

- Rapid updating with DFM allowed modelers to try conceptual alternatives.
- Residual errors in head were reduced from a root-mean-square of 3.0 to 1.3 ft.
- DFM took 75% less time than manual calibration.
- DFM quantified the uncertainty in contaminant breakthrough curves.

The Old Burial Ground is a multiple aquifer system where the heterogeneous distribution of hydraulic conductivity is important for contaminant transport. The DFM numerical grid is a three-dimensional curvilinear grid with an aerial extent of 6760 ft in the x-direction and 5070 ft in the y-direction. The grid is 27x27x21 with uniform spacing along the x and y axes. Hydrostratigraphic picks were used to define the elevation of 21 nodal layers so that each nodal layer was assigned to only one aquifer unit. A Green Clay layer was discretized by the fourth nodal layer from the bottom. The recharge/drain boundary condition was applied to the top nodal layer with the maximum recharge coefficient as a parameter to be calibrated by DFM. The no-flow boundary condition was applied to the interior nodes of the bottom nodal layer. All other boundary nodes were assigned constant head boundary conditions.

The hydraulic conductivity is calibrated separately in three hydrostratigraphic layers: layer 1 is an aquifer above the Green Clay, layer 2 is the Green Clay aquitard, and layer three is an aquifer below the Green Clay. The log of the horizontal conductivity in the aquifer of layer 1 is viewed as the sum of a constant trend plus a spatially correlated random process. The trend and the value of the random process at each node are calibrated. Also, a constant trend for the vertical conductivity anisotropy is calibrated. In the aquifer of layer 2, the same type of model is used for the horizontal conductivity, but the vertical anisotropy is fixed since little data are available for its calibration. The vertical conductivity in the aquitard of layer 3 is a parameter of calibration. Physical model constraints were provided by the VAM3DF model from HydroGeoLogic, Inc.

Heads from 237 observation wells were calibration targets for DFM. Mud fraction data from cores was mapped to horizontal conductivity and vertical conductivity measurements. A total of 880 conductivity measurements were used: 415 horizontal and 465 vertical conductivities. These conductivities plus the other calibrated parameters were used to produce the tritium breakthrough curve to a local stream called Fourmile Branch. The breakthrough curve with 5% and 95% confidence bounds is shown in Figure 17.5.1. Also shown are independent field data that validate the breakthrough curve and the confidence bounds.

17.6 Conclusions

DFM builds on previous hydrogeological work to provide a mature methodology implemented and deployed in software tools for data integration, model calibration, and geostatistical quantification of uncertainty. Key points are the following:

- DFM is the first to solve the causality problem with Kalman filtering to produce a computationally practical spatial state estimation methodology for model calibration.

- DFM provides a solution to the problem often referred to in the hydrogeological literature of combining point data for hydraulic head and conductivity to take advantage of the considerable data worth that these data sets often possess.

- DFM provides a system identification theory for the determination of geostatistical structure based on maximum likelihood methods.

- DFM has been applied to several field-scale groundwater systems. An application is presented at the DOE Savannah River Site where DFM quantified breakthrough curve uncertainty and produced a better data fit with 75% less time spent relative to a conventional calibration.

Solving the causality problem for spatial estimation lays the groundwork for further extensions in hydrogeology and in other fields with similar spatial estimation and identification problems. Some of the possible extensions are the following:

- Integration with management optimization methods for environmental remediation, particularly for quantifying safety margins and managing uncertainty. DFM opens the possibility for real-time monitoring and optimization through rapid model updating for expedited cleanup.

- Oil reservoir management. There are conceptual similarities between oil reservoir management and current environmental applications of DFM that can be exploited.
- Medical imaging. DFM ideas have already been applied to process electroencephalograph data to image electrical activity in the brain. The appropriate quasi-static representation of Maxwell's Laws is identical to steady groundwater flow equations, facilitating DFM extension.

Acknowledgements

I am indebted to Dr. Anthony N. Michel, my major Professor, for his general approach to problem solving, which was an important part of this work and much of what I have done in my career. Tony taught me not to shrink from the complexities of large-scale dynamical systems but to attack them at a fundamental mathematical level where breakthroughs can be made. This specific work follows in the footsteps of departed colleagues and friends Jerry Bierman and Jim Vandergraft. The information filter ideas for DFM came from lively discussions with Jerry Bierman and Bruce Gibbs. Jim Vandergraft provided critical optimization ideas for Bayesian estimation.

Bibliography

[1] A. F. Bennett, *Inverse Methods in Physical Oceanography*, Cambridge University Press, New York, 1992.

[2] D. P. Bertsekas, *Constrained Optimization and Lagrange Multiplier Methods*, Academic Press, New York, 1982.

[3] G. J. Bierman, *Factorization Methods for Discrete Sequential Estimation*, Academic Press, New York, 1977.

[4] G. J. Bierman and D.W. Porter, "Decentralized tracking via new square root information filter (SRIF) concepts," *Business and Technological Systems*, BTS-34-87-32, 1988.

[5] J. Carrera and L. Glorioso, "On geostatistical formulations of the groundwater flow inverse problem," *Advanced Water Resources*, vol. 14, no. 5, pp. 273–283, 1991.

[6] J. Carrera and S. P. Neuman, "Estimation of aquifer parameters under transient and steady state conditions: 1. Maximum likelihood method incorporating prior information," *Water Resources Research*, vol. 22, no. 2, pp. 199–210, 1986.

[7] J. Carrera and S. P. Neuman, "Estimation of aquifer parameters under transient and steady state conditions: 2. Uniqueness, stability, and solution algorithms," *Water Resources Research*, vol. 22, no. 2, pp. 211–227, 1986.

[8] J. Carrera and S. P. Neuman, "Estimation of aquifer parameters under transient and steady state conditions: 3. Application to synthetic and field data," *Water Resources Research*, vol. 22, no. 2, pp. 228–242, 1986.

[9] R. Chellappa and A. Jain, *Markov Random Fields: Theory and Application*, Academic Press, Boston, MA, 1991.

[10] J. J. Clark and A. L. Yuille, *Data Fusion for Sensory Information Processing Systems*, Kluwer, Boston, MA, 1990.

[11] R. L. Cooley, "Incorporation of prior information on parameters into nonlinear regression groundwater flow models," *Water Resources Research*, vol. 18, no. 2, pp. 965–976, 1982.

[12] P. Courtier, J. Derber, R. Errico, J. F. Louis, and T. Vukicevic, "Important literature on the use of adjoint, variational methods and the Kalman filter in meteorology," *Tellus*, vol. 45A, no. 2, pp. 342–357, 1993.

[13] G. Dagan, "Stochastic modeling of groundwater flow by unconditional and conditional probabilities: The inverse problem," *Water Resources Research*, vol. 21, no. 2, pp. 65–72, 1985.

[14] M. M. Daniel and A. S. Willsky, "A multiresolution methodology for signal-level fusion and data assimilation with applications to remote sensing," *Proc. IEEE*, vol. 85, no. 1, pp. 164–180, 1997.

[15] D. B. Duncan and S. D. Horn, "Linear dynamic recursive estimation from the viewpoint of regression analysis," *J. American Statistical Association*, vol. 67, no. 340, pp. 815–821, 1972.

[16] M. J. Eppstein and D. E. Dougherty, "Simultaneous estimation of transmissivity values and zonation," *Water Resources Research*, vol. 32, no. 11, pp. 3321–3336, 1996.

[17] R. A. Freeze, J. Massmann, L. Smith, T. Sperling, and B. James, "Hydrogeological decision analysis: 1. A framework," *Ground Water*, vol. 28, no. 5, pp. 738–766, 1990.

[18] L. W. Gelhar, "Stochastic subsurface hydrology from theory to applications," *Water Resources Research*, vol. 22, no. 9, pp. 1355–1455, 1986.

[19] S. Geman and D. Geman, "Stochastic relaxation, Gibbs distributions, and the Bayesian restoration of images," *IEEE Trans. Pattern Analysis and Machine Intelligence*, vol. PAMI-6, no. 6, pp. 721–741, 1984.

[20] M. Ghil, "Meteorological data assimilation for oceanographers. Part 1: Description and theoretical framework," *Dynamics of Atmospheres and Oceans*, vol. 13, pp. 171–218, 1989.

[21] D. L. Hall, *Mathematical Techniques in Multisensor Data Fusion*, Artech House, Boston, MA, 1992.

[22] M. C. Hill, "A computer program (MODFLOWP) for estimating parameters of a transient, three-dimensional, ground-water flow model using nonlinear regression," *U.S. Geological Survey*, Report 91-484, 1992.

[23] R. J. Hoeksema and P. K. Kitinidis, "An application of the geostatistical approach to the inverse problem in two-dimensional groundwater modeling," *Water Resources Research*, vol. 20, no. 7, pp. 1003–1020, 1984.

[24] A. K. Jain, "Advances in mathematical models for image processing," *Proc. IEEE*, vol. 69, no. 5, pp. 502–528, 1981.

[25] A. G. Journal and C. J. Huijbregts, *Mining Geostatistics*, Academic Press, New York, 1991.

[26] R. E. Kalman, "A new approach to linear filtering and prediction problems," *Trans. ASME, Ser. D: J. Basic Engineering*, vol. 82, pp. 35–45,1960.

[27] P. K. Kitanidis and E. G. Vomvoris, "A geostatistical approach to the inverse problem in groundwater modeling (steady state) and one-dimensional simulations," *Water Resources Research*, vol. 19, no. 3, pp. 677–690, 1983.

[28] L. J. Levy and D. W. Porter, "Large-scale system performance prediction with confidence from limited field testing using parameter identification," *Johns Hopkins APL Technical Digest*, vol. 13, no. 2, pp. 300–308, 1992.

[29] A. Medina and J. Carrera, "Coupled estimation of flow and solute transport parameters," *Water Resources Research*, vol. 32, no. 10, pp. 3063–3076, 1996.

[30] D. McLaughlin and L. R. Townley, "A reassessment of the groundwater inverse problem," *Water Resources Research*, vol. 32, no. 5, pp. pp. 1131–1161, May 1996.

[31] D. W. Porter, "Quantitative data fusion: A distributed/parallel approach to surveillance, tracking, and navigation using information filtering," *Proc. Fifth Joint Service Data Fusion Symposium*, Johns Hopkins University/Applied Physics Laboratory, Laurel, MD, Oct. 1991.

[32] D. W. Porter, B. P. Gibbs, W. F. Jones, P. S. Huyakorn, L. L. Hamm, and G. P. Flach, "Data fusion modeling for groundwater systems," *J. Contaminant Hydrology*, vol. 42, pp. 303–335, 2000.

[33] W. H. Press, S. A. Teukolsky, W. T. Vetterling, and B. P. Flannery, *Numerical Recipes in FORTRAN: The Art of Scientific Computing*, Cambridge University Press, New York, 1992.

[34] Y. Saad, *Iterative Methods for Sparse Linear Systems*, PWS Publishing Company, Boston, MA, 1996.

[35] M. I. Sezan and A. M. Tekalp, "Survey of recent developments in digital image restoration," *Optical Engineering*, vol. 29, no. 5, pp. 393–404, 1990.

[36] M. F. Snodgrass and P. K. Kitanidis, "A geostatistical approach to contaminant source identification," *Water Resources Research*, vol. 33, no. 4, pp. 537–546, Apr. 1997.

[37] N. Z. Sun and W. W. G. Yeh, "A stochastic inverse solution for transient groundwater flow: Parameter identification and reliability analysis," *Water Resources Research*, vol. 28, no. 12, pp. 3269–3280, 1992.

[38] F. C. Van Geer, C. B. M. Te Stroet, and Z. Yangxiao, "Using Kalman filtering to improve and quantify the uncertainty of numerical groundwater simulations 1. The role of system noise and its calibration," *Water Resources Research*, vol. 27, no. 8, pp. 1987–1994, 1991.

[39] J. S. Vandergraft, "Efficient optimization methods for maximum likelihood parameter estimation," *Proc. 24th Conference on Decision and Control*, Ft. Lauderdale, FL, 1985.

[40] P. Whittle, "On stationary processes in the plane," *Biometrika*, vol. 41, pp. 434–449, 1954.

[41] Z. Yangxiao, C. B. M. Te Stroet, and F. C. Van Geer, "Using Kalman filtering to improve and quantify the uncertainty of numerical groundwater simulations 2. Application to monitoring network design," *Water Resources Research*, vol. 27, no. 8, pp. 1995–2006, 1991.

[42] Z. T. Yucel and R. H. Shumway, "A spectral approach to estimation and smoothing of continuous spatial processes," *Stochastic Hydrology and Hydraulics*, vol. 10, pp. 107–126, 1996

[43] S. Zou and A. Parr, "Optimal estimation of two-dimensional contaminant transport," *Ground Water*, vol. 33, no. 2, pp. 319–325, 1995.

Chapter 18

(Control, Output) Synthesis: Algebraic Paradigms*

Michael K. Sain and Bostwick F. Wyman

Abstract: Under broad assumptions, it is known that there is in general no "separation principle" to guarantee optimality of a division between control law design and filtering of plant uncertainty. It is possible, however, to develop parameterizations of nominal (control, output) responses and to examine their capabilities in the feedback situation.

18.1 Introduction

From an intuitive point of view, acceptable response is the hallmark of successful design in a linear multivariable control system. Even if we have perfect plant knowledge and no system disturbances, the development of reasonable specifications and their achievement with available equipment is not a trivial matter. When, however, the plant is unstable, or uncertain, or acted upon by disturbances, the choice of feedback realization for the controller may place the response goals in competition with new goals such as internal stability, sensitivity suppression, and disturbance rejection.

Under broad assumptions, Zames [1] observed that there is in general no "separation principle" to guarantee optimality of a division between control law design and filtering of plant uncertainty. It thus makes sense to develop parameterizations of response and to examine their capabilities in the feedback situation. Because feedback

*This research was supported in part by the Frank M. Freimann Chair in Electrical Engineering at the University of Notre Dame, Notre Dame, IN.

features, such as internal stability, are certainly dependent upon controller configuration, it is desirable that a response parameterization not depend upon controller configuration. In that way, it would permit comparisons and contrasts among specific control arrangements. This suggests that response be studied as a problem in synthesis.

Control synthesis indeed has a rich history. As early as 1951, Guillemin proposed that synthesis of feedback control systems should involve a determination of the closed-loop transfer function from specifications, followed by construction of appropriate compensation networks (cf. [2]). In due course, Truxal [3] discussed the Guillemin method as it related to basic feedback issues of plant pole cancellation, imperfect cancellation, and controller complexity. Not surprisingly, some of the same ideas then appeared in texts on sampled-data control [4] and in works on digital control [5].

As one solves classical transfer function equations for the compensation required by a given closed-loop specification, one of course inverts the plant, thereby obtaining the equivalent series compensator. By 1957, at least for stable plants, authors [6] began to discuss equivalent series compensation as a parameter of feedback synthesis.

The literature of many inputs and many outputs began to follow the trend [7–10]. As part of the general state-space development in the area of control, the problem of synthesizing a given closed-loop command/output-response map became known as "model matching," and was solved in that context by Morse [11]. Subsequently, model matching was studied from an input/output view [12–15], where focus was placed upon the matrix equation

$$[Z_1(s)]\,[Z(s)] = [Z_2(s)]\,, \tag{18.1.1}$$

with $[Z_i(s)]$, $i = 1, 2$, being given rational matrices and with $[Z(s)]$ to be determined. A principal issue was the fact that $[Z(s)]$ might be required to have certain properties, such as being stable or proper, while $[Z_i(s)]$, $i = 1, 2$, might not be so restricted. Investigators then began the use of transfer function rings and subrings [16–18], after which began a gradual coalescence [19] with methods based upon function spaces and operator algebras, as well as generalizations of the rings [20]. In 1977, working outside the ring context in a mixed state-space/transfer function format, Bengtsson [21] solved a broad class of problems concerning feedback realization in a model matching context. Pernebo [22] extended this work on model matching, with the aid of matrices of rings.

Back in 1977, as part of an application study dealing with gas turbine control, Peczkowski and Sain [23] solved a model matching problem using transfer functions. Building upon this experience, Peczkowski, Sain, and Leake [24] proposed in 1979 the total synthesis problem (TSP), wherein both the command/output-response and command/control-response are to be synthesized, subject to the plant constraint. From an algebraic viewpoint, TSP carries with it the idea of tradeoff between control response and output response, one of the important features of approaches based on optimal control. From the outset, TSP was rooted in application studies [25–33].

A useful feature of the TSP idea is that it may be subdivided immediately into a nominal design problem (NDP), which is not dependent upon specific controller structures, and a feedback synthesis problem (FSP), which is. Gejji [34] made the first study of this separation, in a semi-coordinate-free context. Gejji found that the NDP was characterized in terms of the plant structural matrices and a single "good" transfer function matrix.

In this chapter, we present a tutorial study of the NDP and FSP–for the unity feedback case. The treatment is coordinate-free.

18.2 Notation and Preliminaries

Let k be an arbitrary field. Then $k[s]$ is the principal ideal domain of polynomials in s with coefficients in k. As a commutative ring with no zero divisors, $k[s]$ admits the quotient field $k(s)$. Intuitively, $k(s)$ is just the set of rational transfer functions having coefficients in k. Our system functions are to be set up on $k[s]$-modules and on $k(s)$-vector spaces.

Suppose that V_i is a k-vector space, for $i = 1, 2$. Then $V_1 \otimes V_2$ is the tensor product of V_1 with V_2 and may be denoted by $V_1 \otimes_k V_2$ to emphasize the fact that V_1 and V_2 are regarded as k-vector spaces. Observe that $k[s]$ admits the structure of a k-vector space, and choose V_1 to be $k[s]$. Let V be a k-vector space of finite dimension, and choose V_2 to be V. Define the k-vector space $k[s] \otimes_k V$ of polynomials in s with coefficients in V. It turns out that this k-vector space admits the structure of a $k[s]$-module. To see this, write

$$\sum_{i=1}^{n} \{p_i(s) \otimes_k v_i\}$$

to represent a vector. Then scalar multiplication by a polynomial $p(s)$ in $k[s]$ is understood in the manner

$$p(s) \sum_{i=1}^{n} \{p_i(s) \otimes_k v_i\} = \sum_{i=1}^{n} \{[p(s)p_i(s)] \otimes_k v_i\} \ .$$

We write $V[s]$ for this $k[s]$-module. In an entirely similar way, we may develop the k-vector space $k(s) \otimes_k V$ and equip it to be a $k(s)$-vector space, which carries the symbol $V(s)$. Clearly, $V[s] \subset V(s)$; and there is an insertion $i\colon V[s] \to V(s)$, which is a morphism of $k[s]$-modules. Now let W be another k-vector space of finite dimension. Let $p\colon W(s) \to W(s)/W[s]$ be the projection morphism from $W(s)$, regarded as a $k[s]$-module, onto the quotient module $W(s)/W[s]$. By a "transfer function," we mean a morphism $L(s)\colon V(s) \to W(s)$ of $k(s)$-vector spaces. Notice that $L(s)$ is not a matrix, as we are yet in a coordinate-free mode. For specific calculations, bases in V and W may be chosen; in turn, these choices induce bases in $V(s)$ and $W(s)$, and then a matrix for $L(s)$ may be defined. This matrix would be denoted by $[L(s)]$.

Because $V(s)$ and $W(s)$ may be regarded as $k[s]$-modules, $L(s)$ can be regarded

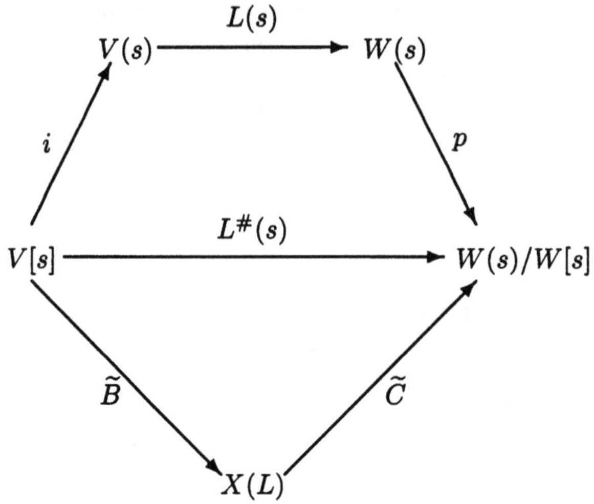

Figure 18.2.1: Realization diagram

as a morphism of $k[s]$-modules. Thus the composition

$$L^{\#}(s)\colon V[s] \to W(s)/W[s],$$

given by $L^{\#}(s) = p \circ L(s) \circ i$, is a morphism of $k[s]$-modules, which we will call the Kalman input/output map associated with $L(s)$. By the "pole module" of $L(s)$, we mean the torsion $k[s]$-module $V[s]/\ker L^{\#}(s)$, denoted by $X(L)$. Onto the pole module there is a controllability epimorphism $\widetilde{B}$ from $V[s]$, while into $W(s)/W[s]$ there is an observability monomorphism $\widetilde{C}$ from $X(L)$, as shown in the realization diagram (Figure 18.2.1). Denote by $m(L)$ the minimal polynomial of $X(L)$. Let $S_g \subset k[s]$ be closed under multiplication in $k[s]$, exclude the zero polynomial, and include the polynomial 1. We shall say that $p(s) \in k[s]$ is a good polynomial if $p(s) \in S_g$. Moreover, we shall say that $L(s)$ is a good morphism of $k(s)$-vector spaces if $m(L) \in S_g$. Alternatively, we can say that $L(s)$ is a good transfer function.

18.3 Problem Statement

In this section, we define in a coordinate-free way the problem of designing simultaneously the complete set of controlled outputs desired from a plant and a corresponding complete set of inputs. The approach will be to characterize these two sets in terms of morphisms of $k(s)$-vector spaces, with each morphism acting upon a space of exogenous requests or commands. Because these morphisms must possess certain "good" qualities as, for example, stability, which may not necessarily be shared by the plant, it is desirable at the outset to clarify carefully just what foundation can be taken as adequate for the task at hand. This requires a basis-independent

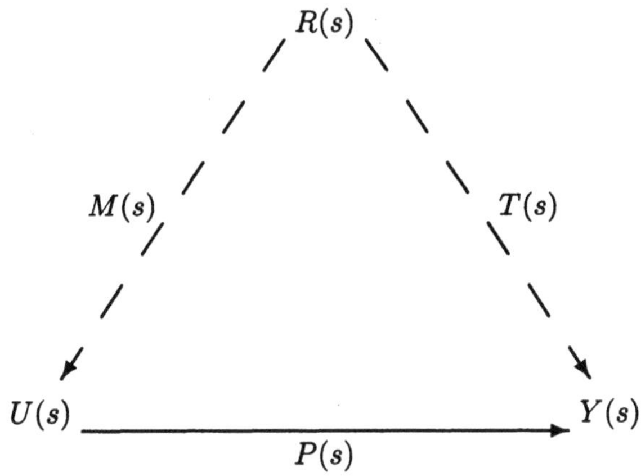

Figure 18.3.1: The NDP

definition of the issues.

Let R be a q-dimensional k-vector space of exogenous vectors, let U be an m-dimensional k-vector space of control vectors, and let Y be a p-dimensional k-vector space of plant output vectors. On these spaces, develop the $k[s]$-modules $R[s] = k[s] \otimes_k R$, $U[s] = k[s] \otimes_k U, Y[s] = k[s] \otimes_k Y$, and the $k(s)$-vector spaces $R(s) = k(s) \otimes_k R$, $U(s) = k(s) \otimes_k U$, $Y(s) = k(s) \otimes_k Y$. Define the plant by a morphism $P(s)\colon U(s) \to Y(s)$ of $k(s)$-vector spaces. As a transfer function, $P(s)$ may be good or it may not. Define the desired plant response to exogenous vectors by the morphism $T(s)\colon R(s) \to Y(s)$ of $k(s)$-vector spaces, and define the controls employed to generate such response by a morphism $M(s)\colon R(s) \to U(s)$ of $k(s)$-vector spaces. Though $P(s)$ is not required to be a good transfer function, it is required that $T(s)$ and $M(s)$ have that property. Further, since $M(s)$ produces the control action which drives the plant $P(s)$, it is of course not independent of the desired plant response $T(s)$. This leads to the first of two basic subproblems to be discussed.

The NDP is to find pairs of good transfer functions $(M(s), T(s))$ such that the diagram of Figure 18.3.1 commutes.

Notice that the NDP is not an exact model matching problem [2, 3, 11] or a minimal design problem [12–15], whose commutative diagram would be that of Figure 18.3.2. In both these diagrams, of course, solid arrows are given morphisms, whereas dashed arrows are to be determined.

Consider next the question of output feedback. By output feedback in the present context we shall mean a morphism $C(s)\colon R(s) \oplus Y(s) \to U(s)$ of $k(s)$-vector spaces. Here $R(s) \oplus Y(s)$ is the $k(s)$-vector (biproduct) space constructed in the usual way from $R(s) \times Y(s)$. Notice that any feedback scheme that is required to be a morphism $R(s) \to U(s)$ when the loop is broken and a morphism $Y(s) \to U(s)$ when there is no exogenous signal must be of this form, according to the following

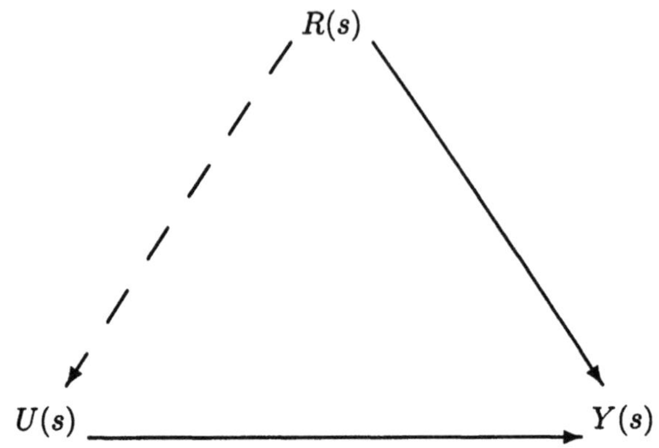

Figure 18.3.2: Model matching

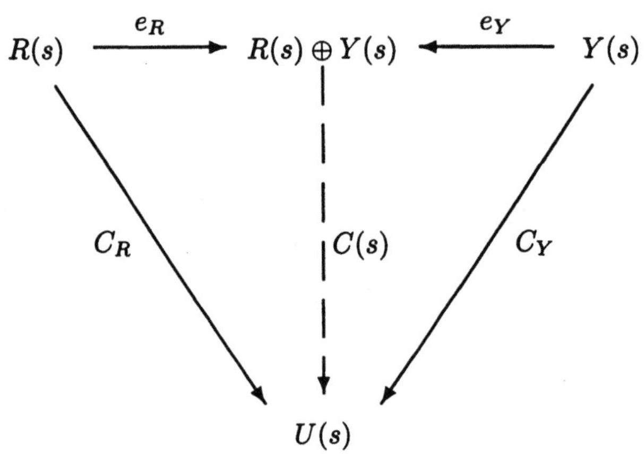

Figure 18.3.3: "Linear" feedback

result.

Proposition 18.3.1 ([35, p. 212]). *If $U(s)$ is a $k(s)$-vector space and*

$$\begin{aligned} C_R\colon R(s) &\to U(s) \\ C_Y\colon Y(s) &\to U(s) \end{aligned}$$

are morphisms of $k(s)$-vector spaces, then there is a unique morphism $C(s)\colon R(s) \oplus Y(s) \to U(s)$ of $k(s)$-vector spaces such that the diagram of Figure 18.3.3 commutes.

Proof. See [35]. The morphisms e_R and e_Y have actions given by

$$e_R(r(s)) = (r(s), 0)\,; \quad e_Y(y(s)) = (0, y(s))\,.$$

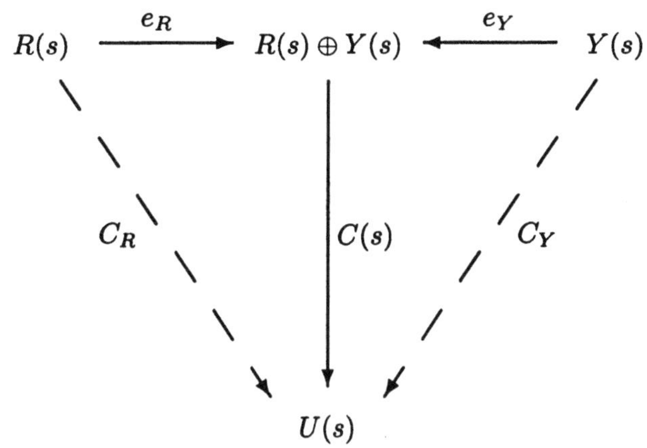

Figure 18.3.4: Synthesis equivalents

Moreover, an output feedback morphism $C(s)$, when given, can always be used to define morphisms C_R and C_Y, as seen in Figure 18.3.4, according to $C_R = C(s) \circ e_R$, $C_Y = C(s) \circ e_Y$. From these facts, it follows that

$$y(s) = P(s)u(s) = P(s)\left\{C_R r(s) + C_Y y(s)\right\},$$

or

$$\left(1_{Y(s)} - P(s) \circ C_Y\right) y(s) = \left(P(s) \circ C_R\right) r(s).$$

Under the technical assumption that $\left(1_{Y(s)} - P(s) \circ C_Y\right)^{-1} : Y(s) \to Y(s)$ exists, we have

$$\begin{aligned} y(s) &= \left(1_{Y(s)} - P(s) \circ C_Y\right)^{-1} \circ P(s) \circ C_R\, r(s) \\ &= P(s) \circ \left(1_{U(s)} - C_Y \circ P(s)\right)^{-1} \circ C_R\, r(s) \end{aligned}$$

which permits the identification

$$T(s) = P(s) \circ \left(1_{U(s)} - C_Y \circ P(s)\right)^{-1} \circ C_R.$$

This leads to the second of the two basic subproblems.

The FSP is to find an output feedback morphism $C(s)$ such that

$$M(s) = \left(1_{U(s)} - C_Y \circ P(s)\right)^{-1} \circ C_R.$$

By itself, FSP is trivial, because of the choice $C_Y = 0$, $C_R = M(s)$. Thus we shall focus on the good FSP (GFSP). By way of example, if $k = \mathbb{R}$ and if S_g is the set of strictly Hurwitz polynomials, then the GFSP may be understood as the FSP with internal stability (FSPIS). Of course, even when $k = \mathbb{R}$, the GFSP and FSPIS are not always the same, because S_g does not have to be the set of strictly Hurwitz polynomials. It may, for illustration, be a subset of those.

18.4 NDP: The Abstract Kernel Problem

In this section, the set of all solution pairs $(M(s), T(s))$ to the NDP will be described as the kernel of an abstract morphism. Some care must be taken to examine the distinction between "polynomial" solutions, "good" solutions, and "arbitrary rational" solutions, because we do not wish to require $P(s)$ to be a good transfer function.

A few additional notations are required. In particular, we need to introduce the localization [36] $S_g^{-1}k[s]$ of the ring $k[s]$. This is done by establishing an equivalence relation $\equiv$ on $S_g \times k[s]$ by a means analogous to that used in developing the quotient field $k(s)$ [35]. $S_g^{-1}k[s]$ is a commutative ring, which satisfies $k[s] \subset S_g^{-1}k[s] \subset k(s)$. Regarded as rings, each of these three sets admits the structure of a $k[s]$-module. Consider the $k[s]$-module $V[s]$. $V[s]$ can be localized as well, by the construction

$$S_g^{-1}V[s] = (S_g^{-1}k[s]) \otimes_{k[s]} V[s] .$$

Then a morphism $J(s)\colon V[s] \to W[s]$ of $k[s]$-modules has a localization

$$S_g^{-1}J(s)\colon S_g^{-1}V[s] \to S_g^{-1}W[s]$$

by the action

$$S_g^{-1}J(s)\left\{\sum_{i=1}^{n}(p_i(s), q_i(s)) \otimes_{k[s]} v_i(s)\right\} = \sum_{i=1}^{n}(p_i(s), q_i(s)) \otimes_{k[s]} J(s)v_i(s) .$$

Some properties of localization should be mentioned. Suppose that $J_1(s)\colon V_1[s] \to V_2[s]$ and $J_2(s)\colon V_2[s] \to V_3[s]$ are morphisms of $k[s]$-modules. Then the sequence

$$V_1[s] \overset{J_1(s)}{\longrightarrow} V_2[s] \overset{J_2(s)}{\longrightarrow} V_3[s]$$

is exact when $\ker J_2(s) = \operatorname{im} J_1(s)$. In such a case, the localized sequence

$$S_g^{-1}V_1[s] \overset{S_g^{-1}J_1(s)}{\longrightarrow} S_g^{-1}V_2[s] \overset{S_g^{-1}J_2(s)}{\longrightarrow} S_g^{-1}V_3[s]$$

is exact also. When applied to the sequences

$$\ker J_1(s) \longrightarrow V_1[s] \overset{J_1(s)}{\longrightarrow} V_2[s],$$
$$V_2[s] \overset{J_2(s)}{\longrightarrow} V_3[s] \longrightarrow V_3[s]/\operatorname{im} J_2(s),$$

this implies that

$$\begin{aligned} \ker S_g^{-1}J(s) &= S_g^{-1}\ker J(s), \\ \operatorname{im} S_g^{-1}J(s) &= S_g^{-1}\operatorname{im} J(s) . \end{aligned}$$

Further, if the $k[s]$-module $V[s]$ is free over $k[s]$, then $S_g^{-1}V[s]$ is free over $S_g^{-1}k[s]$, a fact that follows from the relation of the tensor product to the direct sum.

Next examine the NDP. Denote by $\mathrm{Hom}_k(R, U)$ the k-vector space of morphisms $R \to U$ of k-vector spaces. For simplicity, write

$$H(R, U) = \mathrm{Hom}_k(R, U).$$

Similarly, write

$$\begin{aligned} H(R,U)[s] &= k[s] \otimes_k H(R,U) = \mathrm{Hom}_{k[s]}(R[s], U[s]) \\ H(R,U)(s) &= k(s) \otimes_k H(R,U) = \mathrm{Hom}_{k(s)}(R(s), U(s)), \end{aligned}$$

which are a $k[s]$-module and a $k(s)$-vector space, respectively. Notice that we have identified some naturally isomorphic tensor product structures. Now define $H(R,U)(s)_g = S_g^{-1} H(R,U)[s]$, and observe that $H(R,U)[s] \subset H(R,U)(s)_g \subset H(R,U)(s)$. Given bases in R and U, a morphism $X(s)$ in $H(R,U)(s)_g$ can be expressed as an $m \times q$ matrix with elements in $S_g^{-1}k[s]$. In an exactly analogous way, we can set up $H(R,Y)$ and $H(U,Y)$. Then similar developments follow. Given the plant $P(s)$, define a morphism

$$F\colon H(R,U)(s) \oplus H(R,Y)(s) \to H(R,Y)(s)$$

of $k(s)$-vector spaces by the action $F(M(s), T(s)) = P(s) \circ M(s) - T(s)$. Clearly, $(M(s), T(s))$ satisfies Figure 18.3.1 if and only if $(M(s), T(s))$ is in ker F. Furthermore, since $F(0, -T(s)) = T(s)$, F is epic, with rank pq. Therefore, ker F is an mq-dimensional $k(s)$-vector space. F can be restricted to submodules of its domain. In particular, write

$$\begin{aligned} K(s) &= \ker F \\ K(s)_g &= \ker F | H(R,U)(s)_g \oplus H(R,Y)(s)_g \\ K[s] &= \ker F | H(R,U)[s] \oplus H(R,Y)[s]. \end{aligned}$$

Then $K[s]$ is a free $k[s]$-module of rank mq, $K(s)_g$ is the localization $S_g^{-1}K[s]$ which is a free $S_g^{-1}k[s]$-module of rank mq, and $K[s] \subset K(s)_g \subset K(s)$. A $k[s]$-basis for $K[s]$ gives a $S_g^{-1}k[s]$ basis for $K(s)_g$, so that computations may be done in $k[s]$.

In fact, an explicit description of $K[s]$ may be given. Given $P(s)$, there exist morphisms $D(s)\colon U[s] \to U[s]$ and $N(s)\colon U[s] \to Y[s]$ of $k[s]$-modules, which induce morphisms $U(s) \to U(s)$ and $U(s) \to Y(s)$ of $k(s)$-vector spaces, such that (a) $D(s)$ is invertible on $U(s) \to U(s)$, (b) $P(s) = N(s) \circ D^{-1}(s)$, and (c) there exist morphisms $A(s)\colon U[s] \to U[s]$ and $B(s)\colon Y[s] \to U[s]$ of $k[s]$-modules such that

$$A(s) \circ D(s) + B(s) \circ N(s) = 1_{U[s]}.$$

The pair $(N(s), D(s))$ is called a right coprime factorization [15, 37] for $P(s)$. It is not unique, but any one may serve for the discussion. Given $(N(s), D(s))$, define a morphism

$$\alpha\colon H(R,U)[s] \to H(R,U)[s] \oplus H(R,Y)[s]$$

of $k[s]$-modules by the action $\alpha(X(s)) = (D(s) \circ X(s), N(s) \circ X(s))$.

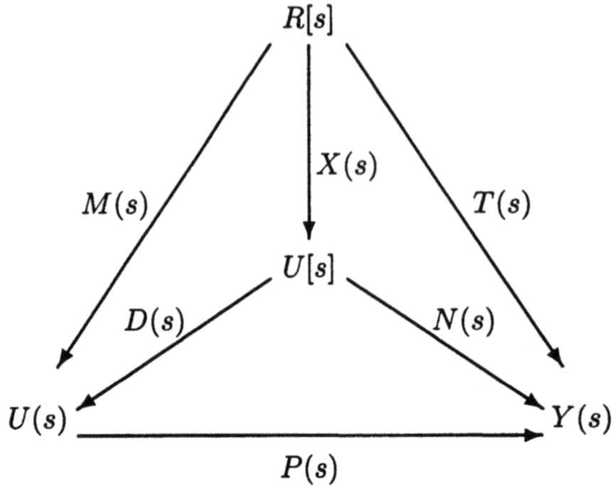

Figure 18.4.1: The NDP at the $k[s]$-module level

Theorem 18.4.1. *The morphism α is monic with image precisely $K[s]$.*

The morphism α defined above can be localized, and since localization of exact sequences gives exact sequences, we have the next result.

Corollary 18.4.1. *The morphism*

$$S_g^{-1}\alpha\colon H(R,U)(s)_g \to H(R,U)(s)_g \oplus H(R,Y)(s)_g$$

defined by

$$S_g^{-1}\alpha(X(s)) = (D(s) \circ X(s), N(s) \circ X(s))$$

is monic and has image precisely equal to $K(s)_g$.

With this background, we can summarize the basic character of the NDP.

Theorem 18.4.2. *The pair $(M(s),T(s))$ is a solution to the NDP if and only if there exists a good transfer function $X_g(s)\colon R(s)_g \to U(s)_g$ such that $M(s) = D(s) \circ X_g(s)$ and $T(s) = N(s) \circ X_g(s)$, where $(N(s),D(s))$ is any right coprime factorization of $P(s)$.*

The good transfer function $X_g(s)$ thus becomes the crucial design parameter in the NDP. The design situation, then, may be sketched as in Figure 18.4.1. In this figure, the viewpoint is at the "computational" level of Theorem 18.4.1 and the morphism α of $k[s]$-modules. Localization of the diagram in Figure 18.4.1 brings us to the diagram of Figure 18.4.2, which is at the level of Theorem 18.4.2. Here we have written $R(s)_g$ for $S_g^{-1}R[s]$, $M_g(s)$ for $S_g^{-1}M(s)$, and so forth. Notice that $S_g^{-1}V(s) = V(s)$, so that $U(s)$ and $Y(s)$ are unchanged. The subscript g on M, X, T, D, and N may be omitted, when in context.

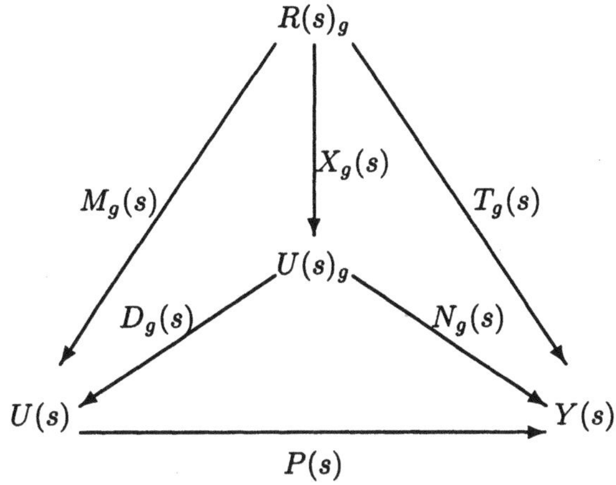

Figure 18.4.2: The NDP at the $S_g^{-1}k[s]$-module level

18.5 The Design Morphism

Consider a good transfer function $X_g(s): R(s)_g \to U(s)_g$. It is clear that one may construct solutions to the NDP according to the scheme indicated in Figure 18.5.1.

Then $(M_g(s), T_g(s))$ generates a good transfer function pair $(M(s), T(s))$ satisfying Figure 18.3.1 by a process of localization up to $k(s)$. It can also be shown that this is the only way that the NDP solution pairs $(M(s), T(s))$ are generated. Thus, in Figure 18.5.2, if $(M(s), T(s))$ is a solution to the NDP, there must exist a good transfer function $X(s)$ such that the diagram commutes.

Remark 18.5.1. We use the same symbol for $D(s): U[s] \to U[s]$ and its localization $D(s): U(s) \to U(s)$ to avoid proliferation of symbols, and we follow the same convention for $N(s)$. The diagrams make the situation clear.

The right-prime pair $(N(s), D(s))$ and the good transfer function $X(s)$ give a specific characterization of solutions to the NDP. In a genuine sense, $X(s)$ contains all the design freedom available. We call $X(s)$ the "design morphism."

18.6 Output Feedback Synthesis (OFS)

By an output feedback, we shall understand a morphism $C(s): R(s) \oplus Y(s) \to U(s)$ of $k(s)$-vector spaces. Let i_R and i_Y be the insertions of $R(s)$ and $Y(s)$ into the biproduct. Then each output feedback determines morphisms $R(s) \to U(s)$ and $Y(s) \to U(s)$ of $k(s)$-vector spaces by

$$C_R(s) = C(s) \circ i_R \quad , \quad C_Y(s) = C(s) \circ i_Y . \tag{18.6.1}$$

Moreover, each pair $(C_R(s), C_Y(s))$ of such morphisms induces a unique output feedback. We shall say that $C(s)$ is an OFS of $M(s)$ if $\{1_{Y(s)} - P(s) \circ C_Y(s)\}$ is

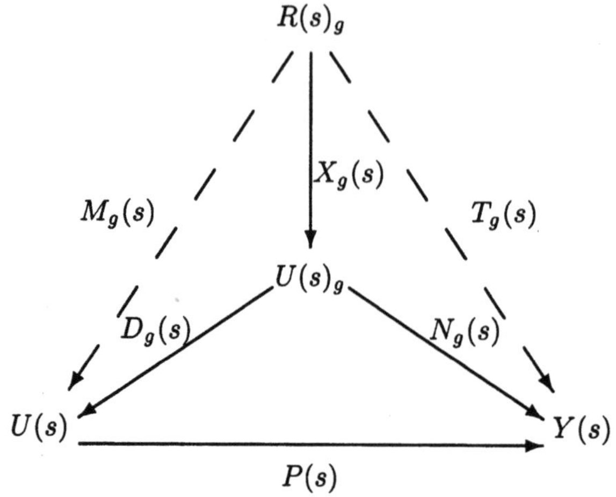

Figure 18.5.1: $X_g(s) \Rightarrow$ NDP solution $(M_g(s), T_g(s))$

an isomorphism and if

$$M(s) = \left(1_{U(s)} - C_Y(s) \circ P(s)\right)^{-1} \circ C_R(s)\,. \tag{18.6.2}$$

By itself, OFS of $M(s)$ poses no problem, as we may choose $C_R(s)$ to be $M(s)$ and $C_Y(s)$ to be the zero morphism.

In order to attack a more interesting problem, both in application and in theory, we choose to constrain $C(s)$. In particular, we specify that

$$C(s)(r(s), y(s)) = G(s)\{r(s) - y(s)\} \tag{18.6.3}$$

for $G(s)\colon R(s) \to U(s)$ a morphism of $k(s)$-vector spaces. $(C_R(s), C_Y(s))$ is given by $(G(s), -G(s))$. We can refer to this well-known case as unity feedback synthesis (UFS). For UFS, $C(s) = 0 \Leftrightarrow G(s) = 0$; thus trivialities do not occur. $G(s)$ determines a UFS for $M(s)$ if $\left(1_{Y(s)} + P(s) \circ G(s)\right)$ is an isomorphism and if

$$M(s) = \left(1_{U(s)} + G(s) \circ P(s)\right)^{-1} \circ G(s)\,. \tag{18.6.4}$$

Remark 18.6.1. A design morphism $X(s)$ may fail to meet the conditions for UFS. Notice also that UFS requires R and Y to be identified, or, if preferred, to be regarded as isomorphic copies of one another.

In what follows, it is useful to have in mind the following lemma.

Lemma 18.6.1. *Let $\mathcal{R}$ be a ring, and let $L_1\colon M_1 \to M_2$ and $L_2\colon M_2 \to M_1$ be morphisms of $\mathcal{R}$-modules. Then*

$$(1_{M_2} + L_1 \circ L_2) : M_2 \to M_2 \tag{18.6.5}$$

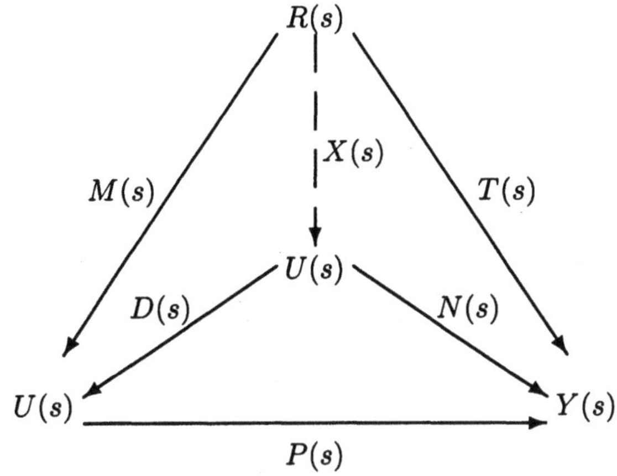

Figure 18.5.2: The NDP solution $(M(s), T(s)) \Rightarrow$ good $X(s)$

has an inverse morphism if and only if

$$(1_{M_1} + L_2 \circ L_1) : M_1 \to M_1 \tag{18.6.6}$$

has an inverse morphism.

Corollary 18.6.1. *In Lemma 18.6.1, when the inverse morphisms exist,*

$$(1_{M_1} + L_2 \circ L_1)^{-1} = 1_{M_1} - L_2 \circ (1_{M_2} + L_1 \circ L_2)^{-1} \circ L_1 . \tag{18.6.7}$$

Corollary 18.6.2. $\left(1_{U(s)} + G(s) \circ P(s)\right) : U(s) \to U(s)$ *is an isomorphism when and only when* $\left(1_{R(s)} + P(s) \circ G(s)\right) : R(s) \to R(s)$ *is an isomorphism.*

Conditions on the design morphism $X(s)$, for a UFS in the NDP, are mild, as shown in the next proposition.

Proposition 18.6.1. *A solution* $(M(s), T(s))$ *to the NDP, characterized by the right-prime pair* $(N(s), D(s))$ *and the design morphism* $X(s)$*, admits UFS if and only if* $\left(1_{R(s)} - N(s) \circ X(s)\right) : R(s) \to R(s)$ *is an isomorphism of* $k(s)$*-vector spaces.*

Proof. In view of Figure 18.5.2, $N(s) \circ X(s)$ is $T(s)$. For sufficiency, choose

$$G(s) = M(s) \circ \left\{1_{R(s)} - T(s)\right\}^{-1}, \tag{18.6.8}$$

which achieves USF. For necessity, observe that

$$\begin{aligned} 1_{R(s)} - T(s) &= 1_{R(s)} - P(s) \circ M(s) \\ &= 1_{R(s)} - P(s) \circ \left\{1_{U(s)} + G(s) \circ P(s)\right\}^{-1} \circ G(s) \\ &= \left\{1_{R(s)} + P(s) \circ G(s)\right\}^{-1}. \end{aligned} \tag{18.6.9}$$

This completes the proof of the proposition.

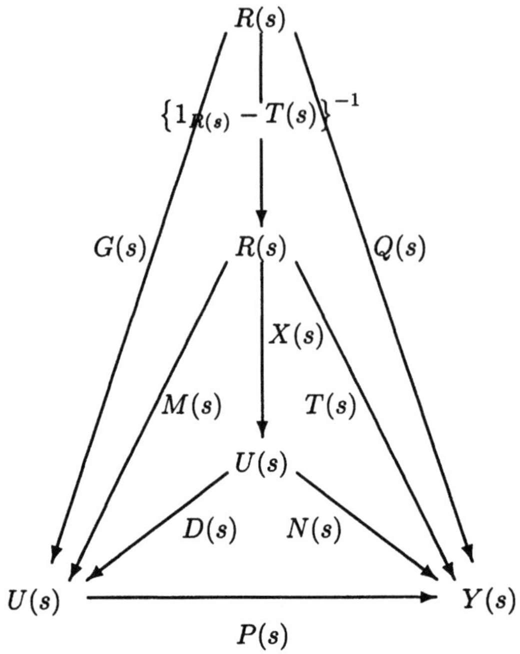

Figure 18.6.1: UFS for the NDP

Remark 18.6.2. The condition of this proposition would typically be met by adding "rolloff" to $[X(s)]$.

Because of Proposition 18.6.1, we can picture the NDP and UFS together in one commutative diagram, Figure 18.6.1. In this diagram, we have denoted the composition $P(s) \circ G(s)$ by $Q(s)\colon R(s) \to Y(s)$.

UFS in the NDP does not yet fulfill application needs. What is needed is a concept of good UFS. Of course, a special case in point is the familiar idea of unity feedback synthesis with internal stability.

18.7 Good Unity Feedback Synthesis (GUSF)

Given $G(s)$, there exist morphisms $\widetilde{D}_G(s)\colon U[s] \to U[s]$ and $\widetilde{N}_G(s)\colon R[s] \to U[s]$ of $k[s]$-modules, with $\widetilde{D}_G(s)$ monic, such that (1) their localizations up to $k(s)$ make the diagram of Figure 18.7.1 commute and (2) there exist morphisms $\widetilde{A}(s)\colon U[s] \to U[s]$ and $\widetilde{B}(s)\colon U[s] \to R[s]$ of $k[s]$-modules with the property $\widetilde{D}_G(s) \circ \widetilde{A}(s) + \widetilde{N}_G(s) \circ \widetilde{B}(s) = 1_{U[s]}$. The pair $\left(\widetilde{D}_G(s), \widetilde{N}_G(s)\right)$ is known as a left-prime factorization of $G(s)$. Using this factorization, we can give an alternate statement of the condition for UFS.

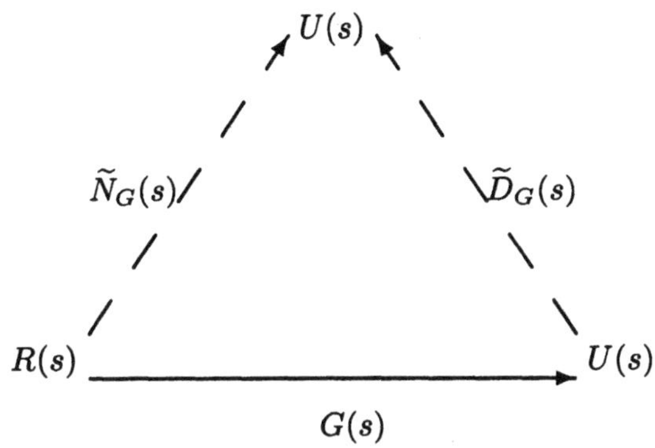

Figure 18.7.1: Left-prime factorization

Proposition 18.7.1. *Suppose that* $(N(s), D(s))$ *is a right-prime factorization of* $P(s)$, *and that* $\left(\widetilde{D}_G(s), \widetilde{N}_G(s)\right)$ *is a left-prime factorization of* $G(s)$. *If* $G(s)$ *generates a UFS for* $M(s)$, *then* $D_i(s)$, *given by*

$$\widetilde{D}_G(s) \circ D(s) + \widetilde{N}_G(s) \circ N(s) \colon U[s] \to U[s], \tag{18.7.1}$$

is a monomorphism of $k[s]$*-modules.*

Proof. Notice that the result is true when and only when localization of the morphism up to $k(s)$ produces an isomorphism $U(s) \to U(s)$ of $k(s)$-vector spaces. For UFS, $\left(1_{U(s)} + G(s) \circ P(s)\right)$ must be an isomorphism. But

$$1_{U(s)} + G(s) \circ P(s) = \widetilde{D}_G^{-1}(s) \circ \left\{\widetilde{D}_G(s) \circ D(s) + \widetilde{N}_G(s) \circ N(s)\right\} \circ D^{-1}(s)\,,$$

and the result follows. □

The manner in which $X(s)$ is constrained by a requirement to achieve UFS can be expressed in terms of $D_i(s)$ and of $\widetilde{N}_G(s)$.

Theorem 18.7.1. *Suppose that the right-prime factorization* $(N(s), D(s))$ *and the design morphism* $X(s)$ *characterize a solution* $(M(s), T(s))$ *to the NDP. If* $G(s)$ *described by its left-prime factorization* $\left(\widetilde{D}_G(s), \widetilde{N}_G(s)\right)$ *generates a UFS for* $M(s)$, *then*

$$X(s) = D_i^{-1}(s) \circ \widetilde{N}_G(s)\,. \tag{18.7.2}$$

Proof. The result follows from the calculation

$$\begin{aligned} X(s) &= D^{-1}(s) \circ M(s) \\ &= D^{-1}(s) \circ \left\{1_{U(s)} + G(s) \circ P(s)\right\}^{-1} \circ G(s) \\ &= D^{-1}(s) \circ \left\{1_{U(s)} + \widetilde{D}_G^{-1}(s) \circ \widetilde{N}_G(s) \circ N(s) \circ D^{-1}(s)\right\}^{-1} \\ &\quad \circ \widetilde{D}_G^{-1}(s) \circ \widetilde{N}_G(s) \\ &= D_i^{-1}(s) \circ \widetilde{N}_G(s)\,, \end{aligned} \tag{18.7.3}$$

which proves the theorem.

Remark 18.7.1. The result does not say that $\left(D_i(s), \widetilde{N}_G(s)\right)$ is a left-prime factorization of $X(s)$.

The idea of the theorem can be reversed, in a sense.

Theorem 18.7.2. *Let $\mathcal{A}(s)\colon U[s] \to U[s]$ and $\mathcal{B}(s)\colon R[s] \to U[s]$ be morphisms of $k[s]$-modules, with $\mathcal{A}(s)$ monic. Further, let $(\mathcal{A}(s), \mathcal{B}(s))$ be a left-prime factorization of the transfer function $\mathcal{A}^{-1}(s) \circ \mathcal{B}(s)\colon R(s) \to U(s)$ constructed after localization up to $k(s)$. If*

$$\mathcal{A}(s) \circ D(s) + \mathcal{B}(s) \circ N(s)\colon U[s] \to U[s], \tag{18.7.4}$$

denoted by $D_{\mathcal{AB}}(s)$, is monic, and if the inverse of its localization up to $k(s)$, when composed with $\mathcal{B}(s)$, is a good transfer function, then the design morphism

$$X(s) = D_{\mathcal{AB}}^{-1}(s) \circ \mathcal{B}(s) \tag{18.7.5}$$

admits UFS with

$$G(s) = \mathcal{A}^{-1}(s) \circ \mathcal{B}(s)\,. \tag{18.7.6}$$

Proof. From Figure 18.6.1, we must have

$$\begin{aligned} G(s) &= M(s) \circ \left(1_{R(s)} - N(s) \circ X(s)\right)^{-1} \\ &= D(s) \circ X(s) \circ \left(1_{R(s)} - N(s) \circ X(s)\right)^{-1} \\ &= D(s) \circ D_{\mathcal{AB}}^{-1}(s) \circ \mathcal{B}(s) \circ \left\{1_{R(s)} - N(s) \circ D_{\mathcal{AB}}^{-1}(s) \circ \mathcal{B}(s)\right\}^{-1} \\ &= D(s) \circ D_{\mathcal{AB}}^{-1}(s) \circ \left\{1_{U(s)} - \mathcal{B}(s) \circ N(s) \circ D_{\mathcal{AB}}^{-1}(s)\right\}^{-1} \circ \mathcal{B}(s) \\ &= D(s) \circ \left\{D_{\mathcal{AB}}(s) - \mathcal{B}(s) \circ N(s)\right\}^{-1} \circ \mathcal{B}(s) \\ &= D(s) \circ \left\{\mathcal{A}(s) \circ D(s)\right\}^{-1} \circ \mathcal{B}(s) \\ &= \mathcal{A}^{-1}(s) \circ \mathcal{B}(s), \end{aligned} \tag{18.7.7}$$

as desired. □

Remark 18.7.2. This theorem assumes only that the composition $D_{\mathcal{AB}}^{-1}(s) \circ \mathcal{B}(s)$ is a good transfer function. In practice, this permits hidden internal behavior which may not be acceptable.

We shall then say that a UFS is good, with acronym GUFS, if $D_i^{-1}(s)\colon U(s) \to U(s)$ is a good transfer function. Theorem 18.7.2 then has an immediate corollary.

Corollary 18.7.1. *If $D_{\mathcal{AB}}^{-1}(s)\colon U(s) \to U(s)$ is a good transfer function, then the construction of Theorem 18.7.2 gives a GUFS.*

In the terminology of Theorem 18.7.2, it is now clear that the family of design morphisms $X(s)$ admitting GUFS can be parameterized in the manner

$$X(s) = \{\mathcal{A}(s) \circ D(s) + \mathcal{B}(s) \circ N(s)\}^{-1} \circ \mathcal{B}(s)\,. \tag{18.7.8}$$

Notice that $1_{U(s)} - X(s) \circ N(s)$ is given by

$$\begin{aligned} & 1_{U(s)} - \{\mathcal{A}(s) \circ D(s) + \mathcal{B}(s) \circ N(s)\}^{-1} \circ \mathcal{B}(s) \circ N(s) \\ = \; & 1_{U(s)} - \left\{1_{U(s)} + D^{-1}(s) \circ \mathcal{A}^{-1}(s) \circ \mathcal{B}(s) \circ N(s)\right\}^{-1} \\ & \circ D^{-1}(s) \circ \mathcal{A}^{-1}(s) \circ \mathcal{B}(s) \circ N(s) \\ = \; & \left\{1_{U(s)} + D^{-1}(s) \circ \mathcal{A}^{-1}(s) \circ \mathcal{B}(s) \circ N(s)\right\}^{-1} \\ = \; & \{\mathcal{A}(s) \circ D(s) + \mathcal{B}(s) \circ N(s)\}^{-1} \circ \mathcal{A}(s) \circ D(s), \end{aligned} \tag{18.7.9}$$

so that $\left\{1_{U(s)} - X(s) \circ N(s)\right\} \circ D^{-1}(s)\colon U(s) \to U(s)$ is a good transfer function. We have established necessity in the following representation theorem.

Theorem 18.7.3. *Suppose that the right-prime factorization $(N(s), D(s))$ and the design morphism $X(s)$ describe a solution $(M(s), T(s))$ to the NDP. Suppose further that $M(s)$ admits UFS. Then $M(s)$ admits GUFS if and only if $Z(s)\colon U(s) \to U(s)$, defined by*

$$Z(s) = \left\{1_{U(s)} - X(s) \circ N(s)\right\}^{-1} \circ D^{-1}(s), \tag{18.7.10}$$

is a good transfer function.

Proof. We need only show sufficiency. Consider the diagram of Figure 18.7.3, where p_R and p_U are the biproduct projections. There exists a unique morphism $\alpha_g(s)$ of $S_g^{-1}k[s]$-modules which makes this diagram commute. Localize $\alpha_g(s)$ up to $k(s)$, and denote it by $\alpha(s)\colon R(s) \oplus U(s) \to U(s)$, where it is a good transfer function. Perform a left-prime factorization $\left(\widetilde{D}_\alpha(s), \widetilde{N}_\alpha(s)\right)$ for $\alpha(s)$, as in Figure 18.7.2. Then the diagrams of these two figures imply that

$$\widetilde{N}_\alpha(s) = \widetilde{D}_\alpha(s) \circ X(s) \circ p_R \tag{18.7.11}$$

$$\widetilde{N}_\alpha(s) = \widetilde{D}_\alpha(s) \circ Z(s) \circ p_U, \tag{18.7.12}$$

so that we may restrict $\widetilde{D}_\alpha(s) \circ X(s)$ and $\widetilde{D}_\alpha(s) \circ Z(s)$ in the manner

$$\widetilde{D}_\alpha(s) \circ Z(s) = \mathcal{A}(s)\colon U[s] \to U[s] \tag{18.7.13}$$
$$\widetilde{D}_\alpha(s) \circ X(s) = \mathcal{B}(s)\colon R[s] \to U[s] \tag{18.7.14}$$

with the former monic because $M(s)$ admits UFS. It is easy to see that

$$G(s) = \mathcal{A}^{-1}(s) \circ \mathcal{B}(s) = Z^{-1}(s) \circ X(s) \tag{18.7.15}$$

is the $G(s)$ required from the assumption that $M(s)$ admits UFS. Indeed,

$$\begin{aligned} G(s) &= M(s) \circ \{1_{R(s)} - N(s) \circ X(s)\}^{-1} \\ &= D(s) \circ X(s) \circ \{1_{R(s)} - N(s) \circ X(s)\}^{-1} \\ &= D(s) \circ \{1_{U(s)} - X(s) \circ N(s)\}^{-1} \circ X(s) \\ &= Z^{-1}(s) \circ X(s)\,. \end{aligned} \tag{18.7.16}$$

Thus it remains to calculate

$$\begin{aligned} \mathcal{A}(s) \circ D(s) + \mathcal{B}(s) \circ N(s) &= \widetilde{D}_\alpha(s) \circ Z(s) \circ D(s) + \widetilde{D}_\alpha(s) \circ X(s) \circ N(s) \\ &= \widetilde{D}_\alpha(s) \circ \{Z(s) \circ D(s) + X(s) \circ N(s)\} \\ &= \widetilde{D}_\alpha(s) \end{aligned} \tag{18.7.17}$$

from which $D_{\mathcal{AB}}(s)$ is monic and the inverse of its localization up to $k(s)$ is a good transfer function. □

Corollary 18.7.2. *If $P(s)\colon U(s) \to Y(s)$ is a good transfer function, then $M(s)$ admits UFS if and only if $M(s)$ admits GUFS.*

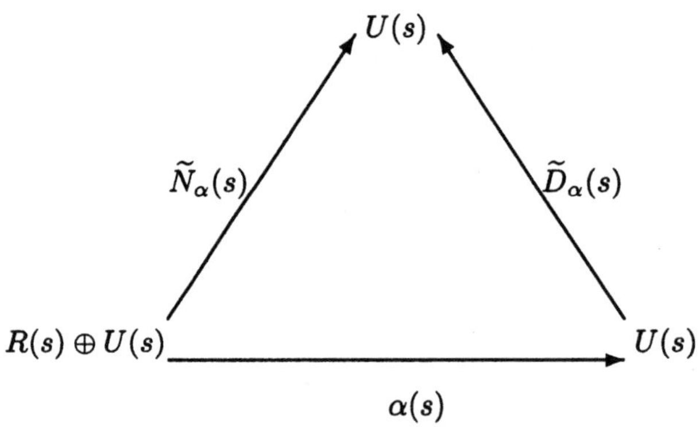

Figure 18.7.2: Left-prime factorization of $\alpha(s)$

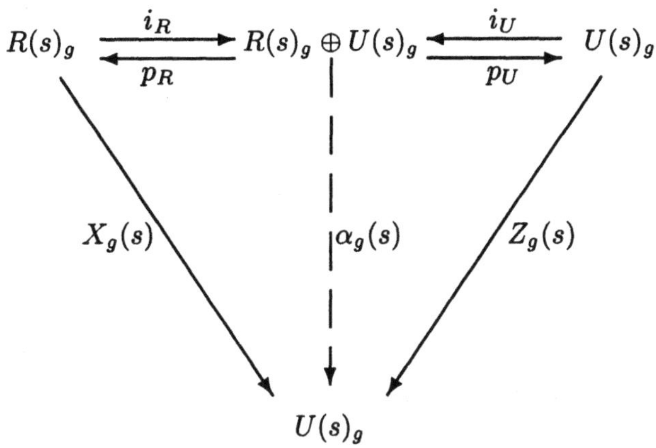

Figure 18.7.3: Module biproduct property

18.8 The Zeros of T(s)

If k were equal to $\mathbb{R}$, the real numbers, and if S_g were the set of nonzero, strictly Hurwitz polynomials, the NDP consists in finding pairs $(M(s), T(s))$ that satisfy the diagram of Figure 18.3.1 and are stable in the sense $m(M) \in S_g$ and $m(T) \in S_g$. In this case, because $M(s)$ is a stable transfer function, we expect that no right-half-plane zero of $P(s)$ can be cancelled by a pole of $M(s)$ and thus that the right-half-plane zeros of $T(s)$ contain those of $P(s)$. Such a statement was established, for example, under rather stringent hypotheses in [42], and related work can be found in [24, 26]. Those results, however, are for the matrix case, and thus are not coordinate-free. In this section, we present a precise and general algebraic result. This result is coordinate-free and makes use of the theory of the zero module introduced in [37].

Intuitively, we show that "the zero module of $T(s)$ maps onto the zero module of $P(s)$ modulo the intersection of the zeros of $P(s)$ with the poles of $M(s)$."

The pole module of $M(s)$ has the description

$$X(M) = \frac{M(s)R[s] + U[s]}{U[s]} \subset \frac{U(s)}{U[s]}.$$

From [37], we have that the zero module of $P(s)$ can be described by

$$Z(P) = \frac{P^{-1}(Y[s]) + U[s]}{\ker P(s) + U[s]}.$$

Let

$$Z_0(P) = \frac{P^{-1}(Y[s]) + U[s]}{U[s]} \subset \frac{U(s)}{U[s]},$$

and write $\pi: Z_0(P) \to Z(P)$ for the natural epimorphism defined by factoring out $\ker P(s)$. Notice that

$$\frac{\ker P(s) + U[s]}{U[s]} \subset Z_0(P).$$

Now $X(M)$ and $Z_0(P)$, as described above, are both submodules of $U(s)/U[s]$. Therefore, their intersection $X(M) \cap Z_0(P)$ can be formed. Moreover, under π, this intersection produces a submodule

$$i(M,P) = \pi\,(X(M) \cap Z_0(P)) \subset Z(P),$$

which has the intuitive interpretation of "the intersection of the poles of $M(s)$ with the zeros of $P(s)$."

It is now possible to state a "pole-zero cancellation theorem."

Theorem 18.8.1. *The morphism $M(s)\colon R(s) \to U(s)$ of $k(s)$-vector spaces induces a morphism $\mu : Z(T) \to Z(P)/i(M,P)$ of $k[s]$-modules. The morphism μ is epic if the rank of $T(s)$ equals the rank of $P(s)$.*

This theorem shows in a coordinate-free way to what extent the zeros of $P(s)$ "appear in" the zeros of $T(s)$. Let us examine one detailed consequence of the theorem. Suppose N is any finitely generated torsion $k[s]$-module. Let m_N be the minimal polynomial of N. The support of N, denoted supp(N), is the set of all zeros of m_N. These zeros need not be elements of k itself. They could lie in an extension field. For example, if k is $\mathbb{R}$, the support is a set of complex numbers. Note that this definition can be extended to modules over arbitrary rings [36]. We shall say that an element in a support is good if it is the zero of some polynomial in S_g. The support itself will be good if each of its elements is good.

The fundamental fact here needed about supports is this: Suppose that $N_1 \subset N_2$ and $N_3 = N_2/N_1$ where N_i, $i = 1,2,3$, are each finitely generated torsion $k[s]$-modules; then supp$(N_1) \subset$ supp(N_2) and supp$(N_3) \subset$ supp(N_2). In fact, supp$(N_2) =$ supp$(N_1) \cup$ supp(N_3). Now supp$(X(M))$ is just the set of poles of $M(s)$, with multiplicities ignored, and supp$(Z(T))$ is just the set of multivariable zeros of $T(s)$, with multiplicities ignored. In particular, if $k = \mathbb{R}$, $M(s)$ is (classically) stable when and only when supp$(X(M))$ is a subset of the open left-half-plane.

Next, the epimorphism μ of Theorem 18.8.1 will be interpreted in terms of supports. The resulting statement, while intuitive, is a severe weakening of the result– because not only module structure but even multiplicity is ignored.

Consider the submodule $i(M,P)$ defined above. Clearly, the properties of support imply supp$(i(M,P)) \subset$ supp$(X(M))$. Furthermore,

$$\text{supp}(Z(P)) = \text{supp}(i(M,P)) \cup \text{supp}(Z(P)/i(M,P)).$$

Now from the construction of μ, $Z(T)/\ker \mu$ is isomorphic to $Z(P)/i(M,P)$, and so supp$(Z(P)/i(M,P)) \subset$ supp$(Z(T))$.

Assume that $M(s)$ is a good transfer function. Then supp$(X(M))$ is good, and this implies that supp$(i(M,P))$ is good. It follows that every element in supp$(Z(P))$ that is not good must occur in supp$(Z(P)/i(M,P))$, and hence also in supp$(Z(T))$.

Corollary 18.8.1. *Suppose that $(M(s), T(s))$ is a solution to the NDP. Then any zero of $P(s)$ that is not good is also a zero of $T(s)$.*

18.9 Conclusions

The TSP has been described. Intuitively introduced by Peczkowski, Sain, and Leake in 1979 [24], TSP consists of the NDP and the GFSP. It has been shown that the NDP can be understood as an abstract kernel problem on localized modules and that design freedom amounts to the choice of a single morphism $X_g(s)$. Further, the idea of plants with a zero module, which is not good, has been discussed, and it has been shown in a coordinate-free way just what type of constraint this imposes on $T(s)$. The treatment considers both control action and plant response at the same time–a departure from the model matching problem. No transformation of rings is used and the approach is coordinate-free.

Various feedback parameterizations have appeared in the literature. For example, Porter [38] discussed a parameter for adaptation, parameter estimation, and sensitivity reduction, and attributed such structures to earlier literature [39]. Moreover, parameterizations of internal stability [40], and their generalizations [20] have been made. Although the design morphism [34] of TSP does not seem to have received prior study, it is to be expected that it should relate to feedback parameters already studied for unity feedback. For example, Desoer and Chen [19] have used the stable parameter of Zames [1] for UFS in the model matching case. When the plant is a good transfer function, and for the case of unity feedback, there exists a good isomorphism relating the Zames parameter to the design morphism. Otherwise, there is only a $k(s)$-relationship.

More importantly, however, model matching does not determine the design morphism associated with TSP, as may be seen easily in Figure 18.4.1. Thus TSP, characterized uniquely by the design morphism, materializes as a problem with fundamental distinctions from model matching.

In the nonlinear case [30, 32, 33], application studies were made. Moreover, Liu and Sung [41] reported nonlinear model matching results for unity feedback with a parameter in the same spirit as the design morphism. Of course, [41] can be reduced to the linear situation. However, TSP and model matching remain distinct problems, as described above.

Dedication

We wish to dedicate this article, which contains not a single ϵ or δ, to Anthony N. Michel, Emeritus Professor at the University of Notre Dame. Good luck, Tony ......

Acknowledgments

We wish to acknowledge our good friend K. D. Pham, who has assisted us greatly with the preparation of the computer files for this chapter; R. R. Gejji, who made the first investigations of the NDP kernel problem and the design matrix and was one of the early workers in polynomial matrix computations; J. L. Peczkowski, or

"Joe P." who was "the man from industry" pioneering these ideas in the context of gas turbine engine control; and P. J. Antsaklis, a virtual wizard with polynomial matrices, who has developed more matrix expressions than we could study in a lifetime of coordinate-free development!

Bibliography

[1] G. Zames, "Feedback and optimal sensitivity: Model reference transformations, multiplicative seminorms, and approximate inverses," *IEEE Trans. Automatic Control*, vol. AC-26, no. 2, pp. 301–320, Apr. 1981.

[2] M. R. Aaron, "Synthesis of feedback control systems by means of pole and zero location of the closed loop function," *AIEE Trans.*, vol. 70, Part II, pp. 1439–1445, 1951.

[3] J. G. Truxal, *Automatic Feedback Control System Synthesis*, McGraw-Hill, New York, 1955.

[4] J. R. Ragazzini and G. F. Franklin, *Sampled-Data Control Systems*, McGraw-Hill, New York, 1958.

[5] G. F. Franklin and J. D. Powell, *Digital Control of Dynamic Systems*, Addison-Wesley, Reading, MA, 1980.

[6] G. C. Newton, Jr., L. A. Gould, and J. F. Kaiser, *Analytical Design of Linear Feedback Controls*, Wiley, New York, 1957.

[7] H. Freeman, "A synthesis method for multipole control systems," *AIEE Trans. Applications and Industry*, vol. 76, Part II, pp. 28–31, Mar. 1957.

[8] R. J. Kavanagh, "Noninteracting controls in linear multivariable systems," *AIEE Trans. Applications and Industry*, vol. 76, Part II, pp. 95–99, May 1957.

[9] H. Freeman, "Stability and physical realizability considerations in the synthesis of multipole control systems," *AIEE Trans. Applications and Industry*, vol. 77, Part II, pp. 1–5, Mar. 1958.

[10] R. J. Kavanagh, "Multivariable control system synthesis," *AIEE Trans. Applications and Industry*, vol. 77, Part II, pp. 425–429, Nov. 1958.

[11] A. S. Morse, "Structure and design of linear model following systems," *IEEE Trans. Automatic Control*, vol. 18, no. 4, pp. 346–353, Aug. 1973.

[12] S.-H. Wang and E. J. Davison, "A minimization algorithm for the design of linear multivariable systems," *IEEE Trans. Automatic Control*, vol. 18, pp. 220–225, 1973.

[13] G. D. Forney, "Minimal bases of rational vector spaces, with applications to multivariable linear systems," *SIAM J. Control*, vol. 13, pp. 493–520, 1975.

[14] M. K. Sain, "A free-modular algorithm for minimal design of linear multivariable systems," *Proc. Sixth IFAC World Congress*, Part IB, Section 9.1, Aug. 1975.

[15] W. A. Wolovich, P. J. Antsaklis, and H. E. Elliott, "On the stability of solutions to minimal and nonminimal design problems," *IEEE Trans. Automatic Control*, vol. 22, no. 1, pp. 88–94, Feb. 1977.

[16] B. D. O. Anderson and N. T. Hung, "Multivariable design problem reduction to scalar design problems," in *Alternatives for Linear Multivariable Control*, M. K. Sain, J. L. Peczkowski, and J. L. Melsa, Eds., pp. 88–95, National Engineering Consortium, Chicago, IL, 1978.

[17] W. A. Wolovich, *Linear Multivariable Systems*, Springer-Verlag, New York, 1974, pp. 250 ff.

[18] A. S. Morse, *System Invariants under Feedback and Cascade Control*, Lecture Notes in Economics and Mathematical Systems, vol. 131, Springer-Verlag, New York, 1975.

[19] C. A. Desoer and M. J. Chen, "Design of multivariable feedback systems with stable plant," *IEEE Trans. Automatic Control*, vol. AC-26, no. 2, pp. 408–415, Apr. 1981.

[20] C. A. Desoer, R.-W. Liu, J. Murray, and R. Saeks, "Feedback system design: The fractional representation approach to analysis and synthesis," *IEEE Trans. Automatic Control*, vol. AC-25, no. 3, pp. 399–412, June 1980.

[21] G. Bengtsson, "Feedback realizations in linear multivariable systems," *IEEE Trans. Automatic Control*, vol. AC-22, no. 4, pp. 576–585, Aug. 1977.

[22] L. Pernebo, "An algebraic theory for the design of controllers for linear multivariable systems-Part I, II," *IEEE Trans. Automatic Control*, vol. 26, no. 1, pp. 171–194, Feb. 1981.

[23] J. L. Peczkowski and M. K. Sain, "Linear multivariable synthesis with transfer functions," in *Alternatives for Linear Multivariable Control*, M. K. Sain, J. L. Peczkowski, and J. L. Melsa, Eds, pp. 71–87, National Engineering Consortium, Chicago, IL, 1978,

[24] J. L. Peczkowski, M. K. Sain, and R. J. Leake, "Multivariable synthesis with inverses," *Proc. Joint Automatic Control Conference*, pp. 375–380, 1979.

[25] J. L. Peczkowski, "Multivariable synthesis with transfer functions," *Proc. Propulsion Controls Symposium*, pp. 111–128, May 1979.

[26] R. J. Leake, J. L. Peczkowski, and M. K. Sain, "Step trackable linear multivariable plants," *Int. J. Control*, vol. 30, no. 6, pp. 1013–1022, Dec. 1979.

[27] M. K. Sain, A. Ma, and D. Perkins, "Sensitivity issues in decoupled control system design," *Proc. Southeastern Symposium on System Theory*, pp 25–29, May 1980.

[28] J. L. Peczkowski and M. K. Sain, "Control design with transfer functions, an application illustration," *Proc. Midwest Symposium on Circuits and Systems*, pp. 47–52, Aug. 1980.

[29] M. K. Sain and A. Ma, "Multivariable synthesis with reduced comparison sensitivity," *Proc. Joint Automatic Control Conference*, vol. 1, Part WP-8B, Aug. 1980.

[30] J. L. Peczkowski and S. A. Stopher, "Nonlinear multivariable synthesis with transfer functions," *Proc. Joint Automatic Control Conference*, Paper WA-8D, Aug. 1980.

[31] M. K. Sain, R. M. Schafer, and K. P. Dudek, "An application of total synthesis to robust coupled design," *Proc. Allerton Conference on Communication, Control, and Computing*, pp. 386–395, Urbana-Champaign, IL, Oct. 1980.

[32] J. L. Peczkowski and M. K. Sain, "Scheduled nonlinear control design for a turbojet engine," *Proc. IEEE International Symposium on Circuits and Systems*, pp. 248–251, Apr. 1981.

[33] M. K. Sain and J. L. Peczkowski, "An approach to robust nonlinear control design," *Proc. Joint Automatic Control Conference*, Paper FA-3D, June 1981.

[34] R. R. Gejji, *On the Total Synthesis Problem of Linear Multivariable Control*, Ph.D. Dissertation, Department of Electrical Engineering, University of Notre Dame, 1980.

[35] S. MacLane and G. Birkhoff, *Algebra*, Macmillan, New York, 1967.

[36] M. F. Atiyah and I. G. MacDonald, *Introduction to Commutative Algebra*, Addison-Wesley, Reading, MA, 1969.

[37] B. F. Wyman and M. K. Sain, "The zero module and essential inverse systems," *IEEE Trans. Circuits and Systems*, vol. CAS-27, no. 2, pp. 112–126, Feb. 1981.

[38] W. A. Porter, "Partial system inverses for adaptation, parameter estimation, and sensitivity reduction," *IEEE Trans. Circuits and Systems*, vol. CAS-25, no. 9, pp. 706–713, Sept. 1978.

[39] J. B. Cruz, Jr., Ed., *System Sensitivity Analysis*, Dowden, Hutchinson and Ross, Stroudsburg, PA, 1973.

[40] P. J. Antsaklis, "Some relations satisfied by prime polynomial matrices and their role in linear multivariable system theory," *IEEE Trans. Automatic Control*, vol. AC-24, no. 4, pp. 611–616, Aug. 1979.

[41] R. Liu and C. H. Sung, "On well-posed feedback systems–In an algebraic setting," *Proc. 19th IEEE Conference on Decision and Control*, pp. 269–271, Dec. 1980.

[42] V. H. L. Cheng and C. A. Desoer, "Limitations on the closed-loop transfer function due to right-half plane transmission zeros of the plant," *IEEE Trans. Automatic Control*, vol. 25, no. 6, pp. 1218–1220, Dec. 1980.

Chapter 19

The Adaptive Dynamic Programming Theorem

John J. Murray, Chadwick J. Cox,* and Richard E. Saeks*

Abstract: The centerpiece of the theory of dynamic programming is the Hamilton-Jacobi-Bellman (HJB) equation, which can be used to solve for the optimal cost functional V^o for a nonlinear optimal control problem, while one can solve a second partial differential equation for the corresponding optimal control law k^o. Although the direct solution of the HJB equation is computationally untenable, the HJB equation and the relationship between V^o and k^o serves as the basis for the adaptive dynamic programming algorithm. Here, one starts with an initial cost functional and stabilizing control law pair (V_0, k_0) and constructs a sequence of cost functional/control law pairs (V_i, k_i) in real time, which are stepwise stable and converge to the optimal cost functional/control law pair, for a prescribed nonlinear optimal control problem with unknown input affine state dynamics.

19.1 Introduction

Unlike the many soft computing applications where it suffices to achieve a "good approximation most of the time," a control system must be stable all of the time. As such, if one desires to learn a control law in real-time, a fusion of soft computing techniques (to learn the appropriate control law) with hard computing techniques (to maintain the stability constraint and guarantee convergence) is required. To implement this fused/hard computing approach to control, an adaptive dynamic program-

*This research was performed in part on the National Science Foundation SBIR contracts DMI-9660604, DMI-9860370, and DMI-9983287.

ming algorithm, which uses soft computing techniques to learn the optimal cost (or return) functional for a stabilizable nonlinear system with unknown dynamics, and hard computing techniques to verify the stability and convergence of the algorithm was developed in [8], where

- the underlying fusion of soft and hard computing concepts was described,
- the adaptive dynamic programming algorithm was formulated,
- a global convergence theorem for the algorithm with a limited sketch of the proof was introduced, and
- several examples of its application in flight control were presented.

The purpose of this chapter is to provide a detailed proof of the adaptive dynamic programming theorem.

The centerpiece of dynamic programming is the Hamilton-Jacobi-Bellman (HJB) equation [2, 3, 7], which one solves for the optimal cost functional $V^o(x_0, t_0)$. This equation characterizes the cost to drive the initial state x_0 at time t_0 to a prescribed final state using the optimal control. Given the optimal cost functional, one may then solve a second partial differential equation (derived from the HJB equation) for the corresponding optimal control law $k^o(x, t_0)$, yielding an optimal cost functional/optimal control law pair (V^o, k^o).

Although direct solution of the HJB equation is computationally untenable (the so-called "curse of dimensionality"), the HJB equation and the relationship between V^o and the corresponding control law k^o, derived therefrom, serves as the basis of the adaptive dynamic programming algorithm [8]. In this algorithm we start with an initial cost functional/control law pair (V_0, k_0), where k_0 is a stabilizing control law for the plant, and construct a sequence of cost functional/control law pairs (V_i, k_i) in real time, which converge to the optimal cost functional/control law pair (V^o, k^o) as follows.

- Given (V_i, k_i); $i = 0, 1, 2, \cdots$; run the system using control law k_i from an array of initial conditions x_0, covering the entire state space (or that portion of the state space where one expects to operate the system);
- Record the state $x_i(x_0, \cdot)$ and control trajectories $u_i(x_0, \cdot)$ for each initial condition;
- Given this data, define V_{i+1} to be the cost to take the initial state x_0 at time t_0 to the final state, using control law k_i;
- Take k_{i+1} to be the corresponding control law derived from V_{i+1} via the HJB equation;
- Iterate the process until it converges.

In Sections 19.2 and 19.3, we will show that (with the appropriate technical assumptions) this process is

- globally convergent to
- the optimal cost functional V^o and is
- stepwise stable; i.e., k_i is a stabilizing controller at every iteration with Lyapunov function V_i.

Since stability is an asymptotic property, technically it is sufficient that k_i be stabilizing in the limit. In practice, however, if one is going to run the system for any length of time with control law k_i, it is necessary that k_i be a stabilizing controller at each step of the iterative process. As such, for this class of adaptive control problems we "raise the bar," requiring stepwise stability, i.e., stability at each iteration of the adaptive process, rather than simply requiring stability in the limit. This is achieved by showing that V_i is a Lyapunov function for the feedback system with controller k_i, generalizing the classical result [7] that V^o is a Lyapunov function for the feedback system with controller k^o.

An analysis of the above algorithm (see Sections 19.2 and 19.3 for additional details) will reveal that a priori knowledge of the state dynamics is not required to implement the algorithm. Moreover, the requirement that the input mapping be known (to compute k_{i+1} from V_{i+1}) can be circumvented by the precompensator technique described in [8] and [9]. As such the above-described adaptive dynamic programming algorithm can be applied to plants with completely unknown dynamics.

19.2 Adaptive Dynamic Programming Algorithm

In the formulation of the adaptive dynamic programming algorithm and theorem, we use the following notation for the state and state trajectories associated with the plant. The variable x denotes a generic state while x_0 denotes an initial state, t denotes a generic time, and t_0 denotes an initial time. We use the notation $x(x_0, \cdot)$ for the state trajectory produced by the plant (with an appropriate control) starting at initial state x_0 (at some implied initial time), and the notation $u(x_0, \cdot)$ for the corresponding control. Finally, the state reached by a state trajectory at time t is denoted by $x = x(x_0, t)$, while the value of the corresponding control at time t is denoted by $u = u(x_0, t)$.

For the purposes of this chapter, we consider a stabilizable time-invariant input affine plant of the form

$$\dot{x} = f(x,u) \equiv a(x) + b(x)u; \; x(t_0) = x_0 \tag{19.2.1}$$

with input quadratic performance measure

$$J = \int_{t_0}^{\infty} l(x((x_0,\lambda), u(x_0,\lambda)))d\lambda$$

$$\equiv \int_{t_0}^{\infty} [q(x(x_0,\lambda)) + u^T(x_0,\lambda)r(x(x_0,\lambda))u(x_0,\lambda)]d\lambda. \tag{19.2.2}$$

Here $a(x)$, $b(x)$, $q(x)$, and $r(x)$ are C^∞ matrix-valued functions of the state that satisfy

- $a(0) = 0$, producing a singularity at $(x,u) = (0,0)$;
- the eigenvalues of $da(0)/dx$ have negative real parts, i.e., the linearization of the uncontrolled plant at zero is exponentially stable;

- $q(x) > 0, x \neq 0; q(0) = 0$;
- $q(x)$ has a positive-definite Hessian at $x = 0$, $d^2q(0)/dx^2 > 0$, i.e., any nonzero state is penalized independently of the direction from which it approaches 0; and
- $r(x) > 0$ for all x.

The goal of the adaptive dynamic programming algorithm is to adaptively construct an optimal control $u^o(x_0, \cdot)$, which takes an arbitrary initial state x_0 at t_0 to the singularity at $(0, 0)$, while minimizing the performance measure J.

Since the plant and performance measure are time-invariant, the optimal cost functional and optimal control law are independent of the initial time t_0, which we may, without loss of generality, take to be 0; i.e., $V^o(x_0, t_0) \equiv V^o(x_0)$ and $k^o(x, t_0) \equiv k^o(x)$. Even though the optimal cost functional is defined in terms of the initial state, it is a generic function of the state $V^o(x)$ and is used in this form in the HJB equation and throughout the chapter. Finally, we adopt the notation

$$F^o(x) \equiv a(x) + b(x)k^o(x)$$

for the optimal closed-loop feedback system. Using this notation, the HJB equation then takes the form

$$\frac{dV^o(x)}{dx}F^o(x) = -l(x, k^o(x)) = -q(x) - k^{oT}(x)r(x)k^o(x) \qquad (19.2.3)$$

in the time-invariant case [7].

Differentiating the HJB equation (19.2.3) with respect to $u^o = k^o(x)$ now yields

$$\frac{dV^o(x)}{dx}b(x) = -2k^{oT}(x)r(x) \qquad (19.2.4)$$

or equivalently

$$u = k^o(x) = \frac{1}{2}r^{-1}(x)b^T(x)\left[\frac{dV^o(x)}{dx}\right]^T \qquad (19.2.5)$$

which is the desired relationship between the optimal control law and the optimal cost functional. Note that an input quadratic performance measure is required to obtain the explicit form for k^o in terms of V^o of (19.2.5), although a similar implicit relationship can be derived in the general case. (See [9] for a derivation of this result.)

Given the above preparation, we may now formulate the desired adaptive dynamic programming algorithm as follows.

Adaptive Dynamic Programming Algorithm.

(1) Initialize the algorithm with a stabilizing cost functional and control law pair (V_0, k_0), where $V_0(x)$ is a C^∞ function, $V_0(x) > 0, x \neq 0$; $V_0(0) = 0$, with a positive-definite Hessian at $x = 0$, $d^2V_0(0)/dx^2 > 0$; and $k_0(x)$ is the C^∞ control law,

$$u = k_0(x) = -\frac{1}{2}r^{-1}(x)b^T(x)\left[dV_0(x)/dx\right]^T.$$

(2) For $i = 0, 1, 2, \cdots$, run the system with control law k_i from an array of initial conditions x_0 at $t_0 = 0$, recording the resultant state trajectories $x_i(x_0, \cdot)$ and control inputs $u_i(x_0, \cdot) = k_i(x_i(x_0, \cdot))$.

(3) For $i = 0, 1, 2, \cdots$, let

$$V_{i+1}(x_0) \equiv \int_0^\infty l(x_i(x_0, \lambda), u_i(x_0, \lambda))d\lambda$$

$$u = k_{i+1}(x) = -\frac{1}{2}r^{-1}(x)b^T(x)\left[\frac{dV_{i+1}(x)}{dx}\right]^T$$

where, as above, we have defined V_{i+1} in terms of initial states but use it generically.

(4) Go to (2).

Since the state dynamics matrix $a(x)$ does not appear in the above algorithm, one can implement the algorithm for a system with unknown $a(x)$. Moreover, one can circumvent the requirement that $b(x)$ be known in Step 3 by augmenting the plant with a known precompensator at the cost of increasing its dimensionality, as shown in [8] and [9]. As such, the adaptive dynamic programming algorithm can be applied to plants with completely unknown dynamics.

In the following, we adopt the notation F_i for the closed-loop system defined by the plant and control law k_i:

$$\dot{x} = F_i(x) \equiv a(x) + b(x)k_i(x)$$

$$= a(x) - \frac{1}{2}b(x)r^{-1}(x)b^T(x)\left[\frac{dV_i(x)}{dx}\right]^T. \tag{19.2.6}$$

To initialize the adaptive dynamic programming algorithm for a stable plant, one may take

$$V_0(x) = \epsilon x^T x$$

and

$$k_0(x) = -\epsilon r^{-1}(x)b^T(x)x$$

which will stabilize the plant for sufficiently small ϵ (although in practice we often take $k_0(x) = 0$). Similarly, for a stabilizable plant, one can "prestabilize" the plant with any desired stabilizing control law such that $d^2V_0(x)/dx^2 > 0$ and the eigenvalues of $dF_0(0)/dx$ have negative real parts and then initialize the adaptive dynamic programming algorithm with the above cost functional/control law pair. Moreover, since the state trajectory going through any point in state space is unique, and the plant and controller are time invariant, one can treat every point on a given state trajectory as a new initial state when evaluating $V_{i+1}(x_0)$, by shifting the time scale analytically without rerunning the system.

The adaptive dynamic programming algorithm is characterized by the following theorem.

Theorem 19.2.1 (Adaptive Dynamic Programming Theorem). *Let the sequence of cost functional/control law pairs (V_i, k_i), $i = 0, 1, 2, \cdots$ be defined by and satisfy the conditions of the adaptive dynamic programming algorithm. Then,*

(i) *$V_{i+1}(x)$ and $k_{i+1}(x)$ exist, where $V_{i+1}(x)$ and $k_{i+1}(x)$ are C^∞ functions with $V_{i+1}(x) > 0, x \neq 0$; $V_{i+1}(0) = 0$; $d^2V_{i+1}(0)/dx^2 > 0$; $i = 0, 1, 2, \cdots$.*

(ii) *The control law k_{i+1} stabilizes the plant with Lyapunov function $V_{i+1}(x)$ for all $i = 0, 1, 2, \cdots$, and the eigenvalues of $dF_{i+1}(0)/dx$ have negative real parts.*

(iii) *The sequence of cost functionals V_{i+1} converge to the optimal cost functional V^o.*

Note that in (ii), the existence of the Lyapunov function $V_{i+1}(x)$ together with the eigenvalue condition on $dF_{i+1}(0)/dx$ implies that the closed-loop system $F_{i+1}(x)$ is exponentially stable [6] rather than asymptotically stable, as implied by the existence of the Lyapunov function alone.

19.3 Proof of the Adaptive Dynamic Programming Theorem

The proof of the adaptive dynamic programming theorem is divided into four steps.

(1) *Show that $V_{i+1}(x)$ and $k_{i+1}(x)$ exist and are C^∞ functions with $V_{i+1}(x) > 0$, $x \neq 0$; $V_{i+1}(0) = 0$; $i = 0, 1, 2, \cdots$.*

By construction $V_{i+1}(x) > 0, x \neq 0; V_{i+1}(0)$, while the existence and smoothness of $k_{i+1}(x)$ follows from that of $V_{i+1}(x)$ since $b(x)$ and $r(x)$ are C^∞ functions and $r^{-1}(x)$ exists.

As such, it suffices to show that $V_{i+1}(x)$ exists and is a C^∞ function. Since $V_{i+1}(x)$ is defined by the state trajectories generated by the ith control law $k_i(x)$, we begin by characterizing the properties of the state trajectories $x_i(x_0, \cdot)$. In particular, since the control law and the plant are defined by C^∞ functions, the state trajectories are also C^∞ functions of both x_0 and t [5]. Furthermore, since $k_i(x)$ is a stabilizing controller and the eigenvalues of $dF_i(0)/dx$ have negative real parts, the state trajectories $x_i(x_0, \cdot)$ converge to zero exponentially [6].

In addition to showing that the state trajectories $x_i(x_0, \cdot)$ are exponentially stable, we would also like to show that the partial derivatives of the state trajectories with respect to the initial condition $\partial^n x_i(x_0, \cdot)/\partial x_0^n$ are also exponentially stable. To this end we observe that $\partial x_i(x_0, \cdot)/\partial x_0$ satisfies the differential equation

$$\frac{\partial}{\partial t}\left[\frac{\partial x_i(x_0, \cdot)}{\partial x_0}\right] = \frac{\partial \dot{x}_i(x_0, \cdot)}{\partial x_0} = \frac{\partial F_i(x_i(x_0, \cdot))}{\partial x_0}$$

$$= \left[\frac{dF_i(x_i(x_0, \cdot))}{dx}\right]\left[\frac{\partial x_i(x_0, \cdot)}{\partial x_0}\right], \quad \frac{\partial x_i(x_0, 0)}{\partial x_0} = 1. \tag{19.3.1}$$

Since $x_i(x_0, \cdot)$ is asymptotic to zero, (19.3.1) reduces to the linear time-invariant differential equation

$$\frac{\partial}{\partial t}\left[\frac{\partial x_i(x_0,\cdot)}{\partial x_0}\right] = \left[\frac{dF_i(0)}{dx}\right]\left[\frac{\partial x_i(x_0,\cdot)}{\partial x_0}\right], \quad \frac{\partial x_i(x_0,0)}{\partial x_0} = 1 \qquad (19.3.2)$$

for large t. As such, the partial derivative of the state trajectory with respect to the initial condition $\partial x_i(x_0,\cdot)/\partial x_0$ is exponentially stable since the eigenvalues of $dF_i(0)/dx$ have negative real parts.

Applying the above argument inductively, we assume that $x_i(x_0, \cdot)$ and

$$\frac{\partial^j x_i(x_0,\cdot)}{\partial x_0^j}, \quad j = 1, 2, \cdots n-1$$

are exponentially stable and observe that $\partial^n x_i(x_0,\cdot)/\partial x_0^n$ satisfies a differential equation of the form

$$\frac{\partial}{\partial t}\left[\frac{\partial^n x_i(x_0,\cdot)}{\partial x_0^n}\right] = \left[\frac{dF_i(x_i(x_0,\cdot)}{dx}\right]\left[\frac{\partial^n x_i(x_0,\cdot)}{\partial x_0^n}\right] + D(t), \quad \frac{\partial x_i(x_0,\cdot)}{\partial x_0^n} = 0 \qquad (19.3.3)$$

where $D(t)$ is a polynomial in $x_i(x_0,\cdot)$ and the trajectories of the lower derivatives $\partial^j x_i(x_0,\cdot)/\partial x_0^j$, $j = 1, 2, \cdots, n-1$. By the inductive hypothesis $x_i(x_0, \cdot)$ and $\partial^j x_i(x_0,\cdot)/\partial x_0^j$, $j = 1, 2, \cdots, n-1$ are all exponentially convergent to zero and, therefore, so is $D(t)$. As such, (19.3.3) reduces to the linear time-invariant differential equation

$$\frac{\partial}{\partial t}\left[\frac{\partial^n x_i(x_0,\cdot)}{\partial x_0^n}\right] = \left[\frac{dF_i(0)}{\partial dx}\right]\left[\frac{\partial^n x_i(x_0,\cdot)}{\partial x_0^n}\right], \quad \frac{\partial^n x_i(x_0,\cdot)}{\partial x_0^n} = 0 \qquad (19.3.4)$$

for large t, implying that nth partial derivative of the state trajectory $\partial x_i(x_0,\cdot)/\partial x_0$ with respect to the initial condition is exponentially stable, since the eigenvalues of $dF_i(0)/dx$ have negative real parts. As such, $x_i(x_0, \cdot)$ and $\partial^n x_i(x_0,\cdot)/\partial x_0^n$; $n = 1, 2, \cdots$ are exponentially convergent to zero.

See [4] for an alternative proof that the derivatives of the state trajectories with respect to the initial condition are exponentially convergent to zero directly in terms of (19.3.1) and (19.3.3).

To verify the existence of $V_{i+1}(x)$, we express $l(x_i(x_0,\cdot), u_i(x_0,\cdot))$ in the form

$$l(x_i(x_0,\cdot), u_i(x_0,\cdot)) = q(x) + k_i^T(x) r(x) k_i(x)$$

$$= q(x) + \frac{1}{4}\left[\frac{dV_i(x)}{dx}\right] b(x) r^{-1}(x) b^T(x) \left[\frac{dV_i(x)}{dx}\right]^T \equiv l_i(x_i(x_0,\cdot)) \qquad (19.3.5)$$

where x denotes $x_i(x_0, \cdot)$ and the notation $l_i(x_i(x_0,\cdot))$ is used to simplify the expression and emphasize that $l(x_i(x_0,\cdot), u_i(x_0,\cdot))$ is a function of the state trajectory.

Now, expanding $q(x)$ as a power series around $x = 0$ and recognizing that $q(0) = 0$ and $dq(0)/dx = 0$, since $x = 0$ is a minimum of the positive-definite function $q(x)$, we obtain

$$q(x) = q(0) + \frac{dq(0)}{dx}x + x^T\frac{d^2q(0)}{dx^2}x + o(||x||^3) = x^T\frac{d^2q(0)}{dx^2}x + o(||x||^3). \quad (19.3.6)$$

As such, there exists K_1 such that $q(x) < K_1||x||^2$ for small x. Similarly, upon expanding $dV_i(x)/dx$ in a power series around $x = 0$, and recognizing $dV_i(0)/dx = 0$ since $x = 0$ is a minimum of V_i, we obtain

$$\frac{dV_i(x)}{dx} = \frac{dV_i(0)}{dx} + \frac{d^2V_i(x)}{dx^2}x + o(||x||^2) = \frac{d^2V_i(x)}{dx^2}x + o(||x||^2). \quad (19.3.7)$$

As such, there exists K_2 such that $dV_i(x)/dx < K_2||x||$ for small x. Finally, since $b(x)r(x)^{-1}b^T(x)$ is continuous at zero, there exists K_3 such that $b(x)r^{-1}(x)b^T(x) < K_3$ for small x. Substituting the inequalities $q(x) < K_1||x||^2$, $dV_i(x)/dx < K_2||x||$, and $b(x)r^{-1}(x)b^T(x) < K_3$ into (19.3.5) therefore yields

$$l(x_i(x_0,\cdot), u_i(x_0,\cdot)) < K_1||x_i(x_0,\cdot)||^2 + K_3K_2^2||x_i(x_0,\cdot)||^2$$

$$= [K_1 + K_3K_2^2]||x_i(x_0,\cdot)||^2 \equiv K||x_i(x_0,\cdot)||^2. \quad (19.3.8)$$

As such,

$$V_{i+1}(x_0) \equiv \int_0^\infty l(x_i(x_0,\lambda), u_i(x_0,\lambda))d\lambda \quad (19.3.9)$$

exists and is continuous in x_0, since the state trajectory $x_i(x_0,\cdot)$ is exponentially convergent to zero.

Finally, to verify that $V_{i+1}(x)$ is a C^∞ function, it suffices to show that trajectories $d^n l_i(x_i(x_0,\cdot))/dx_0^n$ are integrable, in which case one can interchange the derivative and integral operators obtaining

$$\frac{d^n V_{i+1}(x_0)}{dx_0^n} = \int_0^\infty \frac{d^n l_i(x_i(x_0,\cdot))}{dx_0^n}d\lambda. \quad (19.3.10)$$

Now,

$$\frac{dl_i(x_i(x_0,\cdot))}{dx_0} = \frac{dl_i(x_i(x_0,\cdot))}{dx}\frac{dx_i(x_0,\cdot)}{dx_0} \quad (19.3.11)$$

while $d^n l_i(x_i(x_0,\cdot))/dx_0^n$ is a sum of products composed of factors of the form $d^j l_i(x_i(x_0,\cdot))/dx^j$ and $d^k x_i(x_0,\cdot)/dx_0^k$, where every term has at least one factor of the latter type. Since the ith closed-loop system is stable each state trajectory $x_i(x_0,\cdot)$ is contained in a compact set and since $l_i(x_i(x_0,\cdot))$ is a C^∞ function, the derivatives $d^j l_i(x_i(x_0,\cdot))/dx^j$ are bounded on the state trajectory $x_i(x_0,\cdot)$, while we have already shown that the derivatives of the state trajectories with respect to the initial conditions $d^k x_i(x_0,\cdot)/dx_0^k$ converge to zero exponentially. As such, $d^n l_i(x_i(x_0,\cdot))/dx_0^n$ converges to zero exponentially and is therefore integrable, validating (19.3.10) and verifying that $V_{i+1}(x)$ is a C^∞ function.

(2) *Show that the iterative HJB equation*

$$\frac{dV_{i+1}(x)}{dx}F_i(x) = -l(x, k_i(x))$$

is satisfied and that $d^2V_{i+1}(0)/dx^2 > 0; i = 0, 1, 2, \cdots$.

To verify the iterative HJB equation we compute $dV_{i+1}(x_i(x_0,t))/dt$ via the chain rule, obtaining

$$\frac{dV_{i+1}(x_i(x_0,t))}{dt} = \frac{dV_{i+1}(x_i(x_0,t))}{dx}\frac{dx_i(x_0,t)}{dt}$$

$$= \frac{dV_{i+1}(x_i(x_0,t))}{dx}F_i(x_i(x_0,t)) \tag{19.3.12}$$

and by directly differentiating the integral

$$V_{i+1}(x_i(x_0,t)) = \int_0^\infty [l(x_i(x_i(x_0,t),\lambda), u_i(x_i(x_0,t),\lambda))]d\lambda. \tag{19.3.13}$$

Since there is a unique state trajectory passing through the state $x_i(x_0,t)$, the trajectory $x_i(x_i(x_0,t),\cdot)$ must coincide with the tail, after time t, of the trajectory $x_i(x_0,\cdot)$ starting at x_0 at $t_0 = 0$. Translating this trajectory in time to start at $t_0 = 0$ yields the relationship

$$x_i((x_i(x_0,t),\lambda)) = x_i(x_0,\lambda+t),\ \lambda \geq 0 \tag{19.3.14}$$

and similarly for the corresponding control. Substituting this expression into (19.3.13) and invoking the change of variable $\gamma = \lambda + t$ now yields

$$V_{i+1}(x_i(x_0,t)) = \int_0^\infty l(x_i(x_0,\lambda+t), u_i(x_0,\lambda+t))d\lambda$$

$$= \int_t^\infty l(x_i(x_0,\gamma), u_i(x_0,\gamma))d\gamma. \tag{19.3.15}$$

Now,

$$\frac{dV_{i+1}(x_i(x_0,t))}{dt} = \frac{d}{dt}\int_t^\infty l(x_i(x_0,\gamma), u_i(x_0,\gamma))d\gamma$$
$$= l(x_i(x_0,\gamma), u_i(x_0,\gamma))\Big|_t^\infty = -l(x_i(x_0,t), u_i(x_0,t)) \tag{19.3.16}$$

since $l(x_i(x_0,\cdot), u_i(x_0,\cdot)) \equiv l_i(x_i(x_0,\cdot))$ is asymptotic to zero (see (1) above).

Finally, the iterative HJB equation follows by equating the two expressions for $dV_{i+1}(x_i(x_0,t))/dt$ of (19.3.12) and (19.3.16).

To show that

$$\frac{d^2V_{i+1}(0)}{dx^2} > 0,$$

we note that $dV_i(0)/dx = 0$ since zero is a minimum of $V_i(x)$ and, similarly $dV_{i+1}(0)/dx = 0$, while

$$F_i(0) = a(0) - \frac{1}{2}b(0)r^{-1}(0)b^T(0)\left[dV_i(0)/dx\right]^T = 0$$

since $a(0) = 0$. As such, taking the second derivative on both sides of the iterative HJB equation, evaluating it at $x = 0$, and deleting those terms that contain $dV_i(0)/dx$, $dV_{i+1}(0)/dx$, or $F_i(0)$ as a factor yields

$$2\frac{d^2V_{i+1}(0)}{dx^2}\frac{dF_i(0)}{dx} = -\left[\frac{d^2q(0)}{dx^2} + \frac{1}{2}\left[\frac{d^2V_i(0)}{dx^2}\right](b(x)r^{-1}(x)b^T(x))\times \left[\frac{d^2V_i(0)}{dx^2}\right]^T\right]. \tag{19.3.17}$$

Since the right-hand side of (19.3.17) is symmetric, so is the left-hand side. As such, one can replace one of the two terms

$$\frac{d^2V_{i+1}(0)}{dx^2}\frac{dF_i(0)}{dx}$$

on the left-hand side of (19.3.17) by its transpose yielding the linear Lyapunov equation [1]

$$\left[\frac{dF_i(0)}{dx}\right]^T\frac{d^2V_{i+1}(0)}{dx^2} + \frac{d^2V_{i+1}(0)}{dx^2}\left[\frac{dF_i(0)}{dx}\right]$$

$$= -\left[\frac{d^2q(0)}{dx^2} + \frac{1}{2}\left[\frac{d^2V_{i+1}(0)}{dx^2}\right](b(x)r^{-1}(x)b^T(x))\left[\frac{d^2V_{i+1}(0)}{dx^2}\right]^T\right] \tag{19.3.18}$$

where we have used the fact that $d^2V_{i+1}(0)/dx^2$ is symmetric in deriving (19.3.18). Moreover, since the eigenvalues of $dF_i(0)/dx$ have negative real parts, while

$$\frac{d^2q(0)}{dx^2} > 0 \text{ and } \left[\frac{d^2V_i(0)}{dx^2}\right]b(x)r^{-1}(x)b^T(x)\left[\frac{d^2V_i(0)}{dx^2}\right]^T \geq 0,$$

the unique symmetric solution of (19.3.18) is positive-definite [1]. As such,

$$\frac{d^2V_{i+1}(0)}{dx^2} > 0,$$

as required.

(3) *Show that $V_{i+1}(x)$ is a Lyapunov function for the closed-loop system F_{i+1} and that the eigenvalues of $dF_{i+1}(0)/dx$ have negative real parts, $i = 0, 1, 2, \cdots$.*

To show that k_{i+1} is a stabilizing control law for the plant, we show that $V_{i+1}(x)$ is a Lyapunov function for the closed-loop system, F_{i+1}, $i = 0, 1, 2, \cdots$. Since $V_{i+1}(x)$ is positive-definite it suffices to show that the derivative of $V_{i+1}(x)$ along the state trajectories defined by the control law k_{i+1}, $dV_{i+1}(x_{i+1}(x_0, t))/dt$ is negative-definite. To this end we use the chain rule to compute

$$\frac{dV_{i+1}(x_{i+1}(x_0,t))}{dt} = \frac{d[V_{i+1}(x_{i+1}(x_0,t))]}{dx}\frac{dx_{i+1}(x_0,t)}{dt}$$

$$= \frac{d[V_{i+1}(x_{i+1}(x_0,t))]}{dx}F_{i+1}(x_{i+1}(x_0,t)). \tag{19.3.19}$$

Now, upon substituting

$$F_{i+1}(x_{i+1}) = a(x_{i+1}) - \frac{1}{2}b(x_{i+1})r^{-1}(x_{i+1})b^T(x_{i+1})\left[\frac{dV_{i+1}(x_{i+1})}{dx}\right]^T \tag{19.3.20}$$

(where we have used x_{i+1} as a shorthand notation for $x_{i+1}(x_0, t)$) into (19.3.19), we obtain

$$\frac{dV_{i+1}(x_{i+1}(x_0,t))}{dt} = \left[\frac{dV_{i+1}(x_{i+1})}{dx}\right]a(x_{i+1})$$

$$-\frac{1}{2}\left[\frac{dV_{i+1}(x_{i+1})}{dx}\right]b(x_{i+1})r^{-1}(x_{i+1})b^T(x_{i+1})\left[\frac{dV_{i+1}(x_{i+1})}{dx}\right]^T. \tag{19.3.21}$$

Similarly, we may substitute the equality

$$F_i(x_{i+1}) = a(x_{i+1}) - \frac{1}{2}b(x_{i+1})r^{-1}(x_{i+1})b^T(x_{i+1})\left[\frac{dV_i(x_{i+1})}{dx}\right]^T \tag{19.3.22}$$

into the iterative HJB equation obtaining

$$\left[\frac{dV_{i+1}(x_{i+1})}{dx}\right]a(x_{i+1}) = \frac{1}{2}\left[\frac{dV_{i+1}(x_{i+1})}{dx}\right]b(x_{i+1})r^{-1}(x_{i+1})b^T(x_{i+1})\times$$

$$\left[\frac{dV_i(x_{i+1})}{dx}\right]^T - l(x_{i+1}, k_i(x_{i+1})). \tag{19.3.23}$$

Substituting (19.3.23) into (19.3.21) now yields

$$\frac{dV_{i+1}(x_{i+1}(x_0,t))}{dt} = \frac{1}{2}\left[\frac{dV_{i+1}(x_{i+1})}{dx}\right]b(x_{i+1})r^{-1}(x_{i+1})b^T(x_{i+1})\times$$

$$\left[\frac{dV_i(x_{i+1})}{dx}\right]^T - l(x_{i+1}, k_i(x_{i+1}))$$

$$-\frac{1}{2}\left[\frac{dV_{i+1}(x_{i+1})}{dx}\right] b(x_{i+1})r^{-1}(x_{i+1})b^T(x_{i+1})\left[\frac{dV_{i+1}(x_{i+1})}{dx}\right]^T \qquad (19.3.24)$$

while expressing $l(x_{i+1}, k_i(x_{i+1}))$ in the form

$$l(x_{i+1}, k_i(x_{i+1})) = q(x_{i+1}) + \frac{1}{4}\left[\frac{dV_i(x_{i+1})}{dx}\right] \times$$

$$b(x_{i+1})r^{-1}(x_{i+1})b^T(x_{i+1})\left[\frac{dV_i(x_{i+1})}{dx}\right]^T \qquad (19.3.25)$$

and substituting this expression into (19.3.24) yields

$$\frac{dV_{i+1}(x_{i+1}(x_0, t))}{dt} = \frac{1}{2}\left[\frac{dV_{i+1}(x_{i+1})}{dx}\right] b(x_{i+1})r^{-1}(x_{i+1})b^T(x_{i+1})\left[\frac{dV_i(x_{i+1})}{dx}\right]^T$$

$$-q(x_{i+1}) - \frac{1}{4}\left[\frac{dV_i(x_{i+1})}{dx}\right] b(x_{i+1})r^{-1}(x_{i+1})b^T(x_{i+1})\left[\frac{dV_i(x_{i+1})}{dx}\right]^T$$

$$-\frac{1}{2}\left[\frac{dV_{i+1}(x_{i+1})}{dx}\right] b(x_{i+1})r^{-1}(x_{i+1})b^T(x_{i+1})\left[\frac{dV_{i+1}(x_{i+1})}{dx}\right]^T. \qquad (19.3.26)$$

Finally, upon completing the square, (19.3.26) reduces to

$$\frac{dV_{i+1}(x_{i+1}(x_0, t))}{dt} = -q(x_{i+1}) - \frac{1}{4}\left[\frac{d[V_{i+1}(x_{i+1}) - V_i(x_{i+1})]}{dx}\right] \times$$

$$b(x_{i+1})r^{-1}(x_{i+1})b^T(x_{i+1})\left[\frac{d[V_{i+1}(x_{i+1}) - V_i(x_{i+1})]}{dx}\right]^T$$

$$-\frac{1}{4}\left[\frac{dV_{i+1}(x_{i+1})}{dx}\right] b(x_{i+1})r^{-1}(x_{i+1})b^T(x_{i+1})\left[\frac{dV_{i+1}(x_{i+1})}{dx}\right]^T. \qquad (19.3.27)$$

As such,

$$\frac{dV_{i+1}(x_{i+1}(x_0, t))}{dt} < 0 \text{ for } x_{i+1}(x_0, t) \neq 0$$

verifying that $V_{i+1}(x)$ is a Lyapunov function for F_{i+1} and that k_{i+1} is a stabilizing controller for the plant, as required.

To show that the eigenvalues of $dF_{i+1}(0)/dx$ have negative real parts, we note that $dV_{i+1}(0)/dx = 0$ since it is a minimum of $V_{i+1}(x)$, and similarly that

$$\frac{dV_i(0)}{dx} = 0,$$

while

$$F_{i+1}(0) = a(0) - \frac{1}{2}b(0)r^{-1}(0)b^T(0)\left[\frac{dV_{i+1}(0)}{dx}\right]^T = 0$$

since $a(0) = 0$. Now, substituting (19.3.19) for the left-hand side of (19.3.27), taking the second derivative on both sides of the resultant equation, evaluating it at $x = 0$, and deleting those terms that contain $dV_{i+1}(0)/dx$, $dV_i(0)/dx$, or $F_{i+1}(0)$ as a factor, yields

$$2\frac{d^2V_{i+1}(0)}{dx^2}\frac{dF_{i+1}(0)}{dx} = -\frac{d^2q(0)}{dx^2} - \frac{1}{2}\left[\frac{d^2V_{i+1}(0)}{dx^2}\right]b(0)r^{-1}(0)b^T(0)\left[\frac{d^2V_{i+1}(0)}{dx^2}\right]^T$$

$$-\frac{1}{4}\left[\frac{d^2[V_{i+1}(0) - V_i(0)]}{dx^2}\right]b(0)r^{-1}(0)b^T(0)\left[\frac{d^2[V_{i+1}(0) - V_i(0)]}{dx^2}\right]^T. \quad (19.3.28)$$

Now, since the right-hand side of (19.3.28) is symmetric so is the left-hand side and, as such, we may equate the left-hand side of (19.3.28) to its hermitian part. Moreover, since $-d^2q(0)/dx^2 < 0$ while the second and third terms on the right-hand side of (19.3.28) are negative semidefinite, the right-hand side of (19.3.28) reduces to a negative-definite symmetric matrix $-Q$. As such, (19.3.28) may be expressed in the form

$$\left[\frac{dF_{i+1}(0)}{dx}\right]^T\frac{d^2V_{i+1}(0)}{dx^2} + \frac{d^2V_{i+1}(0)}{dx^2}\left[\frac{dF_{i+1}(0)}{dx}\right] = -Q. \quad (19.3.29)$$

Finally, to verify that the eigenvalues of $dF_{i+1}(0)/dx$ have negative real parts we let λ be an arbitrary eigenvalue of $dF_{i+1}(0)/dx$ with eigenvector v. As such, $(dF_{i+1}(0)/dx)v = \lambda v$, while premultiplying this relationship by $v^*d^2V_{i+1}(0)/dx^2$ yields

$$v^*\frac{d^2V_{i+1}(0)}{dx^2}\frac{dF_{i+1}(0)}{dx}v = \lambda v^*\frac{d^2V_{i+1}(0)}{dx^2}v. \quad (19.3.30)$$

Now, upon taking the complex conjugate of (19.3.30) and adding it to (19.3.30), we obtain

$$v^*\left(\left[\frac{dF_{i+1}(0)}{dx}\right]^T\frac{d^2V_{i+1}(0)}{dx^2} + \frac{d^2V_{i+1}(0)}{dx^2}\left[\frac{dF_{i+1}(0)}{dx}\right]\right)v = 2Re(\lambda)v^*\frac{d^2V_{i+1}(0)}{dx^2}v. \quad (19.3.31)$$

Finally, substituting (19.3.29) in (19.3.31) yields

$$-v^*Qv = 2Re(\lambda)v^*\frac{d^2V_{i+1}(0)}{dx^2}v \quad (19.3.32)$$

from which it follows that $Re(\lambda) < 0$, since $d^2V_{i+1}(0)/dx^2 > 0$ (see part (2) of the proof), and $-v^*Qv < 0$.

(4) *Show that the sequence of cost functionals V_{i+1} is convergent.*

The key step in our convergence proof is to show that

$$\frac{d[V_{i+1}(x_i(x_0,t)) - V_i(x_i(x_0,t))]}{dt} = \frac{d[V_{i+1}(x_i(x_0,t))]}{dt} - \frac{d[V_i(x_i(x_0,t))]}{dt} \tag{19.3.33}$$

is positive along the trajectories defined by the control law k_i. Substituting (19.3.5) into (19.3.16) yields

$$\frac{d[V_{i+1}(x_i(x_0,t))]}{dt} = -l(x_i,u_i)$$

$$= -q(x_i) - \frac{1}{4}\left[\frac{dV_i(x_i)}{dx}\right] b(x_i)r^{-1}(x_i)b^T(x_i)\left[\frac{dV_i(x_i)}{dx}\right]^T \tag{19.3.34}$$

(where we have used x_i as a shorthand notation for $x_i(x_0,t)$ and similarly for u_i) while one can obtain an expression for $[dV_i(x,t)/dt]\big|_{x_i(x_0,t)}$ from (19.3.27) by replacing the index $i+1$ by the index i

$$\frac{dV_i(x_i(x_0,t))}{dt} = -q(x_i) - \frac{1}{4}\left[\frac{d[V_i(x_i) - V_{i-1}(x_i)]}{dx}\right] b(x_i)r^{-1}(x_i)b^T(x_i)\times$$

$$\left[\frac{d[V_i(x_i) - V_{i-1}(x_i)]}{dx}\right]^T - \frac{1}{4}\left[\frac{dV_i(x_i)}{dx}\right] b(x_i)r^{-1}(x_i)b^T(x_i)\left[\frac{dV_i(x_i)}{dx}\right]^T \tag{19.3.35}$$

which is valid for $i = 1, 2, 3, \cdots$ after reindexing. Finally, substituting (19.3.34) and (19.3.35) into (19.3.33) yields

$$\frac{d[V_{i+1}(x_i(x_0,t)) - V_i(x_i(x_0,t))]}{dx} = -q(x_i) - \frac{1}{4}\left[\frac{dV_i(x_i)}{dx}\right] b(x_i)r^{-1}(x_i)b^T(x_i)\times$$

$$\left[\frac{dV_i(x_i)}{dx}\right]^T q(x_i) + \frac{1}{4}\left[\frac{d[V_i(x_i) - V_{i-1}(x_i)]}{dx}\right] b(x_i)r^{-1}(x_i)b^T(x_i)\times$$

$$\left[\frac{d[V_i(x_i) - V_{i-1}(x_i)]}{dx}\right]^T + \frac{1}{4}\left[\frac{dV_i(x_i)}{dx}\right] b(x_i)r^{-1}(x_i)b^T(x_i)\left[\frac{dV_i(x_i)}{dx}\right]^T$$

$$= \frac{1}{4}\left[\frac{d[V_i(x_i) - V_{i-1}(x_i)]}{dx}\right] b(x_i)r^{-1}(x_i)b^T(x_i)\left[\frac{d[V_i(x_i) - V_{i-1}(x_i)]}{dx}\right]^T > 0 \tag{19.3.36}$$

for $i = 1, 2, 3, \cdots$.

Since F_i is asymptotically stable, its state trajectories $x_i(x, \cdot)$ converge to zero, and hence so does $V_{i+1}(x_i(x_0, \cdot)) - V_i(x_i(x_0, \cdot))$. Since

$$\frac{d[V_{i+1}(x) - V_i(x)}{dt} > 0$$

on these trajectories, however, this implies that

$$V_{i+1}(x_i(x_0, \cdot)) - V_i(x_i(x_0, \cdot)) < 0$$

on the trajectories of F_i, $i = 1, 2, 3, \cdots$. Since every point x in the state space lies along some trajectory of F_i, $x = x_i(x_0, t)$, however, this implies that $V_{i+1}(x) - V_i(x) < 0$ for all x in the state space, or equivalently, $V_{i+1}(x) < V_i(x)$ for all x; $i = 1, 2, 3, \cdots$. As such, $V_{i+1}(x)$, $i = 1, 2, 3, \cdots$ is a decreasing sequence of positive numbers, $i = 1, 2, 3, \cdots$, and is therefore convergent (as is the sequence, $V_{i+1}(x)$, $i = 0, 1, 2, \cdots$, since the behavior of the first entry of a sequence does not affect its convergence), completing the proof of the adaptive dynamic programming theorem.

Although an initial cost functional of the form $V_0(x) = \epsilon x^T x$ is technically required to initialize the algorithm for a stable plant (to guarantee that $d^2 V_0(0)/dx^2 > 0$), a review of the proof will reveal that one can, in fact, initiate the adaptive dynamic programming algorithm for a stable system with $V_0(x) = 0$.

19.4 Conclusions

Our goal was to provide a detailed proof of the adaptive dynamic programming theorem. The reader is referred to [8] for a discussion of the techniques used to implement the theorem in a computationally efficient manner, and examples of its application to both linear and nonlinear systems.

Bibliography

[1] S. Barnett, *The Matrices of Control Theory*, Van Nostrand Reinhold, New York, 1971.

[2] R. E. Bellman, *Dynamic Programming*, Princeton University Press, Princeton, NJ, 1957.

[3] D. P. Bertsekas, *Dynamic Programming: Deterministic and Stochastic Models*, Prentice-Hall, Englewood Cliffs, NJ, 1987.

[4] A. Devinatz and J. L. Kaplan, "Asymptotic estimates for solutions of linear systems of ordinary differential equations having multiple characteristics roots," *Indiana University Math J.*, vol. 22, p. 335, 1972.

[5] J. Dieudonne, *Foundations of Mathematical Analysis*, Academic Press, New York, 1960.

[6] A. Halanay and V. Rasvan, *Applications of Liapunov Methods in Stability*, Kluwer, Dordrecht, 1993.

[7] D. G. Luenberger, *Introduction to Dynamic Systems: Theory, Models, and Applications*, John Wiley, New York, 1979.

[8] J. Murray, C. Cox, G. Lendaris, and R. Saeks, "Adaptive dynamic programming," *IEEE Trans. Systems, Man and Cybernetics: Part C*, vol. 32, pp. 140–153, 2002.

[9] R. Saeks and C. Cox, "Adaptive critic control and functional link networks," *Proc. 1998 IEEE Conference on Systems, Man and Cybernetics*, San Diego, CA, pp. 1652–1657, 1998.

Chapter 20

Reliability of SCADA Systems in Offshore Oil and Gas Platforms

Kelvin T. Erickson, E. Keith Stanek, Egemen Cetinkaya,
Shari Dunn-Norman, and Ann Miller

Abstract: Supervisory control and data acquisition (SCADA) systems are commonly used in the offshore oil and gas industry for remote monitoring and control of offshore platforms. Using a generalized platform system architecture, the reliability of the entire system is estimated. The outcome of this reliability assessment is an estimate of mean time between failures (MTBF), system availability, and probability of facility damage or pollution release. The reliability was estimated using probabilistic risk assessment. A fault tree was constructed to show the effect of contributing events on system-level reliability. Probabilistic methods provide a unifying method to assess physical faults, contributing effects, human actions, and other events having a high degree of uncertainty. The probability of various end events, both acceptable and unacceptable, is calculated from the probabilities of the basic initiating failure events.

20.1 Introduction

According to ARC Advisory Group [2], a system is classified as a supervisory control and data acquisition (SCADA) system when

> "the system must monitor and control field devices using remote terminal units (RTUs) at geographically remote sites. The SCADA system typically includes

the master stations, application software, remote terminal units and all associated communications equipment to interface the devices. The system must also include the controllers and I/O for the master stations and RTUs and also the system HMI and application software programs. It does not include field devices such as flow, temperature or pressure transmitters that may be wired to the RTU."

In the offshore oil and gas industry, there are four different types of facilities that employ SCADA systems offshore: conventional platforms, subsea systems, pipelines, and mobile drilling vessels.

This paper treats only the reliability of the offshore platform SCADA systems. Results for subsea and pipeline systems are covered in Erickson et. al. [4]. Mobile drilling vessels are excluded since the SCADA system is used strictly for monitoring and there is no remote control of actual drilling operations.

20.2 Conventional and Deepwater Platforms

A conventional platform is defined as a steel structure that consists of topsides and a jacket. The jacket is piled into the seabed and does not require any additional tethers or a mooring system for structural integrity. This type of structure has been used extensively to develop both oil and gas fields in the Gulf of Mexico and offshore California. Conventional platforms may be small (tripods or four pile jackets) but many of these platforms are large (jackets with eight piles or more) and include significant topsides and many wells.

Successful explorations in deepwater and technological advances have fueled trends toward deepwater developments recently. For deepwater, tension leg platforms or guyed towers are used with subsea production elements, such as subsea wells, templates, and manifolds. In these systems, the topside part of the structure is similar to that of conventional platforms.

Wells drilled and completed from conventional or deepwater platforms are tied back directly to the platform, and the produced oil and gas flow directly from the production tubing into process facilities located on the platform.

By surveying a representative sample of offshore operators, two major architectures of SCADA systems on offshore platforms have been identified: (1) distributed programmable logic controller (PLC); and (2) centralized PLC. Only the distributed platform architecture is considered in this paper.

The distributed PLC platform architecture is shown in Figure 20.2.1 and is typical of larger platforms. In this type of system, a separate PLC controls each major unit of the platform. There is a platform communication network that connects the PLCs and the computers used for the human-machine interface (HMI). The HMI/SCADA software uses the communication network to send commands to the PLCs and to receive information from the PLCs. There is generally limited information passing between the PLCs. Each major unit normally has a local operator panel to allow personnel to interact with that unit only. In this type of architecture, the safety system is generally handled by one of the PLCs. Typically, the platform communication network is redundant. If the primary network fails, communication is automatically switched to a redundant communication network. The platform is monitored from an onshore office by a microwave/radio/satellite link. The onshore office may perform some limited control functions, especially when the platform is evacuated due to bad weather.

Each PLC generally works autonomously from the other PLCs and will continue to control even if onshore communication to the PLC is temporarily lost. However, if communication is lost for some significant time, the PLC will shut down the unit.

A deepwater platform has the same basic architecture as shown in Figure 20.2.1. The only difference is that the well SSV PLC and the equipment it controls is replaced by a subsea SCADA system, described in Erickson et. al. [4].

20.3 Fault Tree and Reliability Analysis

The reliability assessment of current SCADA technology has two major directions:

(i) Calculation of a reliability index for the SCADA system as a whole, including sensors, modems, communications channels, servers, and the SCADA workstation. The form of this index is the system unavailability.

(ii) Calculation of the probability of a top event during a given year. This top event is facility damage or a significant oil spill.

The calculations are relatively simple if the system is a series system in a reliability sense. This is true of many electronic or mechanical systems. Once the reliability performance of each component in the chain is found, the overall system performance is easily calculated. For instance, if the availability of each component of the system can be found, the overall system availability is just the product of the component availabilities. Similar simple calculations can be done to find the probability of system failure, or the mean time between system failures. For instance, if λ_i, $i = 1, n$ are the n component failure rates, then the system failure rate is $\lambda_{system} = \sum_{i=1}^{n} \lambda_i$. The system reliability, assuming constant failure rates and no repair, is $R(t) = e^{-\lambda_{system} t}$. The system mean-time-to-failure is $MTTF_{system} = 1/\lambda_{system}$.The component availabilities can be found if the constant repair rates for the n components are known to be μ_i, $i = 1, n$. In that case the availability of the ith component is

$$A_i = \frac{\mu_i}{\lambda_i + \mu_i} = \frac{MTTF}{MTTF + MTTR}$$

and the system availability is $A_{system} = \prod_{i=1}^{n} A_i$.The latter calculation is more challenging but contains the most pertinent information.

Most systems, however, are not simple series systems and therefore more powerful techniques such as fault tree analysis must be used.

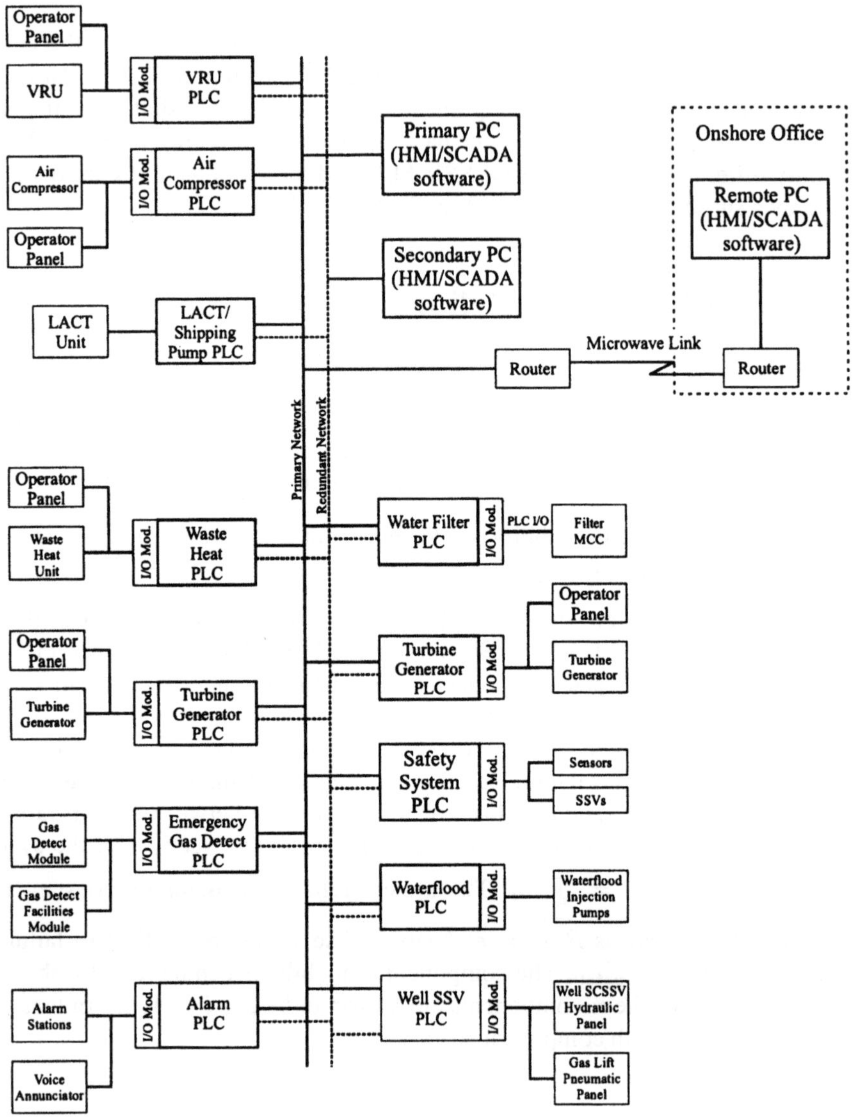

Figure 20.2.1: Distributed PLC platform SCADA system

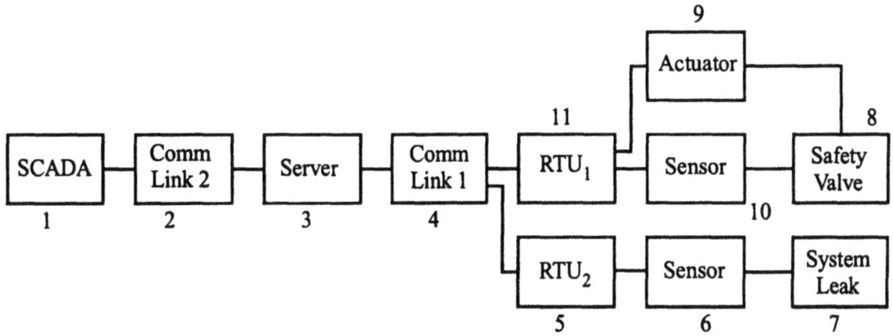

Figure 20.3.1: Simple SCADA system

Probabilistic risk assessment (PRA) was used to assess the effect of contributing events on system-level reliability tree (Billinton and Allan [3], Henley and Kumamoto [6], Shooman [7]). Probabilistic methods provide a unifying method to assess physical faults, contributing effects, human actions, and other events having a high degree of uncertainty. The PRA was performed using fault tree analysis. The probability of various end events, both acceptable and unacceptable, is calculated from the probabilities of the basic initiating failure events.

The fault tree model serves several important purposes. First, the fault tree provides a logical framework for the failure analysis and precisely documents which failure scenarios have been considered and which have not. Second, the fault tree is based on well-defined Boolean algebra and probability theory. The fault tree shows how events combine to cause the end (or top) event, and at the same time defines how the probability of the end event is calculated as a function of the probabilities of the basic events. Thus, the fault tree model can be easily changed to accommodate systems consisting of components from one vendor as well as components from mixed vendors (e.g., software vendors and hardware vendors). The fault tree analysis also illuminates the "weak points" in the design, which will be used to assess trade-offs and to generate recommendations to oil and gas operators.

20.3.1 Example reliability analysis

To illustrate the concept, it will be assumed that a simple SCADA system can be represented as shown in Figure 20.3.1.

Suppose now we postulate the top event to be a system leak that is not mitigated by action of the safety shutoff valve. The fault tree diagram for this system would look like Figure 20.3.2. In Figure 20.3.2,

G1 – Failure to close safety valve in the presence of an oil leak
G2 – Failure to sense an oil leak
G3 – Failure to sense state of an open safety valve
G4 – Failure to close the safety valve.

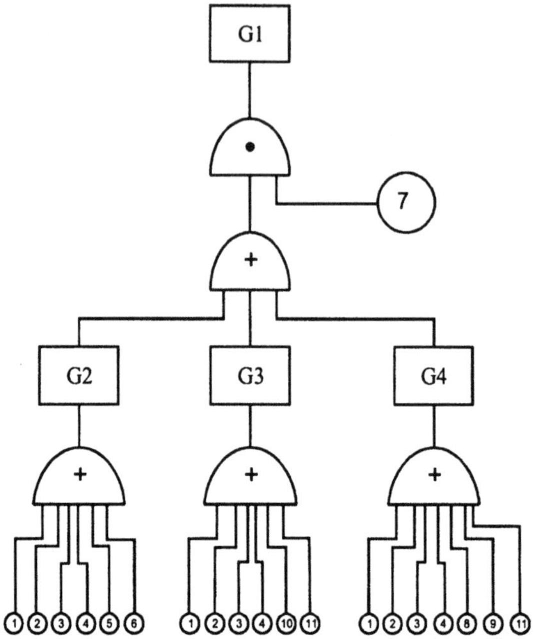

Figure 20.3.2: Fault tree for simple SCADA system

Basic events are

q_1 – Failure of the SCADA system
q_2 – Failure of the communications link 2
q_3 – Failure of the server
q_4 – Failure of communications link 1
q_5 – Failure of RTU_2
q_6 – Failure of the leak sensor
q_7 – Failure of the system: an oil leak
q_8 – Failure of the safety valve to close
q_9 – Failure of the actuator to close the safety valve
q_{10} – Failure of the safety valve position sensor
q_{11} – Failure of RTU_1.

This is an example fault tree diagram. If the safety valve is a fail-safe type valve, the state G4 is not a factor. If the SCADA system is designed to close the safety valve regardless of how its state is sensed, then state G3 is not a factor either.

The simplest approach to solving for the availability of the top event is to draw a reduced fault tree. Since the intermediate states G2, G3, and G4 are all the outputs of OR gates and they all feed into an OR gate, the four OR gates can be replaced by a single OR gate with all the basic events, except number 7, as inputs. The new reduced fault tree is as shown in Figure 20.3.3, where

G5 - Failure of SCADA system or safety value.

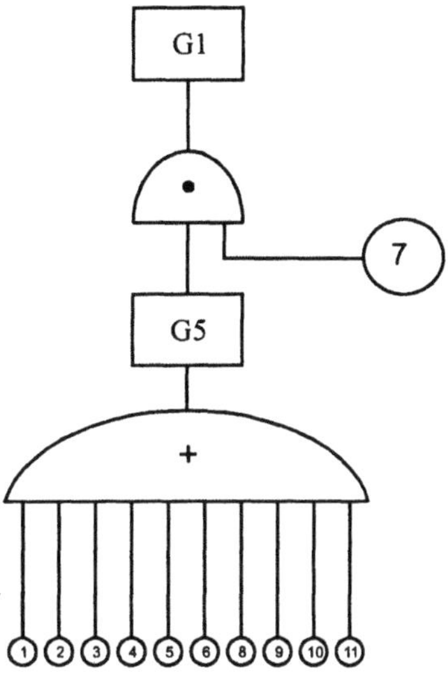

Figure 20.3.3: Reduced fault tree for simple SCADA system

The availability of state G5 is

$$A\,(\text{top event}) = q_1 A\,(\text{top event}\,|q_1 \text{ occurs}) + (1 - q_2)\, A\,(\text{top event}\,|q_1 \text{ doesn't occur})\,.$$

The availability of this top event is

$$A\,(G1) = q_7 \left[1 - \prod_{i=1, i\neq 7}^{11} (1 - q_i)\right].$$

20.3.2 Calculating the availability of the top event

Processing the availabilities of component failures through a series of AND/OR gates in a fault tree is a well-established process. In general, an AND gate with n inputs, each with availability of failure of q_i, results in a new state with availability

$$A\,(\text{output of AND gate}) = q_i\, q_2 \cdots q_n = \prod_{i=1}^{n} q_i.$$

In the case of an OR gate the availability of the new state when there are n inputs, each with availability of failure of q_i, is

$$\begin{aligned} A\,(\text{output of OR gate}) &= 1 - (1 - q_i)\,(1 - q_2) \cdots (1 - q_n) \\ &= 1 - \prod_{i=1}^{n} (1 - q_i). \end{aligned}$$

A program capable of performing these operations for a general fault tree without dependent basic events was available and was used to calculate the availability of the top event in these fault trees.

20.3.3 Failure of the SCADA system

Figure 20.3.4 shows the fault tree for a typical SCADA system as shown in Figure 20.2.1. Using failure rates and repair rates, availabilities and unavailabilities can be found for each subsystem, shown in Table 20.3.1. The communication network failure availability is from one operator. The other availabilities are from SINTEF [8] and Gertman and Blackman [5].

The first column in Table 20.3.1 contains the event numbers (in the circles) that appear in the fault tree of Figure 20.2.1. A short description of the system component that fails occurs in the second column. The third column has the failure rate, in failures per year. This column is derived from the SINTEF [8] tables, which have the failure rate in failures per million hours. The fourth column contains the time to repair, in repairs per year, and is calculated by dividing the number of hours/year (8760) by the hours/repair from SINTEF [8]. The last column, availability, the last column is obtained by dividing the third column by the sum of the third and fourth columns.

The availability of the top event (failure of the SCADA system) can be found by analyzing the fault tree diagram with the basic event data of Table 20.3.1. Calculating this value results in the availability of the top event (SCADA system failed) equal to 1.2×10^{-2}. Note that the communication network dominates the failure rate.

Table 20.3.1: Failure data for basic events in SCADA fault tree

No.	Basic Events	Failure Rates of Basic Events (Failures per Year)	Repair Times for Basic Events (Repairs per Year)	Availability of Failure
1	Remote PC	1.075	2136.6	0.0005
2	Comm. Network	109.1	8760	0.01
3,4	PC	1.075	2136.6	0.0005
5–16	PLC	1.18	7963.6	0.00015

20.3.4 Reliability of the entire platform system

To assess the availability of the entire platform system, the availability of the platform system (not including SCADA) and the subsea system must be included. The fault tree of the platform and subsea systems are developed in Erickson et. al. [4] using the information in [1]. The availability of the system to a failure of the control system that leads to a pollution release or facility damage is about 6.2×10^{-3}. The low availability values for the SCADA system failure and the surface/subsea

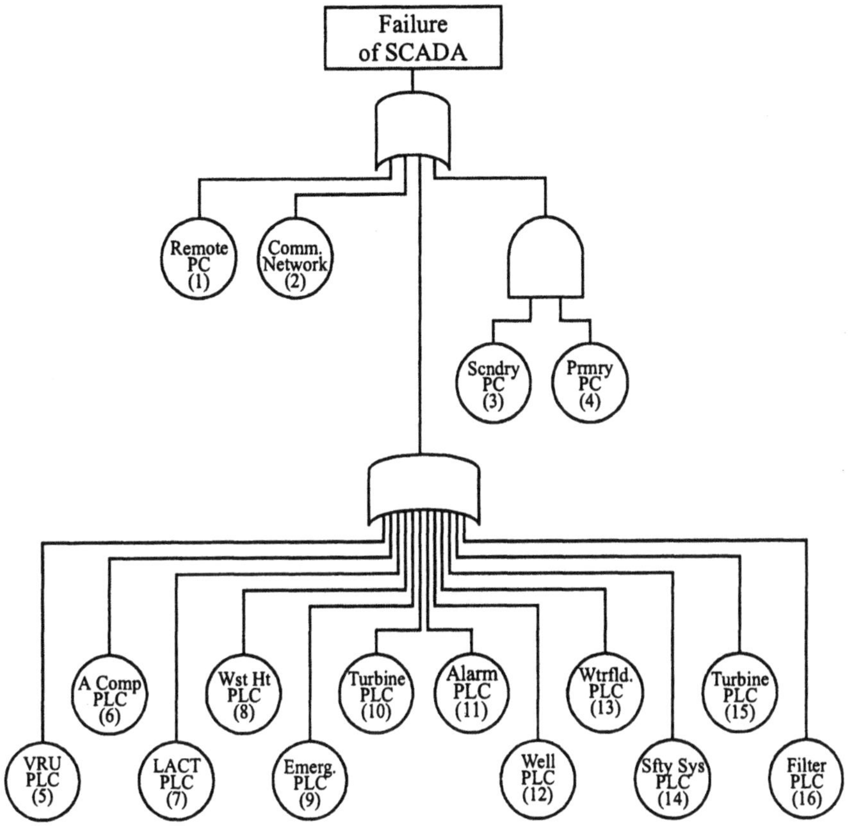

Figure 20.3.4: Fault tree for distributed platform SCADA system

failure make it unlikely that these two events could occur simultaneously. The overall availability of a surface/subsea failure and a SCADA failure is about 8.3×10^{-4}. If such an event took an average of one hour to repair, this would lead to a failure rate of about 8300 failures/10^6 hours or about one failure in 1.4 years.

20.3.5 Operator reliability experience

All of the surveyed operators indicated that distant network communications are the weak link in all of their systems. The conversion of analog microwave to a digital microwave network with a loop architecture has helped. Nevertheless, the operators program their systems to reliably function even when communications is disrupted.

Operators of deepwater subsea systems report one to two failures of the control system per year, which is consistent with the 0.86 year MTBF shown in Table 20.3.2.

Table 20.3.2: Summary of reliability analysis results

System	MTBF (years)	Failure Availability	Failure
Surface	2.5×10^{8}	7.6×10^{-13}	Pollution or facility damage
Subsea	0.86	0.00618	Control
Surface/Subsea	0.86	0.00618	Pollution, facility damage, or control
SCADA	0.09	0.012	Control
Surface/Subsea/SCADA	1.4	0.00083	Pollution, facility damage, or control

20.4 Conclusions

Using a generalized platform system architecture, the reliability of the entire SCADA system used in the offshore oil and gas industry is estimated. The outcome of this reliability assessment is shown in Table 20.3.2, which is corroborated by operator experience. Results of the reliability study indicate that communication system failures are the predominant failure modes in the SCADA systems.

Bibliography

[1] API, "Recommended practice for analysis, design, installation, and testing of basic surface safety systems for offshore production platforms," *Recommended Practice 14C*, 6th ed., 1998.

[2] ARC Advisory Group, *Oil & Gas, Water SCADA Systems Global Outlook*, ARC Advisory Group, Dedham, MA, Mar. 1999.

[3] R. Billinton and R. N. Allan, *Reliability Evaluation of Engineering Systems: Concepts and Techniques*, 2nd ed., Plenum Press, New York, 1992.

[4] K. T. Erickson, A. Miller, E. K. Stanek, and S. Dunn-Norman, *Survey of SCADA System Technology in the Offshore Oil and Gas Industry*, Report to Department of Interior, Program SOL 1435-01-99-RP-3995, Nov. 2000.

[5] D. I. Gertman and H. S. Blackman, *Human Reliability and Safety Analysis Handbook*, Wiley, New York, 1994.

[6] E. J. Henley and H. Kumamoto, *Probabilistic Risk Assessment*, IEEE Press, New York, 1992.

[7] M. L. Shooman, *Probabilistic Reliability: An Engineering Approach*, McGraw-Hill, New York, 1968.

[8] SINTEF, *OREDA–Offshore Reliability Data*, 3rd ed., SINTEF Industrial Management, Trondheim, Norway, 1997.

Chapter 21

Power Control and Call Admission Control for DS-CDMA Cellular Networks*

Derong Liu, Yi Zhang, and Sanqing Hu

Abstract: In this chapter, we develop call admission control algorithms for signal-to-interference ratio-based power-controlled direct-sequence code-division multiple-access cellular networks. We consider networks that handle multiple classes of services. When a new call (or a handoff call) arrives at a base station requesting admission, our algorithms will calculate the desired power control setpoints for the new call and all existing calls. We will provide necessary and sufficient conditions under which the power control algorithm will have a feasible solution. These conditions are obtained through deriving the inverse of the matrix used in the calculation of power control setpoints. If there is no feasible solution to power control or if the desired power levels to be received at the base station for some calls are larger than the maximum allowable power limits, the admission request will be rejected. Otherwise, the admission request will be granted. When higher priority is desired for handoff calls, we will allow different thresholds for new calls and handoff calls. We will develop an adaptive algorithm that adjusts these thresholds in real time as the environment changes. The performance of our algorithms will be shown through computer simulation.

*This work was supported by the National Science Foundation under Grants ECS-9996428 and ANI-0203063.

21.1 Introduction

Call admission control schemes are critical to the success of future generations of wireless networks. On the one hand, call admission control schemes provide the users with access to wireless networks for services. On the other hand, they are the decision-making part of the network carriers with the objectives of providing services to users with guaranteed quality and also achieving resource utilization that is as high as possible. It is therefore conceivable that call admission control policy is one of the critical design parameters in wireless networks [1, 5, 14, 18, 19, 22, 30].

It is known that the capacity of direct-sequence code-division multiple-access (DS-CDMA) networks is limited by the reverse link (up link) rather than the forward link (down link) [7]. Reverse link capacity is limited by the interference received at the base station, which is closely related to traffic characteristics, power control, radio propagation, sectorization, and other factors. Power control is essential to minimize each user's interference on the reverse link in varying radio environments and traffic conditions. Earlier studies [7] have considered power control systems that keep each user's signal arriving at the home base station with the same signal power strength. The base station measures the received power level and compares it with a desired value and then transmits power control bit(s) to the mobile units adjusting their power levels. Networks of this type are referred to as strength-based power-controlled CDMA cellular networks. Several researchers have studied call admission control algorithms for strength-based power-controlled CDMA networks [12, 14]. On the other hand, researchers have argued that the signal-to-interference ratio (SIR) is more important than signal strength in determining channel characteristics (e.g., bit error probability) [13, 26]. SIR-based power control schemes determine the value of the power control bit(s) by comparing the received SIR with the desired SIR threshold. The simulation results of [2] suggest the potential of higher network performance in an SIR-based power-controlled DS-CDMA network for constant bit rate traffic environments. The analysis results of [13] show that the reverse link capacity of SIR-based power-controlled DS-CDMA networks that support ON-OFF traffic is higher than that of the strength-based power-controlled CDMA networks. In [6], [23], and [29], call admission schemes for SIR-based power-controlled DS-CDMA networks have been proposed.

In this chapter, we will develop call admission control algorithms for SIR-based power-controlled DS-CDMA cellular networks that provide multiple classes of services [17, 22]. (e.g., voice, video, and data). The chapter is organized as follows. In Section 21.2, we derive a formula that is used to determine the desired power level (setpoint) for the new call or the handoff call based on the information obtained when it arrives at a base station. We will give necessary and sufficient conditions under which the power control will have a feasible solution. Our results are obtained by deriving the inverse of the matrix used in the calculation of power setpoints. In Section 21.3, we present our call admission control algorithms for SIR-based power-controlled DS-CDMA networks. Our call admission control algorithms will be derived from the viewpoint of controlling the SIR levels for all calls at a base station. Our algorithms will be presented in a basic form where the thresholds are fixed and in adaptive form where the thresholds are adjusted in real time as environment changes. Our adaptive call admission control algorithm will adjust automatically the thresholds of received power strength in order to achieve the overall optimal performance. In Section 21.4, we study the performance of the present algorithms through computer simulation and compare them with existing call admission control algorithms. The simulation studies will show that the present adaptive call admission control algorithm outperforms existing call admission control algorithms. Finally, in Section 21.5, we conclude the present chapter with a few pertinent remarks.

21.2 Calculation of Power Control Setpoints

In the DS-CDMA cellular network model used in this chapter, we assume that separate frequency bands are used for the reverse link and the forward link, so that the mobiles only experience interference from the base stations and the base stations only experience interference from the mobiles. We consider cellular networks that support multiple classes of services. Assume that there are K classes of services provided by the wireless network under consideration, where $K \geq 1$ is an integer. We define a mapping $\sigma : Z^+ \to \{1, \cdots, K\}$ to indicate that the nth user is from the service class $\sigma(n)$, where Z^+ denotes the set of nonnegative integers. We assume that each user in our network may be from a different service class that requires a different qualify of service target (e.g., in terms of different bit error rate for each service class). This includes the case when we allow each user to specify its own quality of service requirements. We assume that traffic from the same service class has the same data rate, activity factor, desired SIR value, and maximum power limit that can be received at the base station.

Consider a base station currently with N active users. The power received at the base station from the nth user (i.e., from the nth mobile station) is denoted by S_n, $n = 1, \cdots, N$. In an SIR-based power-controlled DS-CDMA network [2, 6, 13, 23], the desired value of S_n is a function of the number of active home users and total other cell interference. If we assume that the maximum received power at a base station is limited to H_k for users from service class $k = \sigma(n)$, then S_n is a random variable in the range of $(0, H_k]$.

In CDMA networks, the bit SIR (or the bit energy-to-interference ratio) for the nth user at the base station (in a cell) can be expressed in terms of the received powers of the various users as [7]

$$(E_b/N_0)_n = \frac{S_n W}{I_n R_{\sigma(n)}} \tag{21.2.1}$$

where S_n is the power level of the nth user received at the base station, W is the chip rate, and $R_{\sigma(n)}$ is the data rate of service class $\sigma(n)$. I_n in (21.2.1) indicates the total interference to the nth user received at the base station and is given by

$$I_n = \sum_{i=1, i \neq n}^{N} \nu_{\sigma(i)} S_i + I_n^{other} + \eta_n \tag{21.2.2}$$

where $\nu_{\sigma(i)}$ is the traffic (e.g., voice) activity factor of the ith user from the service class $\sigma(i)$, I_n^{other} is the total interference from neighboring cells, η_n is the background (or thermal) noise, and N is the number of active users in the network. In [26–28], it has been shown that the total interference from neighboring cells I_n^{other} can be expressed as

$$I_n^{other} = f \sum_{i=1, i \neq n}^{N} \nu_{\sigma(i)} S_i$$

where f is called the intercell interference factor with a typical value of 0.55. We can then rewrite (21.2.2) as

$$I_n = (1 + f) \sum_{i=1, i \neq n}^{N} \nu_{\sigma(i)} S_i + \eta_n.$$

To fully utilize the network capacity, calls must be admitted according to the actual network capacity instead of the fixed capacity. Actual network capacity is limited by the amount of interference allowed in the network. In DS-CDMA cellular networks, interference increases as the traffic load increases. However, interference must be limited to meeting the required E_b/N_0 in the link to satisfy the quality of service constraints. For example, the quality of service requirement for voice users with a maximum bit error rate of 10^{-3} can be satisfied by the power control mechanism keeping E_b/N_0 at a required value of 7 dB or higher [7, 14, 23]. If more calls than the actual network capacity can accommodate are admitted into the network, the required E_b/N_0 will not be satisfied. Equation (21.2.1) implies how network capacity must be limited so that the required E_b/N_0 can be ensured. For SIR-based power-controlled cellular networks, the desired power levels received at a base station will reflect the total interference level in the network coupled with the information about the user's service class. If the total interference is high, the required power levels will be high, and vice versa. If the required bit error rate is low, the required power levels must be high to ensure the corresponding SIR.

Assume that there are N active users in a cell at the time when a new call or handoff call requests admission. If this new mobile (henceforth, the 0th user) is accepted by the cell, then it will increase the interference to other active calls in the cell. Other active mobiles must boost their powers to reach the E_b/N_0 requirements. This process continues until a steady state is reached when the requirement $(E_b/N_0)_n \geq \gamma_{\sigma(n)}$ is satisfied for all users, including the new user (i.e., satisfied for $n = 0, 1, \cdots, N$). This implies that at the steady state, we have

$$\frac{S_n W}{R_{\sigma(n)} \gamma_{\sigma(n)}} \geq (1 + f) \sum_{i=0, i \neq n}^{N} \nu_{\sigma(i)} S_i + \eta_n, \; n = 0, 1, \cdots, N. \tag{21.2.3}$$

We can rewrite (21.2.3) in a matrix format as

$$AS \geq b \tag{21.2.4}$$

where $S = [S_0, S_1, \cdots, S_N]^T$,

$$A = \begin{bmatrix} \Delta_0 & -\delta_1 & \cdots & -\delta_N \\ -\delta_0 & \Delta_1 & \cdots & -\delta_N \\ \vdots & \vdots & \ddots & \vdots \\ -\delta_0 & -\delta_1 & \cdots & \Delta_N \end{bmatrix} \tag{21.2.5}$$

with $\Delta_i = W/(R_{\sigma(i)}\gamma_{\sigma(i)})$ and $\delta_i = (1+f)\nu_{\sigma(i)}$ for $i = 0, 1, \cdots, N$, and $b = [\eta_0, \eta_1, \cdots, \eta_N]^T$.

The solution of (21.2.4) that requires the minimum power while satisfying the SIR constraints $\gamma_{\sigma(n)}$ for every user is given by

$$S^* = [S_0^*, S_1^*, \cdots, S_N^*]^T = A^{-1}b, \tag{21.2.6}$$

where A^{-1} is the inverse of matrix A. The solution in (21.2.6) requires that the matrix A be nonsingular and each element of S^* be positive. In this case, we say that the power control algorithm has a feasible solution. We have the following proposition to specify necessary and sufficient conditions under which the matrix A is nonsingular.

Proposition 21.2.1. *The matrix A defined in (21.2.5) is nonsingular if and only if*

$$\sum_{i=0}^{N} \frac{\delta_i}{\Delta_i + \delta_i} \neq 1. \tag{21.2.7}$$

Proof. First we prove that the determinant of matrix A has the following form

$$\det(A) = \prod_{i=0}^{N}(\Delta_i + \delta_i) - \sum_{i=0}^{N} \delta_i \prod_{j=0, j\neq i}^{N} (\Delta_j + \delta_j). \tag{21.2.8}$$

We will prove (21.2.8) using mathematical induction. When $N = 1$, we have

$$\det(A_{(1)}) = \Delta_0\Delta_1 - \delta_0\delta_1 = (\Delta_0 + \delta_0)(\Delta_1 + \delta_1) - \delta_0(\Delta_1 + \delta_1) - \delta_1(\Delta_0 + \delta_0)$$

which implies that (21.2.8) is true for $N = 1$. Now we assume that (21.2.8) is true for $N = k$, i.e.,

$$\det(A_{(k)}) = \prod_{i=0}^{k}(\Delta_i + \delta_i) - \sum_{i=0}^{k} \delta_i \prod_{j=0, j\neq i}^{k} (\Delta_j + \delta_j).$$

Let $N = k + 1$. The matrix $A_{(k+1)}$ is given by

$$A_{(k+1)} = \begin{bmatrix} \Delta_0 & -\delta_1 & \cdots & -\delta_k & -\delta_{k+1} \\ -\delta_0 & \Delta_1 & \cdots & -\delta_k & -\delta_{k+1} \\ \vdots & \vdots & \ddots & \vdots & \vdots \\ -\delta_0 & -\delta_1 & \cdots & \Delta_k & -\delta_{k+1} \\ -\delta_0 & -\delta_1 & \cdots & -\delta_k & \Delta_{k+1} \end{bmatrix} = \left[\begin{array}{ccc|c} & & & -\delta_{k+1} \\ & A_{(k)} & & \vdots \\ & & & -\delta_{k+1} \\ \hline -\delta_0 & \cdots & -\delta_k & \Delta_{k+1} \end{array}\right].$$

Therefore,

$$\det(A_{(k+1)}) = \Delta_{k+1}\det(A_{(k)}) + \sum_{i=0}^{k}(-1)^{(k+1+i)}(-\delta_i)|M_{k+1,i}|$$

where $|M_{k+1,i}|$ is the minor corresponding to the ith element $-\delta_i$ on the last row of $A_{(k+1)}$. Each minor $|M_{k+1,i}|$ has the following form:

$$\begin{vmatrix} \Delta_0 & -\delta_1 & \cdots & -\delta_{i-1} & -\delta_{i+1} & \cdots & -\delta_k & -\delta_{k+1} \\ -\delta_0 & \Delta_1 & \cdots & -\delta_{i-1} & -\delta_{i+1} & \cdots & -\delta_k & -\delta_{k+1} \\ \vdots & \vdots & \ddots & \vdots & \vdots & \cdots & \vdots & \vdots \\ -\delta_0 & -\delta_1 & \cdots & \Delta_{i-1} & -\delta_{i+1} & \cdots & -\delta_k & -\delta_{k+1} \\ -\delta_0 & -\delta_1 & \cdots & -\delta_{i-1} & -\delta_{i+1} & \cdots & -\delta_k & -\delta_{k+1} \\ -\delta_0 & -\delta_1 & \cdots & -\delta_{i-1} & \Delta_{i+1} & \cdots & -\delta_k & -\delta_{k+1} \\ \vdots & \vdots & \cdots & \vdots & \vdots & \ddots & \vdots & \vdots \\ -\delta_0 & -\delta_1 & \cdots & -\delta_{i-1} & -\delta_{i+1} & \cdots & \Delta_k & -\delta_{k+1} \end{vmatrix}.$$

Subtracting the ith row from all other rows, we get

$$|M_{k+1,i}| = (-1)^{(i+k)}(-\delta_{k+1}) \prod_{j=0,j\neq i}^{k} (\Delta_j + \delta_j).$$

Thus,

$$\begin{aligned}
\det(A_{(k+1)}) &= \Delta_{k+1}\det(A_{(k)}) + \sum_{i=0}^{k}(-1)^{(2k+2i+1)}(-\delta_i)(-\delta_{k+1}) \times \\
&\quad \prod_{j=0,j\neq i}^{k} (\Delta_j + \delta_j) \\
&= \Delta_{k+1}\det(A_{(k)}) - \delta_{k+1}\sum_{i=0}^{k}\delta_i \prod_{j=0,j\neq i}^{k} (\Delta_j + \delta_j) \\
&= \Delta_{k+1}\det(A_{(k)}) - \sum_{i=0}^{k}\delta_i \prod_{j=0,j\neq i}^{k+1} (\Delta_j + \delta_j) + \\
&\quad \Delta_{k+1}\sum_{i=0}^{k}\delta_i \prod_{j=0,j\neq i}^{k} (\Delta_j + \delta_j) \\
&= \Delta_{k+1}\prod_{i=0}^{k}(\Delta_i + \delta_i) - \sum_{i=0}^{k}\delta_i \prod_{j=0,j\neq i}^{k+1} (\Delta_j + \delta_j) \\
&= \prod_{i=0}^{k+1}(\Delta_i + \delta_i) - \sum_{i=0}^{k+1}\delta_i \prod_{j=0,j\neq i}^{k+1} (\Delta_j + \delta_j)
\end{aligned}$$

which implies that (21.2.8) is true for all $N \geq 1$.

Matrix A is singular if and only if $\det(A) = 0$, or

$$\det(A) = \det(A_{(N)}) = \prod_{i=0}^{N}(\Delta_i + \delta_i) - \sum_{i=0}^{N} \delta_i \prod_{j=0, j\neq i}^{N} (\Delta_j + \delta_j) = 0$$

which implies

$$\sum_{i=0}^{N} \delta_i \prod_{j=0, j\neq i}^{N} (\Delta_j + \delta_j) = \prod_{i=0}^{N}(\Delta_i + \delta_i).$$

That is equivalent to

$$\sum_{i=0}^{N} \frac{\delta_i}{\Delta_i + \delta_i} = 1$$

which completes the proof.

Remark 21.2.1. Conditions similar to those in Proposition 21.2.1 have been derived in the literature [20, 25, 31] under different network environment settings. Our Proposition 21.2.1 complements those results in [20, 25, 31] under the network environment settings of this chapter.

The next proposition provides necessary and sufficient conditions for S_i^*, $i = 0, 1, \cdots, N$, to be positive.

Proposition 21.2.2. *Each element of S^* in (21.2.6) is positive if and only if*

$$\frac{1}{1 - \sum_{i=0}^{N} \frac{\delta_i}{\Delta_i + \delta_i}} > - \frac{\min_{0 \leq i \leq N} (\eta_i)}{\sum_{i=0}^{N} \frac{\eta_i \delta_i}{\Delta_i + \delta_i}} \tag{21.2.9}$$

when condition (21.2.7) is satisfied.

Proof. If $\sum_{i=0}^{N} \frac{\delta_i}{\Delta_i + \delta_i} \neq 1$, we know that matrix A is nonsingular. Let $A^{-1} = [c_{ij}]$ and $\xi_i = \sum_{j=0}^{N} c_{ij}, i = 0, 1, \cdots, N$. Then, from $A^{-1}A = I_{N+1}$, where I_{N+1} is the $(N+1) \times (N+1)$ identity matrix, it follows that

$$\begin{aligned}
0 &= \Delta_0 c_{i0} - \delta_0(\xi_i - c_{i0}) \\
&\vdots \\
0 &= \Delta_{i-1} c_{i,i-1} - \delta_{i-1}(\xi_i - c_{i,i-1}) \\
1 &= \Delta_i c_{ii} - \delta_i(\xi_i - c_{ii}) \\
0 &= \Delta_{i+1} c_{i,i+1} - \delta_{i+1}(\xi_i - c_{i,i+1}) \\
&\vdots \\
0 &= \Delta_N c_{iN} - \delta_N(\xi_i - c_{iN}).
\end{aligned}$$

From these equalities, we get

$$c_{ij} = \frac{\delta_j \xi_i}{\Delta_j + \delta_j}, \ i \neq j \ \text{and} \ c_{ii} = \frac{1 + \delta_i \xi_i}{\Delta_i + \delta_i}. \tag{21.2.10}$$

Adding these equalities together, we have

$$\xi_i = \frac{1}{\Delta_i + \delta_i} + \xi_i \sum_{j=0}^{N} \frac{\delta_j}{\Delta_j + \delta_j},$$

i.e.,

$$\xi_i = \frac{1}{(\Delta_i + \delta_i)\left(1 - \sum_{j=0}^{N} \frac{\delta_j}{\Delta_j + \delta_j}\right)}, \quad i = 0, 1, 2, \cdots, N. \tag{21.2.11}$$

From (21.2.6), we have

$$\begin{aligned} S_i^* &= A^{-1}b = \sum_{j=0}^{N} c_{ij}\eta_j = \frac{\eta_i(1 + \delta_i \xi_i)}{\Delta_i + \delta_i} + \sum_{j=0, j \neq i}^{N} \frac{\eta_j \delta_j \xi_i}{\Delta_j + \delta_j} \\ &= \frac{\eta_i}{\Delta_i + \delta_i} + \xi_i \sum_{j=0}^{N} \frac{\eta_j \delta_j}{\Delta_j + \delta_j}. \end{aligned} \tag{21.2.12}$$

Clearly, $S_i^* > 0$ is equivalent to

$$\xi_i > -\frac{\eta_i/(\Delta_i + \delta_i)}{\sum_{j=0}^{N} \frac{\eta_j \delta_j}{\Delta_j + \delta_j}}.$$

Combining with (21.2.11), we get

$$\frac{1}{1 - \sum_{j=0}^{N} \frac{\delta_j}{\delta_j + \Delta_j}} > -\frac{\eta_i}{\sum_{j=0}^{N} \frac{\eta_j \delta_j}{\Delta_j + \delta_j}}.$$

For $S_i^* > 0$, $i = 0, 1, \cdots, N$, it is clear that the solution given in (21.2.12) is feasible if and only if

$$\frac{1}{1 - \sum_{i=0}^{N} \frac{\delta_i}{\Delta_i + \delta_i}} > -\frac{\min_{0 \leq i \leq N} (\eta_i)}{\sum_{i=0}^{N} \frac{\eta_i \delta_i}{\Delta_i + \delta_i}}.$$

This completes the proof.

Remark 21.2.2. If every element of b in (21.2.4) has the same value, i.e., if $\eta_i = \eta$ for $i = 0, 1, \cdots, N$, the condition (21.2.9) reduces to

$$\sum_{i=0}^{N} \frac{\delta_i}{\Delta_i + \delta_i} < 1,$$

which has been mentioned by other researchers in [20, 21, 25, 31]. We note that Propositions 21.2.1 and 21.2.2 provide complete proof to the feasibility condition of power control in the CDMA cellular networks studied herein. Equation (21.2.10) provides the inverse of matrix A in an analytical form and (21.2.12) provides the power setpoint calculation for each user.

Remark 21.2.3. When $K = 1$, i.e., when there is only a single class of service in the system, condition (21.2.7) becomes

$$N + 1 = \frac{W}{\nu R\gamma(1 + f)},$$

which is a similar result to the condition given in [7].

We will assume the use of a power control algorithm that achieves the power control setpoint calculated in (21.2.12). We note that many of the power control algorithms provided in the literature (cf. [1, 3, 10]) can be used for this purpose. In the next section, we develop call admission control algorithms based on Proposition 21.2.2 and (21.2.12).

21.3 Fixed and Adaptive Call Admission Control Algorithms

In most existing call admission control algorithms involving power control [1, 3, 6, 10, 12, 14, 23], the speed of call admission control decision process depends on the speed of power vector convergence, for only after the power vector is converged can we make a decision about whether or not to accept the new call. In the present algorithm, we do not need to wait for the power vector to converge to make a decision due to the simple calculation in (21.2.6) [or (21.2.12)]. The special form of matrix A in (21.2.6) renders a simple analytical solution to its inverse as described in the previous section [cf. (21.2.10)]. We note that such a calculation can easily be performed in the processor at a base station.

Our main objective in this chapter is to develop an adaptive call admission control algorithm for SIR-based power-controlled DS-CDMA cellular networks. We first introduce a static call admission control algorithm based on power strength measurements. Equation (21.2.6) or (21.2.12) indicates the minimum power that must be received by the home base station from each mobile station in order to guarantee the minimum SIR given by γ_k, $k = 1, \cdots, K$, where K is the total number of service classes. Assuming that the maximum power limit that may be received by a base station is H_k for service class k, $k = 1, \cdots, K$, then we have the following static admission control scheme:

If condition (21.2.9) is satisfied
    If ($S_n^* \leq H_{\sigma(n)}$ for all $n = 0, 1, \cdots, N$), accept the call;
    Otherwise, reject the call
Otherwise, reject the call.

This algorithm indicates that when condition (21.2.9) is not satisfied, there will be no feasible solution to the power control algorithm and thus the call should be rejected. On the other hand, when $S_n^* > H_{\sigma(n)}$ for some n, we will also reject the new call since otherwise either we would require some mobile stations to transmit more power than they can possibly do or the acceptance of the new call will severely damage the quality of service of existing users, especially those near the cell boundary. The maximum power limits H_k, $k = 1, \cdots, K$, are determined by the power limit of mobile transmitters, cell size, path loss information, and user's service class. They have been used in several previous works on call admission control [7, 14, 23]. We assume that each user may be from a different service class among a total of K service classes. For example, we can choose $H_k = P_k E_k[L]$, $k = 1, \cdots, K$, where P_k is the maximum power that can be transmitted by a mobile in class k and $E_k[L]$ is the expected value of path loss for service class k from the cell boundary to the base station.

The grade of service (GoS) in cellular networks is mainly determined by the new call blocking probability and the handoff blocking probability [34]. The first determines the fraction of new calls that are blocked, while the second is closely related to the fraction of admitted calls that terminate prematurely due to dropout. The static algorithm presented above does not give any priority to handoff calls. Dropping a call in progress is generally considered to be more problematic than blocking a new call and needs to be kept under control. Handoff calls are more important than new calls, and so we need to give priority to handoff calls [9, 19, 23, 24]. The trunk reservation scheme (also called the guard channel scheme) has been extensively studied in the traditional voice-centric cellular networks [9, 19, 23]. The basic idea is to reserve a fixed number of channels at each base station exclusively for handoff calls. In the present approach, we can choose a fixed threshold $T_k < H_k$, $k = 1, \cdots, K$, for new calls to allow higher priority for handoff calls, where we assume that the new user is from a particular service class, i.e., $1 \le \sigma(0) \le K$. Choosing $T_k < H_k$ would usually admit fewer new calls than the case when $T_k = H_k$. The static threshold admission policy that gives the handoff call higher priority is given as follows:

If condition (21.2.9) is satisfied
    If (new call) then
        If ($S_0^* \le T_{\sigma(0)}$ *and* $S_n^* \le H_{\sigma(n)}$ *for all* $n = 1, 2, \cdots, N$*),*
            accept the call;
        Otherwise, reject the call;
    If (handoff call) then
        If ($S_n^* \le H_{\sigma(n)}$ *for all* $n = 0, 1, \cdots, N$*), accept the call;*
        Otherwise, reject the call
Otherwise, reject the call.

$T_{\sigma(0)}$ used in the above algorithm is determined according to the service class of the new user. The GoS metrics are strongly influenced by the call admission control algorithm, which determines whether a new call should be admitted or blocked. Blocking more new calls generally improves the forced-termination probability of the calls that are admitted, and thus there is always a tradeoff. Efficient bandwidth allocation schemes have to ensure that the call dropping probability is maintained at a predefined level while at the same time minimizing the new call blocking probability (or maximizing the bandwidth utilization). For a given set of parameters including traffic statistics and mobility characteristics, fixed call admission control schemes can sometimes yield optimal solutions [4, 19]. All such schemes (cf. [9, 19, 23]), however, by reserving a fixed part of capacity, cannot adapt to changes in the network conditions due to its static nature. This is clearly not suitable in the presence of bursty data and the emerging multimedia traffic. Therefore, an adaptive and dynamic call admission control scheme is essential to the operation of call admission control algorithms in wireless networks [17, 33].

It is noted that new call thresholds T_k, $k = 1, \cdots, K$, in the present algorithm are key design parameters that have a tremendous effect on the performance of wireless networks. In our adaptive call admission control algorithm to be introduced next, the thresholds T_k are adjusted adaptively according to the handoff call blocking rate (for calls in the same class), which is a primary measure of traffic load. Such calculations are restricted to each individual base station, thus eliminating the signaling overhead for information exchange among cells. The main idea is as follows [15, 32]. When a base station experiences a high handoff call blocking rate, we will decrease the corresponding threshold T_k so that handoff calls will be given increasing priority. When a base station maintains the handoff call blocking rate much lower than the threshold over a significant period of time, we can gradually increase the threshold. In doing so, we want to make sure that the handoff call blocking rate is under the maximum allowable level while fully utilizing the network capacity.

We present the following adaptive algorithm for determining automatically the new call threshold T_k to achieve optimal admission control performance.

P_k^h, P_k^l = thresholds for handoff blocking probability for class k with $P_k^l < P_k^h$
P_k^d = handoff call blocking rate for class k
$T_k = T_k^{init} = T_k^M$ for class k (initial value)
$m_k = 0, k = 1, \cdots, K$
While (time increases)
    If (a mobile is handed off to the current cell) then
        Calculate $P_{\sigma(0)}^d$;
        $m_{\sigma(0)} = m_{\sigma(0)} + 1;$
        If (it is blocked) then
            If ($P_{\sigma(0)}^d \geq P_{\sigma(0)}^h$) then
                $T_{\sigma(0)} = \max(\alpha_d T_{\sigma(0)}, T_{\sigma(0)}^m)$
                $m_{\sigma(0)} = 0;$
        If ($P_{\sigma(0)}^d \leq P_{\sigma(0)}^l$) then
            If ($m_{\sigma(0)} \geq M$) then
                $T_{\sigma(0)} = \min(\alpha_u T_{\sigma(0)}, T_{\sigma(0)}^M);$
                $m_{\sigma(0)} = 0.$

where $\alpha_u > 1$, $\alpha_d < 1$, and $M > 0$ are design parameters and T_k^m and T_k^M denotes the minimum and the maximum thresholds for service class k, $k = 1, \cdots, K$, respectively.

In the above algorithm, α_u, α_d, and M may also be chosen differently for different service classes. The admission algorithm will keep a record of the values P_k^l, P_k^h, P_k^d, m_k, and T_k for each service class k, $k = 1, \cdots, K$. This algorithm adjusts the new call threshold $T_{\sigma(0)}$ [for the service class $\sigma(0)$] automatically according to the measured blocking rate of handoff calls in the same service class. It will only decrease the threshold $T_{\sigma(0)}$ when a handoff call is blocked under the condition that $P_{\sigma(0)}^d \geq P_{\sigma(0)}^h$, and it will only increase the threshold $T_{\sigma(0)}$ after a number of consecutive handoff calls indicated by the number M under the condition that $P_{\sigma(0)}^d \leq P_{\sigma(0)}^l$. It tries to make sure that the handoff call blocking rate is below the given threshold $P_{\sigma(0)}^h$. It also tries to reduce the new call blocking rate by incrementing $T_{\sigma(0)}$ when it is observed that $T_{\sigma(0)}$ is lower than needed.

If the decision from the present admission control algorithm is to accept the call request, the power control algorithm will then start to adjust the power levels of all users in the home cell of the new user arrival and in the neighboring cells.

Remark 21.3.1. The development of the present adaptive call admission control algorithm involves the following three steps.

(1) The first step is the simple static call admission algorithm where the idea of performing call admission control based on calculated power control setpoints is first used. Such an approach will in general be more favorable than others, especially in the case when a call needs to be rejected. In the present approach, such a rejection decision is made after calculating the power control setpoints. However, in existing call admission control algorithms involving power control, such a decision can only be made after the actual power control process is converged, which may lead to improper power level assignments [1, 3, 10]. Such a process involves the power adjustment of all users in the cell (and possibly users in the surrounding cells as well) and it takes times and wastes energy (therefore, shortens the battery life of the mobile units). On the other hand, the present approach saves time and does not waste energy for those calls that are destined for rejection.

(2) The second step in the present development is the call admission control algorithm with priorities given to handoff calls. Such an approach has been in general accepted by practitioners and has been employed in actual implementations [9, 18, 19]. The present algorithm assigns a smaller threshold for new call admission than for handoff call admission to guarantee higher priorities for handoff calls.

(3) The third step is the adaptive algorithm that automatically adjusts the threshold values for new calls used in the present call admission control algorithm. The main idea is build upon our previous adaptive guard channel algorithms for call admission control in wireless networks (usually non-CDMA networks) [15, 32]. We note that the present algorithm for adjusting threshold values is simple to implement and efficient in applications (see the next section for simulation results).

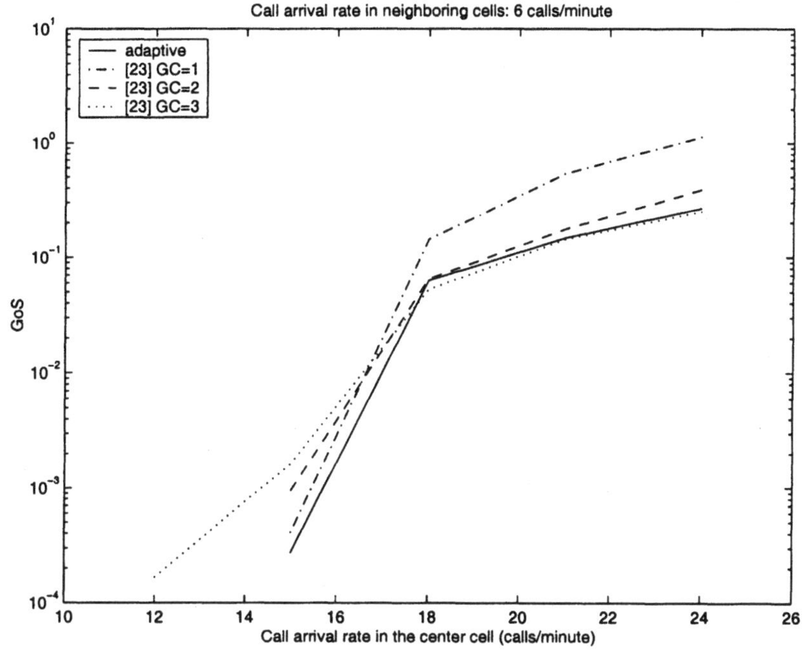

Figure 21.3.1: Neighboring cells with low load

21.4 Simulation Results

According to the digital European cordless telecommunication (DECT) specifications [34], the GoS is defined as

$$\text{GoS} = P(\text{new call blocking}) + w \times P(\text{handoff failure}),$$

where $P(a)$ is the probability of event a. In the present simulation studies we choose $w = 10$ as in the case of [23]. The arrival rate consists of the new call attempt rate λ_c and the handoff call attempt rate λ_h. λ_c depends on the expected number of subscribers per cell. λ_h depends on such network parameters as traffic load, user velocity, and cell coverage areas [8, 9]. In our simulation, we assume that $\lambda_c : \lambda_h = 5 : 1$ [9]. A channel is released by call completion or handoff to a neighboring cell. The channel occupancy time is assumed to be exponentially distributed [8, 9, 16], with the mean value $1/\mu = 3$ minutes .

We first conduct simulation studies for a network with single class of service (e.g., voice). The network parameters used in the present simulation are taken similarly as the parameters used in [13, 23] (cf. Table 21.4.1). The parameters used in the present adaptive algorithm are listed in Table 21.4.2. A larger value of α_u (increment factor) and a smaller value of α_d (decrement factor) make the algorithm more aggressive, i.e., more responsive to changes in traffic conditions. Finally, a smaller value of the parameter M makes the algorithm more sensitive to changes in traffic load condi-

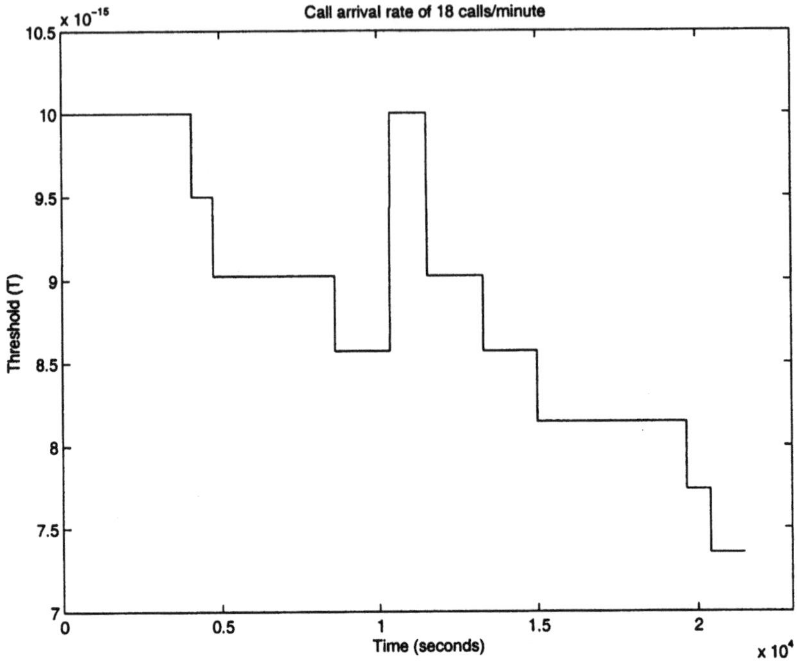

Figure 21.4.1: A plot of the threshold T

tions.

Table 21.4.1: Network parameters

Parameters	Values	Parameters	Values
W	1.2288 Mcps	R	9.6 kbps
η	1×10^{-14} W	H	1×10^{-14} W
E_b/N_0	7 dB	v	3/8

In the following, we conduct comparison studies between the present adaptive call admission control algorithm and that of [23]. We then conduct comparison studies between our fixed and adaptive call admission control algorithms. Using the algorithm in [23], the base station controller reads the current interference from the power strength measurer. It then estimates the current interference margin (CIM) and handoff interference margin (HIM), where CIM < HIM. A total interference margin (TIM) is set according to the quality of service target. If CIM > TIM, reject the call admission request. If HIM < TIM, accept the call request. If CIM < TIM < HIM, then only handoff calls will be accepted.

Figure 21.3.1 compares the present adaptive call admission control algorithm with the algorithm in [23] that reserves 1, 2, and 3 channels for handoff calls, respectively.

Table 21.4.2: Simulation parameters

Parameters	Values	Parameters	Values
α_u	1.1	α_d	0.95
T^m	0	T^M	H
M	3	P^l	$0.9P^h$

The arrival rate in all neighboring cells is fixed at 6 calls/minute. We assume the use of a hexagonal cell structure. The parameter P^h is chosen as 0.001 because of the light traffic load. From Figure 21.3.1, we see that the present algorithm has the best GoS when the call arrival rate is low in the center cell and very close to the best GoS when the arrival rate is high in the center cell. We also see that using the algorithm in [23], when the traffic load changes in the center cell, the values of GoS varies differently for different values of GC (the number of guard channels). In fact, the algorithm in [23] is a kind of GC algorithm used in CDMA systems. Therefore, when the load is low, GC $=$ 1 performs the best, and when the load is high, GC $=$ 3 performs the best. However, our algorithm can adapt to varying traffic load conditions. It has the best overall performance under various traffic loads. Figure 21.4.1 displays the plot of the threshold T for this experiment. The display is for call arrival rate of 18 calls/minute at the center cell. It is clear from the display that the value of threshold T changes often during the whole period of simulation.

Table 21.4.3: Network parameters

Voice users		Data users	
Parameters	Values	Parameters	Values
W_v	4.9152 Mcps	W_d	4.9152 Mcps
R_v	9.6 kbps	R_d	38.4 kbps
H_v	1×10^{-14} W	H_d	1×10^{-13} W
$(E_b/N_0)_v$	7 dB	$(E_b/N_0)_d$	9 dB
v_v	3/8	v_d	1

Figure 21.4.2 compares the present adaptive algorithm with the algorithm in [23] that reserves 1, 2, and 3 channels for handoff calls, respectively. In this case we double the arrival rate in neighboring cells to 12 calls/minute. The parameter P^h is chosen as 0.01 because of the high traffic load. From Figure 21.4.2 we see that the GoS of the present algorithm is higher than that of [23] when the arrival rate is low in the center cell. The reason is that when the interference from neighboring cells is high due to a high arrival rate, our calculation for power control setpoints indicates that the signal strength required in the center cell is sometimes higher than the limit H in order to satisfy the voice quality requirement. In this case, more

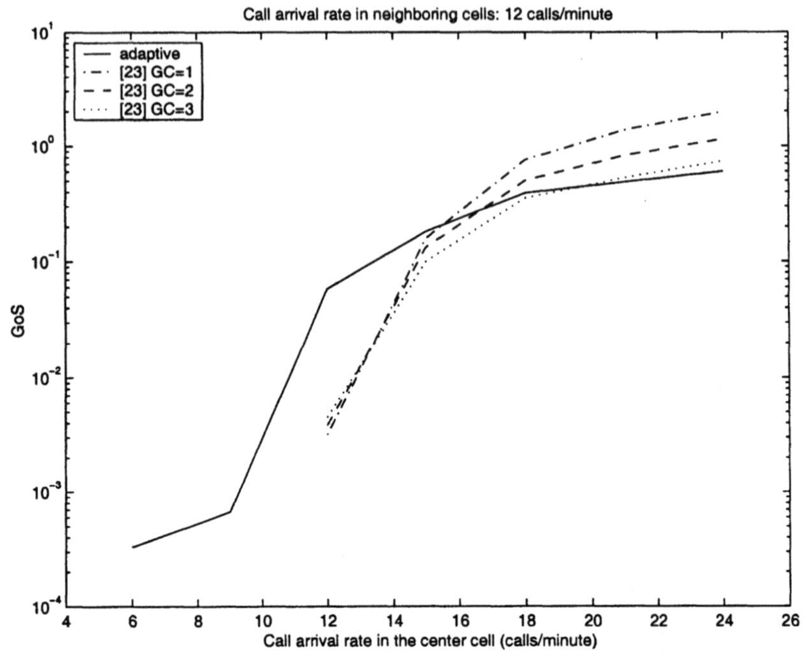

Figure 21.4.2: Neighboring cells with normal load

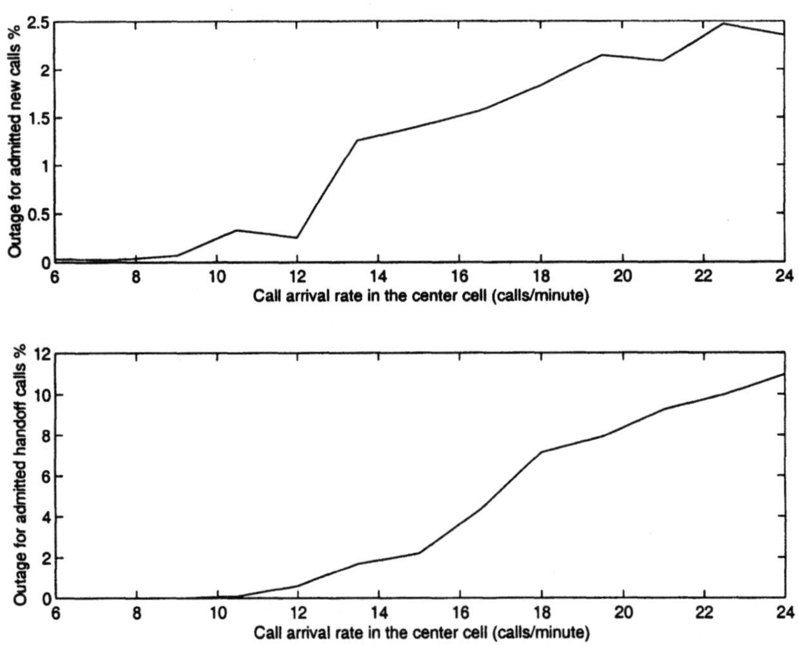

Figure 21.4.3: Outage rate of admitted calls

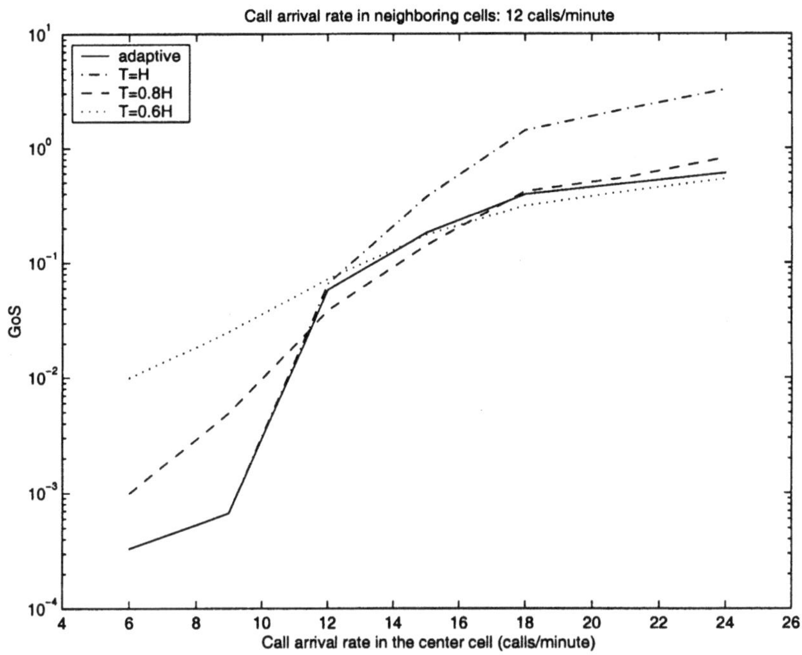

Figure 21.4.4: Comparison between our adaptive and fixed call admission control algorithms

calls are rejected by our algorithm. However, using the algorithm in [23], although the required signal strength is over the limit H, CIM and HIM are still under TIM because of the light traffic condition in the center cell. Thus, using the algorithm in [23], calls are still admitted, and as a result, the quality of service for all calls in the cell will be compromised. For example, there is a new call rejected by our algorithm when the call arrival rate is 10 calls/minute in the center cell. At that moment, the required signal power strength received at the base station is 1.024×10^{-14} W > H, where H is given in Table 21.4.1 as the highest power strength that may be received at the base station. If we accept that call, we would get the E_b/N_0 equal to 6.94 dB at that moment. Since 6.94 dB is less than 7 dB required by the quality of service constraints, the quality of communication cannot be guaranteed using the algorithm in [23]. Figure 21.4.3 shows the actual outage percentage of the admitted new calls and handoff calls using the algorithm in [23] for the same simulation. Here outage means that when a call is admitted, its SIR does not meet the minimum requirement and thus the quality of service will be compromised for all calls in the cell.

Figure 21.4.4 compares our adaptive call admission control algorithm and our fixed algorithm with static thresholds given by H, $0.8H$, and $0.6H$, respectively. The arrival rate in all neighboring cells is fixed at 12 calls/minute. The parameter P^h is chosen as 0.01 because of the high traffic load. From Figure 21.4.4, we see that the present adaptive algorithm has the best GoS for almost all call arrival rates

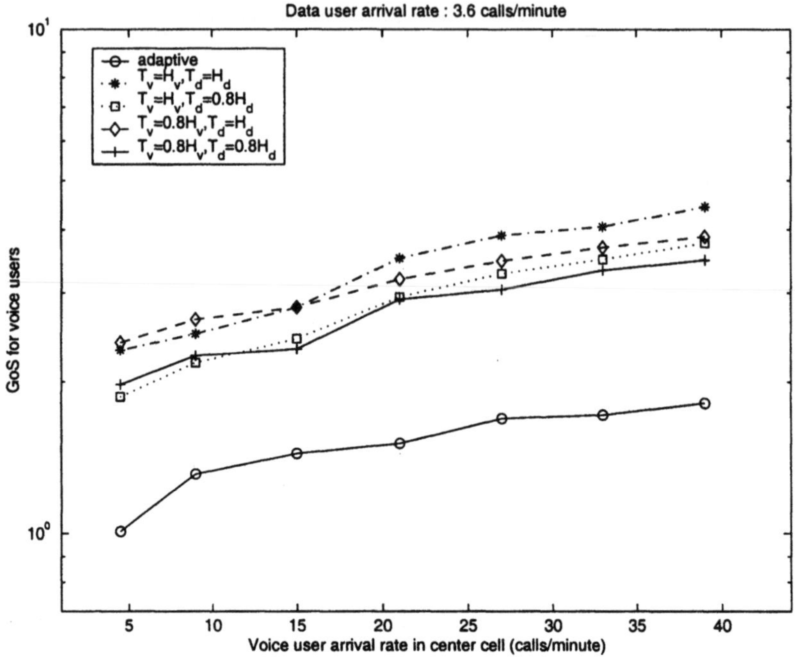

Figure 21.4.5: GoS for voice calls

tested. We can conclude that the present adaptive algorithm performs better than the fixed algorithms because the adaptive algorithm can adapt to varying traffic load conditions.

Finally, we conduct simulation studies for cellular networks with two classes of services. One class is voice service and the other is data service. We note that similar results can easily be obtained for cases with more than two service classes. Network parameters in our simulations are chosen in reference to the parameters used in [11] and [21] (cf. Table 21.4.3). The background noise in this case is chosen the same as in Table 21.4.1. Figures 21.4.5 and 21.4.6 compare our adaptive call admission control algorithm and our fixed algorithm with static thresholds given by H and $0.8H$, respectively. The arrival rates of voice users and data users in all neighboring cells are fixed at 18 calls/minute and 2.4 calls/minute, respectively. For both type of services, the parameters α_u, α_d, M, T^m, and T^M are chosen as in Table 21.4.2. We choose P^h as 0.1 and 0.03 for voice users and for data users, respectively. Figure 21.4.5 shows the results of the GoS for voice users in our simulations and Figure 21.4.6 shows the results of the GoS for data users. The simulation results show that the present adaptive call admission control algorithm performs better than the fixed algorithms for networks with two classes of services.

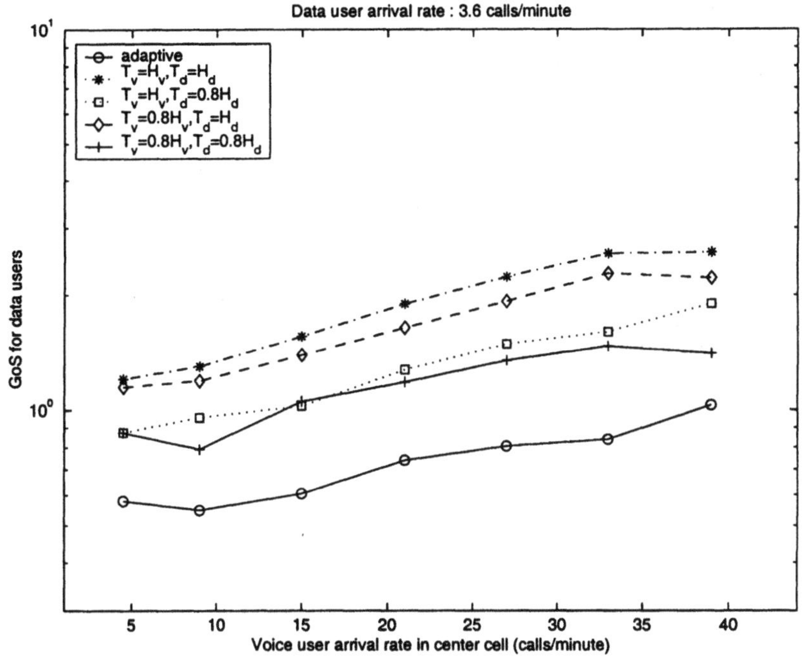

Figure 21.4.6: GoS for data users

21.5 Conclusions

In this chapter, we developed fixed and adaptive call admission control algorithms for multiclass traffic in SIR-based power-controlled DS-CDMA cellular networks. These algorithms were based on the calculation of required power setpoints for mobile stations to be received at a base station in a cell. We established necessary and sufficient conditions under which the power control algorithm will have a feasible solution. These conditions were obtained through deriving the inverse of the matrix used in the calculation of power control setpoints. If there is no feasible solution to power control, or if the calculated power setpoints exceed the maximum power that can be generated by a mobile station or exceed the maximum power allowed in the cell, we simply reject the call. To give handoff calls higher priority than new calls, we use thresholds for new calls that are smaller than the thresholds used for handoff calls. Our adaptive algorithm can search automatically the optimal thresholds for the power levels of new calls. We note that changes in traffic conditions are inevitable in reality. Thus, fixed call admission control policies are less preferable in applications. Our simulation results show that when the traffic load condition changes, the fixed admission policy will suffer either from higher new call blocking rate, higher handoff call blocking rate, or higher interference than the tolerance.

Bibliography

[1] M. Andersin, Z. Rosberg, and J. Zander, "Soft and safe admission control in cellular networks," *IEEE/ACM Trans. Networking*, vol. 5, pp. 255–265, Apr. 1997.

[2] S. Ariyavisitakul, "Signal and interference statistics of a CDMA system with feedback power control–Part II," *IEEE Trans. Communications*, vol. 42, pp. 597–605, Feb.-Mar.-Apr. 1994.

[3] N. Bambos, S. C. Chen, and G. J. Pottie, "Channel access algorithms with active link protection for wireless communication networks with power control," *IEEE/ACM Trans. Networking*, vol. 8, pp. 583–597, Oct. 2000.

[4] S. Choi and K. G. Shin, "A comparative study of bandwidth reservation and admission control schemes in QoS-sensitive cellular networks" *Wireless Networks*, vol. 6, no. 4, pp. 289–305, 2000.

[5] S. K. Das, R. Jayaram, N. K. Kakani, and S. K. Sen, "A call admission and control scheme for quality-of-service (QoS) provisioning in next generation wireless networks," *Wireless Networks*, vol. 6, no. 1, pp. 17–30, 2000.

[6] Z. Dziong, M. Jia, and P. Mermelstein, "Adaptive traffic admission for integrated services in CDMA wireless-access networks," *IEEE J. Selected Areas in Communications*, vol. 14, pp. 1737–1747, Dec. 1996.

[7] K. S. Gilhousen, I. M. Jacobs, R. Padovani, A. J. Viterbi, L. A. Weaver, and C. E. Wheatley III, "On the capacity of a cellular CDMA system," *IEEE Trans. Vehicular Technology*, vol. 40, pp. 303–312, May 1991.

[8] R. A. Guerin, "Channel occupancy time distribution in a cellular radio system," *IEEE Trans. Vehicular Technology*, vol. VT-35, pp. 89–99, Aug. 1987.

[9] D. Hong and S. S. Rappaport, "Traffic model and performance analysis for cellular mobile radio telephone systems with prioritized and nonprioritized handoff procedures," *IEEE Trans. Vehicular Technology*, vol. VT-35, pp. 77–92, Aug. 1986.

[10] D. Kim, "Efficient interactive call admission control in power-controlled mobile systems," *IEEE Trans. Vehicular Technology*, vol. 49, pp. 1017–1028, May 2000.

[11] Y. W. Kim, D. K. Kim, J. H. Kim, S. M. Shin, and D. K. Sung, "Radio resource management in multiple-chip-rate DS/CDMA systems supporting multiclass services," *IEEE Trans. Vehicular Technology*, vol. 50, pp. 723–736, May 2001.

[12] I.-M. Kim, B.-C. Shin, and D.-J. Lee, "SIR-based call admission control by intercell interference prediction for DS-CDMA systems," *IEEE Communications Letters*, vol. 4, pp. 29–31, Jan. 2000.

[13] D. K. Kim and D. K. Sung, "Capacity estimation for an SIR-based power-controlled CDMA system supporting ON-OFF traffic," *IEEE Trans. Vehicular Technology*, vol. 49, pp. 1094–1100, July 2000.

[14] Z. Liu and M. E. Zarki, "SIR-based call admission control for DS-CDMA cellular systems," *IEEE J. Selected Areas in Communications*, vol. 12, pp. 638–644, May 1994.

[15] D. Liu and Y. Zhang, "Adaptive guard channel assignment for cellular networks," *Proc. IEEE CAS Workshop on Wireless Communications and Networking*, pp. 268–273, Notre Dame, IN, Aug. 2001.

[16] J. G. Markoulidakis, G. L. Lyberopoulos, and M. E. Anagnostou, "Traffic model for third generation cellular mobile telecommunication systems," *Wireless Networks*, vol. 4, no. 5, pp. 389–400, 1998.

[17] C. Oliveira, J. B. Kim, and T. Suda, "An adaptive bandwidth reservation scheme for high-speed multimedia wireless networks," *IEEE J. Selected Areas in Communications*, vol. 16, pp. 858–874, Aug. 1998.

[18] S. Ramanathan, "A unified framework and algorithm for channel assignment in wireless networks," *Wireless Networks*, vol. 5, no. 2, pp. 81–94, 1999.

[19] R. Ramjee, D. Towsley, and R. Nagarajan, "On optimal call admission control in cellular networks," *Wireless Networks*, vol. 3, no. 2, pp. 29–41, 1997.

[20] A. Sampath, P. S. Kumar, and J. M. Holtzman, "Power control and resource management for a multimedia CDMA wireless system," *Proc. 1995 Conference on Personal, Indoor and Mobile Radio Communications*, pp. 21–25, Toronto, Canada, Sept. 1995.

[21] A. Sampath and J. M. Holtzman, "Access control of data in integrated voice/data CDMA systems: Benefits and tradeoffs," *IEEE J. Selected Areas in Communications*, vol. 15, pp. 1511–1526, Oct. 1997.

[22] M. Schwartz, "Network management and control issues in multimedia wireless networks," *IEEE Personal Communications*, vol. 2, pp. 8–16, June 1995.

[23] S. M. Shin, C.-H. Cho, and D. K. Sung, "Interference-based channel assignment for DS-CDMA cellular systems," *IEEE Trans. Vehicular Technology*, vol. 48, pp. 233–239, Jan. 1999.

[24] M. Sidi and D. Starobinski, "New call blocking versus handoff blocking in cellular networks," *Wireless Networks*, vol. 3, no. 1, pp. 15–27, 1997.

[25] V. V. Veeravalli and A. Sendonaris, "The coverage-capacity tradeoff in cellular CDMA systems," *IEEE Trans. Vehicular Technology*, vol. 48, pp. 1443–1450, Sept. 1999.

[26] A. J. Viterbi, *CDMA: Principles of Spread Spectrum Communication*, Addison-Wesley, Reading, MA, 1995.

[27] A. J. Viterbi, A. M. Viterbi, and E. Zehavi, "Other-cell interference in cellular power-controlled CDMA," *IEEE Trans. Communications*, vol. 42, pp. 1501–1504, Feb.-Mar.-Apr. 1994.

[28] A. M. Viterbi and A. J. Viterbi, "Erlang capacity of a power controlled CDMA system," *IEEE J. Selected Areas in Communications*, vol. 11, pp. 892–900, Aug. 1993.

[29] M. Xiao, N. B. Shroff, and E. K . P. Chong, "Distributed admission control for power-controlled cellular wireless systems," *IEEE/ACM Trans. Networking*, vol. 9, pp. 790–800, Dec. 2001.

[30] A, Yener and C. Rose, "Genetic algorithms applied to cellular call admission: local policies, " *IEEE Trans. Vehicular Technology*, vol. 46, pp. 72–79, Feb. 1997.

[31] L. C. Yun and D. G. Messerschmitt, "Power control for variable QOS on a CDMA channel," *Proc. IEEE Military Communications Conference*, pp. 178–182, Long Branch, NJ, Oct. 1994.

[32] Y. Zhang and D. Liu, "An adaptive algorithm for call admission control in wireless networks," *Proc. IEEE Global Communications Conference*, pp. 2537–2541, San Antonio, TX, Nov. 2001.

[33] W. Zhuang, B. Bensaou, and K. C. Chua, "Adaptive quality of service handoff priority scheme for mobile multimedia networks," *IEEE Trans. Vehicular Technology*, vol. 49, pp. 494–505, Mar. 2000.

[34] "A guide to DECT features that influence the traffic capacity and the maintenance of high radio link transmission quality, including the results of simulations," *ETSI Technical Report: ETR 042*, July 1992 (available on-line at http://www.etsi.org).

Index

Control Engineering

Series Editor

William S. Levine
Department of Electrical and Computer Engineering
University of Maryland
College Park, MD 20742-3285
USA

Aims and Scope

Control engineering is an increasingly diverse subject, whose technologies range from simple mechanical devices to complex electro-mechanical systems. Applications are seen in everything from biological control systems to the tracking controllers of CD players. Some methods, H-infinity design for example, for the analysis and design of control systems are based on sophisticated mathematics while others, such as PID control, are understood and implemented through experimentation and empirical analysis.

The Birkhäuser series *Systems and Control: Foundations and Applications* examines the abstract and theoretical mathematical aspects of control. *Control Engineering* complements this effort through a study of the industrial and applied implementation of control — from techniques for analysis and design to hardware implementation, test, and evaluation. While recognizing the harmony between abstract theory and physical application, these publications emphasize real-world results and concerns. Problems and examples use the least amount of abstraction required, remaining committed to issues of consequence, such as cost, tradeoffs, reliability, and power consumption.

The series includes professional expository monographs, advanced textbooks, handbooks, and thematic compilations of applications/case studies.

Readership

The publications will appeal to a broad interdisciplinary readership of engineers at the graduate and professional levels. Applied theorists and practitioners in industry and academia will find the publications accessible across the varied terrain of control engineering research.

Preparation of manuscripts

We encourage the preparation of manuscripts in LaTeX for delivery as camera-ready hard copy, which leads to rapid publication, or on a diskette.

Proposals should be sent directly to the editor or to: Birkhäuser Boston,
675 Massachusetts Avenue, Cambridge, MA 02139, U.S.A.

Published Books

Lyapunov-Based Control of Mechanical Systems
M.S. de Queiroz, D.M. Dawson, S.P. Nagarkatti, and F. Zhang

Nonlinear Control and Analytic Mechanics
H.G. Kwatny and G.L. Blankenship

Qualitative Theory of Hybrid Dynamical Systems
A.S. Matveev and A.V. Savkin

Robust Kalman Filtering for Signals and Systems with Large Uncertainties
I.R. Peterson and A.V. Savkin

Control Systems Theory with Engineering Applications
S.E. Lyshevski

Control Systems with Actuator Saturation: Analysis and Design
T. Hu and Z. Lin

Deterministic and Stochastic Time-Delay Systems
E.K. Boukas and Z.K. Liu

Hybrid Dynamical Systems
A.V. Savkin and R.J. Evans

Stability and Control of Dynamical Systems with Applications: A Tribute to Anthony N. Michel
D. Liu and P.J. Antsaklis, editors

Forthcoming books in the Control Engineering series

Stability of Time-Delay Systems
K. Gu, V.L. Kharitonov, and J. Chen

Nonlinear Control of Engineering Systems: A Lyapunov-Based Approach
W.E. Dixon, A. Behal, D.M. Dawson, and S.P. Nagarkatti

PID Controllers for Time-Delay Systems
S.P. Bhattacharyya, A. Datta, and G.J. Silva

Verification and Synthesis of Hybrid Systems
E. Asarin, T. Dang, and O. Maler

Qualitative Nonlinear Dynamics of Communication Networks
V. Kulkarni

MIX
Papier aus verantwortungsvollen Quellen
Paper from responsible sources
FSC® C105338

If you have any concerns about our products,
you can contact us on
ProductSafety@springernature.com

In case Publisher is established outside the EU,
the EU authorized representative is:
Springer Nature Customer Service Center GmbH
Europaplatz 3, 69115 Heidelberg, Germany

Printed by Libri Plureos GmbH
in Hamburg, Germany